ECONOMIC AND SOCIAL COMMISSION FOR ASIA AND THE PACIFIC

REVIEW OF MINERAL RESOURCES POTENTIAL AND POLICY FOR DEVELOPMENT IN NORTH-EAST ASIA

DEVELOPMENT AND MANAGEMENT OF NON-LIVING RESOURCES IN THE COASTAL ZONES OF THE ASIA-PACIFIC REGION

VOLUME 2

UNITED NATIONS
New York, 1997

Front cover photographs

Left upper corner:	Offshore oil and gas drilling in Bohai Gulf of China
Upper centre:	Volcanoes at Kamchatka, Far East of the Russian Federation
Right upper corner:	TAMHAE II, a three-dimensional seismic research vessel of KIGAM of the Republic of Korea
Left lower corner:	Mining pollution control works in the Old Koyama Mine, Japan
Lower centre:	Opencast mining at the Musan Mining Complex of the Democratic People's Republic of Korea
Right lower corner:	Camp for mineral exploration work in Mongolia

ST/ESCAP/1828

UNITED NATIONS PUBLICATION

Sales No. E.98.II.F.54

ISBN 92-1-119843-7

PREFACE

The countries of North-East Asia have a recognized mineral resource base and potential for fuel and non-fuel minerals and have formulated long term national plans and programmes for exploration and development of these resources. Most countries within the subregion have now realized the importance of the development of the mining sector for economic development and social welfare and have formulated or in the process of reviewing their mineral development policies that mainly entail mining legislation and related fiscal regimes including taxation. Consequently, most countries in the North-East Asia region are now devoting much attention in harmonizing their legal and fiscal policies so as to attract private investment.

However, there is a lack of assessment of the resources potential and developmental environment, and a lack of integration of environmental considerations into policy making in the development of the mineral resources at subregional level. Economic and technological cooperation between countries in the North-East Asia is not as yet significant. This may hinder the development process and also cause environmental problems, both locally and subregionally, in the short and long term.

This publication addresses issues related to the development of both fuel and non-fuel land and marine minerals in the North-East Asia. The publication examines the current status and trends of minerals development in the North-East Asian countries in regard to their socio-economic development, reviews the national development policies of these countries (Part One), and the subregional prospects in mineral development and international cooperation activities in support of such minerals development in the North-East Asia (Part Two). The publication incorporates three subregional studies as well as country papers, which have emanated from the execution of the project titled "Review of Mineral Resources Potential of North-East Asia in Support of Sustainable Development of Coastal and Offshore Areas" during 1997. The ESCAP secretariat would like to record its deep appreciation to the Governments of the Republic of Korea and China for providing generous financial and technical support to the project.

The close cooperation, technical and organizational support to the activities of the project which has been demonstrated by the Ministry of Geology and Mineral Resources of China, the Korea Institute of Geology, Mining and Materials (KIGAM) and the Coordinating Committee for Coastal and Offshore Geoscience Programmes in East and Southeast Asia (CCOP) are also highly appreciated. Special thanks should be attributed to Professor Sun Yunsheng, President, Changchun University of Science and Technology for providing excellent host facilities and the chairmanship of the Workshop on Review of Mineral Resources Potential in North-East Asia in February 1997, Dr. Allen L. Clark and Dr. Vladimir Vyssotsky, principal consultants of the project for preparing their studies as well as Dr. Sahng-Yup Kim, Director of CCOP Technical Secretariat for his overall personal contribution to the success of this project.

The ESCAP secretariat views this publication as one of the initial and necessary steps in generating interest among geoscientists, the international mining community and financial institutions in further research of the geology and mineral resources of this subregion and investment promotion in the development of mineral-based industries in the countries of North-East Asia.

CONTENTS

CONTENTS *(continued)*

Page

LIST OF TABLES

PART ONE

PART TWO

CONTENTS *(continued)*

LIST OF TABLES *(continued)*

CONTENTS *(continued)*

LIST OF TABLES *(continued)*

CONTENTS *(continued)*

LIST OF TABLES *(continued)*

LIST OF FIGURES

PART ONE

CONTENTS *(continued)*

LIST OF FIGURES *(continued)*

PART TWO

CONTENTS *(continued)*

LIST OF FIGURES *(continued)*

Page

PART ONE

MINERAL RESOURCES POTENTIAL AND POLICY FOR DEVELOPMENT IN THE COUNTRIES OF NORTH-EAST ASIA

CONTENTS

1. NORTHEAST CHINA'S MINERAL RESOURCES SITUATION AND REGIONAL COOPERATION [1]

ABSTRACT

The northeast China has an area of 1.24 million square km and a population of about 111 million, 12.9 and 9.3 per cent of China, respectively. This area covers 3 provinces (Liaoning, Jilin and Heilongjiang) and the east part of the Inner Mongolia Autonomous Region. Within this area, more than 120 types of mineral resources have been discovered, among which over 100 have proved reserves and 49 occupy the first three places in preserved reserves in China. Magnesite, wollastonite, diatomaceous earth, raw materials for construction, iron, petroleum and coal, can supply both domestic and foreign markets. However, some mineral resources have limited reserves and can not meet domestic demand. This includes copper, lead, zinc, rich iron ore, chromite and bauxite. Abundant mineral resources need to be fully exploited and utilised, particularly focussing on prospecting and exploration for resources in short supply, if the sustainable economic development of northeast China and peripheral regions is to be realised. In the future, activities relating to ore prospecting, exploitation and development of mineral resources should be carried out in order to enhance economic growth, wider cooperation, enhanced technical and financial investment, and environmental protection.

1.0 SUPPLY AND DEMAND TRENDS FOR MINERAL RESOURCES IN CHINA AND NORTH-EAST ASIA

To better understand the mineral resources situation in northeast China, it is necessary to briefy analyse the supply and demand situation for mineral resources in China and North-East Asia (CNA).

With continuous economic development in CNA since the 1980s, the consumption of mineral resources has obviously increased and the mineral industry has further developed. The output of coal, bauxite, aluminium oxide, tungsten, tin, antimony, titanium, rare earth elements, flourite, magnesite, graphite and talc can meet regional demand and supply exports to some extent. Natural gas supply and demand is essentially in balance. However, petroleum, rich iron ore, copper, sulfur, phosphorous and potassium are in short supply, and must be imported from other regions and internationally. Although mineral consumption in CNA has been high in past years, it has not reached its peak and the intensity of per capita mineral consumption is still quite low, with the exception of Japan. Hence the mineral demand in CNA is expected to further expand due to its projected continuous economic development.

Energy consumption in CNA has rapidly increased in the past dozen years. In 1994, China, Japan and the Republic of Korea were the biggest consumers. These three countries will maintain a strong demand for petroleum in the future. Due to the limited proved petroleum reserves in CNA, demand in the region will still depend on import. Currently, 70 per cent of oil imported into CNA comes from the Middle East, and in the next century this figure might rise as high as 90 per cent. Meanwhile, the import of natural gas still shows strong trends. However, coal will be self-sufficient and can be exported to other places in the world.

Early next century, there will be an increasing demand for iron and other steel alloys such as manganese, chromium, nickel, cobalt, tungsten, molybdenum and vanadium, due to the improvement of metallurgical techniques and the growth of steel production in the region. It is estimated that, with the exception of tungsten, the other metals will depend on imports to varying degrees. In addition, the import of copper, lead and zinc will increase, while other nonferrous metals such as tin, rare earth elements and titanium will continue to supply the global market.

1 Sun Yunsheng and Sun Fengyue, Changchun University of Earth Sciences, Changchun, China

For the nonmetallic minerals, agricultural mineral resources like sulfur, phosphorous and potassium, are in short supply, ensuring dependence on other regions of the world. However, the industrial nonmetallic materials are abundant in CNA and will continue to supply the world market.

2.0 MINERAL RESOURCES IN NORTHEAST CHINA

2.1 Introduction of China's mineral resources

Tectonically, China lies astride three mega tectonic zones which are old Asian, Tethyan-Himalayan and the Circum-Pacific. More than 168 types of mineral resources have been discovered, among which 151 have proved reserves. China is one of the few countries in the world with a complete range of mineral types.

In 1993, 45 types of minerals accounted for 99.5 per cent of the total output value in China, which can be classified into 5 categories on the basis of the reserves/annual production ratio and the self-sufficiency rate (table 1).

Mineral resources in China with international superiority are coal, tungsten, tin, molybdenum, antimony, rare-earth elements, rare elements, graphite, magnesite and aluminium, while other relatively small mineral resources are petroleum, natural gas, rich iron, rich manganese, copper, gold, uranium, potash salt, chrome, nickel, platinum group elements, diamond and cobalt.

Table 1. Categories of 45 main types of mineral resources in China

Mineral categories	*Hardly maintain production (reserves/output ratio < 30)*	*Maintain normal production (reserves/output ratio = 30~60)*	*Increase production (reserves/output ratio = 60~100)*	*Extend production scale (reserves/output ratio = 100~200)*	*Rich preserved reserves (reserves/output > 200)*
Not self-sufficient (self-sufficiency rate < 50%)	gold, platinum		chromium		potash salt[1]
Hardly self-suficient (self-sufficiency rate = 50%~90%)		silver, iron*		nickel	cobalt[3]
Self-sufficient (self-sufficiency rate = 100%±10%)	petroleum*, manganese	natural gas, cement limestone*	sulfur*, refractory clay, glass raw material	boron[4], asbestos, coal*, phosphorus	bentonite, gypsum, rock materials, uranium*, bauxite*
More than self-sufficient (self-sufficiency rate = 110%~200%)	lead*	strontianite, barite		graphite	kaolinite, sodium salt*, mirabilite, diatomaceous earth, ilmenite
Large scale exporting (self-sufficiency rate > 200%)		tin, tungsten, zinc, fluorite, talc, antimony		molybdenum, wollastonite	rare earth elements, magnesite

Source: Wen Shicheng, 1996. Characteristics and prospects of mineral resources in China. In: Chinese Institute of Geology and Mineral Resources Information (editor), Geological Sciences and Mineral Resources for the 21st Century, Geological Publishing House, Beijing.

Notes:
* Key mineral resources.
1 Most reserves difficult to utilize.
2 Based on the import and export value from 1989-1993.
3 Mainly associate minerals.
4 Poor ore quality.

2.2 Mineral resources in northeast China

More than 120 types of mineral resources have been discovered in northeast China. Besides the four types of energy resources (coal, oil shale, petroleum, natural gas) these resources include both metallic and nonmetallic minerals.

Metallic mineral resources comprise iron, manganese, chromite, vanadium, titanium, copper, lead, zinc, bauxite, nickel, cobalt, tungsten, tin, bismuth, molybdenum, mercury, antimony, gold, silver, platinum, tantalum, beryllium, zirconium, monazite, xenotime, germanium, gallium, indium, thallium, rhenium, cadmium, selenium and tellurium. Non-metallic mineral resources include: pyrite, sulfur, phosphorous, potash rich rocks, serpentinite (peridotite), limestone for the chemical industry, boron, salt, mirabilite, trona, bromine, arsenic, peat, barite, flux limestone, flux dolomite, molding sand, magnesite, refractory clay, fluorite, siderotil, aluminium-rich raw materials, piezoeletric quartz, silica, smelting quartz, Iceland spar, diamond, mica, asbestos, kaolinite, graphite, gypsum, talc, cement limestone, cement components, cement mixing materials, glass sand, glass dolomite, feldspar, ceramic clay, brick clay, wollastonite, construction marble, construction silicate materials, calcite, molding basalt, perlite, zeolite, vermiculite, diatomaceous earth, bentonite pumice, garnet, jade and stained stone, and agate.

Forty-nine types of mineral resources rank in the first three on China's reserves list. Twenty types are at the first level, including oil shale, petroleum, iron, beryllium, pyrite, boron, flux limestone, magnesite, aluminium-rich materials, Iceland spar, diamond, graphite, talc, wollastonite, calcite, molding basalt, vermiculite, pumice, jade and stained stone, and agate. There are 17 types listed from the second level, which are as follows: coal, natural gas, zinc, molybdenum, tantalum, gallium, sulfur, potash-rich rocks, trona, siderotil, molding sand, smelting quartz, gypsum, cement mixing materials, brick clay, zeolite, and diatomaceous earth. There are a further 12 types listed at the third level: including nickel, gold, germanium, indium, rhenium, bromine, peat, silica, refractory clay, mica, glass dolomite and perlite. These resources are distributed over approximately 20 ore deposition areas.

Economically, 26 types are dominant among the minerals listed, and accordining to their reserves, quality and other characteristics, they can be classified into the following 5 groups.

Mineral resources in Group 1 are large reserves of high quality which include petroleum, iron, magnesite, graphite, talc, molding sand, glass sand, and placer gold.

Group 2 resources have reasonable reserves of rather high quality and make a significant contribution to China's output. They include coal, natural gas, boron and nickel.

Group 3 includes: diamond, wollastonite, zeolite, diatomaceous earth and jade stone. Even though they are rich reserves and good quality, they have not been fully developed to date.

Copper, molybdenum, oil shale, zirconium, tantalum, rare earth elements, and beryllium all belong to the fourth Group, occur in large reserves, but due to economic and technological constraints are not really exploited.

Group 5 includes peat and sillimanite, and both have little or no proved reserves. This group does, however, have a very high development potential due to the existence of a favourable geological setting.

The potential value of the 45 main types of mineral resources in northeast China is up to 1.39×10^{16} Chinese yuan, making up 10.2 per cent of the value of the whole nation's mineral resources (1.3×10^{17} yuan).

Compared to the whole country in average potential value, by land area and population, northeast China has a much greater asset in ferrous metals and metallurgical flux materials, and a slightly bigger asset in nonferrous and precious metals. For energy resources, the asset is slightly less, and is far smaller when it comes to rare elements, rare earth elements, construction materials and chemical industrial raw materials. Hence, ferrous metals and metallurgical flux materials in the north-east are superior resources in China while chemical and other industrial materials are limited ones. Energy resources account for 66 per cent, ferrous metals 15.8 per cent, metallurgical flux materials 8.3 per cent, and the rest 8.3 per cent of the potential value (1.3×10^{16} yuan) of the reserves of northeast China's 45 main types of mineral resources (table 2).

Table 2. Potential value of reserves for various kinds of mineral resources in northeast China

Mineral resources	*Potential value of reserves (10^8 Chinese yuan)*	*Percentage in that of China's same mineral*	*Percentage in that of the total 45 minerals of the area*
Energy	9,161.2	9.4	66.0
Ferrous metals	2,195.1	21.9	15.8
Non-ferrous metals	618.0	13.8	4.5
Precious metals	86.0	17.2	0.6
Rare and rare earth elements	87.6	5.5	0.6
Non-metallic resources for chemical industry	70.5	0.6	0.5
Metallurgical flux non-metallic minerals	1,151.6	65.3	8.3
Construction raw materials and other non-metallic minerals	507.4	6.5	3.7

From the perspective of market supply and demand, magnesite, chromite, wollastonite, diatomaceous earth, construction materials, iron, petroleum and coal can meet the market demand. On the other hand precious metals, rich iron ore, chromite and bauxite are in short supply.

3.0 NATIONAL POLICIES AND PROGRAMMES IN EXPLOITING MINERAL RESOURCES IN ORDER TO MAINTAIN REGIONAL ECONOMIC GROWTH

National policies have a significant influence on mineral resource exploitation and economic growth. China has maintained a continuing interest in the exploration and exploitation of its mineral resources, reflected in continual development of the mining industry. Policies and other measures have ensured reasonable mineral development and sustainable economic growth based around northeast China's mineral potential.

3.1 Attaching importance to mineral prospecting and exploration

Mineral reserves have been expanded through a combination of national policies, mineral prospecting and exploration programmes. In line with the geological prospectivity of the area, programmes have, in recent years, focused on petroleum, gold and non ferrous metals. Target areas have shifted from well mapped areas with relatively developed economies, to more remote places where the geology is poorly understood and the economies primitive. Discoveries have included several large scale copper, lead, zinc and silver deposits in east Inner Mongolia, and a cobalt deposit in southeast Jilin. Key research projects funded by the state during the eighth and nineth five-year periods have had a similar orientation.

3.2 Concern for environment protection during mineral development

In the past, environmental protection received little attention in mineral development, but it is now considered to be an essential component. The new mining law for China includes sections directed towards prevention of pollution and mitigation of environmental damage in mineral development activity. New measures stress a balance of environmental protection and economic benefit.

3.3 Encouragement of international cooperation

There has been little international cooperation in mineral exploration and exploitation. There have, however, been a series of collaborative scientific research projects closely interrelated with mineral exploration and development. These geological projects involved energy, metallic and non-metallic resources. Some programmes are bilateral, others multilateral, involving the Russian Federation, Republic of Korea, Japan, the United States, Mongolia, Australia and the UK as well as China. For this kind of international cooperation work, the state will give financial and policy support priority. For future mineral exploitation, imported technology and capital will be most important.

The new International Centre for Geoscience Research and Education in North-East Asia, founded in 1995, has strengthened the cooperation and exchange of scientific research and education in the region. It is expected that the organization will provide a catalyst for future cooperation in mineral exploration and development among countries in the region.

3.4 Encouraging higher level processing of minerals

Although many mineral products from northeast China have been widely exported, they have brought rather low economic returns, due to a low level of processing. China now encourages enterprises to increase the level of processing of mineral products. This has been a successful policy. Greater economic returns can be expected from coal, petroleum, natural gas, graphite and magnesite with improved processing techniques.

3.5 Attach attention to protecting and reasonably exploiting mineral resources

Mineral resources are limited and non-renewable. In the past, mineral resources were lost in the waste streams of mining, processing and smelting. The new mining law is expected to ensure better protection and efficient utilisation of the country's mineral resources.

4.0 RECOMMENDATIONS ON SUBREGIONAL COOPERATION

North-East Asian countries have a good record of cooperation, with further encouraging prospects for ongoing collaboration in the exploration and development of regional mineral resources. It is recommended that cooperation should be further encouraged in exploration, mining, ore dressing and environment protection in order to support regional economic development.

4.1 Cooperation in mineral exploration

Four suggested cooperation projects are as follows:

1. The study and development of effective prospecting techniques in heavily vegetated areas. It is important to identify effective exploration techniques and methods that can be used in areas such as northeast China's forest or grassland regions.

2. Studies on comparative metallogenic geology in North-East Asian countries.

3. Study metallogenic history, geology and economic potential in northeast China.

4. Study the oil and natural gas potentiality in volcanic areas in northeast China and select new oil prospecting targets. The oil-forming conditions in northeast China are quite favourable. With the increase in oil demand and oil production, a new oil prospecting and exploration base is needed. Attention needs to be given to volcanic areas, as in the past these areas were considered to be completely unprospective. Regional investigation has, however, proved that Meso-Cenozoic volcanic rocks were formed by intermittent eruptions and are intercalated with normal sedimentary layers. Thin coal beds, oil shale and plant and animal fossils have been found in these rocks, all being good indicators of oil and natural gas prospectivity.

4.2 Cooperation in mining and utilization of mineral resources

Some large mineral deposits in northeast China have not been mined due to technical or economic constraints. These constraints urgently require attention and joint exploitation projects may result. Among these deposits are the large Duobaoshan porphyry copper deposit in Heilongjiang province, and Wunuketushan porphyry copper (molybdenum) deposit in Inner Mongolia. Although there are large reserves of potassium rich rocks in northeast China, techniques of exploitation and utilisation are very primitive. Since there is a shortage of supply of both copper and potash in North-East Asia, cooperation for their development would be timely.

4.3 Cooperation in mineral processing

A key factor in mineral utilization, ore processing techniques applied in northeast China require improvement. Many types of deposits are inefficiently utilized due to lack of advanced processing skills, offering a productive field for future cooperation. Sugested projects include:

1. *Separation and efficient utilization of economic elements of certain mineral deposits.* The large Wongquangou Fe-B deposit has not been mined, as there is no proper technique to separate the economic components. Due to low recovery rates by inefficient ore processing methods, excessive waste is generated by the large Errentaolegai manganese-silver deposit in Inner Mongolia.

2. *Processing microfine gold ore.* Several microfine gold deposits have been found in the Laoling belt, in southeast Jilin province. The current recovery rate in ore processing is very low, hindering the development of the gold industry in this region.

3. *Study of processing techniques for nonmetallic minerals.* There are large proved reserves of nonmetallic mineral resources in northeast China. Suitable processing techniques need to be employed to increase their economic return.

4.4 Cooperation in environmental protection for mining and processing activities

The environmental issue is one of the three essential problems to be faced in the next century in order to rectify the environmental pollution caused by mineral exploitation in the past. Cooperation may be carried out in two areas:

1. Control of environmental pollution in mineral development.

2. Forecasting and controlling ground and slope subsidence from mining.

REFERENCES

1. Chinese Institute of Geology and Mineral Resources Information, Chinese Academy of Geological Sciences 1992. Map of China's mineral resources (technical manual attached), Geological Publishing House, Beijing.

2. Chinese Institute of Geology and Mineral Resources Information, Chinese Academy of Geological Sciences 1993. China's mineral resources, Press of China's Construction Material Industry, Beijing.

3. Centre for Resource Studies, Queen's University, Kingston, Ontario, 1988. Prospects for minerals in the 90's, Proceeding No. 21.

4. Wen Shicheng, 1996. Characteristics and mineral prospect of mineral resources in China, in Chinese Institute of Geology and Mineral Resources Information (editor), Geological sciences and mineral resources for the 21st Centure.

CONTENTS

2. THE OIL AND GAS RESOURCE POTENTIAL IN THE NORTHERN CHINA SEAS[1]

ABSTRACT

The northern China seas include the Bohai Sea, Yellow Sea and East China Sea, which encompass a combined area of 1,261,000 square kilometres. About 500,000 square kilometres of this area is available for hydrocarbon exploration.

Since the first commercial discovery of petroleum in 1967, a number of hydrocarbon fields or structures have been discovered in the northern China seas. Hydrocarbon occurrences range in age from Pliocene to Proterozoic, especially Oligocene. The current oil production of the sea area is approximately 5,600 t per day and is only from the Bohai Sea area. Recent assessment of resource potential estimates that there are at least 11.5 billion tonnes of oil and 4,280 billion cubic metres of natural gas that occur in the northern China seas. Proven geological reserves are estimated to reach some 0.6 billion tonnes of oil and 50 billion cubic metres of natural gas primarily within the East China Sea.

The Bohai Sea area is an important offshore production base in China and possesses 6 production fields which occupy 80 per cent of the proven reserve of the sea area. Exploration in the East China Sea has confirmed nine oil/gas fields, one of which has been exploited and another for which a development appraisal has been conducted. Significant commercial oil and gas discoveries have not been made in the Yellow Sea, but geological and geophysical information suggests the possibility of finding commercial oil and gas.

The paper briefly discusses the exploration history, geological setting and hydrocarbon resources of the main basins in the northern China seas. Exploration targets with significant hydrocarbon potential are also highlighted.

1.0 INTRODUCTION

There are two large national corporations engaged in the offshore petroleum exploration, exploitation and production of petroleum in China at present. One is the China National Offshore Oil Corporation (CNOOC) and the other is the China National Star Petroleum Corporation (CNSPC). The CNSPC is in charge of the exploration of the entire Chinese land territory and offshore areas, and is under the Ministry of Geology and Mineral Resources (MGMR).

The northern China seas have a total area of 1,261,000 square kilometres, of which about 500,000 square kilometres are available for exploration, including the Bohai Gulf, Yellow Sea and East China Sea.

Geologically, the northern China seas comprise the Bohai Sea Basin, North Yellow Sea Basin, South Yellow Sea Basin, East China Sea Shelf Basin and Okinawa Trough Basin. Petroleum prospecting within the sea area has, to date, concentrated primarily on the Bohai Sea Gulf Basin, South Yellow Sea Basin and East China Sea Shelf Basin, whereas only geophysical surveys have been conducted in the North Yellow Sea Basin and Okinawa Trough Basin.

The geological history of the sea area indicates that it has all the conditions thought necessary and essential for the generation and accumulation of oil and gas. For example, a comparatively thick sedimentary sequence with good source and reservoir rocks; a sedimentary sequence within a suitable thermal window for hydrocarbon generation; favourable trapping conditions for oil and gas; and commercial accumulations of oil and gas have all been found in rocks ranging in age from Proterozoic to Pliocene.

Current oil production of about 5,600 t per day from the Bohai Sea, and significant discoveries during the last decade in the East China Sea Shelf Basin have opened up a new area in these northern China seas for possible

1 Zeng Guangdong and Jia Jianyi, Shanghai Offshore Petroleum Exploration and Development Institute, Shanghai, China

oil and gas production. To meet demand for East China's energy, accelerating oil and gas exploration and development of the East China Sea Shelf Basin is both pressing and important. Drilling during the late seventies and late eighties did not enable the discovery of significant oil or gas in the South Yellow Sea Basin, but oil shows were seen. Futhermore, the onshore extension of the South Yellow Sea Basin has already become an oil-producing province. If intensively explored, the South Yellow Sea Basin could be very promising.

2.0 BOHAI SEA

The Bohai Sea is a nearly closed shallow shelf covering an area of about 97,000 square kilometres, trending northeastward. The average depth of the Bohai Sea is 18 m with a maximum depth of 78 m. The Bohai Sea, linked with the Yellow Sea through the Bohai Strait, is geologically a part of the Bohai Gulf Basin, located at the joint Liaohe, Jiyang and Huanghau regional depression (figure 1).

2.1 History of exploration

In the early 1960s, the Bohai Sea witnessed the first offshore geological and geophysical survey of China, and the year 1966 was marked by the commencement of offshore drilling in China. In 1967, the first commercial oil well in the sea area was established, and this became the first offshore production area of China in 1976. More than 20 oil-gas and hydrocarbon-bearing structures or fields have been discovered to date.

The Bohai Gulf Basin is well known as an oil/gas-rich basin. It has five oil provinces, namely Huabei, Liaohe, Shenli, Zhongyuan and Dagang, which largely support China's petroleum industry. The Bohai Sea lies nearly at the centre of the basin. The sea area therefore has many similarities in the generation and accumulation of hydrocarbons.

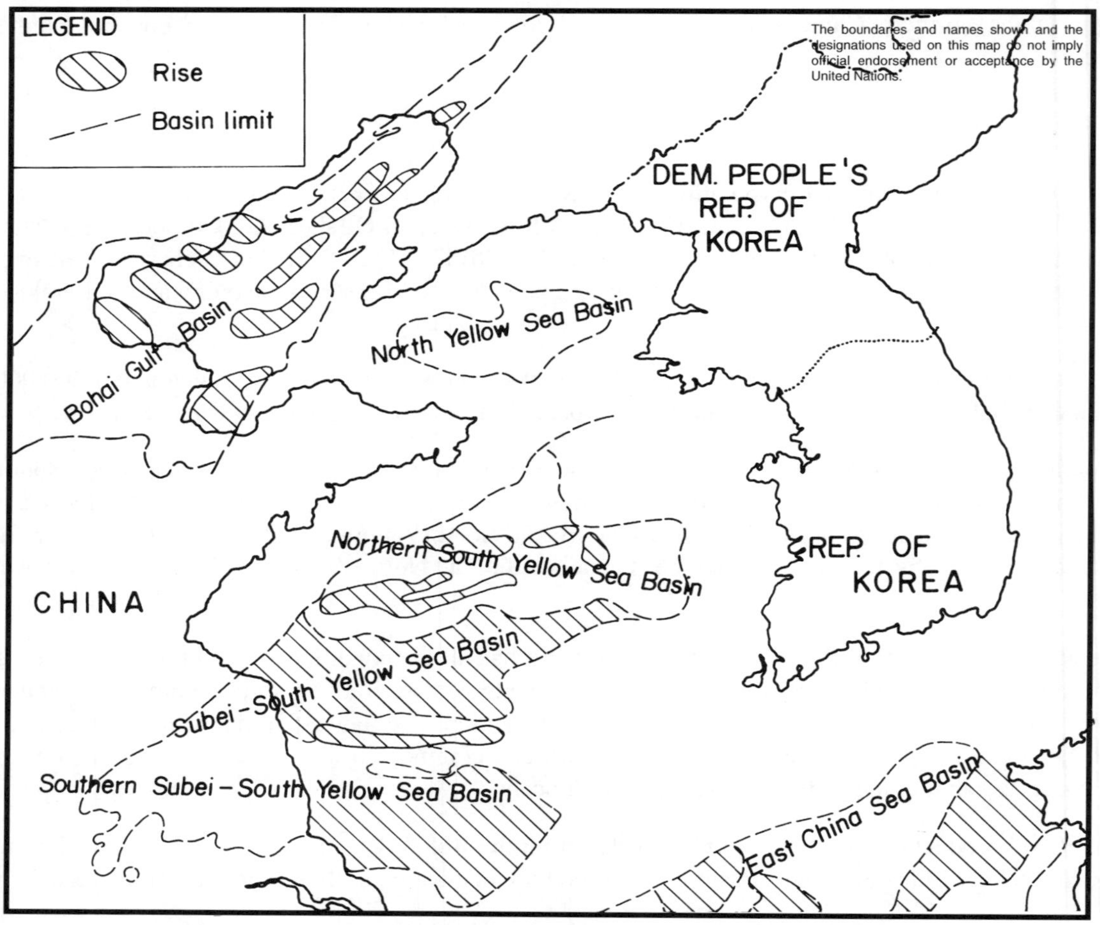

Figure 1. Structural outlines of the Yellow Sea and Bohai Gulf

2.2 Hydrocarbon accumulation geology and resource potential

The evolution of the Sino-Korean block led to the formation of the Bohai Gulf Basin. The Basin was developed in two stages, the platform development stage from the Middle-Upper Proterozoic to Paleozoic period and the rifting development stage of the Meso-Cenozoic period. The break-up of the Sino-Korean block during the Mesozoic resulted in a number of rift depressions filled with coaly and volcanic sequences. Block-faulting activities at the end of the Paleogene were marked by the formation of the Bohai Gulf Basin and characterised by the separation of the rift depressions. The basin eventually became unified in the Neogene.

Due to block-faulting followed by deposition, symmetric rift depressions, or half-rift depressions, were developed on the block-fault bodies. These half-rift depressions controlled deposition throughout the Cenozoic, during which up to 1,000 metres of sediment accumulated.

Under the control of the Paleogene block-faulting activities, the Bohai Gulf Basin was characterised by a series of separated half-rift depressions and lake basins. Each depression and/or lake basin had an independent sedimentary system of the same age.

The geological framework of the lake basins controls the rock type and distribution of reservoirs, whereas, the asymmetric geological base of the rift-lake basins causes the unevenness of sedimentation. Thick sands of sub-lacustrine, alluvial and deltaic fan, and biostromial limestone of oolitic shoals are developed on both the steep and flat flanks of faulted lake-basins respectively. Large-scale sand bodies emerged at the ends of the lake basins, whereas shoe-string sands and turbinate sands are distributed in the central part of lake basins.

Figure 2 shows that the subsidence centres of Bohai Gulf Basin shiftd from the basinal edges to the Bohai Sea area, which led to the thinning of the Eocene Kongdian FM(Ek) and Eo-Oligocene Shahejie FM(Es) formation from the surrounding areas to the Bozhong depression (BZ) – a depression of the central Bohai Sea area. This

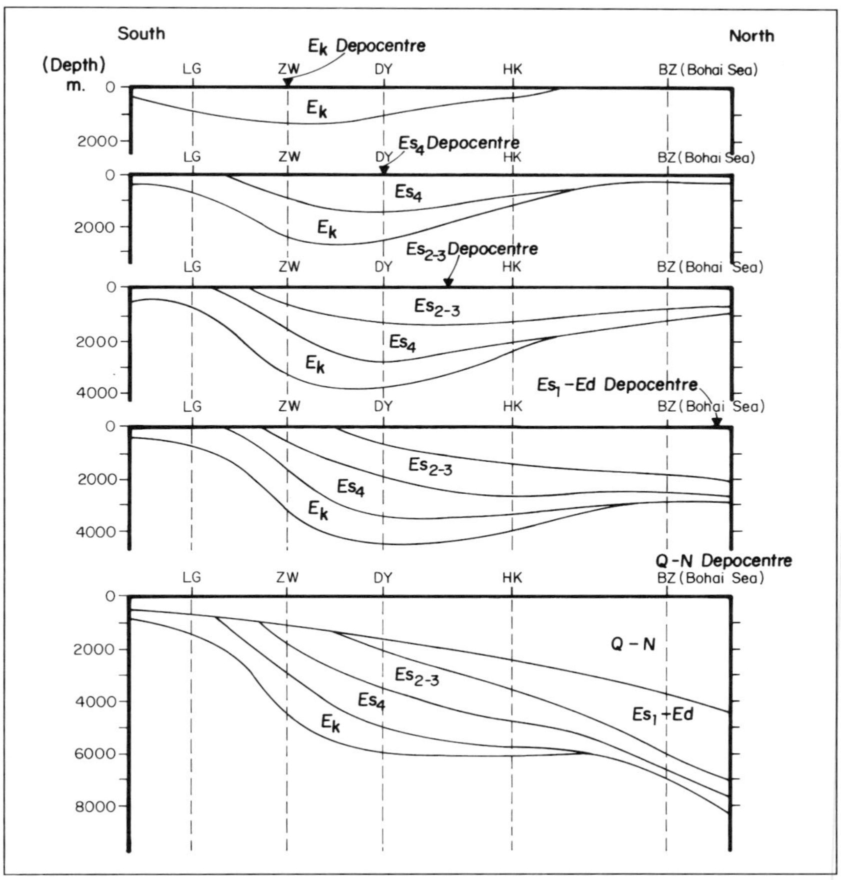

Figure 2. The northwards movement of the depocentre in the Bohai Gulf Basin

resulted in the great thickness of the Oligocene Dongyong FM(Ed) and Neogene. As a result, Ed-Neogene formations serve as the main source rocks and reservoirs, as well as offshore regional covers, of the region.

Because of the opening and block-faulting activities during the Paleogene, a number of fault-bounded rift depressions were formed. Thick Eocene and Oligocene lacustrine sediments, rich in organic matter, were formed in the rift depressions, and are basic units controlling the formation and distribution of hydrocarbons. The rift depression's rich source rocks are characterised by a (1.0-2.7% w) of TOCs, good kerogen type (mainly Type I and Type II) and a high geothermal gradient. As Tertiary sediments are multi-cyclic, delta reservoirs and lacustrine or marine mud provided favourable conditions for the generation, accumulation and preservation of hydrocarbons.

The Bohai Sea region's Paleogene sequence is made up of thick, terrestrial fault controlled deposits which form the main target beds. The Neogene basin is in a united downwarp which generally acts as the cover for oil and gas occurrences. The Cenozoic sedimentary thickness of the Bohai Sea area reaches about 7,000 m and source rocks from the Oligocene Dongying FM(ED) cover an area of about 2,700 square kilometres. The hydrocarbon potential of the area is about 260 x 10^8 t suggesting that the hydrocarbon-generative potential is relatively high. In addition, the sea area has various favourable loci for the accumulation of oil and gas such as drape anticlines, stratigraphic-lithologic composite traps and buried hills. Commercial accumulations of oil and gas have been proven in rocks ranging in age from Proterozoic to Pliocene.

About 0.6 billion tonnes of OOIP and 20 billion m^3 of OGIP (proved plus indicated) are estimated. Currently more than 20 oil fields and hydrocarbon-bearing structures have been found, and five oil fields have been put into production. Two giant-sized fields contain two-thirds of the total geological reserves. Current estimates of potentially undiscovered resources are as follows:

Potential resource:	Oil (10^8 t)	Gas (10^{11} m^3)
Undiscovered resource	49.0	12.8
Economic resource	7.6	2.1

3.0 YELLOW SEA

The Yellow Sea is a half-closed shelf sea with an area of 412,000 square kilometres and an average water depth of 42 m. The sea can be geologically divided into northern and southern sections.

3.1 History of exploration

The north Yellow Sea is an uplift region, wheras the south Yellow Sea is a downwarping region. To date the north Yellow Sea has been studied only by a regional geophysical survey, however, the south Yellow Sea underwent geophysical prospecting in 1968 and during 1974-1979, 7 dry wells were drilled. Several dry wells were subsequently drilled by a foreign oil firm in the late eighties. Some oil shows were seen in two wells. Only the south Yellow Sea is discussed in this paper.

The south Yellow Sea Basin is actually linked with the north Jiangsu Basin, so both of them constitute a united body – the Subei-south Yellow Sea Basin, which is composed of the two basins separated by a central uplift.

3.2 Hydrocarbon accumulation geology and resource potential

The Subei-south Yellow Sea Basin is tectonically located in the north Yangtze block and belongs to an intercontinental faulted downwarp basin. The basin was formed during the late Mesozoic and underwent an evolutionary process from fault to downwarp during the Meso-Cenozoic age. The tectonic evolution of the Subei-south Yellow Sea Basin involved the break-up of the Late Proterozoic Sino-Korean block which resulted in the formation of aulacogen and marginal rifts in the Early Paleozoic to Early Mesozoic; the basement decoupling of the Jurassic to Cretaceous and the Tertiary conversion from rift to downwarp. Since the basin was developed on a marginal subsidence zone between a craton and an active belt, rifting and downwarping alternately emerged. Two-stage rifting activities, namely Paleo-Eocene rifting and Oligocene rifting, were very important in the basin. The sites and conditions for hydrocarbon accumulation are closely related to the development of rift-downwarp.

As a result of a complete rift-downwarp (beginning-developing-rifting-downwarping) during the Tertiary, the basin is characterised by a Neogene downwarp underlain by a Paleogene separated half-graben.

The basin has experienced 6 periods of tectonic movement, resulting in numerous local structures. Seismic data indicate that 191 local structures are found in the sea area, especially in the northern south Yellow Sea. Depositionally, a suite of fluvial-lacustrine fans were deposited in the half-grabens of the basin. Source rocks from the basin are the 5,000-8,000 m thick Paleo-Eocene Funing Group, which is also an important source-reservoir-seal combination in the basin. Onshore data show that the Late Paleozoic Qinglong Group is of hydrocarbon-generative potential and may also be a prospective target for moderate-small sized fields. The northern south Yellow Sea Basin may be even more prospective. The estimates of potential resources areas are as follows:

Potential resource	Oil (10^8 t)	Gas (10^{11} m^3)
Undiscovered resource	5.5	1.0
Economic resource	0.9	0.2

4.0 EAST CHINA SEA

The East China Sea is a marginal sea, with a broad shelf area of 752,000 square kilometres and an average water depth of 349 m. The sea extends like a fan in a NNE direction. The west of the East China Sea is a large stretch of shelf, making up two-thirds of the sea area, whereas the east portion consists of continental slope, sea troughs, island arcs and trenches. In the inner shelf there are many inter-island reefs, beaches and troughs. The outer shelf has a smooth surface and along its outer margin occur islands such as Diaoyu Island and the Chiwei Island.

4.1 History of exploration

Specialists pointed out very early that the East China Sea is the richest source of hydrocarbons in the China seas. In 1974, exploration teams entered the sea area to conduct petroleum exploration and research studies. In 1980, a two year comprehensive geophysical survey was undertaken, covering 600,000 square kilometres with grid sizes of 40 x 80 kilometres. Oil and gas exploration, including drilling and modern seismic, was carried out for important sags and favourable structures. Thirty-two self-operating wells (29 by CNSPC, 2 wells by CNOOC and one well by a small oil company) have been drilled in the East China Shelf Basin to date. Of these, 17 commercial oil/gas wells have been completed, yielding a drilling success ratio of 53.1 per cent 9 fields, namely, DH, BYT, CX, TWT, WYT, DQ, KQT, YQ and CHX, have been proved.

Although the East China Sea has still only had a relatively low level of exploration, more than twenty years of hard work has led to a substantial and distinct outcome of hydrocarbon exploration. For example, the PH field, which is of moderate size, is under construction, and the BYT field is under appraisal. Discovered reserves and exploration costs indicate that the Xihu depression of the East China Sea Shelf Basin is a low-risk and high-yield area for hydrocarbon exploration. Significant oil and gas discoveries have not been made in other depressions due to the extremely low levels of exploration.

4.2 Hydrocarbon accumulation geology and resource potential

The two Cenozoic basins, which developed in the East China Sea with differences in geological age, type, architecture and genetic mechanisms, are:

- the East China Sea Shelf Basin, a marginal rift basin that developed on continental crust and formed during two episodes of rifting of the continental crust as it moved oceanward; and

- the Okinawa Trough Basin, a back-arc basin developed at the transitional zone and formed by the rifting of the continental crust as a result of the subduction of the Pacific plate under the East Asia plate.

The East China Sea Shelf Basin was filled with Neogene and Paleogene deposits and overlapped on the reformed Mesozoic Basin, whereas the Okinawa Trough Basin is Neogene in age.

Due to the evolutionary differences of the two basins, the sediment-filling ages are newer from west to east producing a "South-Marine and North-Terrestrial" framework. As a result of the conversion of the extensive to compressive, structural traps are different, not only from south to north, but also from east to west.

The above mentioned geological features complicate hydrocarbon exploration in the East China Sea and indicate that new occurrence models will be needed for different areas.

There are two major views with respect to the Pre-Mesozoic history of the East China Sea, one, that the East China Sea was an old block of Tathys composed of the Mid-Late Proterozoic metamorphic rock supported by the metamorphic rock of age 1860 Ma in the LF-1 well. The other view sees the East China Sea as a Palaeozoic folded belt based on the extrapolation of the small distribution of late Palaeozoic metamorphic outcrops in the adjacent areas.

It has been proposed that the Pre-Cenozoic history of the East China Sea is similar to that of the marginal areas of east China, and others generally think that the Mesozoic evolution may be correlated with that of Tathys.

The continental crust of the East China Sea was reformed by three-stage pulling-apart in the Cenozoic age. The reformations lay a foundation of hydrocarbon formation in the different periods.

The first pull-apart, along the east flank of Zhe-Min Uplift Folded Zone at the end of the Mesozoic age, created a series of half-rift depressions with east-fault and west-overlap (figure 3), namely the Changjiang, Qiantangjiang, Oujing, Nanredao and Penxi Depressions. From north to south these constitute the so-called west depression zone of 6,000-8,000 m thick marine Paleocene clastics. Geochemical data indicate that the sequence has hydrocarbon potential, and that some oil from LF-1 well was derived from the Paleocene sequence. It is clear that the west depression zone is the target to search for Paleocene hydrocarbons in the East China Sea.

The second pull-apart, at the end of the Paleocene, resulted in the formation of the half-rift depressions along the central East China Sea Shelf Basin: from north to south, these are Fujian, Xihu, Jilong and Xinzhou depression, which form the east depression zone. A 10,000 m thick coal-bearing formation exists in this depression zone, characterised by the previously described "south-marine and north-terrestrial" deposition. All the wells in the depression zone have to a greater or lesser degree, hydrocarbon shows, and commercial hydrocarbon has been discovered in ten of these wells. Oil/gas from the Oligocene and Eocene reservoirs are mainly derived from the Eocene, and as a result the Eocene sequence is undoubtedly the most important target in the East China Sea. The Eocene sediments deposited in the west depression zone were uplifted and seriously denuded as a result of the Yuquan movement at the end of the Eocene. These residual Eocene strata in the zone are also an important target in the East China Sea. The estimated hydrocarbon resources for the Eocene make up about 80 per cent of the whole resource in the East China Sea.

The two above mentioned riftings are probably related to the shifting of the Chinese landmass towards the east as west China was compressed by the Indian plate subducting towards the north and the Eurasian plate shifting southward, from the beginning of the Mesozoic and Cenozoic. Meanwhile, the Pacific plate underwent a shearing movement. All the events caused the east part of the Chinese landmass to creep towards the southeast, resulting in a pull-apart of the eastern part of China.

The third pull-apart occurred after the Mid-Late Miocene in the Okinawa Trough Basin, which is located between the continental shelf and Ryuku Archipelago. This third pull-apart was either caused by the directional change of the Pacific plate, or by conversion from a passive margin into an active margin. The subduction of the Pacific plate made the Okinawa Trough change into a back-arc basin. Seismic data indicates that there is a nearly 10,000 m thick Mid-Late Miocene to Quaternary sedimentary sequence deposited in these rift areas, which are about 800 kilometres in length and 40-80 kilometres in width. The maximium thickness of the sediments is located on the shelf slope instead of the present centre of the Okinawa Trough Basin.

In summary, the west depression zone (the depression formed during the first pull-apart stage), is the main area to search for the Paleocene oil-gas field. The east depression zone, namely the depressions formed in the second pull-apart stage, is the favourable area to explore for Eocene and Oligocene oil-gas fields. The Okinawa Trough Basin (the depression dominating the third stage), is a promising area for prospecting for Late Miocene oil-gas fields.

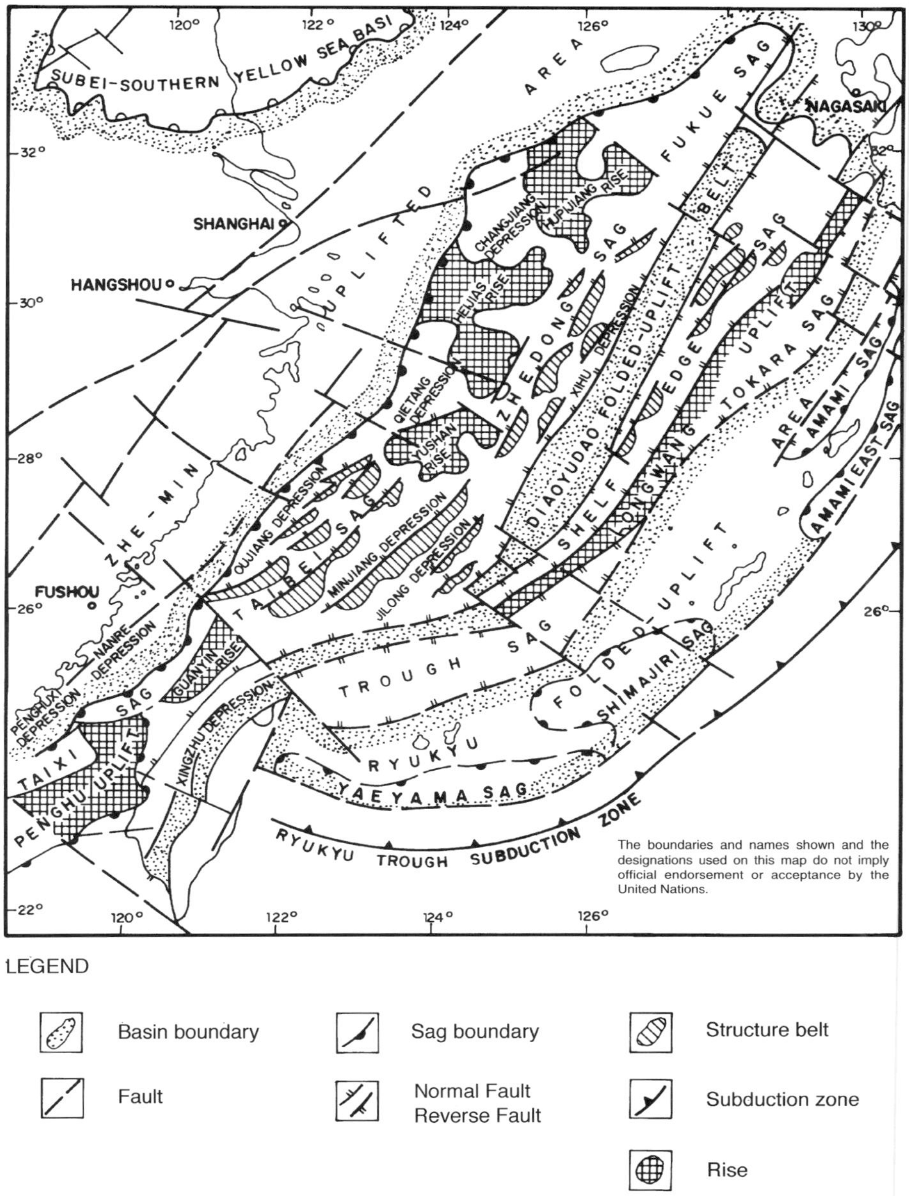

Figure 3. The structural division of the East China Sea

Based on present knowledge, exploration techniques and economic conditions, exploration activities should be concentrated on the east depression zone, with the goal of finding moderate to large fields, and strengthening exploration in the west depression zone area with the hope of discovering additional areas of commercial hydrocarbon. Detailed exploration of the Okinawa Trough Basin will be necessary in the future. Current estimates of potential resources for the East China Sea Shelf Basin are as follows:

Potential resources	**Oil (10^8 t)**	**Gas (10^{11} m^3)**
Undiscovered resources	60.0	29.9
Commercial	9.3	4.8

REFERENCES

1. Li, B.Q. et al., 1991. The Basic Patterns of Oil-Gas Accumulation in Bohai Offshore Area and the Forecasting of Prospective Districts: Oil-Gas Accumulation and Distribution of China, September 1991, Petroleum Industry Press (in Chinese with English abstract).

2. Hu, J.Y. et al., 1990. Geological Texture and Hydrocarbon Accumulation of the Bohai Bay Basin: Meso-Cenozoic Sedimentary Basin of China, 1990, published by Petroleum Industry Press (in Chinese).

3. Zhang, Y.CH. et al., 1990. Tectonic Evolution of Subei-South Yellow Sea Basin: Meso-Cenozoic Sedimentary Basin of China, 1990, Petroleum Industry Press (in Chinese).

4. Zhao, J.H. and Liu, SH. SH., 1995. "Relation between Cenozoic Geological Framework and Oil-Gas Exploration in the East China Sea", in Proceedings of International Symposium on Petroleum Geology in the East China Sea, Shanghai, November 1995.

5. Li, B.Q. et al., 1996. "Hydrocarbon Resource Estimate of East China Sea and Adjacent Regions (1996)" (in Chinese).

6. Gong, Z. SH. and Lin, D.H., 1989. The situation analysis of offshore oil and gas exploration: China Offshore and Gas, v. 3, No. 1, 5-10 (in Chinese).

7. Gong, Z. SH., 1992. The prospects for oil and gas development of China offshore: China Offshore Oil and Gas, v. 6, No. 2, 1-10 (in Chinese).

CONTENTS

3. MONGOLIA'S MINERAL RESOURCES[1]

1.0 INTRODUCTION

Mongolia is a large (1.565 million square kilometre), sparsely inhabited country, sandwiched between the Russian Federation and China in central Asia, with a small population of 2.4 million and a poorly developed economy.

The economy, which is now undergoing the change to a market economy, is expected to be led to recovery by its mineral resources. The mineral industry accounts for 35 per cent of gross industrial output and 65 per cent of Mongolia's exports. The intensive development of the mineral resources is now vital to the growth of Mongolia.

The Mongolian government has a general policy of promoting the mining sector, and the necessary legislative and administrative framework for mineral development has recently been formulated. For the long term, there is a clear need, in order to promote the mineral industry, to develop the national mineral policy and legislation so that it is suitably attractive to foreign investors.

2.0 MINERAL RESOURCES AND MINING

Mongolia is well endowed with mineral resources (figure 4). The country has deposits of fluorite, copper-molybdenum, coal, gold, silver, tin, tungsten, rare earths, lead, zinc, and phosphorite as well is numerous occurrences of other commodities.

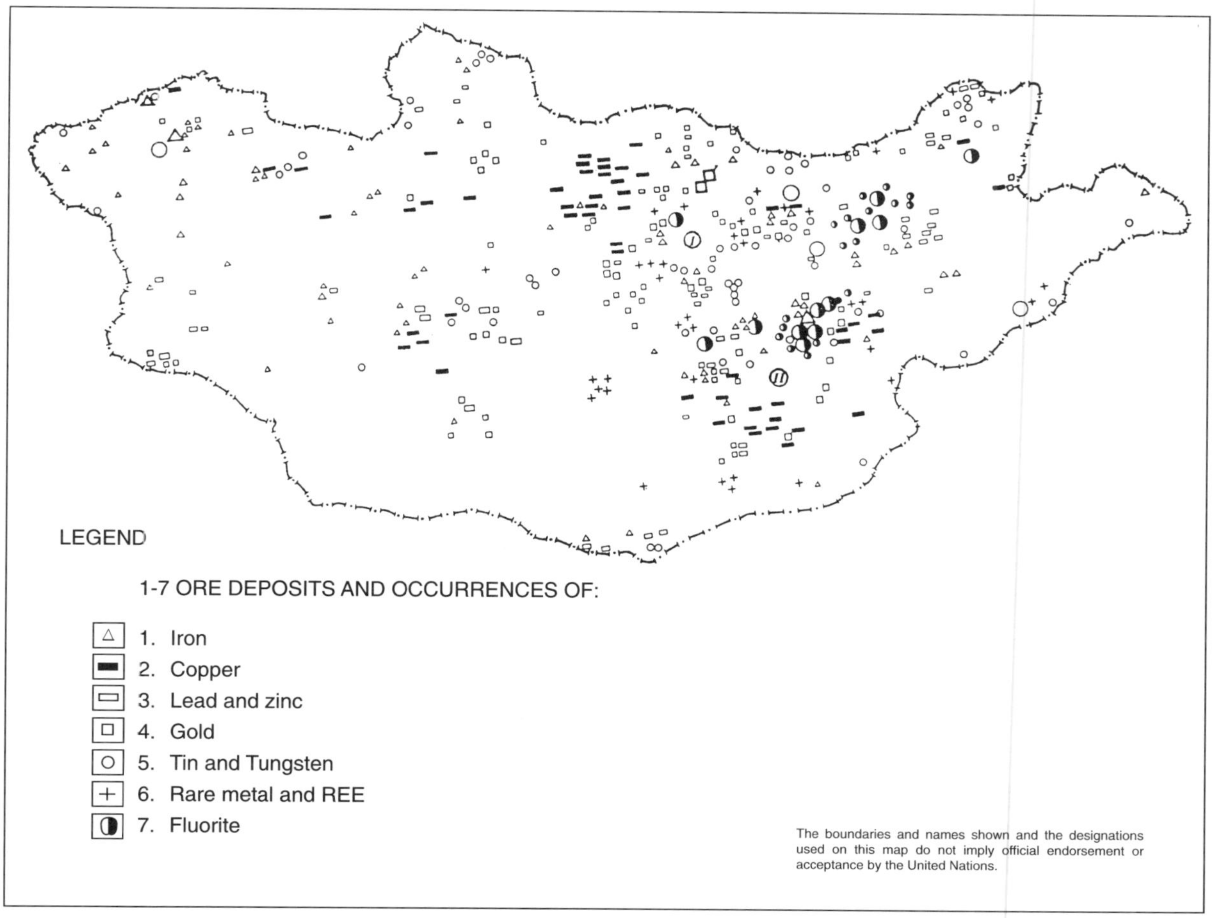

Figure 4. Distribution of ore deposits and occurrences in the territory of Mongolia

---

[1] Ochir Gerel, Mongolian Technical University, Ulaanbaatar 46, Mongolia

Mongolia is the world's largest producer of acid and metallurgical grade fluorite, and produces 15 per cent of the world's fluorspar. In the four year period 1988-1991 Mongolia produced 1,212,400 tonnes of fluorspar. Copper and molybdenum are derived from the Erdenet mine, one of the largest porphyry copper deposits in Asia, and the country's principal earner of foreign exchange. In 1988-1991, Mongolia produced 645,000 tonnes of copper concentrate and 5,957 tonnes of molybdenum concentrate. Fifteen coal mines are active and in 1988-1991 produced 15,024,300 tonnes of coal.

Mine production fell during the 1991-1995 period of political and economic reconstruction. Mongolia's geology and mineral resources have been documented comprehensively since 1939, in a series of collaborative projects carried out with the former Soviet Union and other COMECON countries.

The market for Mongolia's mineral products contracted significantly with the demise of the Soviet system. The loss of major financial aid and technical assistance from the COMECON countries has created an urgent need to obtain new foreign investment to develop the mining sector. In early 1990, geological exploration for oil recommenced through joint ventures with western companies and the Russian Federation. The interest of international mining companies in mineral prospecting, exploration, and mining of hardrock gold, copper, lead, zinc, uranium, tin, and silver continues to grow. Gold is now the most readily saleable commodity, given Mongolia's remote, landlocked position. Bulk-mineable gold deposits are particularly attractive exploration targets. Over 50 Mongolian private companies are working on placer gold mining projects. Gold production in 1996 reached 6 tonnes. Presently there are a total of 200 deposits of coal, copper, molybdenum, placer gold, placer tin, tungsten, fluorite and various industrial minerals being mined by 158 mining enterprises (tables 3, 4 and 5).

3.0 EXISTING MAJOR MINERAL OPERATIONS

3.1 Copper-molybdenum

The mining sector in Mongolia is dominated by the Erdenet copper-molybdenum mine and processing plant, located 350 km northwest from Ulaanbaatar. Erdenet was formed in 1971 as a joint Mongolian-former Soviet Union venture (51 per cent Mongolia, 49 per cent Russia). Present output is 20,000 mt per year of ore, containing 0.8 per cent copper and 0.01 per cent molybdenum. Since 1974 the processing plant has regularly produced approximately 350,000 t of copper concentrate (35 per cent Cu) and 1,500-2,000 t of molybdenum (47 per cent Mo). Copper reserves in the Erdenet area are sufficient for 60 years production. The following types of copper-molybdenum mineralization are known in Mongolia: copper-porphyry, skarn copper, nickel and copper in gabbroids, native copper, sediment-hosted (sandstone- and shale-hosted) deposits, massive sulphide and vein copper (figure 5).

3.2 Fluorite

Mongolia is the world's third largest producer of fluorspar after the Peoples Republic of China and Mexico. Total revenue generation from fluorite mining and processing is estimated at US $33 million per year, which represents 14 per cent of present mineral sector revenues. There are seven main fluorspar mining areas, all concentrated in the Middle Gobi area, 350 km southeast of Ulaanbaatar. Six of the mines are operated by a Mongolian-Russian joint venture (Mongolrostsvetmet).

3.3 Gold

Gold mining in Mongolia is small scale, currently restricted to simple alluvial operations. Prior to 1993, it provided generally less than 1 ton per year and contributed 3 per cent of the sector's revenues. The three main alluvial operations are located at Tolgoit, Shariin gol and Duvunt. A Government programme to promote and encourage gold mining began in 1993, and presently 50 Mongolian private companies are actively undertaking placer gold mining operations at North Khentei, Bayankhongor and Eastern Mongolia. Gold production from alluvial operations in 1993 was 1.2 t, in 1994 1.7 t, in 1995 4.0 t, and in 1996 6.0 t. Two hard rock operations at Bumbat (Zaamar area) and Olon-Ovoot (South Gobi) are operated by foreign and Mongolian companies (figure 6).

Table 3. Geological epochs and mineralization in Mongolia

Stage (geodynamical regime)	*Earth crust development*	*Metallogenic zone, belt*	*Known mineralization*	*Possible mineralization*
Precaledonian	Preriphean basement		Fe, ceramic pegmatites with garnet, apatite-magnetite ores, gold	Rare-metal pegmatites and metasomatites
Late Riphean-early Cambrian	Oceanic	Khentei	Fe, Au	Massive sulphides, polymetal
		Ider-Jida, Lake, Bayankhongor, Central-Mongolian	Au, Cu, Ni, nephrite	Cr, nephrite, massive sulphides, polymetal, Co, Ni, Cu
	Transition	Khuvsgul, Zavhan	Au, phosphorites, nephrite, Mn, Fe	Cr, massive sulphides, polymetal
Early Paleozoic accretion of earth crust	Oceanic	South Mongolian	Au	Massive sulphides with Cu, Zn, Co and Ni
	Transition	Khentei, Govi-Altai-Suchbaatar	Fe	Massive sulphides with Cu, Co and Ni
	Continental	North Mongolian (Ider-Khentei, Lake, East Mongolian)	Fe, Zn, Cu, Au, Pb, Ag, graphite	
Middle Paleozoic accretion of earth crust with continental rifting	Oceanic	South Mongolian, Khangai-Khentei	Massive sulphides, polymetals with Au, Cr, Fe, Mn, magnesite, chrysotile-asbestos, talc	Massive sulphides with Cu, Co and Ni
	Transition	South Mongolian, Govi-Tianshan, Govi-Altai-Suchbaatar	Polymetals, Cu, Au, Ag, graphite	
	Continental	North Mongolian Ider-Selenga	Fe	
		Mongol-Altai	Fe, polymetals, Cu, W, Mo, Au, rare elements	Mo, W, Sn, rare elements
		Khuvsgul	Fe, Al, V, graphite	

Late Paleozoic continental, accretion and rifting	Oceanic	Suliinkher	Cr	Massive sulphides with Ni
	Transition	Totoshan		
		North Mongolian	**Cu**, **Mo**, W, Au	Rare metals
	Continental	Central Mongolian	**Cu**, Mo, **Pb**, Au, Ag	
		Mongol-Altai	W, Mo, Au, Pb, Zn, Cu	
		South Mongolian	**Cu**, Mo, REE, Nb, Zr, Au, Ta, Sn	
Early Mesozoic reworking of continental crust	Continental	Khentei	**Au**, W, Sn, Ta	
		Khangai-East Mongolian	**W**, **polymetals**, Cu, **Ta**, **Au**, Fe	
		South Eastern	W, Mo, polymetals, Cu	
Late Mesozoic reworking of continental crust	Continental	Internal	**Sn, W, Au,** polymetals, Cu	Fluorite, Hg
		Closed	**W**, **Au**, polymetals, Cu, Sn, **Ta, Be, fluorite**	
		Govian	**REE, Be, Ta**	
		North Western	**Ag**, Bi, As	
Cenozoic	Continental	Khangai-Khuvsgul	Gem stones	Rare elements
		Dariganga	Gems stones	Rare elements

Table 4. Mongolia's main mineral production in the last 15 years

Commodity, product	*Units*	*1980-1984*	*1985-1989*	*1990-1995*
Copper, concentrate	1,000 ton	1,229.5	1,773	1,935.6
Molybdenum concentrate	ton	12,067.4	16,141	24,095
Fluorspar, ore	1,000 ton	3,319.9	3,023.8	3,280
Fluorspar, concentrate	1,000 ton	–	342.5	601.3
Gold, metal	ton	5,000	5,000	9,000
Coal	1,000 ton	24,005.3	37,998.6	35,930.4
Tin, concentrate	ton	308	814.6	396.8
Tungsten, concentrate	ton	112.9	194.1	–
Lime	1,000 ton	429.1	544.5	416

Table 5. Mongolia's mineral production 1994-1995

Commodity	*Units*	*1994*	*1995*
Cement	1,000 ton	85.8	108.8
Coal	1,000 ton	4,980	5,003
Anthracite and bituminous coal	1,000 ton	1,430	1,445
Lignite and brown coal	1,000 ton	3,550	3,558
Copper-mine production	ore 1,000 ton conc. 1,000 ton met. 1,000 ton	20,377.3 408.8 120.16	20,050.3 415.7 121.9
Fluorspar	ore 1,000 ton conc. 1,000 ton	383.2 88	526.9 122.5
Gold	kg	1,742	4,080
Molybdenum	1,000 ton	3.956	3.531
Tin – mine production	ton	53.4	59
Tungsten	ton	16.67	61.6
Salt	1,000 ton	67.924	86.826

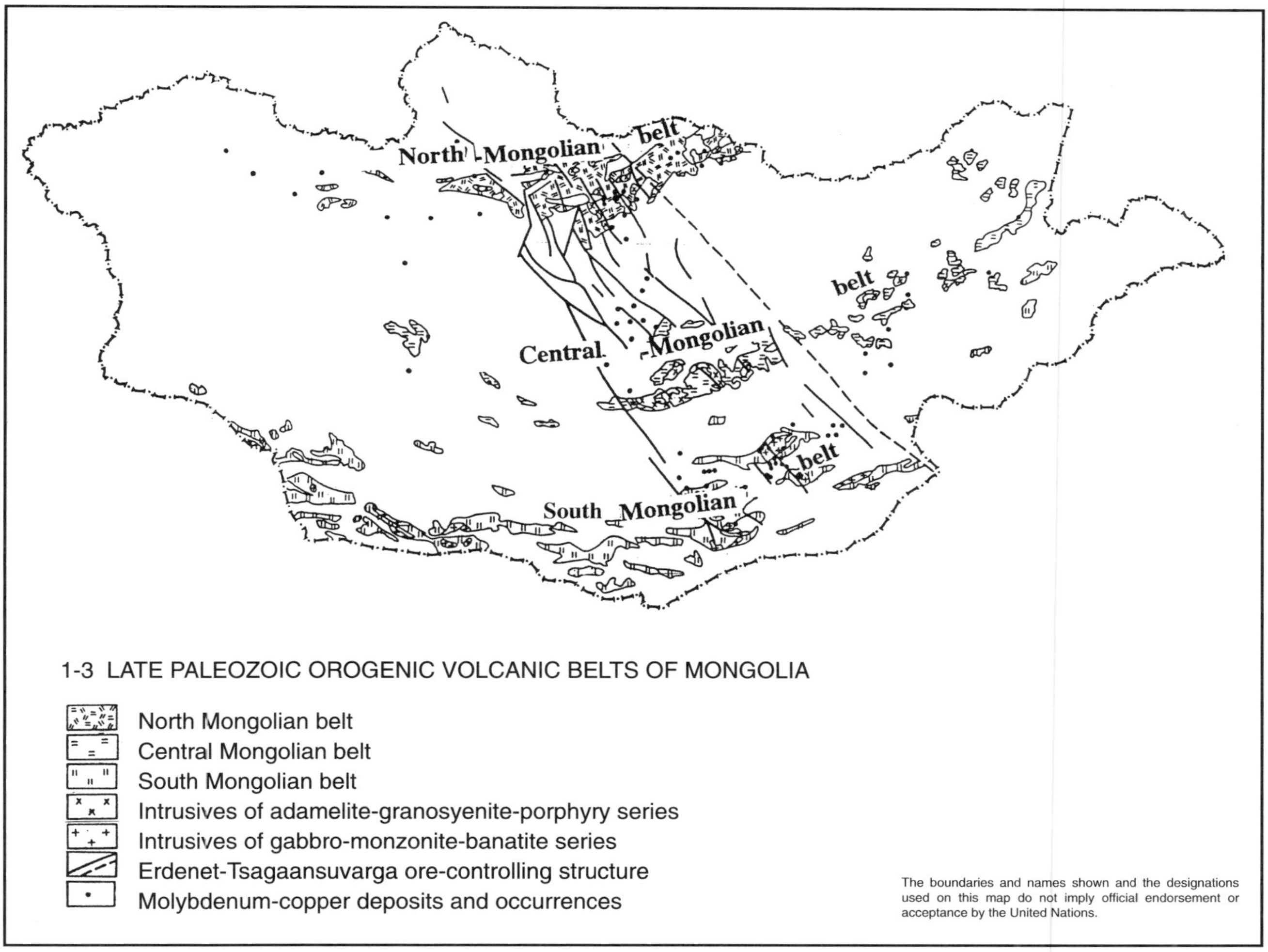

Figure 5. Molybdenum-copper metallogenic belts of Mongolia

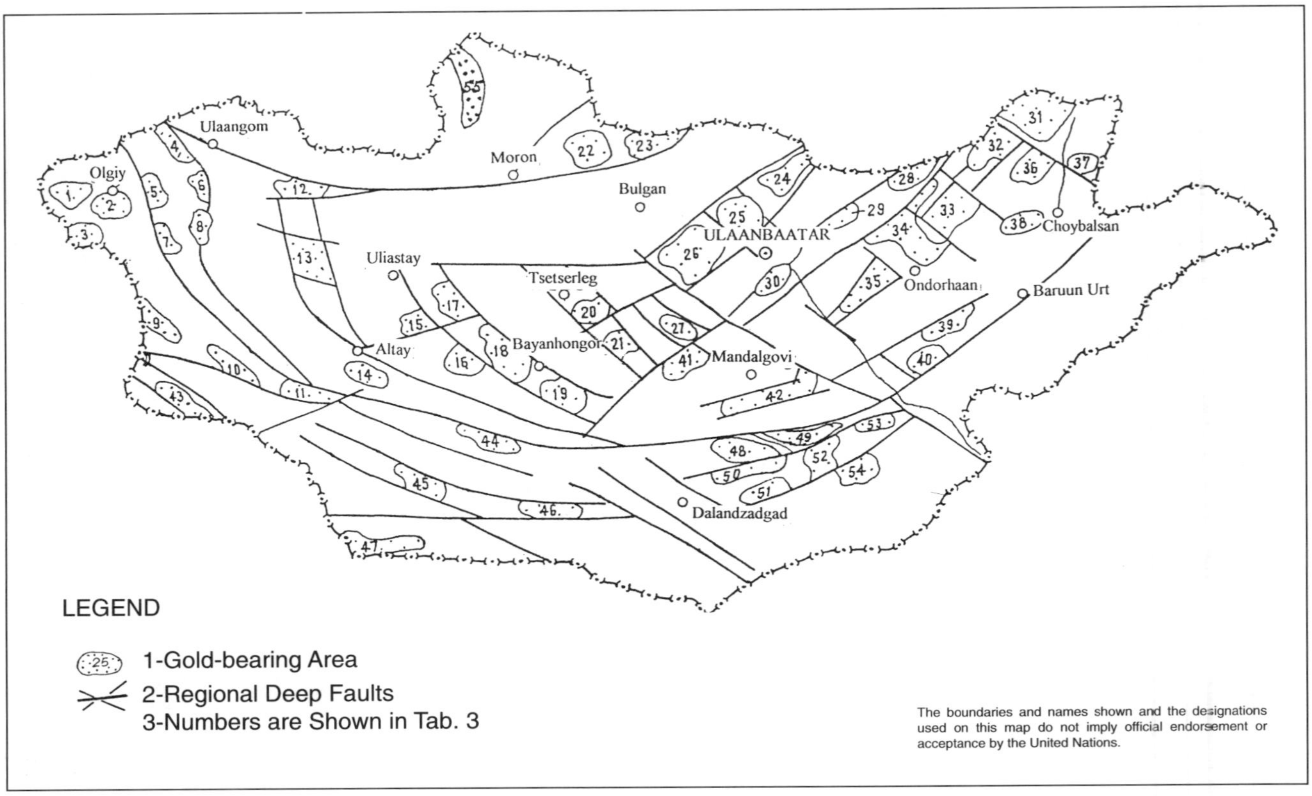

Figure 6. Distribution of gold mineralization in Mongolia

3.4 Coal

Coal mining in Mongolia began in 1912 and increased dramatically through the 1970-1980s in response to the fuel demands of the power sector. More than 70 per cent of present coal production is from the mines at Baga Nuur (4 million tonnes per year) and Sharyn Gol (1.5 mt). Of the 15 operating mines, all but one are open pit mines.

3.5 Tin and tungsten

Other mineral exports of economic note are tin and tungsten, which together contribute only 0.2 per cent of mineral sector revenues, amounting to less than US$ 1 million per year. Tin has been mined by dredge at Modot, by Russians in 1943 and in 1970s by a Mongolian/Czech joint venture. A new placer tin mine is under construction at Janchivlan near Ulaanbaatar, to be operated as a Mongolian venture. Historical production of tungsten concentrates reached 125 tonnes in the late 1970's (Tumentsogt, Burentsogt and Yuguzer mines) but production has declined due to market conditions and was less than 50 tonnes in the 1980's. Production in 1990 was 45 tonnes of concentrate. A Mongolian-Hungarian joint venture developed the Tsagaan Davaa tungsten operation, 100 km northwest of Ulaanbaatar, but because of depressed tungsten prices the operation was stopped in 1990.

Several other mineral operations for lime, building and facing stones, gemstones, and salt, exist mainly to satisfy local needs.

4.0 MINERAL POTENTIAL

Mongolia has a great potential for future mineral discoveries and development. In Mongolia, there are currently more than 6,000 deposits and occurrences of about 80 different minerals, such as copper, lead, zinc, iron ore, tin, tungsten, gold, silver, gemstones, graphite, coal, oil, marble, phosphorites and rare earths (figure 4). The mining industry accounts for 18 per cent of Mongolia's industrial product and over 60 per cent of Mongolia's total export earnings. However, bearing in mind the currently depressed mineral market, the relatively early stage of exploration, the difficult geographic conditions and infrastructure deficiencies in many areas, the potential for economic mineral development is more limited.

Developing an export oriented mining industry with foreign investment, is a crucial element in Mongolia's development strategy. The government has prioritized potential mineral development areas and opportunities based on market conditions, price forecasts and the availability of infrastructure. Short and medium-term mineral prospects are quite well defined. The main issue the government needs to currently address to facilitate the flow of foreign private capital investment into the sector, is the question of the government's equity involvement in new mineral ventures with foreign interests. In particular, the government should be encouraged to permit 100 per cent foreign investment in future hard rock mines.

Recent exploration has identified the potential for expansion of alluvial gold operations and the development of hard rock open pit and underground gold mines. More than 10 foreign companies such as RTZ, BHP, Molopo, Quincunx, Mongolian Gold and Golden Tiger have commenced prospecting and gold mine development. This represents the first entry of western capital into Mongolia's mineral sector.

The Erdenet mine produces some 350,000 tonnes of copper concentrates, representing 112,000 tonnes of contained copper metal per year. The government correctly sees as a priority, a feasibility study for the development of an in-country smelting operation to gain the value-added benefit of exporting refined copper, whether in cathode or blister form.

The main porphyry copper type deposits for future development are in the South Gobi belt, and include such deposits as Tsagan Suvargaa, Shuten and Kharmagtai. Other first priority mineral developments are lead zinc deposits such as the Ovoo skarn deposit with 12 per cent zinc, and the Ulaan-Tsav hydrothermal polymetallic deposit in east Mongolia (figure 5).

4.1 Oil

As a result of oil exploration from 1931-1969, and more recently in the 1990s, several oil fields and oil seepages have been discovered: Zuubayan, Tsagaan Els, Uhin, Baruunbayan and Tamsag (figure 7). About 80 per cent of oil exploration work and oil development was oriented to the Zuunbayan and Tsagaan Els fields in the East Gobi basin, which are being developed by Bayan Oil Co., an American (Nescor Energy Corporation) and Mongolian joint venture. The Zuubayan oil field produced a total of 550,000 tonnes of oil, and has total reserves of 6.336 million tonnes with total recoverable reserves estimated at 1.759 million tonnes. The Tsagaan Els oil field's recoverable oil reserves are estimated at 950,000 tonnes.

After more than 20 years of dormancy, a new period of Mongolian oil exploration history commenced when democratic change brought in an open-door economic policy in the 1990's. The foreign investment law, announced on May 1, 1990, permitted foreign investments in all economic sectors, particularly the mineral sector. The Mongol Petroleum Co. (Mongol Gazryn Tos-MGT) was established in May 1990 for the development of the oil and gas resources of Mongolia.

During the last few years the former MGT has entered into joint venture agreements with various technical contractors to assemble, evaluate, and market Mongolian geological and geophysical data.

Twenty two contract areas have been defined around Mongolia's Mesozoic sedimentary basins.

4.2 Coal

Coal is a leading mineral and the principal source of heat and electrical energy in Mongolia.

The country's coal production started in 1912. In 1989 the output reached 9.0 mt, 90 per cent of which was mined by opencut methods: 80 per cent of the produced coal is from the Baganuur (4 mt), Sharyn Gol (2 mt) and Aduunchuluun (0.4 mt) open pits. The Naliah (0.5 mt) underground mine was closed in 1993 due to technical and economic problems. To date, over 40 coal deposits have been found and exploited. Of these 18 are being actively

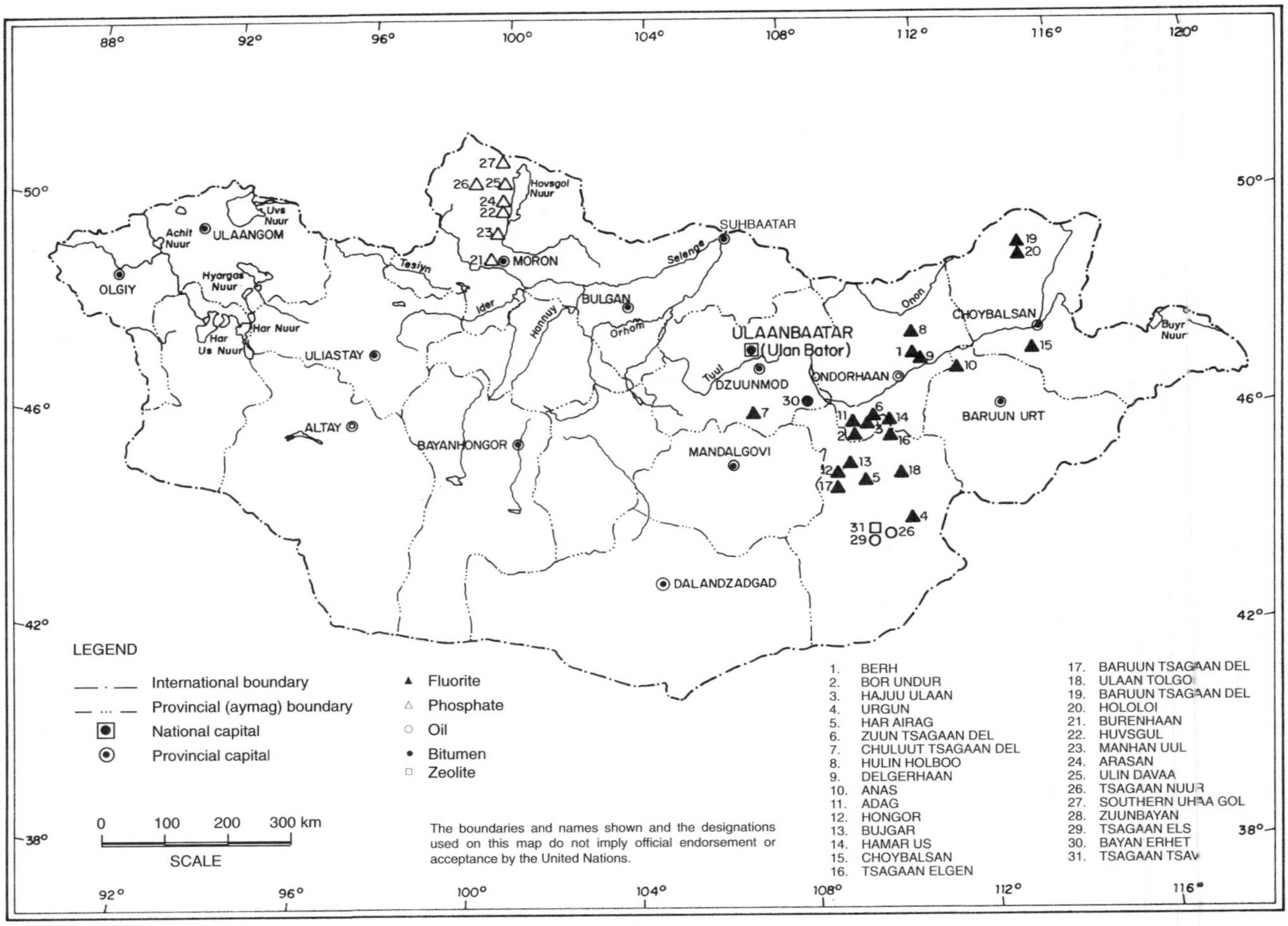

Figure 7. Distribution of non-metallic deposits of Mongolia

mined and 9 have been prepared for development. As of 1995, identified coal reserves amounted to 10,000 mt, including 3,500 mt of proved reserves of which over 1,000 mt are coking coal.

Coal deposits are irregularly distributed. Over 90 per cent of overall reserves are concentrated in the central economic region. Coal resources were formed during four coal-forming epochs: the Carboniferous, Permian, Jurassic and Cretaceous. The deposition of coal in the Carboniferous and the Permian took place in the superimposed basins of the post folding development of the region while the Jurassic and the Cretaceous coals were in superimposed basins resulted from tectonic rejuvenation. The bulk of the coal mined in Mongolia is for power generation plants.

Carboniferous coalbearing formations are developed in the Mongolian Altai and Harhiraa basins which are characterised by grey terrigenous sediments which fill the basins. The surface area of these basins is 60,000 square kilometre (Hahiraa basin) and 37,000 square kilometre (Mongolian Altai) and undiscovered resources in the above basin are estimated to be 26 billion tonnes and 1.5 billion tonnes respectively. Carboniferous coals constitute high calorific power-generating fuels and are also mined on a small scale for local needs. The principal mining areas include the Hartarvgati, Nuurst hotgor, Huden, Hundlun and Zeegt [of which] the former is best studied (figure 8).

Late Permian sediments, with potentially economic coal, are developed mainly in south Mongolia where they fill east-west trending inter-montane troughs. Cross-sections show that thick marine terrigenous sequences give way, in the upper part, to the coastal plain and lagoon facies of a sea retreating to the south. Deposits of Permian coals form the South Gobi coal basin, including the Gurvantes, Nomgon and Urlan nuur coal depressions and the Tavantolgoi, Gurvantes, Narinshuhait and other deposits.

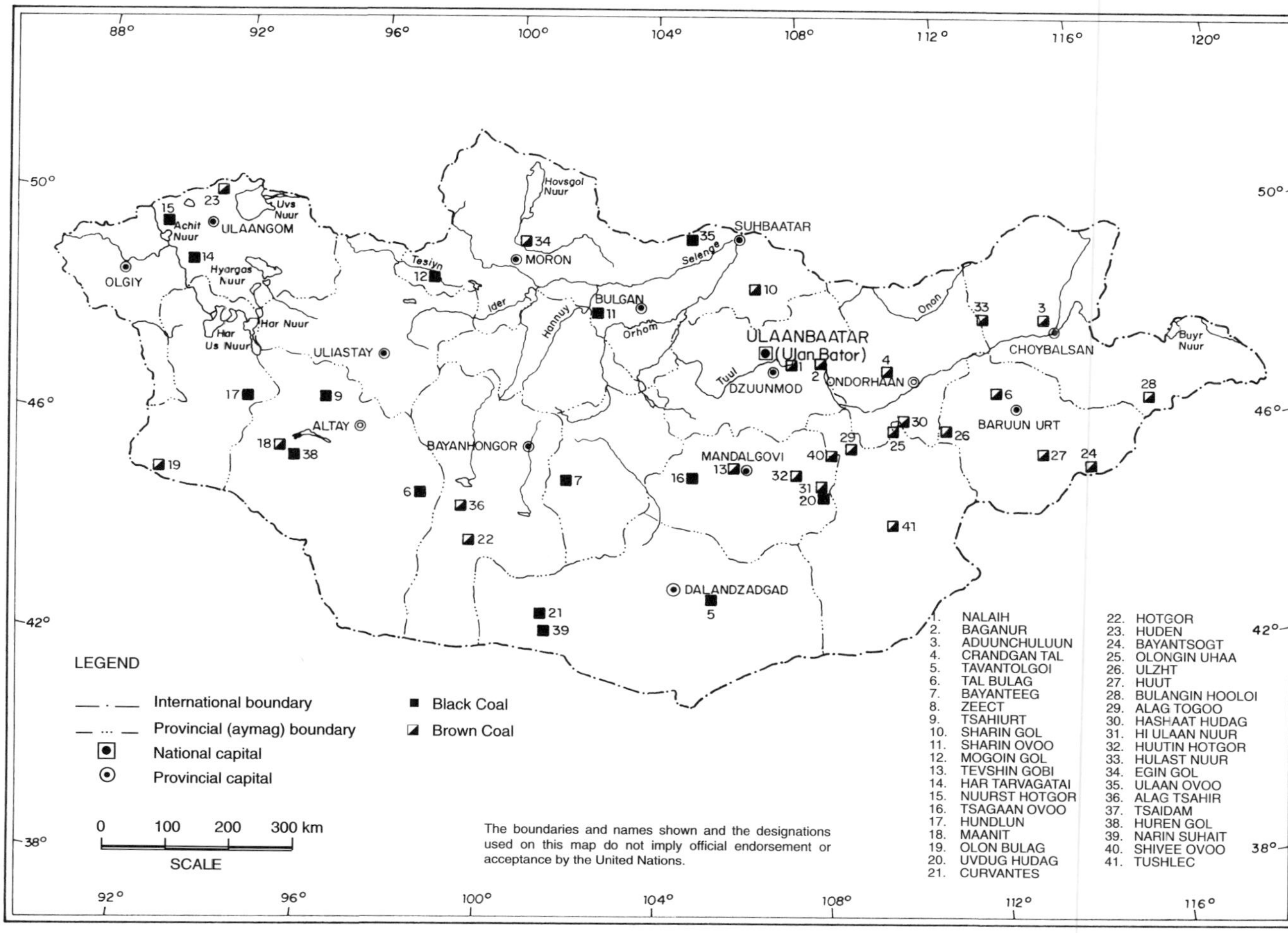

Figure 8. Distribution of major coal deposits of Mongolia

The Tavantolgol deposit is unique in terms of the size of the reserves and quality of the coal. It is located 90 km east of Dalanzadgad, the administrative centre of the South Gobi province. The central part of the deposit has been studied to a depth of 340 m. Other areas of the Ulaan nuur basin (Uhaa hudag, Eastern and Bortolgoi localities) have also been evaluated to a depth of 340 m. Overall resources of coals from the Ulaan nuur basin, including the Tavantolgoi deposit and Uhaa hudag, Bortolgoi and eastern areas, to a depth of 940 m (maximum depth of coal seams), are estimated to be 7 billion tonnes, including 2.5 billion tonnes amenable to coke production and 1.5 billion tonnes of self-coking coal. On this tonnage 3.5 billion tonnes can be mined by the open pit method. The overall coal reserves of the Tavantolgoi deposit proper are estimated to be 5 billion tonnes of which 2.8 billion tonnes are amenable to open pit mining. The proven coal reserves, to a depth of 300-340 m, amount to 1.2 billion tonnes, and the probable reserves 300 mt (totalling 1.5 billion tonnes), of which 800 mt and 200 mt, respectively, are charge coking and 500 mt are self-coking.

The economic potential of coal in Jurassic basins is well known. The basins occur in the western and central parts of the country in the Sharyn gol, Bayanteeg, Mogoin gol, Saihan Ovoo and Ulaan Ovoo deposits. The Sharyn gol deposit consists of Middle Jurassic terrigenous rocks and the thickness of coal seams ranges from 4.0 to 34.0 m. The deposit, exploited since 1964, has a current production of 2 mt/year.

Cretaceous deposits of coal and oil shales were formed in the central, south and eastern portions of Mongolia.

Mongolia's large proven coal reserves could sustain a considerable increase in coal production. Among the present day, highly mechanized coal-producing enterprises, 84 per cent are operating with the technical assistance of the former Soviet Union. There is considerable potential to broaden the production of brown and hard coals including coking coals, due to both the short distances from railroads and because of the large coal-producing targets: Baganuur, Shivee Ovoo and others. The bulk of the coal reserves prepared for development (95 per cent) are amenable to modern open pit mining.

One of the problems characterizing the coal industry of Mongolia is the limited market, namely the power-generating industry. The possible production of liquid synfuels from coal, gasification, manufacturing of metallurgical coke and other possible products remain unsatisfactorily studied. The use of coal without the prior extraction of by-products is comon, although the possiblity of commercial concentrations of valuable by products, germanium in particticular, has been shown to occur in the Mesozoic coals of eastern Mongolia. It has also been found that oxidized coals may be used in agriculture for enhancing soil fertility and coal ash may be used in cement production.

4.3 Gold

More than 40 favourable ore-bearing structures (figure 6) have been discovered in the territory of Mongolia. Some of these structures form elongated ore zones, such as North Hentii and Bayankhongor (Sillitoe et al., 1996). Gold deposits and occurrences are classified into 5 groups and several types (Dejidmaa, 1996). The major groups include Archean, Phanerozoic-mesothermal, epithermal, intrusion related, gold in ancient sea floor hydrothermal systems and ancient placers.

Archean to Early Proterozoic metamorphic terrains which belong to granite-greenstone belts are few in number. The Baidrag zone is the most prospective for the discovery of Archean-early Proterozoic lode gold deposits. Few silicate-oxide iron and gold-sulphide-quartz replacement occurrences have been discovered. The Kharaat uul gold occurrence is represented by pyritized, silicified replacement of gneisses and amphibolites. Host gneisses are pyritized, silicified, sericitized and chloritized, and cut by quartz-pyrite veins intensively oxidized at the surface. The grade of gold varies from 0.3 to 10 g/t on the surface.

Phanerozoic mesothermal gold deposits and occurrences are distributed widely in ancient tectonically active continental slope areas of Late Proterozoic and Early Paleozoic (Ordovician-Devonian) age. They are presented by turbidite-hosted (black and green schist) mesothermal veins and disseminated veining classes. The veins are low sulphide gold-quartz occurrences whereas the disseminated veining is a high sulphide type. There are a few occurences of the gold-quartz type mineralisation within the Upper Proterozoic and some occurences of disseminated sulphide-quartz veining in the Vendian-Early Cambrian within the Baidrag-Burdyn gold district. Deposits such as the Olon Ovoot and other occurrences of Devonian age are distributed in south Mongolia, related to an active continental margin and related spatially to concordant dykes and sills of diabase, gabbro and diorites.

Epithermal gold occurences in volcanic terrains are distributed in southern and eastern Mongolia within East-Mongolian Late Mesozoic and South Mongolian Late Paleozoic volcano-plutonic belts. They are subdivided into adularia-sericite and alunite-kaolinite classes.

Intrusion related gold deposits and occurrences are subdivided into five classes and several types. The intrusion hosted/disseminated class are related to porphyry type deposits. Skarn deposits and occurrences are represented by gold-copper and gold-copper-iron types distributed throughout central and eastern Mongolia (Late Mesozoic, Early Mesozoic and Late Mesozoic). Stockwork, disseminated, and replacement in non-carbonate rocks, are represented by widely spread gold-scheelite stockwork and gold-sulphide-quartz veining replacement types.

Vein gold deposits are the best studied and exploited in Mongolia. The gold content of known deposits varies from a few hundred kilograms up to 30 tonnes and grades range from 1.5 g/t up to 26 g/t, with proved reserves from 4t up to 15t (Tsagaan tsahir uul). Mineral composition ranges widely, becoming more complex from older to younger. Wall rock alteration is developed very widely.

Known gold deposits and occurrences developed in *ancient seafloor hydrothermal systems* which occurred in western Mongolia in the Late Proterozoic-Early Cambrian "Lake" zone and include volcanogenic gold-rich massive sulphides and metalliferous sediments. The *ancient gold placer* group has been insufficiently studied. However, this group is a primary source for several recent gold placer deposits.

4.4 Copper

Porphyry copper-molybdenum mineralization is located within three metallogenic belts (figure 5) North Mongolian, Central-Mongolian and South-Mongolian (Sotnikov et al., 1984, Gerel, 1990). These metallogenic belts coincide with sublatitudinal Late Paleozoic (south and central Mongolia) and Late Paleozoic-Early Mesozoic (north Mongolia) volcanic belts. These belts, controlling copper mineralization, preserve the main orientation of Hercynides and are superimposed on the Hercynides structures, indicating a genetic relationship between the mineralization and deep faults. Paleozoic accretionary wedges and magmatic arcs are believed to host the main porphyry copper deposits of Tsagaan Suvarga and Kharmagtai in the South Mongolian belt; Erdentiin Ovoo in the North Mongolian belt and Bayan Uul in the Central Mongolian belt. These four porphyry-copper deposits display different characteristics and erosion levels, but all are associated with magmatic complexes, including trachyandesitic volcanics of elevated alkalinity (Sillitoe et al., 1996). The Bayan uul porphyry copper prospect is dominated by pyrite-rich sericitic and quartz-tourmaline alteration representing a shallow erosion level (Koval et al., 1988). The even shallower, epithermal parts of the porphyry copper systems are also preserved in the South Mongolian belt where lithocaps, characterized by pyrite-rich quartz-alunite and quartz-pyrophyllite alteration of trachyandesitic volcanic rocks, occupy extensive areas at Shuteen and in the Ikh Shanhai range. The total reserves of Tsagaan Suvarga deposit have been estimated at 1,280,000 t at 0.53 per cent Cu, and a total Mo content of 43,600 t at 0.018 per cent Mo.

4.5 Lead and zinc

In the 1970s and 1980s, large and medium scale deposits and a number of promising ore occurrences, such as Ulaan, Muhar, Tsav, Tumurtin Ovoo, Bayan uul, Altantolgoi, Mongon Ondor and others were discovered and explored in the eastern part of the country. The deposits were mainly formed in the Mesozoic and are concentrated within the Central Mongolian volcanic belt, usually occurring in marginal parts of Late Mesozoic depressions, filled with Upper Jurassic-Lower Cretaceous trachybasalt-trachyrhyolite and latite series volcanics (Gerel, 1990). The known base metal deposits belong to four types: *mineralized fluid-explosive pipes, veins in various rocks, mineralized shear zones in rocks of various composition, and aposkarn metasomatic bodies in limestones*.

Total reserves of the main metals from the Ulaan deposit are 830,000 t lead, 1,815,000 t zinc at average contents of 1.0-1.45 per cent and 0.4-4.44 per cent respectively. These reserves include 430,000 t lead at average content of 1.1 per cent and 757,000 t zinc, at average content of 1.92 per cent which are amenable to open pit mining. Silver content ranges in the ore bodies from 30 to 58 g/t. Potential resources from deep horizons are estimated to be 300,000 t lead and 600,000 t zinc.

4.6 Rare metal

Rare metal mineralization in Mongolia is represented by three occurrences of rare metals (Sn, W) and rare elements (Ta, Nb, Zr, Be, Li, Cs, Sr and REE) which can be divided into two groups: *magmatic and hydrothermal* (Kovalenko & Yarmolyuk, 1995).

The magmatic group includes rare metal granites (Ta bearing lithium-flourine granites and Nb-REE-Zr-bearing peralkaline granites), pegmatites, ongonites, cesium glasses, berillium tuffs, Nb-REE-Zr pantellerites and comendites, rare metal albite-bearing nepheline syenites, REE-bearing carbonatites and magnetite-apatite alkaline volcano-plutonic rocks. The hydrothermal group comprises hydrothermal deposits of Sn, W and Mo with a significant enrichment of rare elements (Ta, Li, Be) and fluorine.

Rare metal mineralization in both groups is closely related to igneous rocks of a relatively narrow compositional range including peralkaline granites, lithium-fluorine granites, lithium-flourine granites, leucogranites, nepheline and pseudoleucite syenites, and corresponding volcanic rocks. Four metallogenic epochs of magmatism and rare metal mineralization are distinguished: (1) Middle Paleozoic, (2) Late Paleozoic, (3) Early Mesozoic, and (4) Late Mesozoic.

Prognostic resources of tin are estimated to be 30,000 t in ore with an average Sn contact of 0.4-0.5 per cent. The major deposit types include the Narsan Hundlun cassiterite-silicate type which contains 40,000 t of tin, in ore averaging 0.6 per cent; the stockwork system of the Ortsog deposit which has reserves of 175,000 t of WO_3 and 26,000 t of Mo; and the Ongon-Khirhan deposit with 22,000 t of tungsten averaging 0.14 per cent.

Prognostic resources of the Khalzan Buregtei deposit (western Mongolia) are estimated to be 600,000 t for niobium oxides, 35,000 t for tantalum oxides, 4 million tonnes for zirconium oxides and more than 100,000 t for yttrium oxides. Prognostic resources of REE deposit of Mushgai hudag (South Gobi) are estimated to be 3 million tonnes, the average content of ore being 1.5 per cent.

4.7 Silver

Silver is an accompanying element in the composition of gold-sulfide-quartz, polymetallic ore deposits. Basic silver resources are in the Asgat (western Mongolian) and Mongon Ondor (eastern Mongolia) deposits. Both occur as silver sulphides in mineralized zones within the terrigenous sediments. The Asgat deposit's reserves were estimated at 2,600 t silver, 83,400 t copper, 37,000 t antimony and 4,000 t bismuth (evaluated and explored resources). The Mongon Ondor deposit reserves were estimated as 280,000 t of lead, 230,000 t of zinc, 24,000 t for tin and 3,212 t of silver (evaluated and prognostic resources).

4.8 Iron

A number of iron deposits with small and medium-scale reserves, and numerous occurrences, have been discovered in Mongolia. These are located in the northern part of the country. The most widespread deposits and occurrences are related to Proterozoic and Vendian-Cambrian metamorphic volcano-sedimentary and less frequently in Devonian cherty terrigenous rocks. Skarn-magnetite types are characterized by a high content of iron: 50.4-54.76 per cent (Tomortei), 34.57-57.06 per cent (Bayangol), 52.27-54.76 per cent (Tomortolgoi) with proved reserves of 137 mt-158 mt and probable reserves from 51 mt up to 165 mt for the Bayangol zone.

4.9 Uranium

Uranium exploration in Mongolia started immediately after World War II and was conducted jointly by the geological organisations of Mongolia and Russia until 1966. Numerous uranium occurrences were discovered in lignite deposits, and after 1966, exploration became more systematic and about 650,000 square kilometres or 40 per cent of the country was covered by airborne radiometric surveys at scales of 1:25,000 and 1:1,000,000. A metallogenic appraisal of undiscovered uranium resource potential was done in an area of 500,000 square kilometres and drilling, trenching and underground exploration was conducted in an area of about 50,000 square kilometres. Based on the results of these investigations, four uranium-bearing provinces were defined: Mongol-Priargunian, Gobi-Tamsag, Hentei-Daurian and North Mongolian districts. Within these provinces six uranium deposits, about 100 uranium occurrences and 1,400 mineral showings and radioactive anomalies were identified (figure 9).

4.10 Phosphorite

Mongolia possesses significant reserves of marine sedimentary-metamorphosed sheet phosphorites. These are concentrated in the Huvsgul phosphorite-bearing basin. Occurrences of phosphorites similar to those from the Huvsgul basin have also been found in the carbonate sequences of the Zavhan zone of Riphean-Vendian, but to date they have not been investigated. The Huvsgul basin covers a band 300 km long, 7-170 km wide, elongated in a north-south direction along the western coast of Huvsgul Lake and from the Murun River in the south to the Russian frontier in the north. The Huvsgul basin was formed in the Pre-Huvsgul epicontinental trough on the Tuva-Mongolian Precambrian microcontinent. The trough is filled by Vendian-Cambrian marine volcanosedimentary rock units (figure 7).

The phosphorite potential occurs within two strategic levels with more economic concentrations confined to the lower level of phosphate deposition. The content of phosphate minerals in ores varies from 10-98 per cent, and comprises 40-98 per cent in standard ores. The P_2O_5 content ranges from 5-34 per cent in ores and those with P_2O_5 less then 16 per cent are considered substandard. The other ores' useful component include fluorine (1-2 per cent). For most of the ore reserves, the P_2O_5 content is in the range 17-21 per cent. The ore bodies could be mined by open pit method.

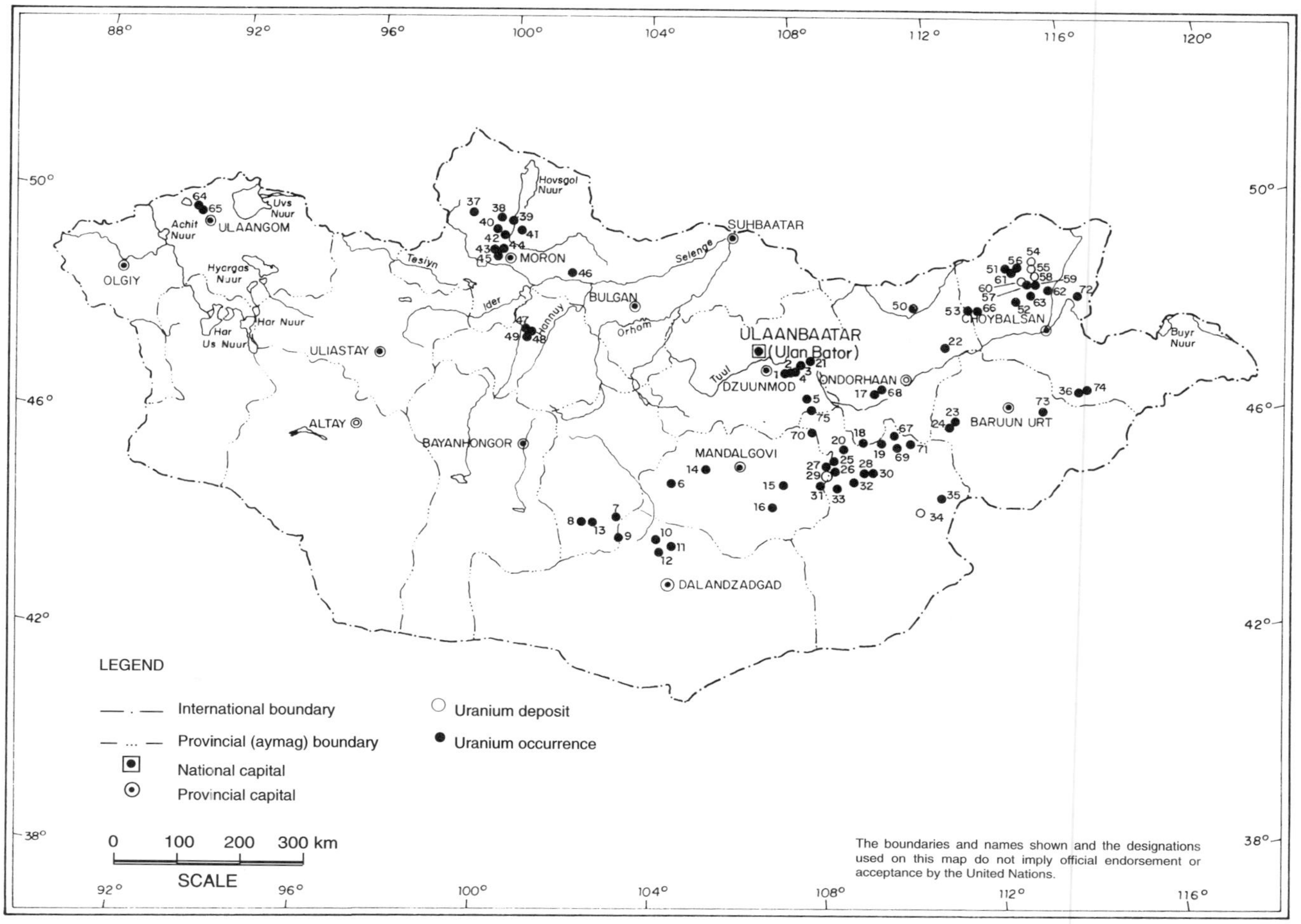

Figure 9. Distribution of uranium deposits and occurrences in Mongolia

4.11 Zeolite

There are 10 natural zeolite deposits and occurrences of industrial importance which are concentrated in the eastern Mongolian volcanic belt (figure 7). They belong to a distinctive type of zeolite, of igneous-limonitic genesis, dating back to the Upper Jurassic and early Cretaceous periods. The main mineral of the zeolite group is clinoptilolite, seldom chabasite, heulandite, analcime and ferrierite. The Tsagaantsav zeolite deposit has about a 160 mt of reserves and is composed of over 80 per cent zeolite ranging in thickness from 3-35 m.

4.12 Asbestos

More than 30 chrysolite-asbestos and amphibole-asbestos deposits and occurrences are known in Mongolia. These are mainly related to ultramafic intrusions accompanying regional deep fracture zones.

4.13 Magnesite

Most of the magnesite deposits and occurrences of economic potential in Mongolia are located in the Manlai and Ikh Bogd deep fault zones, occurring as the weathering product of ultramafic intrusions.

5.0 MONGOLIAN GEOLOGICAL SURVEY'S POLICY

At present 80 per cent of the country has been mapped at a scale of 1:200,000 and about 9 per cent at a scale of 1:50,000. Approximately 3,000 occurrences of various minerals have been defined, however very few have been economically evaluated. Existing geological, geochemical and geophysical information requires modern interpretation using modern methods of remote sensing. Airborne geophysical surveys are required and exploration of the southern part of the country is considered particularly important.

In 1995, the Mongolian Parliament passed the Mineral Law of Mongolia and at present all geological exploration work (regional mapping, prospecting and feasibility study) is conducted in accordance with this Law and its accompanying regulations.

Every year the state budget allocates about US$1.0-1.7 million for geological exploration. The funding of detailed prospecting for all types of mineral deposits is now conducted by private companies.

The major responsibilities of the Geological Survey fall into five general categories:

- Mineral policy and planning research to define the most effective contribution that mineral resources can make to the economy of the country.
- Organization and regulations of all exploration activities in the country as determined by the Mineral Law and regulations.
- Information capture and management of all geological, mining and prospecting data related to mineral resources.
- Organization of the geoscientific surveying of all geological and mineral resources and production maps and reports on such surveys for domestic and international promotion.
- Distribution of state budget for regional geological exploration, and the issuing of exploration licenses for all types of geological exploration.

In connection with market demand, the Geological Survey is keeping to the following policy:

- To continue mineral exploration with the objective of discovering bulk mineable gold deposits (ancient conglomerates, sedimentary hosted, black shales, copper-porphyry systems, carlin type, etc.).
- Commencing regional geophysical and geochemical exploration in the southern and western portions of Mongolia to define geologically favourable structures for oil concentration.
- To commence regional uranium exploration in eastern and southern Mongolia to determine areas for further detailed prospecting.
- To acquire modern analytical equipment, and undertake technical renovation of the state exploration organization and training of Mongolian specialists abroad.
- Propose to the Mongolian parliament a package of amendments and changes to the Mineral Law with the purpose of attracting foreign investment and promoting national exploration and mining companies to develop the Mongolian mineral resources.

REFERENCES

1. Gerel, O., 1990. Petrology, geochemistry and mineralization of subalkaline magmatism in Mongolia, Irkutsk, 40 p. (in Russian).

2. Gerel, 0., 1995. Mineral Resources of the western part of the Mongol-Okhotsk foldbelt. Resource Geology Special Issue, No. 18, 1), p. 151-157.

3. Dejidmaa, G., 1996. Gold metallogeny of Mongolia. Mongolian Geoscientist, No. 1, Project report of Institute of geology and Mineral resources, p. 6-30.

4. Jargalsaihan, D., Kazmer, M., Baras, Z. and Sanjadorj, S., 1996. Guide to the Geology and Mineral Resources of Mongolia. Geological Exploration, Consulting and Services Co. Ltd., Ulaanbaatar, p. 329.

5. Koval P.V., Ariunbileg S., Libatorov Yu. I., and Maximyuk I.Ye., 1988. The Bayan-ula molybdenum-copper- porphyry prospect, central Mongolia, and its relation to magmatism: International Geology Review, v. 30, p. 900-911.

6. Kovalenko, V.I. and Yarmolyuk, V.V., 1995. Endogenous rare metal ore formations and rare metal metallogeny of Mongolia. Economic Geology, v. 90, pp. 520-529.

8. Kovalenko, V.I., Zaitsev, N.S., Yarmolyuk, V.V. et al., 1984. Main geodynamic stages of Mongolia territorial evolution and their metallogenic features, in Endogenous ore formations of Mongolia. Moscow, Nauka, p. 7-41 (in Russian).

9. Sillitoe H. Richard, Ochir Gerel, Gunchingiin Dejidmaa et al., 1996. Mongolia's gold potential. Mining Magazine, July, p. 12-15.

10. Sotnikov V.I., Berzina A.P., Bold D., 1984. Regularities of spatial distribution of copper-molybdenum mineralization in Mongolia, in: Endogenic ore-bearing formations of Mongolia. Nauka, p. 89-101 (in Russian).

CONTENTS

4. ENVIRONMENTAL POLICY OF MONGOLIA[1]

1.0 INTRODUCTION

Mongolia is a country with ancient traditions of nature protection. Official documents such as Chinggis Khaan's "Ikh Zasag" (Great Governance) issued in XIII century and "Khalkh Juram" issued in XVI century contained provisions to protect nature by declaring some beautiful natural places as protected areas in which tree cutting, animal hunting and land disturbance were prohibited.

Bogd Khaan mountain, one of the oldest protected areas in the world was declared a protected area in 1778. Its administration office was set up in 1911.

Ancient Mongolian national concepts and policies of nature protection were based on the philosophy of the interdependent development of human beings, nature and the economy in terms of space and time. With this philosophy, Mongolians paid great respect to the earth as the life sustaining source, and created and developed a specific nomadic lifestyle and culture in harmony with nature. Consequently, Mongolia remains one of the world's few countries, that has kept nature relatively undisturbed.

However, the traditional Mongolian lifestyle that was developed in harmony with nature over many centuries was greatly disturbed and changed under the more than 200 years of rule of the ancient Manj chin state of Mongolia, European socialist ideas, technical revolution and the centrally planned economic system of the last 60 years.

There now even appears to be a tendency towards ecological imbalance due to environmental degradation and the improper use of natural resources, which has taken place as a result of mismanagement, urbanization and industrialization in Mongolia. For example, half of the total cropland and one-third of the total pastureland has been degraded.

There also has been a reduction of 20 per cent in soil fertility and animal resources and a 33 per cent reduction in forest resources. About 5 million ha. of land in the Gobi has been affected by sand shift, more than 300 lakes, rivers, streams and springs have disappeared and more than 100 animal and plant species are threatened with extinction.

Due to the contamination of air, water and soil cover in developed areas, thousands of people drink water of a quality that does not meet acceptable standards, and air pollution, in cities such as Ulaanbaatar and Darhan, exceeds permissible levels by 3-5 times.

The depletion of natural resources and the degradation of the environment have affected the economy of the country, and was indeed a key reason for the socio-economic crisis of the 1990s. There has been a tremendous reduction in timber, crops, coal, cement, brick and chalk production. Many enterprises have stopped or will have to stop their operations, due to the economic crisis.

These impacts are due to the exploitative nature of natural resource utilization and the non-integrated approach to environmental, social and economic issues.

These circumstances have greatly affected Mongolia's traditional respect for nature and state policies on human and livestock settlement in a life-sustaining manner.

2.0 POLICY ASPECTS

The transition to democracy and the market economy, which started at the beginning of the 1990s, has provided a great opportunity to re-assess the historic path we passed through and to follow the concept of environmentally sound and sustainable development in the future.

---

[1] B. Ganbaatar, Department of International Cooperation and Projects, Ministry of Nature and the Environment of Mongolia
D. Gantigmaa, Department of Legal Affairs and Cooperation, Mongolia

Mongolia's new constitution, adopted in 1992, grants the right of citizens to live in a healthy and safe environment, and declares that Mongolian land territory and natural resources are the property of its citizens, thus being placed under state protection. The Constitution also states that the maintenance of ecological balance is one of the important factors to ensure the national security of Mongolia.

The adoption of a package of environmental laws including the Land Law, Special Protected Areas Law, Underground Resources Law, Mineral Resources Law, Environmental Protection Law, and laws on water, forests, hunting, plants and air, are all important steps towards the creation of a legal framework for environmental protection.

The above laws were defined on the basis of a detailed assessment of the quantity and quality of natural resources, and the life-sustaining capacity of nature. They reflect many important issues, such as fees for natural resource utilisation, proper and efficient use of natural resources, the User Pays-Polluter Pays Principle, and the active participation of citizens, economic entities, and public and private sectors in environment protection activities.

In addition to the above, laws recently adopted include laws on fees for use of water, forest, plant resources and for animal hunting. Draft laws on fees for land use and for mineral resource use have also been drafted. It is estimated that 4-5 billion tugrigs of income for state and local budgets could be generated from natural and mineral resources use.

Actions have also been taken to develop ecological norms and standards, determine natural carrying capacity and to conduct environmental impact assessments for development projects before they are implemented.

A solid foundation has been established for further development of bilateral and multilateral cooperation in the field of environment protection. Currently 11 environmental projects are being implemented in Mongolia with the assistance of UN agencies and donor countries. The projects include the Mongolia Biodiversity Project, Mongolia's Action Programme-21, and Strengthening of the Environment Management Capabilities of the Ministry of Nature and the Environment.

Mongolia is now a signatory to the following international conventions: Convention on Biological Diversity; UN Framework Convention on Climate Change; Vienna Convention for the Protection of the Ozone Layer; Convention to Combat Desertification in Those Countries Experiencing Serious Drought and/or Desertification, Particularly in Africa; Convention on International Trade in Endangered Species of Wild Fauna and Flora; and the Basel Convention on the Control of Transboundary Movements of Hazardous Wastes and Their Disposal.

There remains a great need for time and resources to stop environmental degradation, rehabilitate the natural environment and to create an ecological base for sustainable development. There are many obstacles to resolving these acute problems, as well as in drawing public attention to the protection of the environment, particularly during the current period when the country faces financial constraints, an unstable economy, poverty, unemployment and weak discipline. Ecological violence may become a real problem due to lack of decisive steps being taken towards maintaining ecological balance.

Examples of improper use of natural resources are: exports of timber, animals and plants; over-mining of gold and construction materials (without conducting environmental impact assessments); and contamination of beautiful natural areas by mismanaged tourism. Overconcentration of economic entities and organizations in the capital city and other large settlement areas reduces ecological carrying capacity around them. There is also extensive migration of people from the countryside to settlement areas, making remote territories empty and affecting the development of some provinces.

The frequency of natural disasters is likely to increase in Mongolia due to climate change, foreign trade imbalance and the growing number of adverse impacts from human activities. Potential for disasters can be observed in the annual increase of deserted steppe areas, crops fertilised with chemicals that can pollute the environment, the spreading of insects and diseases as well as the effects of snow, floods, and forest fires.

In response to these concerns, the Parliament of Mongolia promotes the development and implementation of ecologically sound policies. While promoting these objectives, Mongolia attempts to enrich its traditions based on the concept of the interdependent development of human beings, nature and the economy in new historic conditions. This integrated economic and ecological policy for Mongolia is directed towards the implementation of the concept of sustainable development in harmony with nature, which was promoted by the United Nations Conference on Environment and Development in 1992. This policy is also reflected in Mongolia's proposal to include the whole territory of Mongolia into the world biosphere zone.

In line with these trends, the Government of Mongolia aims to create a basis for a relatively self-sustaining economy which meets both economic and ecological requirements.

The Government considers the environment as a priority area and works to ensure maximum productivity from natural resources without damaging their quality and value. By achieving this objective the national capacity to protect the environment should be enhanced.

Attention must be paid to developing a new network and planning criteria for strategic policies capable of handling ecological issues through comprehensive legal, economic, technical and organizational methods and mechanisms. Actions must be taken to provide a background of state policy which will reflect priority for ecological policies within the framework of industry, infrastructure, society, science and technology, justice and external relations. Policies need to be supported by government initiatives for policy implementation and coordination of inter-sectoral issues.

An important area for awareness raising in the policy renewal process, is the role of natural resource management in development. This issue may be tackled by using a portion of the derived income from activities utilising natural and mineral resources, for the protection or effective rehabilitation of natural resources.

It is necessary to make environmental laws easily understandable to implement, and to develop them in conjunction with other laws regulating economic activities.

Significance is attached to the introduction of economic and technical planning methods and principles for environment protection activities. Projects and measures will ideally be implemented in a step-wise manner to determine ecological consequences, to properly refect real values of natural resources in national calculations, apply and develop environmentally sound technology, standards and ecological norms. A methodology will be developed to establish national ecological standards based on consistency, international criteria, and technological models for: a) saving resources; b) evaluation of costing of natural resources; c) determining environmental quality and pollution status, and; d) environmentally sound processing of natural and mineral resources. Some projects and measures are being implemented to enhance capacity of environmental management, finance and human resources.

The Ministry of Nature and the Environment has been instructed to both develop objectives and obligations which provide the state, citizens, economic entities and organizations with strategic planning, ecological policies and legal directives; and to perform intersectoral coordination. These objectives are to be initiated through three agencies: the Environment Protection Agency; the Hydrometeorology and Environment Monitoring Agency; and the Land Management Agency; all of which come under the Ministry of Nature and Environment.

A programme of decentralisation and strengthening of local administrations is expected to enhance the capacity of local authorities and their professional organizations to deal with, and take full responsibility for, environment protection activities.

To ensure equal participation of government, citizens, non-governmental organizations, economic entities and other organizations, it is proposed to set up environment units in ministries and local administrations. This will complement the creation of ecological databases in economic entities and organizations and assist in the planning of environment activities. Some government duties will be transfered to non-govenment organizations.

Actions are required to: develop methods and mechanisms for increasing fees for natural resource use; to enable a better integration of environment policy, management, budget and finance; to promote the use of fee incomes for environment protection measures; and to encourage ecological investments by tax reduction and foreign aid programmes.

Legal and administrative mechanisms for imposing fines for damages caused to the environment by economic entities, organizations and citizens will be strengthened. Actions for rehabilitating degraded environments, and for using natural resources efficiently and responsibly will be encouraged by introducing environmentally sound technology. Considering the importance of human factors in environmental protection activities, measures are necessary to increase science, information, awareness and training among government staff and the general public. Attempts are also being made to create a more favorable legal environment for citizens to own land and use its renewable resources without any charges.

One idea being considered as part of Mongolia's policy on sustainable development, is the concept of socioeconomic priority areas being coordinated at a regional level. Geography and the specific characteristics of Mongolia dictate that the background for sustainable development must be created at the regional level. With this concept, the country would be divided into macro regions, and sustainable development programmes and projects devised for each region. Appropriate mechanisms would be implemented to ensure provinces cooperate with one another and sustain ecosystems inside the region.

The Government of Mongolia is also attempting to accelerate the development of foreign relations and to benefit from this as much as possible in terms of protecting and maintaining Mongolia's pristine environment. To date the following environmental issues have been identified, for resolution within the framework of external aid and international environmental cooperation:

- comply with international conventions on the environment and sustainable development and conclude and implement bilateral agreements;
- use development aid through projects and measures to rehabilitate degraded environments, evaluate natural resources, and produce market compatible products;
- encourage foreign trades which meet requirements to protect people from toxic chemicals, conserve gene pools of animals and plants, ensure ecological safety, and sustain ecological carrying capacity;
- develop scientific and technical cooperation in the field of introduction and transfer of environmentally sound technology, re-introduction of rare animals, tree planting, expanding of special protected areas network, combating and prevention of natural disasters and hazards, weather forecasting, ecologically clean production and environmental management, and use of information systems, and
- develop international cooperation to conduct joint studies on Mongolia's specific nomadic traditions and customs which were in full harmony with nature, raising public awareness of them amongst other nations, and developing environmentally sound economic activities and lifestyles.

Mongolia follows the new concept of external relations, based on a suitable combination of national interest with that of the international community, proper and effective coordination of foreign loans and aid and provision of multilateral diplomatic activities. Mongolia will maintain its proposal to include its territory within the world biosphere reserve and will be happy to cooperate with nations who support this proposal.

Mongolia considers its pristine and relatively untouched nature and traditional nomadic lifestyle and culture as part of the common heritage of the global community, and will do its best to contribute to the common interest of world nations on tourism, consumption of ecologically clean products, equal sharing of natural resources, and selection of a sustainable development path.

It is hoped that Mongolia's great expectations for the protection of its natural and environmental heritage will be realised, as the world community enters the 21st century promoting the concept of sustainable development. We greatly appreciate and warmly welcome international negotiations and decisions, particularly at the World Economic Summit, which ensures ecological safety throughout the world.

3.0 EIA ENFORCEMENT IN MINERAL RESOURCES DEVELOPMENT

Mining is one of the core sector industries which play a positive role in the progress of the country's development. However, when compared to some industries such as thermal power plants and the chemical industries, its environmental impact is seen to be, to a certain extent, unavoidable. Concern about this impact has reached such a level that many question whether a industry such as mining should be encouraged. In several fragile and eco-sensitive areas it is often questioned whether mining should be permitted. In recent years, even the country's highest courts of law have queried the impact of mining and questioned whether mining is done scientifically.

Mining in both developed and developing countries has drawn the attention of world bodies who are sceptical about safe and eco-friendly mining. The problem is, perhaps, not so much economic, but whether proper techniques and management practices are being utilised to reduce the impact, while optimising mineral production.

Fortunately several countries, including some of the developing ones, have taken measures to mitigate the impact, and consequently have made mining more viable and acceptable. Overall, however, although the developed world has achieved limited success in this direction, the developing world is still struggling to do so.

Mongolia is one such developing country, where the techniques of assessment have been well established and it has inched a step ahead through the implementation of the "Environmental Management Plan" (EMP). For a mining project an "EMP" includes, in brief, the baseline data, the mining method, its estimated impacts and mitigation measures, and the landuse plan after the mining is over.

In Mongolia, Environmental Impact Assessments (EIA) for a mining project are submitted in the form of an EMP and are judiciously scrutinised by an interdisciplinary expert Bureau of the Environmental Protection Agency of the Ministry for Nature and the Environment.

Gold mining is one of the main sectors of Mongolia's mining industry, promoted since 1994 by the Government of Mongolia through its "Gold" project. At present about 110 companies are in operation, and about 90 of them are active in gold mining.

Mongolia produces 0.5 per cent of the world's gold, and has the potential to expand its mining and export. Deposits such as Zaamar, Tolgoit, Bugant, Hailaast, Dubunt and Boroo are being extracted and further investigated. Another 10 areas in the region of Mongol Altai and Trans Altai Gobi including Onon, Ulz, Zabhan, Hubsgul, Bayanhongor and Umnugobi are being prepared for extraction.

In 1992, 770 kg of gold was extracted. During 1994-1996 gold extraction was raised from 1,750 to 6,000 kg. The main task of the government by the year 2000, is to raise gold extraction to 10 tonnes by establishing modern companies that employ high technology and techniques.

According to the Mongolian Law on Environment Protection (1995), the term *environmental impact assessment* means the prior identification of any possible adverse effects from production or service activities by citizens, economic entities or organizations to human health and the environment, and determination of the measures to minimise and mitigate such adverse impacts. Specifically, the law states that, "impact assessments shall be conducted for the development of proposals and programmes, as well as for establishing contracts for the operation, initiation, and expansion of production or services which may have adverse environmental impacts. Citizens, economic entities and organizations implementing proposals, programmes or contracts shall comply with the requirements determined by the impact assessment."

Major features of the current period have been the establishment of legal regulations and procedures for EIA. These include:

- EIA Procedures for proposed projects (Resolution No.: 121 of the Government Cabinet of Mongolia);
- Joint Decree of the Minister for Nature and Environment and the Chairman of the National Development Board on Guidelines for implementation of EIA Procedures;
- Decree of the Minister for Nature and the Environment on Guidelines and Methodology of EIA Procedures.

During this period (after adoption of Resolution No. 121) about 300 general assessment and 20 detailed assessments were conducted. General assessments are conducted by the above mentioned Bureau of

Environmental Protection Agency, MNE, and detailed assessments are carried out by private consultants licenced by the Ministry.

A general EIA is carried out by the MNE or relevant local Government during the first stage of project identification, on the basis of data available from the preliminary survey made prior to the investment and is aimed at identifying the project's environmental impacts and negative consequences, and to decide whether further detailed assessments are necessary.

On the basis of the conclusions of the general EIA, the following decisions are taken:

i) To provide approval for implementation of projects without further detailed assessments, if the project impact and consequences meet the requirements of existing environmental and hygienic standards and regulations;

ii) In case of limited negative impacts of a project, the project's implementation will be accepted with specific conditions added as management and organizational measures;

iii) In case of the projects impact being defined as negative, a decision for a detailed assessment of environmental impact of the project will be made.

General EIAs are carried out by a unit of the Environmental Protection Agency, a Bureau with a full time staff of two. The basic methodology is to assess the technological aspects of projects in line with existing environmental standards. Additionally, the team interacts with other departments and obtains information on local environmental conditions for the proposed project. Based on the above investigations the unit identifies whether a detailed EIA is required for the project or whether to limit it to a conditional approval. Results of the assessment are sent to project proponents and the respective local Government agencies. The General EIA is carried out during the stage of project formulation and identification.

The "detailed EIA" is conducted during the feasibility study and work design stages of the project with the aim of identifying the negative and positive impacts on the environment (based on subsequent methodology and recommendations), their consequences and to assess hazards or damages caused by the project.

The governmental bodies related to EIA may conduct the sectoral EIA. However, the sectoral EIAP is not developed as yet.

The "detailed EIA" shall include an investigation to identify and determine the adverse impact of the project on the environment and human health. The extent of these adverse impacts will be defined and mitigative measures to decrease adverse impacts to an acceptable level will be recommended during project implementation.

The "detailed EIA" may be partial or complete, depending on the significance and the extent of the predicted impacts and their effect on and damages to the environment and human health. The type of detailed EIA depends upon the following: if the predicted impacts affect one or few environmental and human health criteria, then a focused EAI is required; if the predicted impacts affect most environmental and human health criteria, and have potential to cause severe damage, a full EIA is required.

The Project Proponent is responsible for contracting a licensed organization to conduct/implement the Detailed EIA on the basis of the General EIA recommendations.

In general, the practical application of the EIA procedure is very poor for the following reasons:

- institutional problems associated with uncertainties regarding the duties of local and central government as well as inter-agency coordination;
- economic difficulties have weakened government capacity to implement established regulations and to make available the required specialists at the appropriate levels; and
- lack of information and a database essential for EIAs.

At present the Ministry of Nature and the Environment is reviewing and developing EIA regulations, and preparing a draft law on EIAs and the accompanying regulatory framework.

CONTENTS

5. THE SUPPLY-DEMAND SITUATION AND DEVELOPMENT POLICY FOR METALLIC MINERAL RESOURCES IN THE REPUBLIC OF KOREA[1]

The Republic of Korea is a small nation with poor natural resource endowments. However, the Republic of Korea is a major consumer of resources, with mineral consumption approximately 1.2 per cent of GNP. Total mineral consumption increases on the average about 10 per cent every year with economic growth increases about 9 per cent. Approximately 60 per cent of total mineral consumption, from 1990 to 1995, was for metallic minerals such as iron ore, copper, lead and zinc concentrates.

Since the middle 1970's domestic mineral production has been insufficient to meet the needs of the Republic of Korea's industry, and dependance on mineral imports has increased yearly to meet the increasing consumption resulting from rapid economic growth and welfare promotion. To meet this increasing demand for mineral resources, the government of the Republic of Korea is continuously undertaking exploration for domestic mineral resources and has emphasized overseas minerals development in resource-rich countries.

Overseas mining projects have a high risk in that such projects characteristically require long lead times, a large amount of capital investment, social costs for infrastructure, environment protection and so on. To decrease the mining investment risk and promote resource development activities abroad the government has established: tax and insurance incentives for overseas mining investment; special funds for loans for exploration and development, with favorable rates and networks to analyze the resource potential; and mining legislation, foreign investment procedures, dispute settlement methods and tax policies of resource-rich countries.

1.0 DEMAND AND SUPPLY

1.1 Non-fuel minerals

As noted, the consumption of non-fuel minerals has continuously increased in line with economic growth and total consumption was approximately 1.2 per cent of GNP in 1995. The consumption rate increases about 10 per cent every year while economic growth increases about 9 per cent. The consumption rate in the period 1990-1995 was 11.7 per cent which was much higher than the 7.4 per cent rate of economic growth shown in table 6.

Metallic mineral consumption from 1990 to 1995 was approximately 60 per cent of total mineral consumption, as shown in table 7. The major metallic commodities are iron ore, copper, lead and zinc concentrates. Recently, precious metal consumption has increased rapidly in parallel with the growth of per capita income.

Some metallic commodities are imported in the form of ingots as there is no domestic smelting capacity and no such ores in the Republic of Korea. Aluminum is a representative example in that the Republic of Korea consumes more than about 8 million tonnes of aluminum ingot per year, however, bauxite is imported only to make refractory bricks.

Domestic mineral exports were an important source of foreign exchange in the 1960's. However, because of changes in the policy related to heavy industry and chemicals in the 1970's, the role of domestic minerals changed from foreign exchange earners to providers of raw materials to meet local demand. Nevertheless, supply of domestic minerals was inadequate for local industry from the middle of the 1970s.

Imports of metallic ores, excluding gold, was about 2 billion US dollars in 1995 while mineral exports were very poor, as shown in table 8.

1.2 Metal

Metal consumption also increased with industrialization and improvement of the living standard. Per capita consumption in 1994 was 724.1 kg which represented an increase of about 4.5 times from that of 1977 and an

[1] Dr. Kalip-Chin Chung, Director, Computer and Information Dept., KIGAM, the Republic of Korea

Table 6. Economic growth and mineral consumption in the Republic of Korea, 1980-1995

Period	*1980-1995*	*1980-1985*	*1985-1990*	*1990-1995*
Growth (%)	8.6	8.0	10.5	7.4
Consumption (%)	10.5	9.6	9.7	11.7

Source: Ministry of Trade, Industry and Energy (MOTIE).

Table 7. Consumption trend of non-fuel minerals in the Republic of Korea, 1990-1995

(Units: Billion Won)

Year	*1990*	*1991*	*1992*	*1993*	*1994*	*1995*
Metallic Ore	1,162	1,333	1,496	1,529	1,529	1,970
Non-Metallic	622	834	892	948	1,039	1,129
Total	1,784	2,167	2,388	2,475	2,568	3,099

Source: Ministry of Trade, Industry and Energy (MOTIE).

Table 8. Trade trend of mineral resources in the Republic of Korea, 1990-1995

(Unit Million $)

Year	*1990*	*1991*	*1992*	*1993*	*1994*	*1995*
Mil. Ton	21.4	28.4	32.4	36.0	34.0	35.4
Mil. $	560.4	828.4	880.1	887.5	796.9	850.7

Source: Ministry of Trade, Industry and Energy (MOTIE).

increase of 44.6 per cent since 1990. Base metals and steel had a similar consumption trend. However, some commodities have shown a decreasing trend in Intensity of Use analysis, as shown in figure 10. This means that the economy of the Republic of Korea is in the transition stage from a resource-intensive based industry to a knowledge-intensive industry.

1.2.1 Iron Ore

Iron ore is mainly used in steel production. There is one steel manufacturing company in the Republic of Korea with a crude steel production capacity of 21 million tonnes per year. This production capacity will be expanded to 24 million tonnes with the completion of a new steel manufacturing factory in the near future. The Republic of Korea required about 35 million tonnes of 56-67 per cent Fe grade iron ores in 1995 (table 9). The completion of the expansion programme will create an additional iron ore demand of about 5 million tonnes/year. Most iron ore is imported from Australia, Brazil, Peru and India through long term contracts. More than 50 per cent of iron ores are imported from Australia. The domestic supply of iron ore is about 0.2-0.3 million tonnes, produced from 2 iron ore mines.

1.2.2 Copper

Copper consumption in 1995 was about 530,000 tonnes, up from 140,000 tonnes in 1980 and 350,000 tonnes in 1990, and representing an increase of about 52 times over that of 1970. This rapidly increasing demand has resulted from rural electrification and industrialization projects initiated in the 1970s. Copper demand in 2010 is predicted to reach about 830,000 tonnes/year.

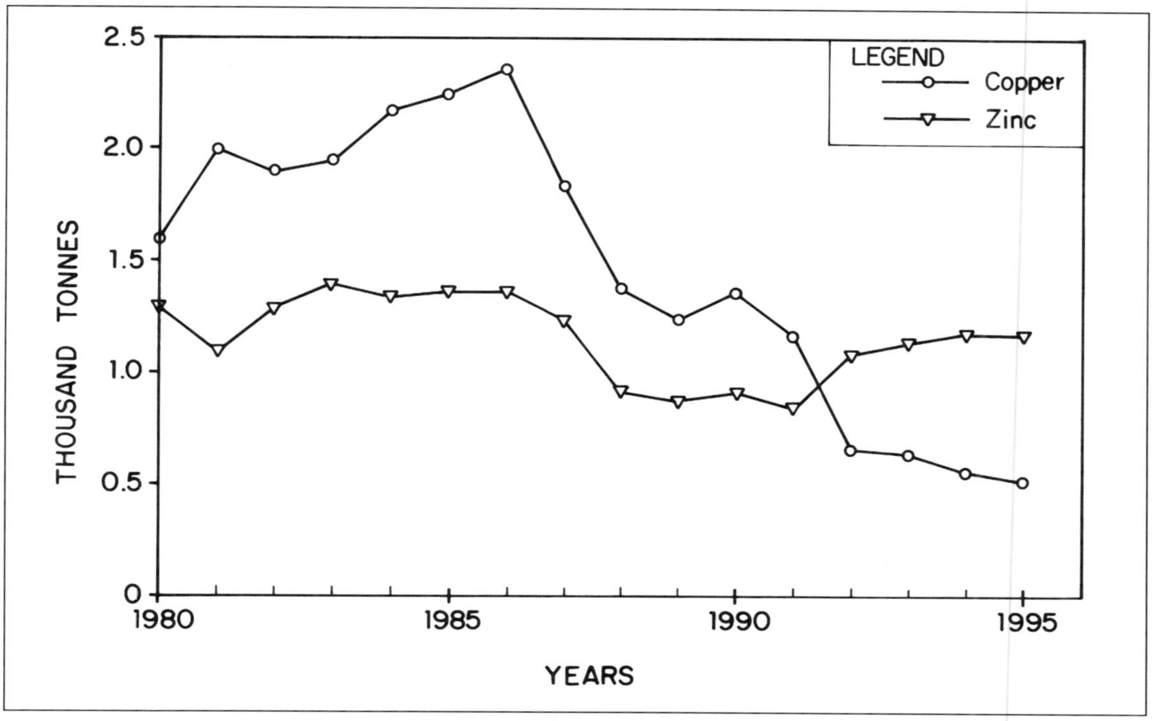

Source: Korea Institute of Geology, Mines and Materials (KIGAM), 1995.

Figure 10. Demand for copper and zinc in the Republic of Korea, 1980-1995

Table 9. Import amount of iron ore in the Republic of Korea, 1990-1995

Year	*1990*	*1991*	*1992*	*1993*	*1994*	*1995*
Mil. Ton	21.4	28.4	32.4	36.0	34.0	35.4
Mil. $	560.4	828.4	880.1	887.5	796.9	850.7

Source: Ministry of Trade, Industry and Energy (MOTIE).

There were several secondary copper refineries in the Republic of Korea until the early 1980s, however most were closed in the 1980s owing to the internationally low price of electrolytic copper. At present, there is only the LG metal company which has a smelting capacity of 150,000 tonnes and an expansion plan for 300,000 tonnes by 2001. LG company annually imports about 450,000 tonnes of copper concentrates from Papua New Guinea, Canada and Indonesia through long term contracts.

The Intensity of Use trend for copper has decreased from a peak of 2.35 in 1986 (table 10). This indicates that the rate of copper consumption will slow year by year when compared with GNP. It is expected that copper demand will depend heavily on the level of substitution in the future.

Table 10. Intensity of use trend for copper in the Republic of Korea, 1985-1995

(Unit: Ton/Mil. $ of GDP)

Year	*1985*	*1986*	*1987*	*1988*	*1989*	*1990*	*1991*	*1992*	*1993*	*1994*	*1995*
IU	2.24	2.35	1.83	1.37	1.24	1.37	1.17	0.66	0.64	0.56	0.51

Source: Korea Institute of Geology, Mines and Materials (KIGAM).

1.2.3 Zinc

The increasing demand for zinc is closely related to the increasing requirements for zinc-galvanized steel used in the automobile industry. Zinc demand in 1995 was about 340,000 tonnes which represents an increase of about 20,000 tonnes every year since 1990. The zinc demand in 2000 and 2010 will be 420,000 and 820,000 tonnes respectively forecasted by regression analysis and intensity of use analysis (tables 11 and 12).

To meet zinc demand until 2000, Korea has developed a self-sufficient capacity within the facilities of both the Korea Zinc and Youngpung Companies which will have smelting capacities of 200,000 and 90,000 tonnes respectively. Zinc ores have traditionally been imported under long term supply contracts. As a result of zinc supply problems, the Republic of Korea has tried to diversify supply sources through joint development projects in Canada, Australia, Namibia and China, Furthermore, some companies are planning to establish new smelters in foreign countries, where zinc ores can be secured at reasonable prices.

1.2.4 Lead

Statistical data compiled by the Korea Non-ferrous Refining Association shows that lead demand in 1995 was about 230,000 tonnes, of which 56 per cent was supplied by Korea Zinc Company, who produces about 130,000 tonnes of lead ingot (Table 13). The remaining lead demand is imported from international markets, as shown in table 13. Approximately 163,000 tonnes of lead ore was consumed in 1995 of which about 88 per cent was imported from foreign countries and 12 per cent supplied by domestic mines.

More than 80 per cent of lead demand is used for making automobile batteries and is forecast to increase to about 300,000 tonnes in 2000, an increase of 7 per cent per year.

Table 11. Demand trend for zinc in the Republic of Korea, 1990-1995

(Unit: Ton)

Product	*1990*	*1991*	*1992*	*1993*	*1994*	*1995*
Galvanizing	155,532	165,001	166,955	192,179	218,144	216,965
Alloying	29,434	26,100	19,931	17,737	14,948	26,091
Diecasting	21,329	24,839	24,281	26,903	31,696	37,415
Metal Chem.	7,415	7,755	5,895	3,302	3,767	5,866
Other	14,069	14,447	13,334	12,955	10,746	16,812
Export	55,639	27,557	52,880	42,572	79,160	34,739
Total	283,418	265,699	283,276	295,648	358,461	337,880

Source: Korea Non-ferrous Refining Association.

Table 12. Demand forecast for zinc ingot in the Republic of Korea, 1995-2010

(Unit: Thousand Ton)

Year	*1995*	*1997*	*1999*	*2000*	*2005*	*2010*
Demand	338	340	380	420	500	820

Source: Korea Institute of Geology, Mines and Materials (KIGAM).

Table 13. Supply and demand of lead in the Republic of Korea, 1989-1995

(Unit: Ton)

Year	*1989*	*1990*	*1991*	*1992*	*1993*	*1994*	*1995*
Demand	137,495	137,684	142,815	154,323	162,647	187,052	229,772
Import	92,467	97,056	107,869	99,395	86,898	116,820	120,288
Export	3,817	4,295	5,950	8,612	14,424	16,560	20,491
Production	48,678	44,778	40,554	63,377	89,976	96,457	129,744

Source: Korea Non-ferrous Refining Association.

2.0 MINERAL RESOURCES DEVELOPMENT POLICY

2.1 Domestic minerals development

The Republic of Korea is implementing dynamic policies for domestic supply of mineral resources to maintain and expand the local production base for mineral resources so that it can supply urgently needed mineral resources in the case of emergency. In particular, the research vessel Tanhae No. II will begin marine exploration on the continental shelf off the Republic of Korea and for manganese nodules in the Pacific beginning in March 1997. The main components of this national policy are:

i) continuous exploration to secure mineral resources

ii) high technology development to increase the add-value of mineral commodities

iii) development of exploitation technology relevant to the different environments

iv) assistance for land reclamation and environment protection of abandoned mining areas.

2.2 Overseas minerals development

As domestic mineral production is inadequate to provide industry with sufficient mineral resources, there is a need to secure a stable international supply of minerals on a long-term basis. The government of the Republic of Korea has been undertaking a great effort to raise self-supplying capability through active overseas mineral development projects, which will not only contribute to supply security but also consolidate economic cooperation with host countries. In 1980, the Republic of Korea set up development targets for strategic resources in the year 2000 (as shown in table 14). Strategic resources are mineral resources essential for national economic development and for which the import value is more than US$100 million per year and import dependence is more than 90 per cent.

In 1978, the Republic of Korea enacted the Overseas Resources Development Law to promote overseas resources development activities. In 1982, the Law was revised from a “permission system” to a “report system” which may be interpreted as a prelude to the liberalization of nearly all types of resource development activities abroad. The revised Law establishes tax and insurance incentives for overseas mining investment, a special fund for loans at a favorable interest rate for exploration and development of energy and mineral resources under the jurisdiction of foreign states and the establishment of networks to analyze resource potential, mining legislation, foreign investment procedures, dispute settlement methods and tax policy of resource-rich countries.

Table 14. The development target of overseas resources of the Republic of Korea in 2000s

Item	*Copper ore*	*Iron ore*	*Zinc ore*	*Uranium*
Target (%)	20	20	20	30

Source: Ministry of Trade, Industry and Energy (MOTIE0.

To date, about 10 successful international mining projects have been undertaken, such as Los Palembras Copper in Chile, while 7 projects such as Klongton fluorite in Thailand have failed. The projects proposed in host states include 6 mines such as lead and zinc in Canada, talc in China, silica in Viet Nam and others. The mineral resources imported through overseas minerals development projects are far behind planned targets.

3.0 ENVIRONMENT PROTECTION FOR THE MINING INDUSTRY

Mining activities have the potential to disturb the environment by altering land forms and ecosystems, disrupting the hydrological cycle and discharging wastes into air and water. However, there have been relatively few studies which attempt to identify and quantify the environmental impacts from mining activities. Therefore, it is difficult to determine the scale and cost of environmental problems associated with the mining industry at present.

Despite the absence of detailed environmental data related to minerals development, the government of the Republic of Korea pays close attention to the environmental impacts from mining activities and has adopted new legislation for strong control of environmental problems. Environmental criteria have already become an important consideration for the mining industry in the Republic of Korea.

The Mining Law of the Republic of Korea has an environmental goodwill principle and a general reference to an environmental protection obligation. According to Korean Environmental Laws, miners are not required to complete an environmental impact assessment or to minimize negative impacts from mineral development. Environmental protection, generally included as part of specific conditions in the mining license, is as follows:

i) to submit a separate environmental impact statement produced by an environmental expert

ii) to submit a comprehensive environmental management plan

iii) to submit bonds or guarantees for environmental liability

iv) to carry out environmental audits by government agency

v) to restore mining areas by refilling, landscaping and re-planting.

As a protective measure, the government of the Republic of Korea is carrying out a basic survey of acid mine drainage (AMD) and is building a data base for closed mines. The current national trend is to emphasize recycling of materials rather than mineral resources development.

4.0 CONCLUSION

As a result of heavy industry and chemical industry policy since the 1970's, the economy of the Republic of Korea has switched to resource intensive industries. To meet the rapidly increasing requirements for mineral resources, the Republic of Korea imports from foreign countries such as Australia, Brazil and Indonesia. To secure supply sources on a long-term basis and at a reasonable price, the government of the Republic of Korea has emphasized not only domestic mineral exploration but also investment in overseas mineral development projects. To decrease the environmental impacts from mining activities, the Republic of Korea is also developing exploitation technologies friendly to the environment, which can be used in international technical cooperation.

REFERENCES

1. Korea Institute of Geology, Mines and Materials (KIGAM), a Study on the Safety Supply and Demand Methods for Strategic Mineral Resources (Aluminum, Zinc, Copper and Iron Ore), 1993-1996.

2. KIGAM, a Legal Study on Overseas Mining Investment into Resource-rich Countries (Canada, Australia, Indonesia, Russia, China, Brazil and Viet Nam), 1990-1996.

3. KIGAM, A Study on the Medium and Long Term Policy for Overseas Resources Development, 1995.

4. Ministry of Trade, Industry and Energy (MOTIE), Annual Report for Supply and Demand Situations of Mineral Commodities, 1985-1995.

5. Korea Non-ferrous Refining Association, Korean Non-Ferrous Metal Bulletin (Quarterly), 1985-1995.

CONTENTS

6. A REVIEW OF ENERGY RESOURCES OF THE REPUBLIC OF KOREA[1]

1.0 STABLE SUPPLY OF PETROLEUM

The Republic of Korea is forced to import crude oil to meet its entire domestic oil needs. Crude oil is being imported mostly by five oil refining firms, and whenever it is deemed necessary, the Korea Petroleum Development Corp. imports crude oil for stockpiling.

Under the Petroleum Business Act, crude oil refiners are required to register with the Ministry of Trade, Industry and Energy with the qualifications of their refining and stockpiling facilities.

Crude oil imports to the Republic of Korea reached 11 million barrels in 1965, and since then annual imports have risen sharply to about 625 million barrels in 1995, 57 times the 1965 figure, and largely influenced by the rise in domestic oil demand.

Keeping in mind the first and second worldwide oil crises, the government is pushing ahead with a top priority directive designed to ensure a stable supply of crude oil. Under the policy directive, the government is exerting its utmost efforts to maintain an optimum level for crude oil imports – the ratio of crude oil imports under a long-term contract basis to crude oil imports on a spot basis, and to diversify import sources of crude oil. In spite of the government's policy of diversifying crude oil sources, with the intention of avoiding over-concentrated dependence upon the Middle East, the Republic of Korea's import dependency on the Middle East increased to 77.8 per cent in 1995. Stepped-up efforts were made to diversify import sources through the expansion of self-developed crude oil on a long term basis by participating actively in new oil-field development projects in Viet Nam, China, Africa and Latin America. Since 1988, the Republic of Korea has been importing self-developed crude oil through its participation in overseas development projects. In 1995, imports of self-developed crude oil stood at 7.5 million barrels, or 1.2 per cent of the Republic of Korea's overall crude oil imports.

The domestic supply of petroleum products in the Republic of Korea is based on the domestic refining system, which means that imported crude oil is processed into petroleum products to meet domestic petroleum needs. The importation of petroleum products is based on maintaining a balance of the domestic supply-demand situation, and consequently, imports of petroleum products are limited mostly to low-sulfur fuel oils (which are in supply shortages), naphtha (a raw material for petrochemicals), LPG and similar products. In 1995, the overall imports of petroleum products reached 205 million barrels, up to 7.7 per cent from a year earlier. The domestic supply of low-sulfur fuels has increased, thanks to the expansion of desulfurizing facilities at domestic oil refineries.

The five refiners, two LPG importers, Korea Electric Power Corp. (KEPCO) and a number of trading companies are all importing products. In April 1995, the government abolished the prior export-import approval system for petroleum products, except for LPG, and as a result, any enterprise with storage capacity may import petroleum products.

Until 1990, the Republic of Korea's exports of petroleum products was very small, and on a processing contract basis. With the rapid expansion of oil refining capacity during the 1991-1992 period, exports of petroleum products began to rise drastically. Export schemes under processing contracts had been fully replaced by direct exports on a normal letter of credit basis. Export products are widespread, ranging from gasoline to asphalt. Overall exports of petroleum products in 1995 reached 115 million barrels, up to 28 per cent from a year earlier.

Domestic consumption of petroleum in the Republic of Korea hit the 200 million barrel mark in 1986, thirteen years after passing the 100 million barrel level in 1973. Annual domestic consumption began rising from 300 million barrels in 1990, to 400 million in 1991, 500 million barrels in 1992, 600 million barrels in 1994, and to 677 million in 1995.

---

[1] Bok-Jae Lee, Korea Energy Economics Institute, the Republic of Korea

Petroleum consumption in 1995 stood at 677 million barrels, up 9 per cent from a year before. Gasoline, kerosene and diesel led petroleum consumption during the year. Consumption of gasoline went up by 16.2 per cent due to an increasing number of passenger cars, and consumption of kerosene and diesel increased at rates of 31 per cent and 10.8 per cent respectively. On the other hand, consumption of LPG which continued to mark an upward curve for cooking and heating, showed signs of setbacks to massive inflows of LNG.

Demand by the industrial sector for petroleum products in 1995 posted a 2.6 per cent rise, and the transportation sector demand rose by 13.8 per cent because of an increasing number of motor vehicles, expanding preferences for recreation and leisure, and growing volumes for physical distribution, thus playing a lead role in expanding petroleum consumption.

Crude oil distillation capacity of the Republic of Korea reached 1.7 million barrels per day as of 1995, and 2.44 million barrels per day at the beginning of 1997.

2.0 DEREGULATION OF THE PETROLEUM INDUSTRY

For the past two decades, the domestic oil-refining industry has been running along a path of stable growth, and has been sheltered by a high, thick wall of government protection. Strict restrictions have been imposed on prices of oil products, which have blocked oil-refining firms from plunging into a self-destructive price-cut war, and thus enabled them to earn steady profits.

The local oil refining market has remained closed to both domestic oil-refining aspirants and foreign competitors in a bid to stem excessive inter-industry competition, which could destabilise the Korean economy.

With price liberalization and the market opening up, however, the good old days seem to be over and the domestic oil-refining industry is expected to undergo major changes during the next five years.

The industry will be forced into fiercer competition as more players are expected to muscle in on the market with liberalized prices. The coming five years will be a crucial time when a rough, high tide of price liberalization and market opening may put their very survival at stake. With changes in the market situation, the industry will face its toughest ever challenge in the coming years.

2.1 Accelerated deregulation

Price restrictions on petroleum products were lifted at the beginning of 1997. New players, both domestic and foreign, will be allowed into the oil-refining business in 1999. These deregulation measures herald the beginning of a seismic upheaval which oil-refining firms, who are already locked up in stiff competition to clinch a bigger share of the market, are expected to go through over the coming years.

During the first six months of 1997, oil companies have been required to report their price changes to the government, but that reporting requirement will be abolished thereafter.

The government eliminated a prior approval system for imports and exports of crude oil and petroleum products, permitting all eligible businesses to trade via cross-border transactions. The licensing system for oil companies and gas stations has been changed to a registration system. All would-be sellers of petroleum products will have appropriate facilities to do business, but they will be required to maintain a certain quality level for their products.

With the opening of the petroleum market to domestic and foreign aspirants in 1999, several domestic business groups are expected to enter the industry, destroying the current market situation, where five oil companies are locked in hot competition.

2.2 New strategies for survival

Industry analysts say that to stem the tide of competition and survive well into the next century, domestic oil companies should change their basic conception of business.

Under the protection of the government, local oil companies have served as the supplier of oil products rather than the seller. Now, they are feeling anew the importance of their role as the seller, and are drawing up new strategies for survival that focus on product and service differentiation and an overseas push.

Last year, the five oil companies – Yukong, LG-Caltex, Hanwha Energy, Ssangyong, and Hyundai – conducted massive restructuring and adopted aggressive marketing strategies. In addition to its core strategy to diversify into five business lines, which include gas and fine chemicals, Yukong is determined to internationalise its operations. As part of its push to diversify production bases, the company plans to build an oil-refining plant in China, on which it will spend $1.5 billion. The compnay also intends to actively participate in developing overseas oil fields and purchasing deposits of existing oil fields in an effort to broaden its import sources.

Hyundai Oil Co. is pushing for a comprehensive, three-stage plan aimed at developing itself into a corporation dealing with all kinds of energy resources. Through this three-stage plan the company seeks to increase its daily refining capacity to 810,000 barrels by 2000, while focusing on exporting its products to Japan and South-East Asian countries.

The goal of Ssangyong Oil Refining Co. is to become a highly competitive international oil company by pursuing a production and marketing strategy to link the domestic market with overseas markets. Toward that objective, the company will strive to keep production costs as low as possible, while making its facilities more sophisticated in order to more effectively cope with changing market conditions. Greater efforts will also be made to expand overseas markets based on its exported products.

To promote productivity and increase profits, LG-Caltex Oil Refining Company plans to invest a huge amount of capital to upgrade its facilities by the year 2000.

3.0 EXPLOITATION POLICY OF PETROLEUM AND COAL

Governments of countries with natural resources want to see that their resources are developed and exploited in a way that is consistent with the national interest, and that the government secures a fair share of the economic rent arising from the exploitation of the resource.

In the Republic of Korea, ownership of onshore and offshore natural resources is vested in the state. In the following section, we examine the nature of the Korean government's involvement in the exploitation of petroleum and coal resources of the Republic of Korea.

3.1 Petroleum

Oil and gas exploration and development in the Republic of Korea and offshore is carried out under concessions issued by MOTIE (Ministry of Trade, Industry and Energy). Concession contracts cover issues such as exploration period, production period and exploration requirements. They also cover economic terms such as the environmental protection charge (this is a refundable guarantee which must be paid by any company involved in exploration on the continental shelf), corporation tax and, where appropriate, the special bonus. The latter is an additional fee payable by companies wishing to explore in areas with high levels of prospectivity. To date no such fee has been levied.

There are no restrictions on who participates in exploration. However, the government has to date only released exploration acreage to PEDCO who, if they wish, can then seek other Korean or overseas partners. Other participants joining a project do so on the basis of a production sharing agreement. Under this agreement, shares of any eventual production are allocated on the basis of the proportion of total costs paid by each participant. In addition, the production sharing agreement deals with issues such as procurement of domestic facilities and manpower. Here, foreign participants are asked to use their best endeavours, where possible, to use equipment and manpower from Korean sources.

The government imposes a 15 per cent royalty on any of the Republic of Korea oil and gas production. Normal corporate taxes also apply. There are no requirements that any oil or gas found be sold in the Republic of Korea. Currently there are no producing fields in the Republic of Korea. PEDCO also participates in overseas exploration and at present has shares in four producing blocks. Another of PEDCO's objectives is to encourage the private sector. It does this in a number of ways. One of these is by providing technical and manpower assistance free of charge to the Republic of Korea (and overseas) explorers. It also provides loans to the Republic of Korea explorers. These loans need only be repaid if a commercial discovery is made. Funds for these loans come from the Petroleum Business Fund (PBF), as do PEDCO's operating costs. In addition, operators are exempt from a range of taxes on expenses related to the exploration and production process, including VAT and customs duties.

3.2 Coal

Coal, in the form of anthracite, is the major indigenous energy resource in the Republic of Korea, and has a relatively low calorific value and a high ash content. In 1993, anthracite reserves were estimated to be around 1.6 billion tonnes, of which 0.74 billion tonnes are recoverable. Ownership of the reserves is shared between the government, with 18 per cent, and private mine operators, with 82 per cent, as of the end of 1993.

For energy security reasons, the Government of the Republic of Korea, in the past actively promoted the expansion of the domestic anthracite mining industry. Consequently, the coal sector was provided with a number of different kinds of financial support, including direct subsidies and low interest loans. The government's stated aim was to enforce a policy for the development and maximum utilisation of domestically produced anthracite. The government is also involved in the health and safety aspects of coal mining.

In November 1950, the government established the Dai Han Coal Corporation (DHCC) as a wholly owned government enterprise. In 1992, the five mines owned and operated by the DHCC produced about 3.62 million tonnes of anthracite, or just over 31 per cent of all the anthracite mined in the Republic of Korea, the remaining production was from privately owned mines.

The government has also provided strong support to private sector companies involved in the anthracite industry. In June 1967, the Korea Mining Promotion Corporation (KMPC) was founded. This corporation provides assistance to private sector anthracite miners for the operation and management of their mines. The KMPC's main means for assisting the private sector mining operators is through the provision of low interest loans. The KMPC also provides a range of other services to miners, such as assistance in mineral related R & D. The minerals analysis services are provided on a cost recovery basis, other services are provided free of charge.

The government is also in a position to influence the profitability of mines by means of its control over the price that mine owners receive for their product. The government maintains a ceiling on the price of domestically mined anthracite for social reasons.

The Government has taken active steps to promote the rationalisation of the Republic of Korea's coal industry. In 1987, the Coal Industry Promotion Board (CIPB) was established to assist in the process of encouraging uneconomic mines to close down. As part of this process, the government provides a grant to mines opting to close down (approximately $10 per ton of production in 1988). It also pays benefits to miners who lose their jobs as a result of mine closures. As a result of the government's programme, 246 private coal mines (out of a total of 356) with a combined production capacity of some 11.75 million tonnes have closed down since 1987. The total cost of the government's intervention in the coal sector in 1993 was about $460 million. The trend towards subsidies, which are both greater in total terms and in terms of subsidies per ton, and which gave rise to concern in the first report, is continuing.

4.0 ENERGY AND ENVIRONMENT

The relationship between energy and environmental interests, if any, has never been good. Virtually all forms of pollution are caused by the consumption of energy, and the environment is worse off for it. With the rapid pace of industrialisation the world over, the environment has suffered. From the depletion of the ozone layer to the greenhouse effect, nature and the inhabitants of the earth have not fared well.

It has always been acknowledged that sources of energy must be improved so as to emit as little pollution as possible and auto makers, along with oil refining companies, are doing their share in this regard. One of the first oil refining companies to come up with some type of environmentally-friendly fuel in the Republic of Korea, was LG-Caltex Oil Refinery, which introduced the nation's first registered trademark for petroleum products.

Techron, marketed here under license of Chevron of the United States since the beginning of 1995, is an ambitious line of products that provide improved fuel injection for maximum combustion and greater mileage.

On the heels of LG-Caltex Oil's success, other oil refining companies are attracting consumers with environmentally-friendly, yet efficient, fuel.

Yukong introduced "Enclean", a name derived from the combination of the terms clean and energy, clean environment and clean engine. Marketing under the themes of clean up, power up, keep up and drive up, "Enclean" boasts some of the highest combustion rates in the industry. Hanhwa Energy introduced Emax gasoline to the market. Other refining companies also introduced environmentally friendly petroleum products.

CONTENTS

7. ODA PROGRAMME BY THE JAPANESE GOVERNMENT AND TECHNICAL COOPERATION ACTIVITIES IN ASIA AND THE PACIFIC REGION BY METAL MINING AGENCY OF JAPAN[1]

1.0 INTRODUCTION

This paper is presented in four sections:

1. An overview of the ODA programme of Japan.
2. The technical cooperation programme by the Japanese Government.
3. A case study of the technical cooperation for mineral exploration by Metal Mining Agency of Japan (MMAJ) in the Asian region, and
4. The future for MMAJ's activities in the ODA technical cooperation programme.

2.0 OVERVIEW OF JAPAN'S ODA PROGRAMME

Development assistance over the past half century has attained remarkable achievements in some sectors and countries. Mankind, however, is still confronted with a number of serious problems with which the international community needs to grapple through the global partnership, including environmental destruction and poverty within the developing countries.

However, the reason that Japan helps the developing countries is not merely because "it is desirable". It is also because ODA's contribution to global peace and prosperity benefits Japan and the Japanese people. In particular, tackling the following problems through development assistance leads to protecting the life of the Japanese people.

- *Environmental problems:* Acid rain could fall on Asian countries including Japan as a result of the industrialization of the Asian countries.
- *Infectious diseases:* They may cross borders and spread easily throughout the world.
- *Oppression of life due to poverty:* This could become the cause of wars and terrorism where the Japanese people could become victims.

Additionally, the growing economies of developing countries benefit the rest of the world through expanding trade and investment. The Japanese economy has also become greatly interdependent with those of growing Asia as can be seen from the following:

1. Japan exports more to Asian countries ($193 billion) than to the U.S. ($120.9 billion).
2. In the manufacturing sector, Japan's direct investment in 1995 in Asia ($6.5 billion) is larger than that in North America ($4.2 billion).
3. Japan depends on the developing countries for more than 40 per cent of its imports.
4. Japan depends on imports for more than 90 per cent of its energy resources (a large part of which are from developing countries).

The Japanese government would like to make a positive contribution, as one of the developed countries, towards peace, a stable world and the sound development of the world economy, by assisting the developing countries to make the transition to market economics and finding solutions to environmental, refugee and other serious global problems.

---

[1] Hiroshi Kobota, Metal Mining Agency of Japan

3.0 ODA AND JAPAN'S INTERNATIONAL CONTRIBUTION

According to OECD statistics for 1996, Japan's GDP is about $4,960 billion which is 22.3 per cent of the total of all OECD member countries and second to that of the U.S. ($6,980 billion, 31.4 per cent). This is greater than the total GDP of all developing countries (about $3,780 billion in 1993) where 80 per cent of the world's population live.

Japan's economic development was not achieved by its own eftorts alone. It was made possible through the strong support provided by the international community. Now that it has amassed great economic power, Japan must move over to the side that supports the international community, taking positive leadership as well as responsibility.

When Japan's historical, political and economic situations are taken into account, ODA, utilizing its economic and technological power, constitutes the most important pillar of Japanese international contribution.

In regarding ODA as the pillar of Japan's international contribution, Japan is required to "undertake a reasonable burden" commensurate with its economic strength. As a percentage of Japan's GNP, which indicates its economic power, Japan's ODA (net disbursement) during the past 20 years has been around 0.3 per cent. Although the Nordic countries show a much higher level, Japan's contribution equals the average for DAC member countries.

The 1995 yen value for ODA extended by Japan increased 0.6 per cent over the year before to ¥1,395.4 billion. In dollar terms, this represented a significant increase of 9.3 per cent, the increase being largely a result of the appreciation of the yen against the dollar. The dollar value of DAC member countries' ODA spending also kept pace with 1995 expenditure at $59.21 billion (up 0.1 per cent from the year before). With ODA spending increased to $14.49 billion (excluding assistance to Eastern Europe) Japan was the largest donor for the fifth year in a row.

4.0 RESULTS ACHIEVED BY JAPAN'S ODA

1995 ODA spending was at almost the same level as for the previous year.

Japan's grant share, which shows the ratio of grants (such as grant aid, technical cooperation, and financial contributions to international organizations) to the total ODA spending by Japan, on a commitment basis, stood at 46.6 per cent in 1995, exactly the same level as the 1993/1994 average. This compared with the DAC average of 77.1 per cent. Meanwhile Japan's grant element, which is an indicator showing the degree of concessionality of aid measured on the basis of interest rates and periods of repayment, stood at 78.7 per cent in 1995, also at the same level as the 1993/94 average (compared with the DAC average of 90.8 per cent).

A major part of Japan's ODA has been extended to Asian countries with which it traditionally has close ties, and in 1995 this represented approximately 54 per cent of ODA spending. In 1995, 44.5 per cent was used for constructing economic infrastructure, a sector to which Japan has traditionally given high priority. More importantly, Japan's ODA disbursements to the social development field (education, health care, etc.), has been increasing, reflecting the growing attention towards this sector of the international community in recent years. In 1995, Japan's ODA disbursements to the social sector increased by 3.5 per cent points over that of 1994, to a total of 26.7 per cent.

5.0 JAPAN'S MINERAL RESOURCE DEVELOPMENT AND TECHNICAL COOPERATION THROUGH MMAJ

Japan is one of the biggest consumers of mineral resources in the world. However, Japan depends on imports for the supply of almost all these resources. As a result, a major necessity for Japan is to secure not only domestic resources, but also to cooperate with other countries in order to secure a stable, global supply of these resources, on a long-term basis. Additionally, there are many developing countries with rich mineral resources, that have not yet been fully developed due to insufficient resources and infrastructure.

The Metal Mining Agency of Japan (MMAJ), in cooperation with the Japan Cooperation Agency (JICA), is promoting and assisting in the field of mineral exploration, as part of the official development assistance (ODA) programme of the Japan government.

The Metal Mining Agency of Japan (MMAJ) has provided technical cooperation since fiscal 1970, in response to requests from developing countries, as part of the Japanese Government's ODA programme. The MMAJ assists those countries in developing mineral resources by transferring advanced exploration technology through the programme. The MMAJ has been greatly contributing to the economic advancement of developing countries, as well as securing a worldwide supply of mineral resources. The MMAJ has been involved in a total of 134 projects, including some which have resulted in mine development. Basic feasibility reports were provided for use in further resource surveys in various countries. We have achieved steady results in all of these projects.

In fiscal year 1995, 24 exploration projects (8 new and 16 continuing ones) were carried out in 24 countries, as well as 9 evaluation surveys for prospective mineralization and 2 follow-up surveys on past completed projects.

6.0 TECHNICAL COOPERATION FOR MINERAL EXPLORATION

Responding to requests from developing countries, MMAJ conducts all surveys and studies required, from regional geological surveys to feasibility studies for actual mine development, providing full survey reports of such studies, as well as implementing technology transfer to counterpart personnel in the course of the cooperation (figure 11).

Technical cooperation for mineral exploration comprises two categories, the General Programme and Special Programme. Request for MMAJ cooperation project needs to go through several steps and procedures (table 15, 16, 17).

6.1 General Programme

Basic Programmes: exploration is conducted to delineate high potential areas for ore deposits through various survey techniques. The MMAJ carries out topographic mapping, geological, geochemical, geophysical, drilling, tunneling surveys, etc. According to the requests of recipient countries, the MMAJ performs the necessary surveys from the reconnaissance to the evaluation stage (15 projects in a year).

Feasibility Study (Regional Development Planning): The MMAJ provides the counterpart country with the most effective development plan for a newly discovered ore deposit through various surveys, or the most effective rehabilitation plan for an existing mine. These include drilling and tunneling surveys in order to determine the reserves and grade of the ore deposit. If necessary, the MMAJ will carry out ore dressing tests and finally will prepare a preliminary feasibility study report including economic evaluations. (3 projects in a year)

Supra-Regional Survey for Mineral Resources: In view of the importance of systematic and basic surveys for mineral resources development, MMAJ conducts cooperation projects to prepare basic maps for mineral resources development upon request from the concerned governments. These surveys combine a comprehensive variety of tests with surveys that included satellite image analysis, geological and geochemical surveys, as well as geophysical surveys, if necessary (2 projects in a year).

6.2 Special Programme for MMAJ Activities

Integrated Metal Exploration in China: this programme is currently contributing to the integrated development of metal resources in China, through mineral exploration and technical cooperation, in the field of separation and recovery technology of metals.

Mineral Exploration for countries Transitional to Market Economies: Based on requests from countries which transfer to market economies, following re-evaluation of the existing mines and other mineral resources of those countries, drilling and other surveys are carried out in order to secure new mineral reserves. This will help these countries carry out resource development projects in the future.

Survey Stage	S – 1 (Reconnaissance Survey)	S – 2 (Regional Survey)	S – 3 (Semi-Detailed Survey)	S – 4 (Detailed Survey)	S – 5 (Evaluation)
Survey Method	• Photogeologic Study * Airborne Geophysical Survey • Photogrammetric Map	• Regional Geological and Geochemical Survey * Regional Ground Geophysical Survey • Reconnaissance Drilling	• Semi-Detailed Geological and Geochemical Survey • Semi-Detailed Ground Geophysical Survey • Semi-Detailed Drilling	• Detailed Geological and Geophysical Survey • Detailed Ground Geophysical Survey • Detailed Drilling/ Tunneling	• Ore Dressing Test • Pre-F/S
Objective	Regional Appraisal Selection of Favourable Area	Appraisal of Favourable Area Selection of More Favorable Area	Appraisal of More Favourable Area Selection of Target Area	Appraisal of Target Area	Preliminary Evaluation of Ore Body
Explanatory Diagram	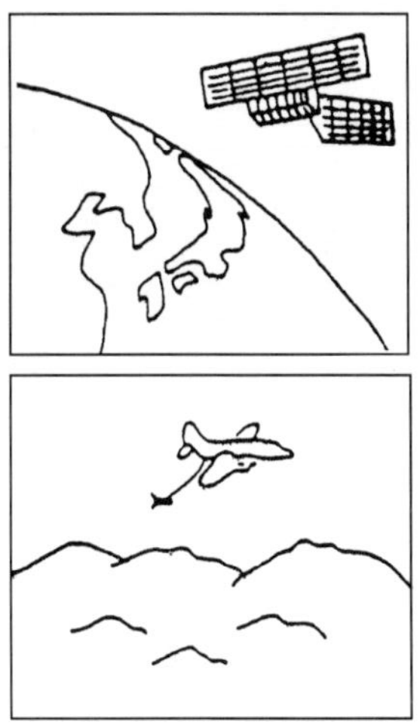				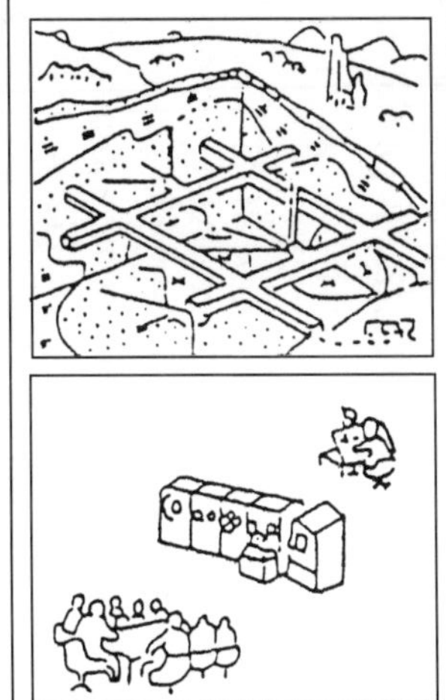
Area	30,000 km^2	1,000 km^2	500 km^2	100 km^2	10 km^2

|—— Mineral Exploration ——| |— Preliminary F/S and Development Planning —|

Figure 11. Schematic mineral exploration programme of the Metal Mining Agency of Japan

Table 15. Steps and procedures for MMAJ's cooperation project

Steps	*Procedures*
1. Preliminary Contact Counterpart ↔ MMAJ	Exchange of information on the technical cooperation project should be communicated directly, or indirectly, from either side through various channels.
2. Project-Finding Mission Partner ↔ MMAJ/JICA	Based on the preliminary information, a mission will be sent to the partner country for the purpose of promoting mutual understanding of the mineral exploration project, and of discussing the possibility of establishing the cooperation project.
3. Official Request Government of Partner Country ↔ Embassy of Japan	This is an essential step expected by any developing country commencing a new cooperation project. The request should be made in a written form by the government organization in charge and sent to the Embassy of Japan through an appropriate channel.
4. Budget for the Project (Government of Japan)	A draft of the budget for the following year is made in July, finalized at the end of the year, and presented to the National Diet of Japan. The annual budget is finalized at the end of March.
5. Consultation on the S/W Counterpart Organization ↔ Japanese Delegation	The Japanese delegation (preliminary survey team), which consists of the staff of MITI, JICA, and MMAJ will be sent to exchange ideas with the counterpart organization about the general plan, project area, period of the project, field work and contribution from both sides. The Scope of Work will be made in written form between the counterpart organization and the delegation. (See Appendix 1, Scope of Work)

Table 16. Contents to be filled in official request for MMAJ's cooperation project

An official request from the government of a partner country is submitted to the Japanese Government through diplomatic channels with terms of reference and relevant information and data. There is no official format for the request for the Mineral Exploration Project.

It is important to make clear the content of the requests to be investigated by the Japanese Government. Attached information and data are indispensible in promoting the procedure.

Principal matters to be described in the Terms of Reference are as follows;

1. Title of the Project
2. Purpose of the Project
3. Target minerals
4. Project areas with maps
5. Survey method and schedule
6. Geological information and data obtained by previous work
7. Outline of existing survey/exploration results

Table 17. The Scope of work by MMAJ/JICA, and the recipient country's organizations

S/W (Scope of Works) should be agreed upon and exchanged after discussion between the Japanese (Preparatory survey team), dispatched by MMAJ and JICA, and the Recipient Country's organization(s) concerned. The Survey is conducted according to this S/W, after signatures are obtained. The contents of the S/W are usually as follows; 1. Objectives of the Survey 2. Outline of the Survey; survey period, area, methods, etc 3. Undertaking of the receipient government and JICA, MMAJ Minutes of Meeting (M/M) are prepared for the first year Survey and other necessary matters, after the discussions concerning detailed survey planning.

Deep-sea Minerals Survey: MMAJ conducts surveys for deep-sea mineral resources (manganese nodule, cobalt-rich manganese crusts and hydrothermal mineral deposits) on the deep seabed within South-Pacific Ocean countries' exclusive economic zones by a survey vessel, HakureiMaTu No. 2, employing state-of-the-art undersea exploration equipment and methods.

Follow Up: In order to obtain comments on the finished project and to receive possible requests for future projects from the counterpart country, a follow up session is held in order to carry out any future projects more effectively. Where a counterpart country follows up the finished project, the MMAJ sends some experts with the necessary equipment to the country in order to assist them.

6.3 Mineral Exploration Programme

The general components of routine mineral exploration are as follows:

Area of the Project

The area of the project may range from a small area to a larger area, for example from a few square kilometres to 10,000 square kilometres, in accordance with the stage of exploration, or the purpose of the project.

Period of the Project

As a rule, each project may be carried out within three years. Extending the period of the project might be possible in special cases.

Survey Stage

The project may include one or more of the stages.

Provision of equipment

The Government of Japan can provide the necessary equipment for the field survey and laboratory investigation of the project to the counterpart organization. The annual budget for provision is about US$35,000 (C.I.F.).

Report

Results and conclusions of the annual field survey works will be reached through discussions between experts of the counterpart organization and the staff of the survey team of MMAJ-JICA. Annual and final reports will be printed by JICA and presented to the partner Government through diplomatic routes.

Acceptance of participants to Japan

In order to discuss the results obtained in the field survey, and to prepare the draft report and/or decide the survey plan for the following year, as mentioned above, JICA may annually accept one or more participants of the counterpart organization to visit Japan under the framework of the Technical Cooperation Plan designed by the Government of Japan. The period of the stay in Japan will be, in general, about one month.

Project Budget

The project budget is annually requested by the National Diet of Japan and covers, in principle, the cost of field activities, office and laboratory works of Japanese survey teams. Expected contributions from the counterpart organization only cover the costs incurred by their own scientists and engineers.

Project manager and field coordinator

To execute the mineral exploration project, MMAJ appoints a responsible person from the organization as Project Manager, and the counterpart organization appoints a responsible person as Project field coordinator. Field surveys should then be conducted by the joint team with the chief of the survey team being the field coordinator.

A detailed work plan for the project will be determined annually by mutual consultation between the counterpart organization, MMAJ and JICA.

7.0 MMAJ AND THE FUTURE

Asian economies are growing dramatically, attracting keen worldwide interests. During the 12-year period from 1980 to 1992, per capita GNP increased at an annual rate of 8.5 per cent in the Republic of Korea, 6.0 per cent in Thailand, 4.0 per cent in Indonesia, and 7.6 per cent in China.

However, Asian countries are facing various problems, including a widening income disparity and environmental problems, which arise in the course of economic growth. There is a growing tendency also among these countries, to seek a more balanced, people-centered development. MMAJ attaches importance to this tendency and would like to make a contribution for solving those problems through integrating the development of mineral resources and protection of the environment.

Until now, MMAJ has mainly been carrying out the survey of mineral resources, but in the future we have to consider the balance between development of mineral resources and environmental protection, and cooperate with developing countries to ensure the sustainable development of mineral resources.

APPENDIX

REPORT ON MINERAL EXPLORATION: SUPRA-REGIONAL SURVEY IN CENTRAL SABAH, MALAYSIA FEBRUARY, 1994

JAPAN INTERNATIONAL COOPERATION AGENCY, METAL MINING AGENCY OF JAPAN

ABSTRACT

The Government of Malaysia and the Government of Japan agreed on a four-year mineral exploration project, starting from 1990, covering the central part of Sabah. The Scope of Work for this project was signed by both governments on Ist August 1990. Objectives of this project are to identify the mineral potential and to obtain useful data for future development of the mineral resources in the area.

The surveys conducted in this project are collection and compilation of existing data, satellite image analyses, geological survey, geochemical survey, heliborne geophysical survey and investigation of known mineral showings. The survey conducted for each phase is as below:

Phase I: Collection and compilation of existing data, satellite image analyses, orientation geochemical and heliborne geophysical surveys.

Phase II: Regional geochemical survey and investigation of mineral showings in the Segama and Semporna areas and heliborne geophysical survey.

Phase III: Semi-detailed geochemical surveys for eight selected areas in the Segama and Semporna areas, and regional geochemical survey and investigation of mineral showings in the Kinabalu and Labuk areas.

Phase IV: Semi-detailed geochemical survey in ten selected areas in the Kinabalu and Labuk areas, semi-detailed geological survey in one area in the Segama area, investigation of the Tampang area in the Kinabalu area and overall data analyses and interpretation.

Results from the semi-detailed survey in eighteen selected areas have delineated five areas as potential areas for mineral resources. Name of these areas and the targets are as follows:

1. Area T (southern margin of the Labuk area): gold and/or porphyry copper deposits.
2. Area N (west of Telupid in the Labuk area): lateritic nickel deposits
3. Area Q (along S. Karamuak in the Labuk area): vein and/or disseminations of copper deposits.
4. Area B (along S. Danum in the Segama area): vein and/or disseminations of copper deposits.
5. Area K (east of Ranau in the Kinabalu area): lateritic nickel deposits.

Among these potential areas, Area T is the most significant. Some mineralized samples collected from Area T indicate more than 10 g/t Au (maximum 18.4 g/t Au, 931.4 g/t Ag). As only a semi-detailed survey was carried out and in order to delineate more accurately the mineralized zones, detailed exploration work should be conducted for these areas.

During this survey numerous basic data were obtained. These data will assist in carrying out further exploration work more effectively and to identify new potential areas. These data should be treated as follow:

1. The satellite images should be used for a geological survey and exploration work in order to understand the large scale structure.

2. The heliborne survey data should be used to understand the hidden geological structure. It is also possible to delineate a large scale altered zone.

3. Some geochemical anomalous zones which are not covered by the semi-detailed survey should be examined in the future work.

4. Many anomalous zones were obtained by the semi-detailed geochemical survey, but some anomalous zones show no clear relationship with mineralization. These anomalous zones should however be examined to understand the nature of the anomalies.

All data obtained in this survey were input in magnetic tape, so anybody could access any part of the data for re-examination and further investigation. Because these data are the basic data for the exploration work in this area, the data should be used to assist in carrying out further exploration work more effectively.

CONTENTS

8. NATIONAL POLICIES, PROGRAMMES AND PROJECTS FOR ASSESSMENT AND DEVELOPMENT OF LAND AND MARINE MINERAL RESOURCES IN THE FAR EASTERN REGION OF THE RUSSIAN FEDERATION[1]

Step by step, overcoming the difficulties of the transition period, Russia builds a new economy. The country is establishing a home market and searches for a place in the international market. Mineral resources are a primary focus. Russia has unlimited possibilities for domestic and foreign investors in geological exploration in promising regions, industrial development of explored deposits and reconstruction of operating mines.

The Far Eastern region is the richest mineral resources region of the Russian Federation. The leading role of this region results from it containing more than 35 per cent of Russian gold reserves (62 per cent production), more than 95 per cent of Russian proven reserves and production of tin, 43 per cent of tungsten, (37 per cent of production), 8 per cent of lead (36 per cent of production), 4 per cent of zinc (11 per cent of production), 88 per cent of antimony (100 per cent of production), 8 per cent of iron, 63 per cent of mercury, 9 per cent of sulfur and the largest portion of reserves and production of diamonds.

A national programme for the assessment of so called prognostic resources in the Far Eastern region has established that the prognostic resources are more than 8 times proven reserves. The evaluation of prognostic resources is based on a comparison between object parameters and geological and geochemical standards which have been developed on the basis of the results of special geological-geochemical modelling of different types of mineral deposits.

1.0 GOLD

Gold deposits are the most important resource for the Far Eastern region of the Russian Federation. It is known that the principal sources of gold are endogenic gold deposits in bedrock (gold-quartz, goldsulfide lodes, etc.) and placers. Gold is also recovered as an important by-product from copper, nickel, lead and zinc ore (mainly copper-porphyry, copper-nickel, copper-pyrite types of endogenic deposits). Table 18 lists data on the different sources of total gold production within the Russian Federation Far East Region. In Russia, about 85-90 per cent of gold is extracted from placer deposits, demonstrating the considerable untapped potential of endogenic deposits of gold.

Table 19 lists data for gold production in different districts of the Far East Region of the Russian Federation, highlighting the area's importance as a producer of gold in the Russian Federation, accounting for about 73 per cent of the country's output. Table 20 outlines proved reserves and prognostic reserves of different types of gold resources (ore deposits of gold, placers and complex deposits).

Tables 18 and 20 show that, throughout the Russian Federation, placers are the principal sources of gold production, while at the same time the reserves of other types of deposits (endogenic gold deposits and complex ones) account for more than 80 per cent of proved reserves (more than 90 per cent of prognostic resources). The same situation is characteristic of the Far East Region of the Russian Federation in areas such as the Khabarovsk and Magadan districts. In the Magadan district, for example, the production of gold from placers exceeds 90 per cent while the prognostic resources are less than 6 per cent. In Yakutia, 69 per cent of gold production is from placers with prognostic resources of 13 per cent.

These figures highlight that the Russian Federation, and its Far East Region in particular, are very favourable for wide scale production of gold both from placer and endogenic types of deposits. The investments of domestic and foreign companies, will, it is hoped, contribute to the rapid growth of gold production in the Russian Federation Far East Region.

[1] E.N. Isaev and S.V. Grigorian, the Russian Federation

Table 18. Russian gold production in 1993

Type of deposit	*Prognostic resources, per cent*	*Reserves, per cent*	*Production, per cent, (t)*
Total	100	100	100 (167.14)
Endogenic	75	52.1	16.5 (27.53)
Placer	9	19.9	71.4 (119.36)
Complex	16	28	12.1 (20.25)

Table 19. Gold production in the Far East Region of the Russian Federation

Districts	*Production*	
	t	*% of Russian production*
Yakutia	24.39	20.4
Magadan	27.75	23.2
Amur	11.6	9.7
Chukotka	15.18	12.7
Khabarovsk	8.07	6.8
Total	86.92	72.8

Table 20. Reserves and prognostic resources of gold in the Far East Region of the Russian Federation

Districts	*Ore deposits*		*Placers*		*Complex deposits*		*Total*	
	Reserves	*Resources*	*Reserves*	*Resources*	*Reserves*	*Resources*	*Reserves*	*Resources*
Russia-total	100	100	100	100	100	100	100	100
Yakutia	15	8.5	22.0	10.7	0.2	3.1	12.3	7.9
Magadan	10.8	11.2	15.3	5.8	–	10.8	8.6	10.7
Chukotka	7.0	6.3	9.7	4.9	–	–	5.6	5.2
Khabarovsk	4.8	4.8	3.7	8.6	–	–	3.3	4.5
Amurskaya	2.4	5.2	9.9	26.9	–	–	3.1	6.3
Kamchatka	2.2	3.8	–	–	–	2.2	1.1	2.8
Korakskaya	2.1	3.9	0.4	2.2	–	–	1.8	3.1
Primorye	0.05	1.4	0.6	0.9	0.2	2.7	0.2	1.8
Sakhalin	–	1.9	0.2	0.9	–	–	0.03	1.5
Juish autonomous	–	–	–	0.4	–	–	0.02	0.05

Considering the present day economic conditions of the Russian Federation, an important priority of natural mineral resources policy is to alert foreign mining companies to the potential for focussing their mining activities on the Russian Federation Far East Region. Success in this regard, and the subsequent successful development of mineral deposits in the region, requires the expansion of exploration activities, mainly aimed at transferring the region's predicted resources into proved reserves to assist future profitable exploitation.

The prospectivity of the region for gold in general and for bedrock gold deposits, in particular, is illustrated in table 21, which lists the proved reserves of some bedrock gold deposits of the region in question.

In the last few years, since the collapse of the USSR, a distinct reduction of gold production in the Russian Federation has taken place (tables 22 and 23).

2.0 TIN

The Russia's Far East Region is the major tin producing region in the Russian Federation, with more than 85 per cent of all Russian reserves and more than 93 per cent of production. The following types of tin deposits are characteristic of the region;

i) tourmaline-cassiterite (Deputatskoye, Dekhdardakh, Valkumeiskoye, Solnechnoye, Festivalnoye, etc.);

ii) quartz-cassiterite (Jultin, Svetloye, etc.).

3.0 TUNGSTEN

The Far East Region is one of the leading producers of tungsten within the Russian Federation, comprising more than 37 per cent of production and more than 43 per cent of prognostic resources.

Vostok II is the major active tungsten deposit in the region. The deposit is a skarn type and is confined to the contact of granodiorites of Cretaceous age, and sedimentary rocks of upper Paleozoic age. The ore mineralization is superimposed on steeply dipping skarn zones.

4.0 LEAD AND ZINC

The Nickolaevskove and Smirnovskoye skarn lead-zinc deposits account for the bulk of lead-zinc production in the Far East Region. The region produces more than 36 per cent and 11 per cent of Russian production of lead and zinc respectively.

5.0 DIAMONDS

Diamond deposits are primarily confined to ultrabasic rocks commonly in the form of intrusive necks resulting from high temperature and pressure prevalent in the plutonic depths.

The Russian Federation is the world's greatest source of diamonds. The Yakutian diamond bearing kimberlite pipes (Mir International, Yubileinaya, Aikhal, Udachnaya) are the major source of Russian diamonds (99.7 per cent of production and 84.3 per cent of reserves of the Russian Federation).

6.0 OTHER MINERALS

Placer deposits of platinum discovered within the the Kamchatka and Khabarovsk districts, are particularly worthy of mention, with production reaching 5.7 tonnes.

In conclusion, the attraction of both domestic and foreign investments to the mineral development programmes in the Far East Region would undoubtedly have a positive influence in securing a prominent position in the international markets for the Russian Federation's mineral exports.

Table 21. Reserves of major bedrock gold deposits of the Russia's Far East Region

District	*Deposit*	*Reserves, t*
Yakutia	Nezhdanenskoye Kyuchusskoye	475 157
Magadan	Maiskoye Kubaka Evenskoye Karalveemskoye Shkolnoye	277 92 52 39 29
Kamchatka	Ametistovoye Rodnikovoye Agginskoye Asochinskoye	97 40 30 30
Khabarovsk	Khakanja	50
Amurskaya	Pokrovskoye Berezitovoye	59

Table 22. Gold production in the Russian Federation, 1983-1992

Years	*Gold Production, t*			
	Total	*Bedrock*	*Placer*	*Complex*
1983	172.26	28.00	118.25	26.51
1984	172.34	28.73	118.08	25.53
1985	175.39	27.93	121.61	25.25
1986	180.48	28.63	125.88	25.97
1987	184.20	31.66	126.63	25.91
1988	188.55	31.25	131.44	25.86
1989	186.39	31.19	129.18	26.02
1990	179.75	30.72	124.28	24.75
1991	178.39	28.54	126.74	23.11
1992	167.24	27.53	119.36	20.25

Table 23. Gold production in the Russian Federation, 1991-1994

Deposit	*1991*	*1992*	*1993*	*1994*
Gold deposits (bedrock and placer)	133.7	126.02	136.23	131.9
Gold bearing complex deposits	10	8.5	7.96	4.9
Total	143.7	134.52	144.19	136.8
Mining production	178.39	167.24	–	–

Considering the present day economic situation of the Russian Federation, the important priority of the nation's mineral resources policy is to alert foreign mining companies to the opportunities for locating their mining activities in the Russian Far East Region. It is evident that the successful development of mineral deposits requires the extension of exploration activities in the region's economically prospective regions, ensuring that potential deposits move quickly from the category of resources into proved reserves for future profitable exploitation.

REFERENCES

1. Benevolsky, B.I., 1995. Gold of Russia: Prospects of the Use and Replacement of Mineral Resources. (in Russian). Publ. "Geoinformmark", Moscow, 90 p.

2. Geological Survey and Development of Mineral Base, 1993. Publ. TSN I GFI (in Russian), Moscow, 618 p.

CONTENTS

9. THE ASSESSMENT AND DEVELOPMENT OF FUEL MINERAL RESOURCES IN THE FAR EASTERN (OKHOTSK) REGION OF THE RUSSIAN FEDERATION AND KEY ISSUES FOR SUCH DEVELOPMENT TO MEET THE DEMANDS OF COUNTRIES OF THE NORTH-EAST ASIA[1]

INTRODUCTION

The Sea of Okhotsk region is one of the most geologically complex regions of the world, including elements of ancient, Precambrian-Paleozoic consolidations to the recent active volcanic island arc zones. Over much of it, the mosaic structure of tectonic blocks of various ages is submerged by the Okhotsk Sea. Hydrocarbon prospecting is carried out mostly on Sakhalin Island and in West Kamchatka, and has led to the discovery of 75 oil and gas fields. Sixty-six of these fields have been discovered on North Sakhalin and the adjacent shelf; 2 gas fields, on the western Sakhalin coast; 3 gas fields, in the south of the island (Aniva district); and 4 gas and condensate fields have been discovered in southwest Kamchatka (figure 12).

The principal exploration targets are the following sedimentary basins: North Sakhalin, Aniva, West Kamchatka, Makarov, Bilibin, Shelikhov, Gizhiga, Penzhina, Tinro, Central Okhotsk, North Kuril, and South Kuril, covering 1.85 million km^2, confined to the Sea of Okhotsk megabasin; and the Ishikari-West Sakhalin and North Japan basins of the Sea of Japan megabasin, covering 0.3 million km^2.

1.0 GEOLOGICAL STRUCTURE OF OKHOTSK REGION

1.1 Structural Elements, Geological Sequences

The Okhotsk region is at the North-East Asian continental margin, in which many changes to its passive and active regimes of development took place over time.

The passive margin encompasses the main part of the region – Sakhalin Island. Structures within the shelf and continental slope formed in the course of rifting with the pulling apart of the continental blocks, as a result of the Shelikhov, Bilibin, Tinro, North Sakhalin, and Ishikari-West Sakhalin rifts (figures 13, 14).

The central part of the Sea of Okhotsk is occupied by the Central Okhotsk cratogenic block, cross-cut by a series of regional faults forming narrow "gaping" grabens.

The active margin includes the Kuril-Kamchatka island arc system, limited by the Kuril trench on the side of the Pacific thalassogene. The West Kamchatka trough is an accretionary structure, formed by convergence of continental blocks. As the oceanic plate plunged beneath the island arc, the South Kuril forearc trough was formed on the southeastern margin of the region.

In the western and northern onshore parts of the Sea of Okhotsk region are the folded complexes of the Mesozoides and the volcano-plutonic belt that complicates them. Along the margin of the Okhotsk volcanogene, the north Okhotsk rift zone extends in the sub-latitudinal direction, which is the offshore part of the Okhotsk-Anadyr rift zone of the latest Cretaceous to earliest Paleogene age.

The structures of the Mesozoides of the continental margin and the older Paleozoic blocks extend over much of the Sea of Okhotsk area. In the northeastern sector of the Sea of Okhotsk, the Paleozoic block probably underwent active subsidence along a system of convergent listric faults forming a deep triangular Tinro rift trough, filled with a thick (up to 10 km) sedimentary sequence. In the southern part of the Sea of Okhotsk, the South Okhotsk deep basin of similar structure is recognized, submerged to a depth of 5 km, while in the northwest is the Deryugin trough. Here active crustal spreading was accompanied by pre Cenozoic deep-sourced basalt extrusions.

---

1 V.I. Vyssotski and R.D. Rodnikova, Petroleum Institute, JSC "VNIIZARUBEZHGEOLOGIA"

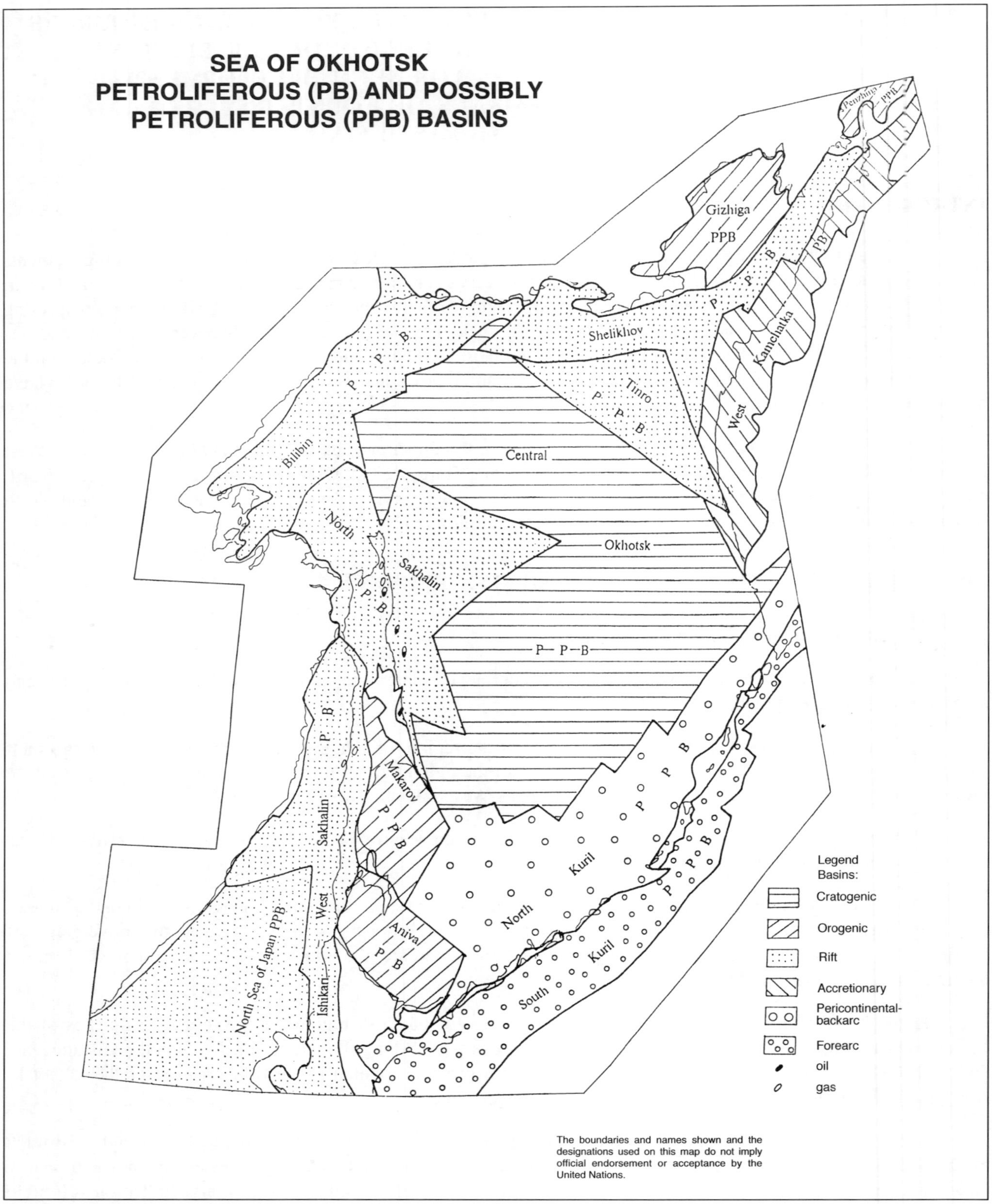

Figure 12. Petroliferous (PB) and possibly petroliferous (PPB) basins of the Sea of Okhotsk

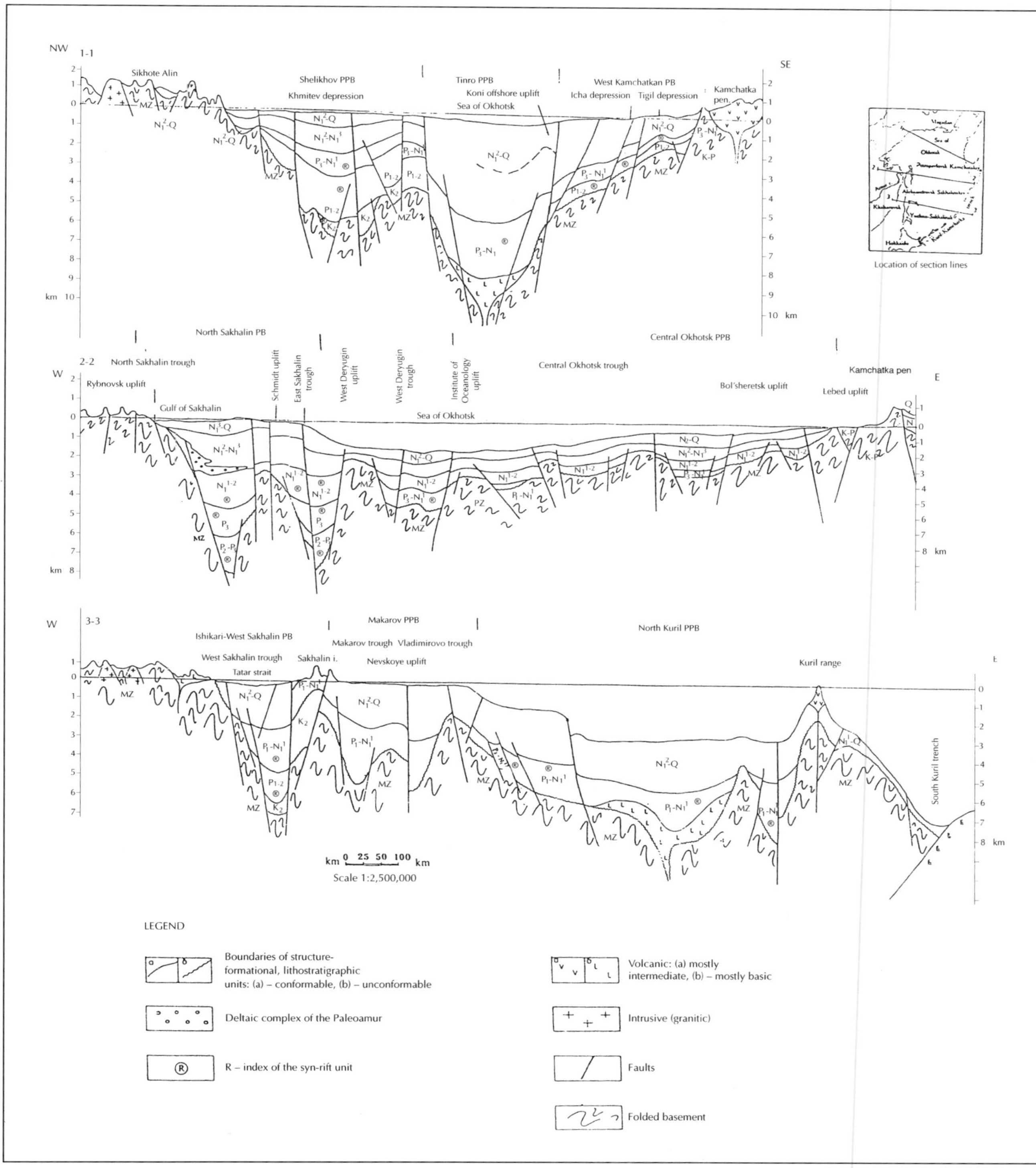

Figure 13. Geological cross-sections in the Sea of Okhotsk region

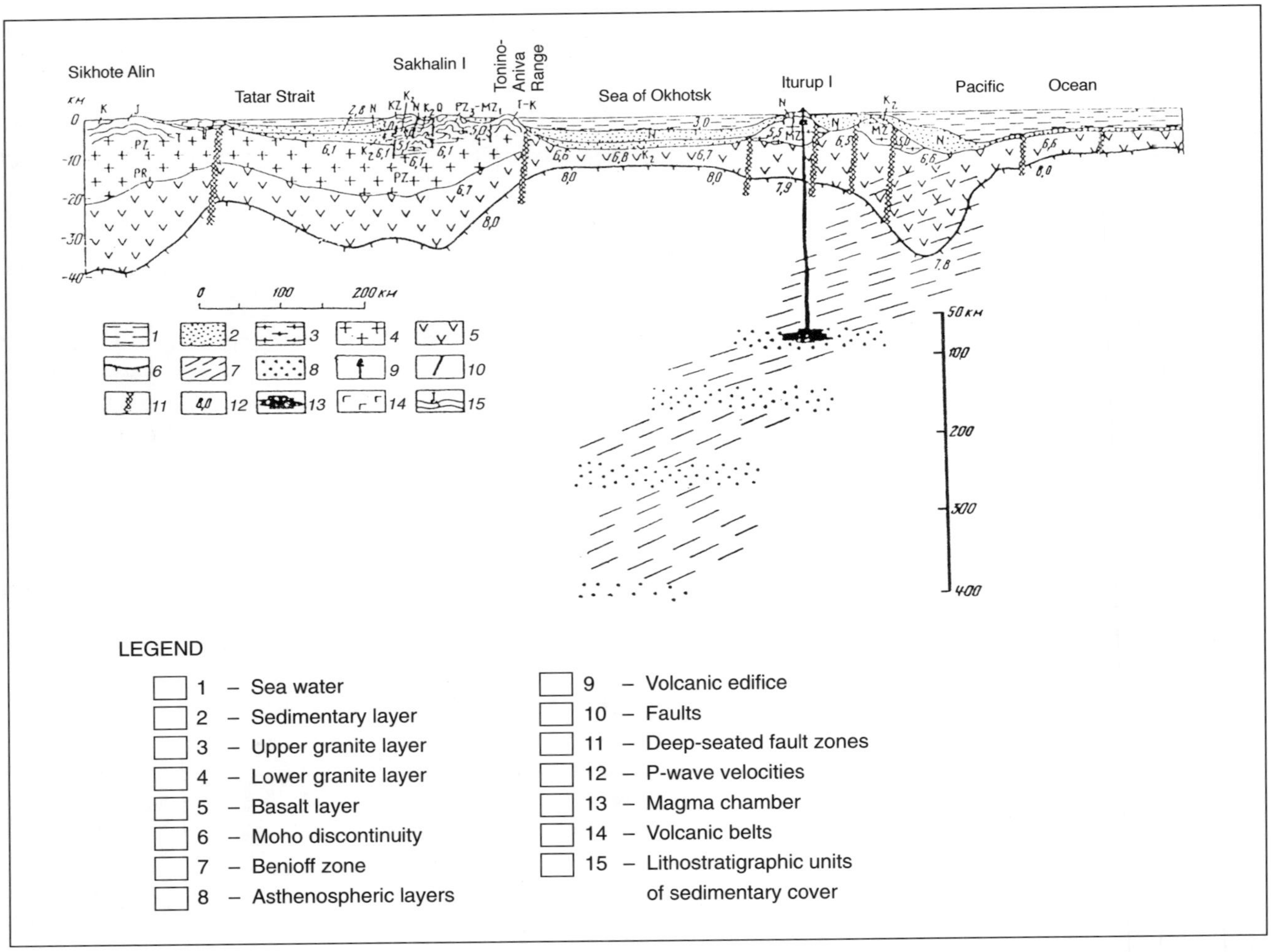

Source: Data compiled by A.G. Rodnikov on the basis of geophysical data of the Shmidt Institute of Earth Physics, Sakhalin Multidisciplinary Research Institute, Institute of Oceanology and the Geological Survey of Japan.

Figure 14. Geological cross-section of the Earth's crust along the Sikhote Alin-Sakhalin-Kuril Island-Pacific Ocean line

Sakhalin Island is a continental block within which the strike of the major tectonic elements are largely inherited from the ancient structural pattern.

The folded basement of the island is composed of the Paleozoic-Lower Mesozoic metamorphic rock complexes which lie at considerable depth, and outcrop at the surface only in isolated, fault-bounded blocks that divide the vast Sakhalin sedimentary depression into a number of submeridional troughs: West Sakhalin, Alilva, Makarov (Tym-Poronay), North Sakhalin, and East Sakhalin.

The *Kamchatka Peninsula* sequence, belonging to the active margin of the Kuril-Kamchatka island arc system, consist of: Paleozoic gneisses, crystalline schists and phyllites, and metamorphosed eruptive and pyroclastic rocks, exposed on the surface in small isolated blocks; Lower Mesozoic to Lower Cretaceous intensely dislocated terrigenous rocks, forming (jointly with the Paleozoic rocks) the basement of the long West Kamchatka trough, extending along the western coast of the peninsula and bounded by the Central Kamchatka Cenozoic volcanic rise; and the Upper Cretaceous-Cenozoic deposits, most fully developed in the depression structures.

The *Kuril island arc* consists of two ranges, the volcanic arc rise, represented by the islands of the Greater-Kuril Range (Kunashir, Iturup, Urup, etc.), and the outer non-volcanic arc, including the islands of the Lesser Kuriles – Shikotan, Polonskiy, Zelenyy, etc. Structures of the Kuril Island Arc extend into the Nemuro Peninsula of Hokkaido Island, and its northeastern segment is represented by the submarine Vityaz Ridge.

The Kuril trench separates the Pacific oceanic plate from the Eurasian plate.

1.2 Deep structures

These major structural elements of the Sea of Okhotsk region are the surface reflection of deep subsurface processes that have taken place during the recent stage of tectonogenesis. Figure 15 shows the geological-geophysical cross-section of the crust of the region along the Sikhote Alin – Sakhalin – Kuril Islands – Pacific Ocean line.

The earth's crust in this region tapers from 37-34 km to 15-10 km, during the transition from continent to ocean. The greatest thickness of sedimentary cover is recorded in the offshore zones adjacent to North Sakhalin, the Tatar Strait, the northern Sea of Okhotsk, and the Tinro trough.

In general, the region is characterized by high heat flow (HF) values, with an average value of 90 mW/m^2, while the average global HF value is 50 mW/m^2. [24]

1.3 Geodynamics of sedimentary basins

The high tectonic mobility of the Sea of Okhotsk region, caused primarily by the heterogenity of the crustal structure, has existed at least since the Paleozoic. Tectonics which took place at the end of the Paleozoic period, led to the formation of tectonic zones over a considerable part of the Sea of Okhotsk which, during the Early Mesozoic, were the sites for the development of the Sikhote Alin volcano-plutonic belt on the continental margin. Also marked by magmatism, in all probability, are the marginal parts of the lower-order pull-apart structures.

The Mesozoic rifts were filled by deep marine deposits. In the marginal parts of blocks, olistostromes are represented. The boundary between the Lower and Upper-Cretaceous is characterized by a manifestation of tectonic movements (Yanshanian), resulting in extensive regression and high volcanic activity.

In the Upper Cretaceous, marine sedimentation took place in the regions of Sakhalin, West Kamchatka, and the Kuril Islands. The final episodes of the Cretaceous manifested themselves as a geocratic epoch of drying over much of the Sea of Okhotsk Region, with fold dislocations taking place.[6]

The beginning of the Cenozoic stage corresponded to regeneration of rifting processes in the region. This was probably caused by the influence of astenospheric plumes within the Sea of Okhotsk region of pericratonic Eurasia. Two trends of tectonic movements during that period determined the formation of the structure of present-day sedimentary basins, for example: (1) the subduction of the Pacific plate in the region of the Kuril-Kamchatka trench and the beginning of active margin development; and (2) the dispersed rifting in the zone of pericratonic subsidence. Rifting centres coincided with the weakened deep-seated fault zones separating crustal blocks with various ages of consolidation.

Rift development continued until the end of the Early Miocene.

The middle Miocene-Pliocene post-rift subsidence encompassed practically the whole region except its northern and western Mesozoic folded areas, where sediment accumulation took place in linear zones bounded by deep-seated faults. During the post-rift period the sedimentary basins acquired their present-day pattern. In the zone of the Kuril-Kamchatka trench, an accretionary prism was formed. Compression processes also manifested themselves in West Kamchatka in connection with the ongoing rifting in the adjacent Tinro zone.

Rifting processes were the primary determinate in the formation of the structure of the sedimentary basins. Rift-related troughs, as a rule, were filled with thick sedimentary sequences, including source rocks, reservoirs, and seals, and are therefore favorable for HC prospecting.

Ongoing and recent tectonic movements, especially in the southeastern zone, expressed primarily by active volcanic and seismic activity, continue to exert an influence on the geodynamic regime of the sedimentary basins, with which HC prospecting is connected.

2.0 PETROLEUM ZONING

Analysis of the geodynamic settings of the sedimentary basins of the region allows recognition of the following types of the related petroliferous and potentially petroliferous basins:

Cratogenic – Central Okhotsk PPB, confined to the fault-bounded submerged Paleozoic-Mesozoic central block of the Sea of Okhotsk;

Orogenic – Aniva PB, Makarov PPB, Penzhina PPB, and Gizhiga PPB basins, formed during the Cenozoic stage of orogenesis on the folded Mesozoic-Paleozoic basement;

Rift – North Sakhalin PB, Ishikari-West Sakhalin PB, Shelikhov PPB, Bilibin PPB, Tinro PPB, North Japan PPB;

Accretionary – West Kamchatka PB,

Pericontinental-Backarc – North Kuril PPB;

Forearc – South Kuril PPB.

The basins are filled with Upper-Cretaceous and (mostly) Cenozoic deposits, the thickness of which reaches 10-12 km.

The principal parameters characteristic of the PBs and PPBs of the region are listed in table 1.

2.1 Sea of Okhotsk Megabasin

The Sea of Okhotsk megabasin includes the Central Okhotsk, Aniva, Makarov, Penzhina, Gizhiga, North Sakhalin, Shelikhov, Bilibin, Tinro, West Kamchatka, North Kuril, and South Kuril basins, with a total area of 1,855.6 km^2 (table 24).

2.1.1 Cratogenic Central Okhotsk PPB

Recognized in the central part of the Sea of Okhotsk.

Geological structure: A submerged cratogenic Sea of Okhotsk block, deformed by tectonic processes. The basin is fault-bounded.

The basement of the basin consists of blocks of various ages of consolidation, including the (mostly) Paleozoic, in the region of the Institute of Oceanology rise, and Mesozoic over much of its territory.

The thickness of *sedimentary fill* within the narrow graben-shaped troughs reaches 3-4 km, and its age is probably Neogene-Quaternary. In some grabens Oligocene deposits are present. On the rises the sedimentary cover is Pliocene-Quaternary.[9]

Source rocks are probably the Oligocene-lower Miocene deposits.

Reservoirs are possibly sandstone horizons developed throughout the sequence.

Seals correspond to mudstone varieties.

Traps are anticlinal structures, basement highs, sedimentary complex and horizon pinchout zones.

2.1.2 Orogenic (Intermontane) Basins

2.1.2.1 Aniva PB

Recognized in southern Sakhalin Island, including the offshore area of Aniva Bay.

Table 24. Characteristics of Petroliferous (PB) and Possibly Petroliferous Basins (PPB) of the Sea of Okhotsk Region

PB, PPB	*Basin Area, 1,000 km^2*		*Chronostratigraphic Age (Range)*		*Maximum Sediment Thickness, km*	*Number of Discoveries*		*Geodynamic Type*
	Total	*Off-shore*	*Sedimentary cover*	*Folded basement*		*Oil*	*Gas*	
1	*2*	*3*	*4*	*5*	*6*	*7*	*8*	*9*
SEA OF OKHOTSK MEGABASIN								
Central Okhotsk PB		544.5	P_3-Q	PZ, MZ	4	–	–	Cratogenic
Aniva PB	72.5	40.5	K_2, P_3-Q	MZ	6	–	3	Orogenic
Makarov PPB	54.7	42.4	K2, P3-Q	PZ, MZ	6	–	–	Orogenic
Penzhina PPB	9.3	5.2	KZ	MZ	3	–	–	Orogenic
Gizhiga PPB	60.2	58.1	P_3-Q	MZ	4	–	–	Orogenic
North Sakhalin PB	195.5	159.2	K_2, P_2-Q	MZ	10-12	22	44	Rift
Bilibin PPB	130.3	124.2	K_2, P-Q	MZ	8	–	–	Rift
Tinro PPB		67.2	P_3-Q	MZ, K_2-P_2, basalt	10	–	–	Rift
West Kamchatka PPB	115.4	39.6	K_2, P-Q	MZ	4	–	4	Accretionary
North Kuril PPB	266.4	255.6	P_3-Q	MZ, KZ	7	–	–	Pericontinental-Backarc
South Kuril PPB	219.6	207.3	K_2, P-Q	MZ, KZ, K_2-P_2 basalts	6	–	3	Forearc
Shelikhov PPB		120	K_2, P-Q	MZ	7	–	–	Rift
SEA OF JAPAN MEGABASIN								
Ishikari-West Sakhalin PB	125.8	88.6	K_2, P-Q	J-K_1	9	31	12	Rift
North Sea of Japan PPB	199.5	187.6	N-Q	MZ basalt basalt	5	–	–	Rift

Geological structure. The basin is confined to an orogenic graben-shaped subsidence zone bounded to the north and south by the uplifts of the Paleozoic-Mesozoic basement.

The *sedimentary fill* is Oligocene-Quaternary with a thickness of up to 6 km, composed primarily of coal-bearing, sand-shale deposits.

Oil/gas source rocks are Oligocene-lower Miocene silt-shale deposits; gas source rocks are the middle-upper Miocene coal-bearing deposits.

Reservoirs are sandstones, tuffs, and tuffaceous breccias.

Seals are mudstone intervals.

Traps for HC accumulation are anticlines. Non-anticlinal pools sealed by thrust planes are also possible. Three small gas fields have been discovered in the basin, Lugovskoye Yuzhnoye, Lugovskoye Vostochnoye, and Zolotorybinskoye. The Lugovskoye Yuzhnoye and Lugovskoye structures are rootless brachyform anticlines in the zone adjacent to the Central Sakhalin fault.[16] Pay zones are tuffstones of the Maruyama Formation at a depth of 960-1,415 m.

All discoveries are small. Ultimate proved gas reserves are 0.6 Bm^3 (as of 01.01.1995). The Lugovskoye Vostochnoye field is currently producing a cumulative gas production reaching 0.02 Bm^3.

2.1.2.2 Makarov PPB

The basin is recognized in the area of Terpeniya Bay on Sakhalin Island and the adjacent offshore territory, including the Tym-Poronay depression.

Geological structure. The basin is confined to the Makarov and Vladimirovo graben-shaped orogenic troughs.

The basement of the basin is composed of Mesozoic folded complexes exposed in the surrounding structures, and dredged on the slopes and in submarine mounts of the Terpeniya Bay depression.[21] According to absolute datings, they mostly correspond to the Jurassic and Cretaceous systems.

The sedimentary fill of the Oligocene-Quaternary age reaches a maximum thickness of 4-6 km. Over much of the basin, the Upper Cretaceous formations occur as the acoustic basement.

Oil and gas *source rocks*, in all probability, are the Oligocene-lower Miocene siltstone-mudstone deposits; gas-prone source rocks are the middle-upper Miocene coal-bearing deposits.

Reservoirs are recognized throughout the sequence.

Seals are provided by mudstone intervals.

Traps are related to anticlinal folds with a height of 150-200 m. The structures identified by seismic surveys are characterized by wide gentle crests; the limbs are cross-cut by fault systems. Folds also drape the pre-Cenozoic horsts. The presence of non-anticlinal pools is inferred, related to pinchouts of lithological complexes sealed by thrust planes.

2.1.2.3 Penzhina PPB

Within the region under consideration, the basin occupies 4,100 square kilometres onshore and 5,200 square kilometres offshore in the Penzhina Bay *(Penzhinskaya Guba).*

Geological structure. The basin is confined to a Cenozoic orogenic trough in the Mesozoides zone, the uplifts of which define its boundaries. The trough is separated from the Shelikhov rift by the NW-EW trending Talovka high.

The basement of the basin is composed of pre-Upper Cretaceous folded volcanoclastic formations intruded by granitoids.

The sedimentary cover, with a thickness of up to 3 km, is composed of terrigenous rocks.

2.1.2.4 Gizhiga PPB

The basin covers much of the Shelikhov Bay.

Geological structure. The basin is confined to the Gizhiga trough and the Yama graben-shaped depression.

The northern boundary of the basin coincides with a NE-trending regional fault; the southern, with the uplifts of the Mesozoic volcano-plutonic *basement*, buried beneath the Cenozoic cover.

The sedimentary cover is composed of Oligocene-Quaternary deposits with a maximum thickness of 4 km.

Source rocks, in all probability, are associated with the Oligocene-lower Miocene complex.

2.1.3 Rift Basins

2.1.3.1 North Sakhalin PB

The basin was found to occur in North Sakhalin Territory and the adjacent offshore area of the Sea of Okhotsk and Sakhalin Bay.

Geological structure. The basin is confined to an equant pericratonic rift zone of subsidence, including the North Sakhalin and East Sakhalin troughs, separated by the Shmidt block uplift, and the Central Deryugin trough limited by the East Deryugin high (figure 13).

The basement of the basin is composed of Triassic-Lower Cretaceous cherty volcanics, in some areas Upper Cretaceous volcanic and volcanic-sedimentary folded complexes; buried to a depth of 5-12 km, and on the inner uplifts, to 1.5-3 km.

The sedimentary fill are Eocene-Quaternary deposits with a thickness of up to 8-12 km, including Upper Cretaceous deposits in some parts of the basin (figure 15).

The upper units of the *sedimentary cover* are characterized by the development of folded structures, as a rule along fault zones, where their height reaches 1.0-1.5 km.[22, 23] The lower units are distinguished by the high degree of dislocation, development of sub-parallel tilted blocks, forming half-grabens with a thickness of 3-5 km.

Source rocks in the basin are the Oligocene-Miocene shales, cherty mudstones, and coal-bearing terrigenous deposits accumulated during the rifting phase, and by the beginning of the termination.

The North Sakhalin Basin shows the highest values of emigration among the Sea of Okhotsk region sedimentary basins.

Reservoirs within the basin's discovered fields are composed of sandstones and siltstones. Open porosities are 6-31 per cent. The reservoir of the Okruzhnoye field is classified as a fractured-porous type in which fluid migration takes place, both at the extent of pressure differences, and under the influences of capillary forces.

HC *seals* are mudstone seals or members with silty-sandstone interbeds. In terms of continuity, the seals are most commonly semi-regional and local, with a thickness from 10 to 80-90 m, rarely more than 100 m.

Traps are numerous anticlines, as a rule linear in shape and confined to fault zones. Also common are gentle dome-shaped and coffer folds. The size of the structures range from 5 to 300 km^2, the height, 100-600 m. Stratigraphic (lithological) and conformity traps are widespread, as are traps with fault-dependent closure. A group of recent discoveries have been related to the horsts of middle Miocene shale-sand deposits of the Dagi Formation that are buried beneath the lower Okobykay mudstone member.

Exploration Results. By the end of 1992, a total of 2,292 wells had been drilled in North Sakhalin, of which 1,939 wells were productive. The average flow rate is 2.5 t/d (in 1985 the flow rate was 3.8 t/d).

The first discovery in the basin was made in 1923 at Okha (Okhinskoye). Offshore exploration of Northeast Sakhalin has been carried out jointly with the Japanese consortium SODEKO since 1976, within the framework of a Soviet-Japanese agreement, signed January 1975. The activities of 1976-1982 resulted in the discovery of the Chayvo and Odoptu-More fields. By 1995, 66 onshore and offshore fields had been discovered, including 11 oil, 10 gas, 28 gas/oil and oil/gas, 7 gas/condensate, and 10 oil/gas/condensate. The discoveries are grouped by petroleum accumulation zones, structurally corresponding to anticlinal zones. The richest petroleum accumulation zones were identified in the vicinity of Piltun-Chayvo and Baikal-Pomyr. They account for practically all the relatively large discoveries of North Sakhalin and are fed from synclinal zones characterized by a great thickness of Oligocene-Pliocene sedimentary cover (up to 10-11 km).

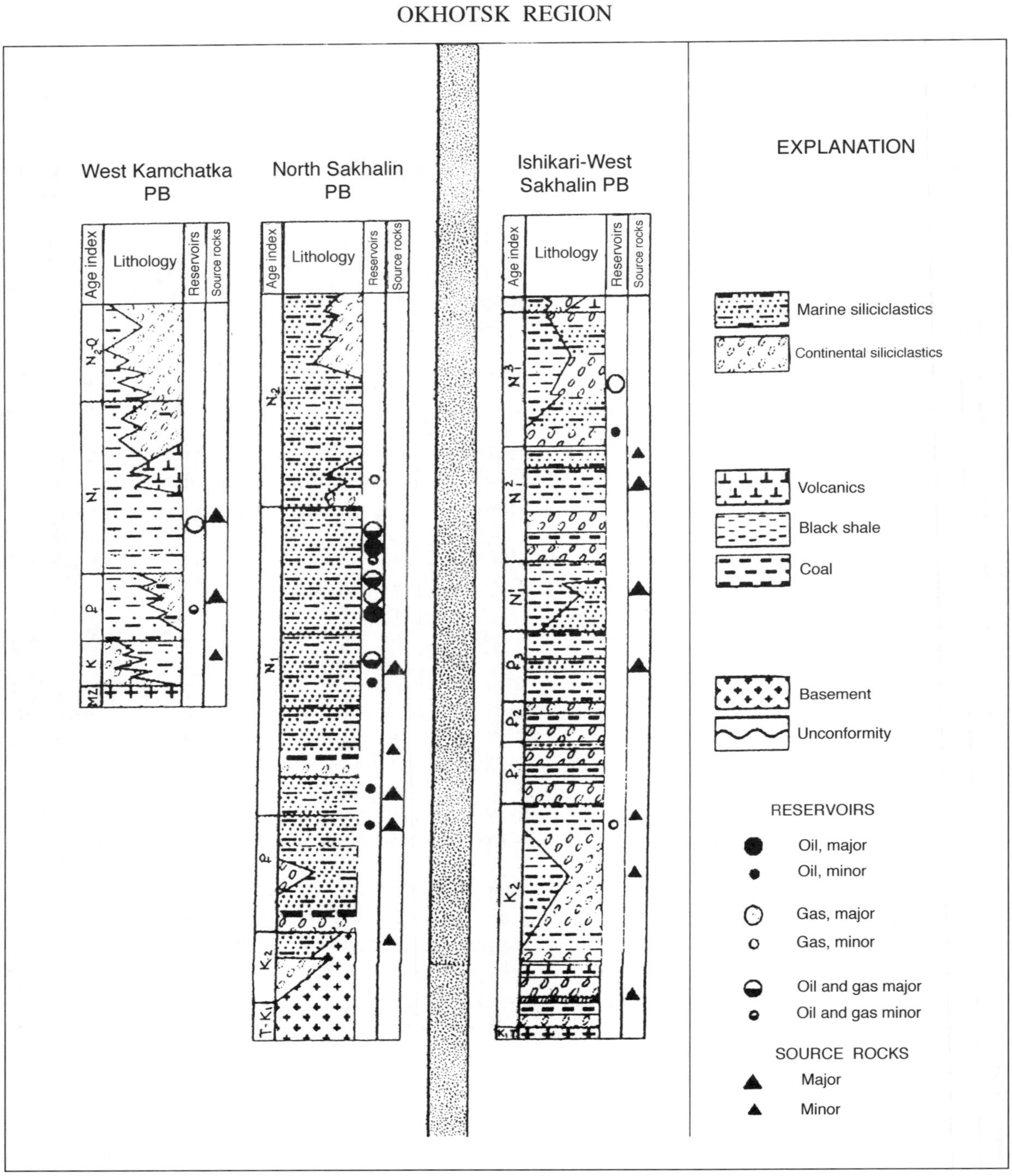

Figure 15. Stratigraphic sections of the Okhotsk region

Pay zones are terrigenous and cherty-terrigenous deposits within three units: lower Nutovo-Okobykay (N_1^{2-3}), Dagi-Uyni (N_1^{1-2}), Daekhurie (N_1-P_3). The age of reservoirs increases from north to south. Predominately petroliferous are the deposits of the first two units at a depth of 3 km. The deepest (gas/condensate) pools were discovered in the Ust Evay field in the deposits of the Dagi Formation at a depth of 4.8 km.

Offshore deposits were explored with sea depths from 10 to 50 m.

Reservoirs in most of the fields are stacked, with the number of reservoir units reaching 33 (Kydylanyi). As a rule the fields are characterized by complex tectonic structures (numerous transverse and diagonal normal and reverse faults, lengthwise faults and thrust faults) and complex lithology (strong lithological discontinuity of the strata, abrupt pay zone thickness variation, pinchouts of reservoir units or the whole reservoir sequence). As a result, in addition to the full-shaped anticlinal pools, are pools with cross-cut shapes (fault- or lithology-dependent closure); anticlinal pools, split into blocks; pools lithologically or tectonically bounded on the limbs and periclines; and subthrust pools. In a number of fields, supra-unconformity sealing was reported. The pools are subdivided into bedded (stratabound), massive-bedded, and lenticular. The most numerous are bedded pools, related to the sequences where sandstones and siltstones are interbedded with mudstones.

Porous reservoirs predominate. Related to a fractured-porous reservoir is the oil pool in the Pilenga Formation of the Okruzhnoye field (4 per cent of the North Sakhalin proven oil reserves).

Net pays of reservoir units are a few meters to a few tens of meters; in bedded-massive reservoirs, which can be as high as 300 m. The pool heights sometimes reach 400-500 m, but are most commonly within 40-100 m and considerably less than the trap heights.

The pay zone depth of occurrence ranges from a few meters (sometimes outcropping on the surface – the Okha and Katangli fields) to 4,850 m (Ust Evay field); but most of the pools occur at depths less than 2,500 m. The total thickness of the productive interval of the sequence often exceeds 2 km, for example at Gilyako-Abunan (2.16 km), Volchinka, Sabo Maloye (2.15 km), and Kydylanyi (2.32 km).

Three deposits are classified as large: Lun (oil, more than 27 MMt; gas 291.3 Bm^3); Odoptu More (oil, 38.1 MMt; gas, 62.1 Bm^3); and Piltun-Astokh (oil, 66.4 MMt; gas, 78.1 Bm^3). (figure 16)

Seven fields are classified as moderate in size: Chayvo (oil 2.9 MMt; gas, more than 28 Bm^3); Arkutun-Dagi (oil, 9.3 MMT; gas, 23.2 Bm^3); Okha Tsentralnaya (oil, 21.2 MMt); Mongi (oil, 16.8 MMt); Ekhabi (oil, 16.1 MMt; gas, 2.3 Bm^3); Kolendo (oil, 15.5 MMt; gas, 5 Bm^3), and Mirzoyev (oil MMt, gas, 9.3 Bm^3).

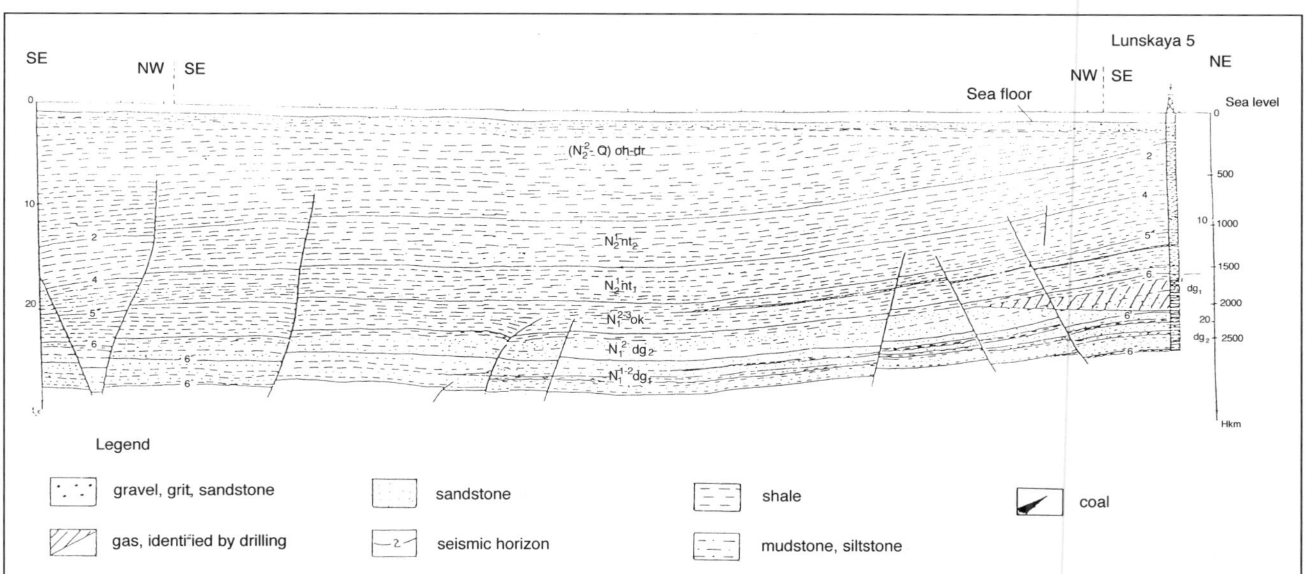

Figure 16. Lithofacies profile along seismic cross-sections (Lun anticline)

The initial proved oil reserves of the North Sakhalin PB as of January, 1993, are 302.6 MMt (including 33.0 MMt of condensate); dissolved gas, 54.4 Bm^3.

The first large oil field in Sakhalin, the Okha field, was brought onstream in 1923. By 1940, annual oil production reached 600 Mt. During the 1960s, about 20 small and moderate discoveries were made. During 1961-1967, annual production increased from 1.7 MMt to the peak value of 2.7 MMt.

Until recently, gas was produced mostly for local consumers. Annual production since 1961 has ranged from 0.38 to 0.8 Bm^3. In recent years it has increased to 1.2-1.6 Bm^3/yr (from 38 producing fields). Condensate is produced in limited amounts. Cummulative production of oil as of 1993 reached 94.5 MMt; gas, 37.7 Bm^3. The bulk of oil production is obtained from the Okha Tsentralnaya, Mongi, and Mirzoyev, yielding 0.3 MMT oil each, jointly accounting for 55 per cent of total annual production. The oil is sweet, low-paraffinic, with a specific gravity of 0.8-0.9 g/cm^3 (26-45 degrees API) and viscosity between 100 and 3,000 cst (centistokes).

Gas, as a rule, is associated with oil and condensate. The main oil/gas fields are Mirzoyev, Ust Evay, Mongi, and Volchinka, which in 1992 yielded 1.5 Bm^3, i.e. more than 85 per cent of Sakhalin's total gas. Condensate content in the Sakhalin gas fields ranges from 9 to 160 kg/m^3.

Gas is produced by proper gas wells, providing for 90 per cent of total production, jointly with oil. By the beginning of 1993, well stocks of Sakhalinmorneftegaz included 80 gas wells, of which 51 were producing. The average daily gas yield of the gas wells increased from 54 thou eu m in 1985 to more than 85 thou m^3 in 1992. Dissolved gas production in 1992 was 0.3 Bm^3.

2.1.3.2 Shelikhov PPB

The basin is located in the northern Sea of Okhotsk.

Geological Structure. The basin consists of a linear system of pericontinental rifts forming a long zone parallel to the northern Sea of Okhotsk coast.

The basement of the basin is Mesozoic, composed of a folded volcanic-sedimentary complex cut through by granitoid intrusions.

The sedimentary cover reaches 7 km in thickness.

Source rocks are Paleogene-Miocene mudstones and cherty mudstones accumulated in rift depressions under a thermodynamic regime favourable for HC generation.

Well Magadanskaya 1 in the interval 500-2,500 m penetrated a sequence of muddy diatomites, gaize and very muddy siliciliths (figure 17).

Reservoir horizons occur mostly at the top of the sequence, and are confined to alluvial fans, deltas and turbidites, widespread in the Oligocene-Miocene part of the sequence.

Seals are mudstone intervals.

Traps for hydrocarbons are various uplifts. In the Oligocene-lower Miocene syn-rift complexes, faulline anticlines, gravity folds, and basement onlap zones are widely developed. Non-anticlinal pools are found in the deltaic formations and fans. In the upper complexes (upper Miocene-Pliocene), gentle anticlines are recognized at the background of the general subhorizontal bedding. Angular uncomfortities at the base of the middle Miocene-Pliocene complexes, may however, cause differences in structural patterns.

2.1.3.3 Bilibin PPB

The basin is located in the northwestern Sea of Okhotsk.

Geological Structure. The basin is confined to a pericontinental rift trough limited to the north and west by the block uplifts of Mesozoides and the Sikhote Alin volcano-plutonic belt.

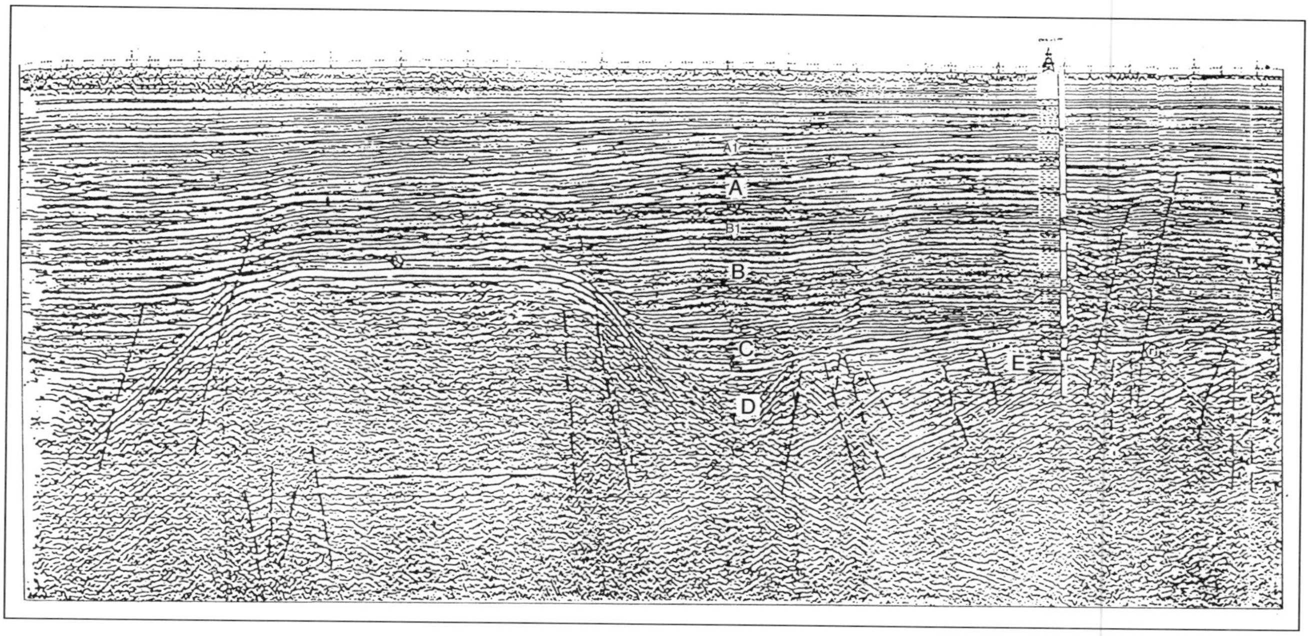

Figure 17. Seismic profile along the drilling well on Magadanskaya line

The basement of the basin consists of the folded complexes of Mesozoides complicated by magmatites.

The *sedimentary cover* of Late Cretaceous-Neogene-Quaternary age has a total thickness of 6-8 km.

Spatial distribution of the sedimentary thickness is controlled by a system of faults, often parallel to the shoreline or submeridional, forming a system of isolated grabens and half-grabens.

Possible *source rocks* are syn-rift and post-rift pelagic mudstones and cherty mudstones, and deltaic complexes.

Seals are confined to mudstone intervals reaching considerable thickness in the pre-Pliocene part of the sequence.

Traps are provided by uplifts of various genesis, and non-anticlinal lithological lenses in the zones of deltaic deposits and fans. Not inconceivable is the presence of fault-sealed pools confined to fault zones. Horst-type uplifts have been identified with a height up to 2-4 km and 120-330 km across, representing basement highs covered by deposits of the upper units.

2.1.3.4 Tinro PPB

The basin is confined to the Tinro sea basin.

Geological Structure. The basin's boundaries coincide with a pericratonic rift trough which is the depocentre of the North Okhotsk area of Cenozoic extension and subsidence.

The Oligocene-lower Miocene part of the *sedimentary sequence* is represented (in grabens) by deep marine cherty mudstones. Development of the middle Miocene-Quaternary deposits was controlled by intense subsidence along listric faults. The total thickness of the sedimentary cover reaches 8-10 km.

Possible oil and gas *source rocks* are middle Miocene rocks. The Oligocene-lower Miocene unit occurs in severe thermobaric conditions, and probably generates gas and condensate (lower gas window).

Possible *reservoirs* are sandstone horizons and fractured siltstones and cherts in the vicinity of faults.

Structural *traps* are probably confined mostly to the fault zones on the flanks of the rift.

2.1.4 Accretionary west Kamchatka PB

The basin occupies the western coast of the Kamchatka Peninsula and the adjacent offshore area.

Geological Structure. The basin is confined to a vast West Kamchatkan subsidence zone limited to the East by the Central Kamchatka volcanic rise. The basin comprises the Palana, Voyampolka, Tigil, Icha, Kolpakovo, Kol, Pustaya Reka, and Parapol troughs filled with Cenozoic deposits with a thickness of more than 4 km.

The basement of the basin consists of folded Mesozoic complexes; over much of the basin the basement includes Upper Cretaceous formations.

The sedimentary cover comprises six structural-stratigraphic units[2], which are as follows; Upper Cretaceous, Paleocene-Eocene, Oligocene-Lower Miocene, Middle Miocene-Upper Miocene, Upper Miocene-Pliocene and Pliocene. The lithofacial scheme of the upper Paleocene-lower Oligocene deposits of the southern Kolpakovo trough shows a wide development of coal in the Paleogene and upper Miocene sequence of the basin, and volcanics in the Oligocene-middle Miocene part of the sequence (figure 18).

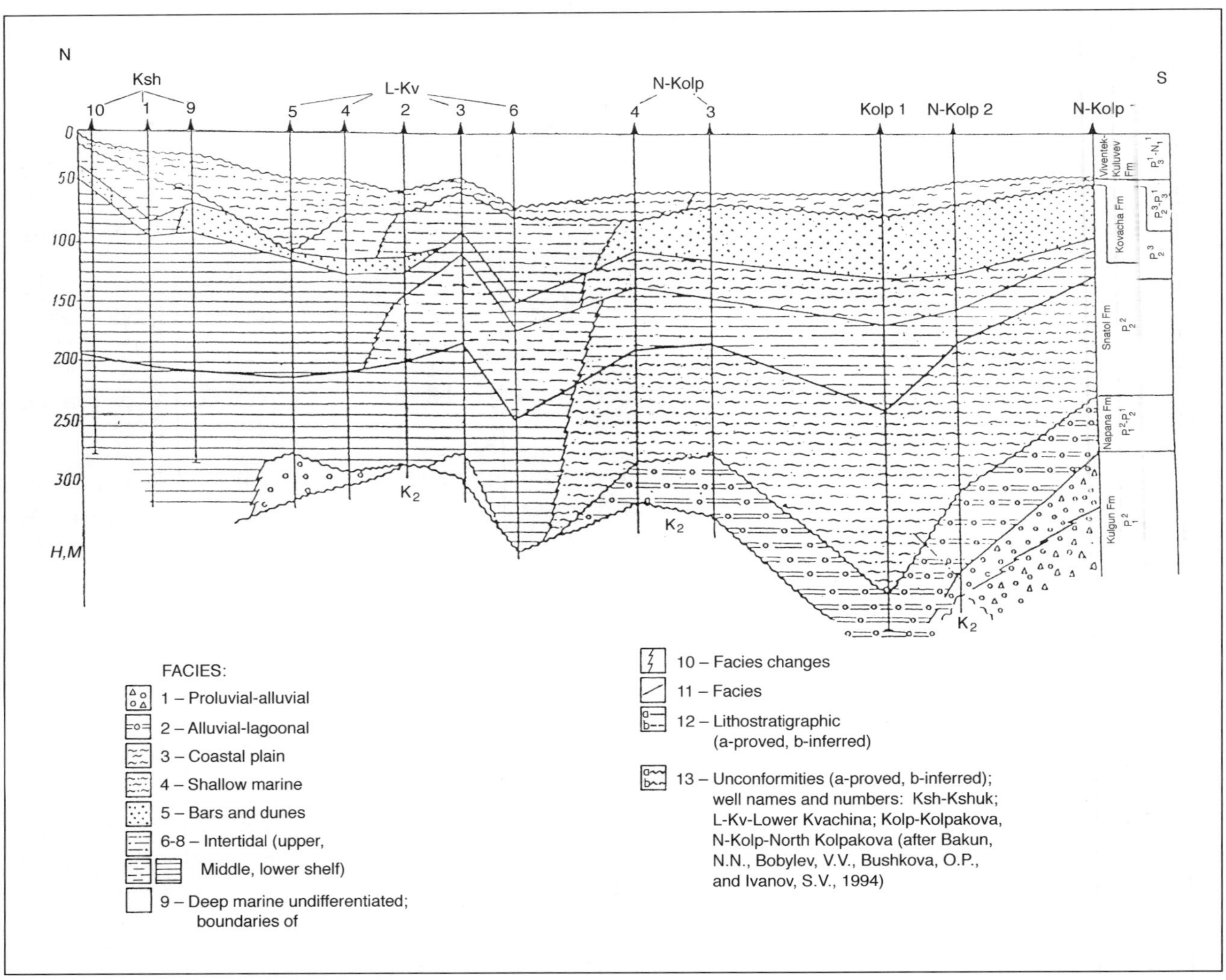

Figure 18. Lithofacies correlation of the Upper Paleocene-Lower Oligocene deposits in the southern part of the Kolpakova depression

Reservoirs are most widely developed in the pre-Oligocene and middle Miocene-Pliocene parts of the sequence. As a rule, they are composed of sandstones interbedded with conglomerates, siltstones, tuffstones, and tuffaceous siltstones.

Seals are provided by mudstone intervals, most common in the Oligocene-lower Miocene part of the sequence.

Traps are numerous uplifts, often complicated by thrusting, typical of accretionary zones. HC discoveries are possible in the subthrust zones of the basement.

Exploration Results. Four gas/condensate fields have been discovered in the Kolpakova trough. Reservoirs are sandstones, siltstones, and tuffstones of the Utkholok, Etolon, and Ermanov formations of Neogene age and the Eocene tuffstones and tuffaceous siltstones of the Snatol and Kovachi formations. Pay zones occur at depths between 116 and 2,450 m.[16] Classified as mean-size is the Nizhniy Kvakchik field (proven gas reserves, 10.3 Bm^3). Other discoveries are small. Non-commercial oil flows in the West Kamchatka basin were obtained from the Middle Kunzhik, Veyber, Liman, Gavan, and Rossoshina structures. Numerous oil and gas shows were obtained from deposits of various ages from the Cretaceous to, and including the Neogene.

Ultimate proven reserves of the West Kamchatka up to the end of 1992, were about 14 Bm^3 gas and more than 0.5 MMT condensate. Total reserves of the Kolpakovo Severnoye and Kshuk fields ($C_1 + C_2$ categories) are estimated at 10 Bm^3 gas and 0.6 MMT condensate.

2.1.5 Pericontinental-Backarc North Kuril PPB

The basin is located in the southern Sea of Okhotsk of the Greater Kuril Islands.

Geological Structure. The basin is confined to a graben-like subsidence zone lengthening northwest of Kunashir, Iturup, Urup, and Simushir islands, also including the Atlasov and Golygin troughs located near the southern end of Kamchatka Peninsula.[15]

The sedimentary cover is Oligocene-Quaternary, reaching a thickness of 4-7 km. The sequence contains deep marine mudstones, cherts, sand-shale, and volcanic-sedimentary deposits. The basin is subdivided into a series of depocenters, separated by horst-type basement highs.[20]

Possible *source rocks* are Oligocene-lower Miocene mudstones and cherty mudstones.

Geochemical parameters of the sequence were studied only in the Golygin trough. Well Krestovskaya 1 (TD 3,550 m) penetrated tuffaceous and cherty rocks of the Voyampolka and Kovran groups.

Reservoirs occur throughout the Neogene sequence. Open porosities of tuffaceous breccias range from 8.2 per cent to 23.2 per cent (average 16.8 per cent). Open porosity of the psephytic varieties of rhyolite tuffs ranges between 6.0 to 20.4 per cent; psammitic, 10.9-27.6 per cent. Permeability of the psephytic rhyolite tuffs varies from 0 to 16 md; that of the psammitic tuffs, from 0 to 8.6 md. Average open porosity of psammitic andesite-basalt tuffs is 5.6 per cent.

Average open porosity of the Hot Beach Formation tuffs is 24 per cent; permeability, 36 md.

Open porosity of the middle-upper Miocene tuffs (Lovtsovskaya Formation of Kunashir Island, Rybakovskaya Formation of Iturup Island) ranges from 32 to 65 per cent (average 46 per cent); permeability, from 0 to 1,140 md.

Possible *seals* are mudstone intervals mostly in the pre-Pliocene sequence, which are the most lithified.

Traps are probably most widely developed on the slopes of the deep basin and the linear subsidencezone and are related to faults.

2.1.6 Forearc South Kuril PPB

The basin is located to the SSE of the Greater Kuril Islands.

Geological Structure. The basin is confined to a sedimentary lens on the southeastern slope of the Kuril arc, including the Lesser Kuril Islands rise, and the southeastern depression of Hokkaido Island.

The basement over much of the basin and on Hokkaido is composed of folded Paleozoic-Mesozoic formations. Within the trench the sedimentary cover lies upon the partially oceanized basement. In the western part of the basin the sedimentary fill includes Upper Cretaceous-Cenozoic coal-bearing (Cretaceous), sand-shale, and volcanoclastic deposits with a total thickness of up to 6 km.

Source rocks on Hokkaido Island are the Cretaceous coal-bearing rocks and Paleogene mudstones.

Reservoirs are sandstones, gritstones, and conglomerates developed throughout the sedimentary sequence. Open porosity of the volcanic conglomerates, gritstones, and sandstones of the Matakotan Formation ranges from 10 to 12 per cent and permeabilities about 1 md. Open porosity of the volcanic sandstones of Malokurilskaya Formation ranges from 6.5 to 13.0 per cent, permeability is close to zero on the Lesser Kuril Islands and 0.5 md on Shikotan Islands. On Hokkaido Island, sandstone reservoirs with permeabilities between 50 and 1,000 md are widespread. Their thickness is not more than a few tens of meters.

Seals are provided by mudstone and volcanic rock seams.

On Hokkaido Island gas accumulations were discovered in the upper Miocene sandstones (Ikeda field); upper Pliocene (Tokati field), and Anthropogene (Kushiro field). Coal gas was obtained from the Oligocene and Upper Cretaceous sandstones and offshore from the Cenozoic deposits.

Traps are structural – gentle anticlinal folds complicated by faulting (Tokati, Ikeda) – and stratigraphic, related to pinchout zones (Kushiro, Tokati). The pools are bedded anticlinal (Ikeda), fault-sealed and lithographically sealed (Taihaie, Tokati, Kushiro). The pools occur at a depth of 35-1,000 m. Gas fields are small and not producing.

2.2 The Sea of Japan Megabasin

With the region under consideration the megabasin includes the Ishikari-West Sakhalin and North Sea of Japan, with a total area of 325.4 thousand square kilometres.

2.2.1 Rift Basins

Geological Structure. The basin is confined to a rift formed on the extension of the Sea of Japan's deep basin. The basin is fault-bounded. To the west it is surrounded by the structures of the East Sikhote Alin volcano-plutonic belt; to the east, the uplifts are confined to the Central Sikhalin overthrust [5].

The northern boundary of the basin is drawn along the strike-slip fault complicating the Rybnovsk uplift; to the south the basin is closed by sediment pinchout on the southern shelf of Hokkaido Island (outside the region under consideration).

The central part of the Tatar Strait is a large extension zone. Accumulation of the Paleogene-lower Miocene deposits in the Ishikari-West Sakhalin basin took place in conditions of rift development with intense subsidence and activation of the permeable lithospheric zones with high heat and fluid conductivity, providing for a high heat flow.

Geophysical data indicate that the pre-middle Miocene deposits in the central Tatar Strait are strongly reworked by thermodynamic processes and are therefore included in the granite layer of the crust (figure 19).

The post-rift (post-middle Miocene) sag of the basin was also characterized by avalanche sedimentation; active ascending movements, accounting for the principal structural uplifts, began only in the Pliocene.

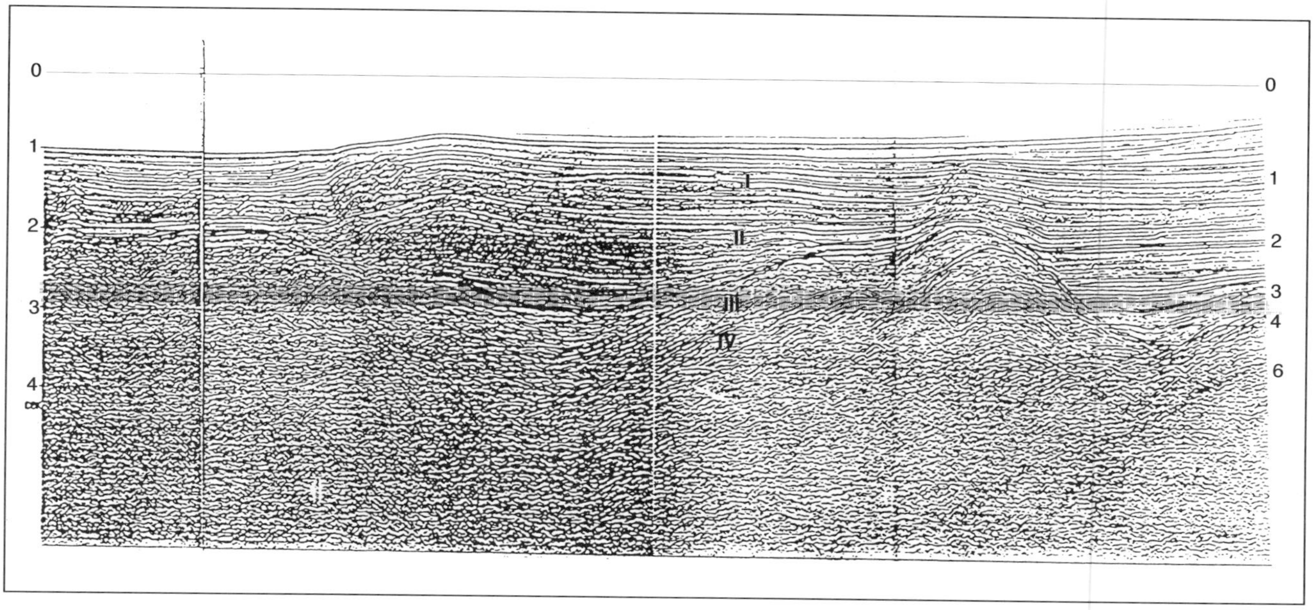

Figure 19. Tatar Strait. Fragment of time-section along seismic profile 177812, showing abrupt thickening of sediments towards Sakhalin

The basement of the Ishikari-West Sakhalin basin is composed of Triassic-Lower Cretaceous volcanic-sedimentary complexes and volcanics.

In *the sedimentary cover,* four structural-stratigraphic units, separated by unconformities, are recognized: Upper Cretaceous, Paleocene-Eocene, Oligocene-Lower Miocene and Middle Miocene-Quaternary.

In the West Sakhalin part of the basin, the stratigraphic range of the sedimentary sequence increases from NE to SW along with a dramatic increase in the proportion of shallow marine and marine facies, the most favorable for petroleum generation, reduced lithification of rocks and the extent of folding, increasing hydrogeological openness of the structure and improvement of geochemical parameters.

Source rocks recognized in the basin are argillites, shales, siltstones, and cherts of Late Cretaceous, late Eocene, early Oligocene, early-middle Miocene, and the earliest upper Miocene age.

The Upper Cretaceous, muddy-terrigenous deposits have exhausted their liquid generation potential over much of the basin, but possess a considerable gas generation potential. Besides, these rocks are very important not only as the source of a combustible fuel, but also as the source of a solvent fluid for the hydrocarbons of the superjacent strata to stimulate realization of their oil generative potential.

The Paleogene and lower Miocene terrigenous and volcanic-muddy strata, which are within the oil window, show a sufficient oil generative potential to produce oil pools.

The sequence of the basin includes two coal-bearing sequences, the Lower Doue and the Upper Doue formations, which occur between the oil generative source rocks. Coal seams are active gas generators, facilitating liquid emigration. In addition, the coal seams can generate highly paraffinic oils.

Reservoirs are widely developed throughout the sequence.

Volcaniclastic rocks include reservoirs with high porosities (tuffstones, tuffs) but insufficient permeabilities. The best reservoir properties were reported in the Campanian-Danian (porosity, 2.0-28.0 per cent; permeability, 0.02-37.0 md); middle Eocene (4.7-28 per cent, 8-49 md); middle-upper Oligocene (4.5-2.7 per cent, 0.1-34 md); and upper Miocene-Pliocene porous clastic reservoirs (20.4-28.8 per cent, 10-100 md). Lower-middle Miocene porous-fractured and fractured reservoirs with a thickness between 90 and 300 m show porosities of 3.1-36 per cent and permeabilities, 9.8-45.0 md.

Seals are upper Senonian-lower Campanian argillites; upper Eocene-Lower Oligocene, lower Miocene, and middle Miocene mudstones and siltstones; and lowest upper Miocene mudstones. In the Pliocene sequence, mudstone caprocks are practically absent. Numerous oil and gas shows and non-commercial oil and gas flows in exploratory wells have been encountered in the basin. Exploration targets are the numerous *local structures* identified by seismic surveys. On the shelf, linear N-S trending faulted anticlines with dip angles of 16-35 degrees on the limbs are predominant.[7] In deeper offshore zones, large brachyanticlines and dome-shaped anticlines have been identified.

Exploration Results. Within the territory of the basin, numerous oil and gas shows and non-commercial HC flows have been obtained from deposits of various age.

In the Sakhalin part of the Ishikari-West Sakhalin basin two gas fields have been discovered, the Pole Shakhty Uglegorskaya (Uglegorsk Colliery field) (onshore) and Izylmetyevsk (offshore). Pay zones are composed of the Miocene tuffstones and siltstones of the Maruyama Formation at a depth of 1,238-1,510 m. The discoveries are small. Ultimate gas reserves at the end of 1992 were 6.8 Bm^3, including 3.8 Bm^3 in the Izylmetyevsk field.

On Hokkaido, the first Ishikari oil field was discovered in 1903. By 1994, 31 oil and 10 small gas fields were discovered. Some of the fields are already depleted.

The principal reservoirs are lower Miocene-Pliocene sandstones. Small pools have been discovered in the upper Oligocene fractured argillites. The fields are confined to gentle symmetrical faulted anticlines with bedded anticlinal pools, often fault-sealed. Pay zones occur at depths between 70 and 2,405 m.

Oil and gas are produced in small volumes. Cumulative production has reached about 0.2 MMt oil and 0.4 Bm^3 gas. Ultimate recoverable HC resources of the Hokkaido shelf are estimated at about 20 MMtoe.

2.2.2 PPB in the north of the Sea of Japan

The basin is located in the north of the Sea of Japan.

Geological structure. The basin is confined to a Cenozoic backarc rift in the transitional zone between the Japan island arc and the deep sea basin.

The basement is composed of tholeiitic basalts (deep sea basin) and the folded complexes on the extension of the Mesozoides of the volcano-plutonic belt of the western surrounds of the basin. The Moneron horst separates it from the Ishikari-West Sakhalin basin. The northern boundary is defined by a fault-line basement high.

The sedimentary fill of Neogene-Quaternary age reaches a thickness of 5 km. It is composed mostly of terrigenous and volcanic-sedimentary rocks.

3.0 PETROLEUM PROSPECTS

Petroleum exploration in the Sea of Okhotsk region was carried out mostly in North Sakhalin and the adjacent shelf. Exploration volumes in the other parts of the island, West Kamchatka and northern (Magadan) shelf were much less.

Half of the region under consideration is actually not covered by exploration. The above data on the 14 basins of the region give grounds to consider it as promising for HC prospecting.[1, 22] According to recent estimates the initial potential HC resources in place are more than 12 billion toe.

Studies of the geological features of the sedimentary basins indicate a correlation between the geological paramenters, petroleum generation volumes, and the HC potential of the basins with their geodynamic regime.

Petroleum prospects for the region are related to terrigenous, cherty-terrigenous, and vocanoclastic deposits of Cenozoic age, mostly Neogene, and to a lesser extent with the Upper Cretaceous.

Exploration targets are reservoir units composed of sandstones, siltstones, conglomerates, tuffstones, and tuffs. Of special interest in the region are non-traditional reservoirs composed of turbidites, weathered rocks, volcano-plutonic rocks, and basement rocks. Pay zones probably do not occur deeper than -4-5 km.

Traps are related to numerous uplifts, including those developed in the sedimentary cover in response to vertical and horizontal movements, and some basement highs. Deep-seated faults are associated with zones of uplifts of various length and height, particularly promising for HC prospecting. Non-anticlinal traps are widespread. The region is characterized by wide development of traps related to subthrust and wrench zones, and possible discompaction zones not only in the sedimentary cover, but also in the basement, and the regional structures confined to extensional rifts.

The high petroleum prospects of the region are caused by wide development of rift structures in zones of continental and transitional crust (intracontinental and pericontinental rifts), where not only marine and deltaic, but also lacustrine and paludal continental mudstones are charcterized by high HC generation potential. HC migration in the rifts is restricted by the linearity of the structures, which reduces petroleum exploration areas, and provides for the concentration of activities on the promising objects. The faulted contacts between source rocks and reservoirs resulting from syn- and post-sedimentary faulting and the unconformities provide additional HC migration pathways, including downward migration into older deposits and even the fractured basement rocks.

4.0 ASPECTS OF OIL AND GAS EXPLORATION AND DEVELOPMENT IN THE SEA OF OKHOTSK REGION

Allowing for much of the region being covered by sea, the current estimates of the region's resources may be considerably amended with increasing density of its geophysical coverage.

Geological prerequisites based mostly on seismic data (figure 20) suggest that the initial total HC resources in place are about 12-15 Btoe. Resource development is most active on Sakhalin shelf. S.M. Bogdanchikov and E.G. Koblov (Sakhalinmorneftegaz)[3] recognize 33 petroliferous and possibly petroliferous zones (figure 21), and according to their estimate, the total in situ HC resources of Sakhalin are 7.8 Btoe, including 3.8 Bt oil. 2.9 Bt oil and 3 Tm^3 gas lie on the Sakhalin shelf. Oil is predominant in onshore resources; offshore, oil and gas proportions are approximately equal. Recoverable gas resources on Sakhalin are estimated at 935 MMt, including 295 MMt onshore, and 640 MMt offshore. Oil recovery factors in offshore fields (using improved oil recovery techniques) reach 0.43-0.55 (average 0.35). In the offshore pools they are estimated at 0.1-0.42 (average 0.225).

52 per cent of total onshore resources of Sakhalin Island are concentrated in three of the 25 petroliferous zones-Okha-Ekhabi, Paramay-Chayvo, and East Dagi. Another 25 per cent are associated with the Espenberg, Volchinka-Sabo, and Lun-Pogranichnaya zones. These six zones are the principal exploration targets on the island. Offshore, three zones of high resource concentration are recognized, the Odotu, Nyyvo, and Chayvo. They jointly account for 69 per cent of the shelf's resources and the largest oil and gas pools.[10, 13, 14]

The Decree of the President of the Russian Federation N 539 of June 1, 1992, "On the Urgent Activities of the Development of New Large Gas Discoveries on Yamal Peninsula, in the Barents Sea, and on the shelf of the Russian Federation" stipulates preparation and confirmation by the Government of the Russian Federation of a programme for the development of gas/condensate fields on Sakhalin shelf and expansion of the gas pipeline network of Sakhalin Island, and the Russian Far East region.

According to the decisions of the Russian Government and local authorities, international offshore Sakhalin exploration and development projects (summary 6) are presently underway. The first project – Sakhalin 1, participated by Sodeco (Japan) and Exxon (USA), Rosneft-Sakhalinmorneftegas, implies development of the Odoptu, Chayvo, and Arkutun-Dagi fields, total ultimate recoverable resources of which are estimated at 290 MMt oil, 33 MMt condensate, and 425 Bm^3 gas. It will include 6 platforms, 8 offshore production units, and 611 wells. Production start-up is due onstream in 2000 from the Arkutun-Dagi field. Gas production from the Chayvo field is due in 2005; from the Odoptu field, in 2001. Peak oil and condensate production of about 23 MMt/yr is expected to be reached by 2001 (figure 22).

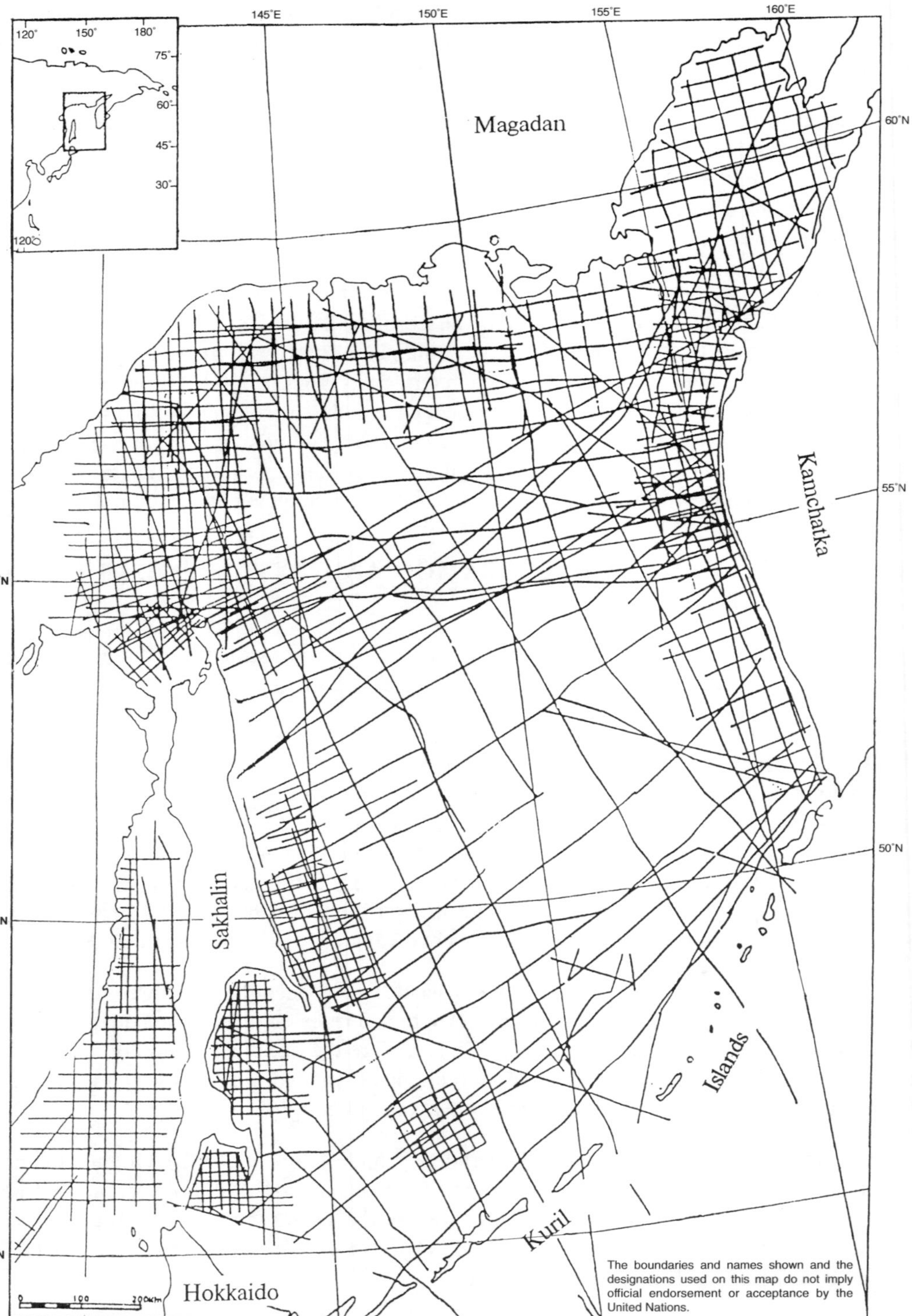

Figure 20. Seismic coverage of the Sea of Okhotsk

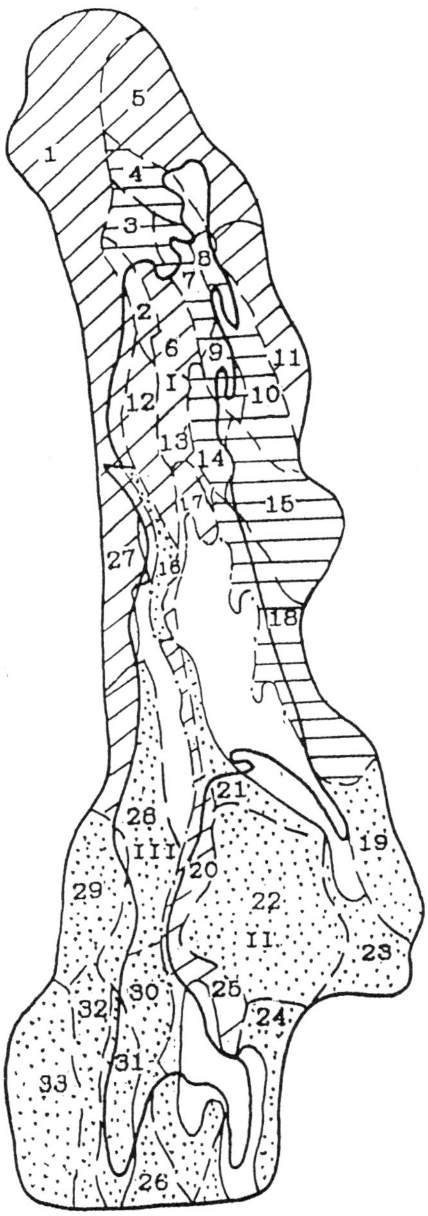

Figure 21. Petroliferous regions and zones of the Sakhalin Oblast

PROJECT SAKHALIN-I

Projection: UTM, Central meridian 141°00 E
Spheroid: Krassovsky

Sakhalin Offshore Well & Seismic Line Location Map

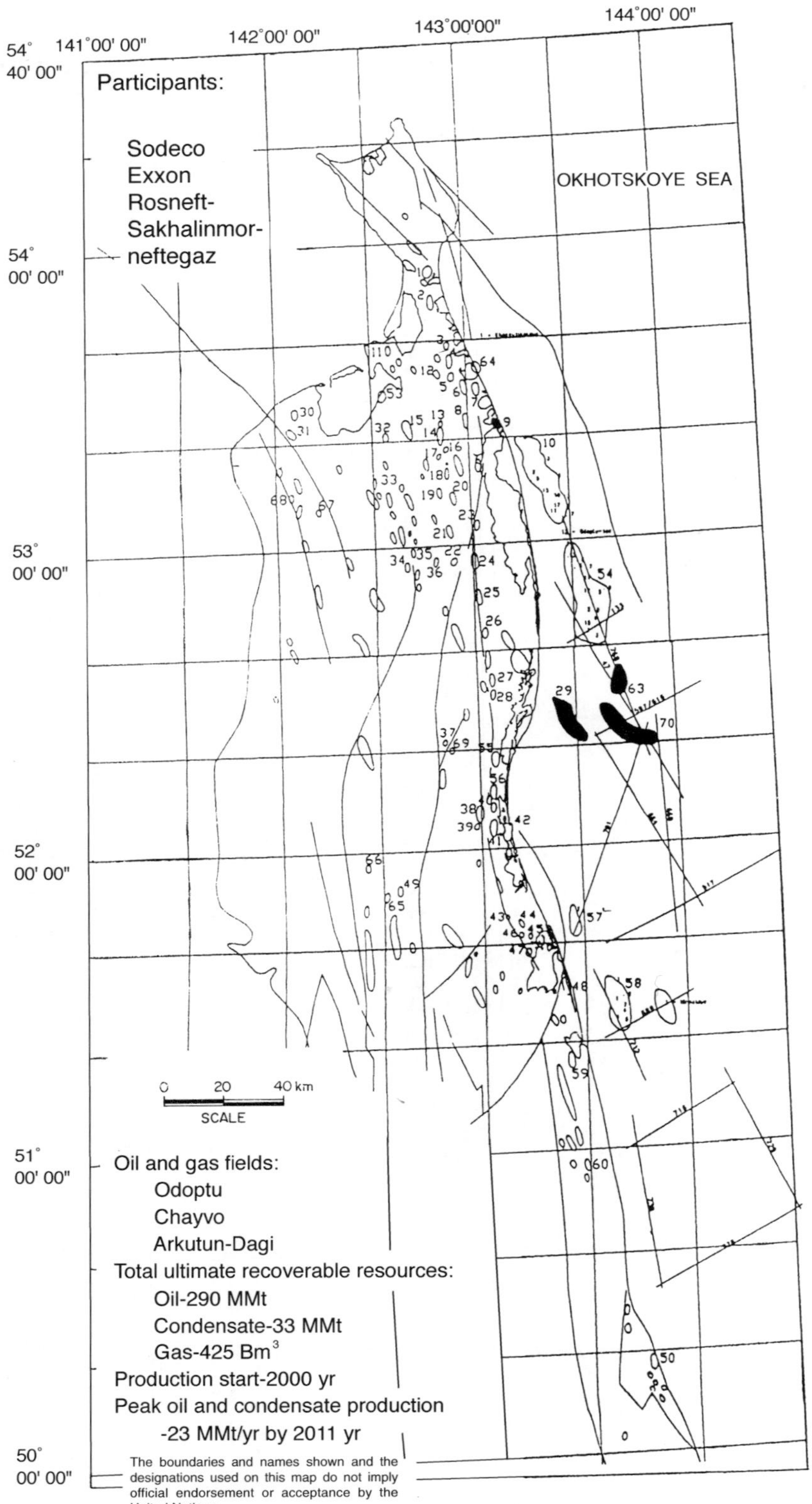

FIELDS:
1 - North Kolendo
2 - Kolendo
3 - North Okha
4 - Okha (Central Okha, Tsentralnaya Okha)
5 - South Okha
6 - Ekhabi
7 - East Ekhabi
8 - Tungor
9 - Odoptu
10 - Odoptu-Sea (Odoptu More)
11 - Nkrasovka
12 - Gilyako-Abuman
13 - Abanovskoye
14 - Nelma
15 - Volchinka
16 - Erri
17 - West Erri
18 - South Erri
19 - West Sabo
20 - Sabo
21 - Maloye Sabo
22 - South Kenlga
23 - Kydulanyi
24 - Mukhto
25 - Paromay
26 - Piltun
27 - Goromay
28 - North Boatasino
29 - Chayvo
30 - Astrakhanovskoye
31 - Uzlovoye
32 - Shkhunnoye
33 - North Glukharka
34 - Mostovoye
35 - Gyrgylani
36 - Kraplvnenskoye
37 - South Val
38 - East Dagi
39 - South Dagi
40 - Nizhneye Dagi
41 - Mongi
42 - Ust-Tomi
43 - Lmchinskoye
44 - Uyglekuty
45 - Katangli
46 - West Kntangli
47 - Pribrezhnoye
48 - Nabil
49 - Verkhne-Tatamskoye
50 - Okruzhnoye
51 - South Lugovskoye
52 - East Lugovskoye
53 - East Baykal'skoye
54 - Piltun-Astokhskoye (Piltun-Astokh
55 - Ust Evay
56 - Name R.S. Mirzoyev
57 - Veninskoye (Veni)
58 - Lunskoye (Lun)
59 - Ufskoye
60 - Polyarninskoye
61 - Izylmetevskoye
62 - Zolotorybnoye
63 - Arkutunskoye (Arkutun)
64 - East Kayganskoye
65 - West Kayganskoye
66 - Verkhne-Nyshskoye
67 - Berezovskoye
68 - Tsentralno-Langryyskoye
69 - Sredny Askasay
70 - Daginskoye (Dagi)

Figure 22. The Odoptu, Chayvo and Arkutun-Dagi oil and gas fields-Project Sakhalin-I

Sakhalin 2 project – participants are Marathon, Mc Dermott (US), Mitsui, Mitsubishi (Japan), Shell Development Sakhalin B.V. (Netherlands), and Royal Dutch, forming the Sakhalin Energy Investment Consortium. The consortium is presently carrying out extension and development of the Piltun-Astokh and Lun fields, the reserves of which are estimated at more than 1 Bt oil and 400 Bm^3 gas. Oil production is expected to peak at 10 MMt/yr; gas at 16 Bm^3 (figure 23).

The *Sakhalin 3* project encompasses the development of the Ayashkay and East Odoptu and Kirinsk offshore blocks. Summary resources more than 1 Btoe (figure 24).

Under the *Sakhalin 4* project, the Sakhalin Oblast authorities and Roskomnedra offered a sector of the Sakhalin Bay, the Shmidt and Astrakhanovka blocks, and the Veni area (block 3B). Summary resources more than 1 Btoe (figure 25).

Project Sakhalin 5 implies East Shmidt area (resources near 0.5 Btoe) (figure 26).

Project Sakhalin 6 implies Pogranichny area (resources near 0.6 Btoe) (figure 27).

The Russian Federation Government Petroleum Law of 30th December 1995 confirmed a production-sharing agreement. This law confirms the list of oil and gas fields and areas for exploration and development with the participation of foreign companies.

There are 6 fields in Sakhalin Island in Development (figure 28):

1. Central Okha: 7.9 million tonnes oil (current recoverable reserves)
2. Katangli: 5.1 min.t
3. Uyglekuty: 3.2 min.t
4. West Sabo: 3.2 min.t
5. Nabil: 1.4 min.t
6. East Ekhabi: 0.9 min.t

Annual production on every field is estimated at 100-200 thousand tonnes. Production may increase fourfold after technical reconstruction.

An Australian group including Stirling and Hardman Resources NL have established the Samir Consortium with Vostokgeologia and the Sakhalin Potential Company, and were awarded a license for the development of the Prizalivnyy block at a distance of 14 km from the Lun field. The Mezhdurechenskoye oil field is located within the block.

In March 1994, the Stirling group was also awarded a license for petroleum exploration in a prospect covering 1,000 sq km in the Kholmsk district of the Ishikari-West Sakahalin basin.[26]

On the near-Magadan shelf, the Ministry of Gas Industry have shot 30 thousand km of seismic profiles at a depth of 50-120 km. A total of 130 structures, with a total prospective area of 23 thousand km^2, were identified, and more than 120 geophysical anomalies, presumably related to HC pools, were reported. Three oil columns are predicted in the intervals 900-1,200, 1,700-2,000, and 2,700-3,200 m. According to Ye. M. K hartukov[8], the total resources of this part of the region are 8-10 Btoe. Favourable conditions explain the high interest of foreign companies to the near-Magadan shelf.

Potential recoverable liquid resources of the West Kamchatka basin are estimated at 500 MMt; gas, 750-1,000 Bm^3. Late in 1993, the Kamchatkan Committee on Geology and the Regional Soviet announced a bidding round for the development of the Kolpakovo Severnoye and Kshuk gas/condensate fields. However, the legal regime of HC exploration and development in the Okhotsk region has not been finally settled.

On December 10, 1996, the Government of the Russian Federation decreed submitting the draft federal law on the list of subsurface areas permitted for bidding on production sharing terms for the confirmation of the State Duma.

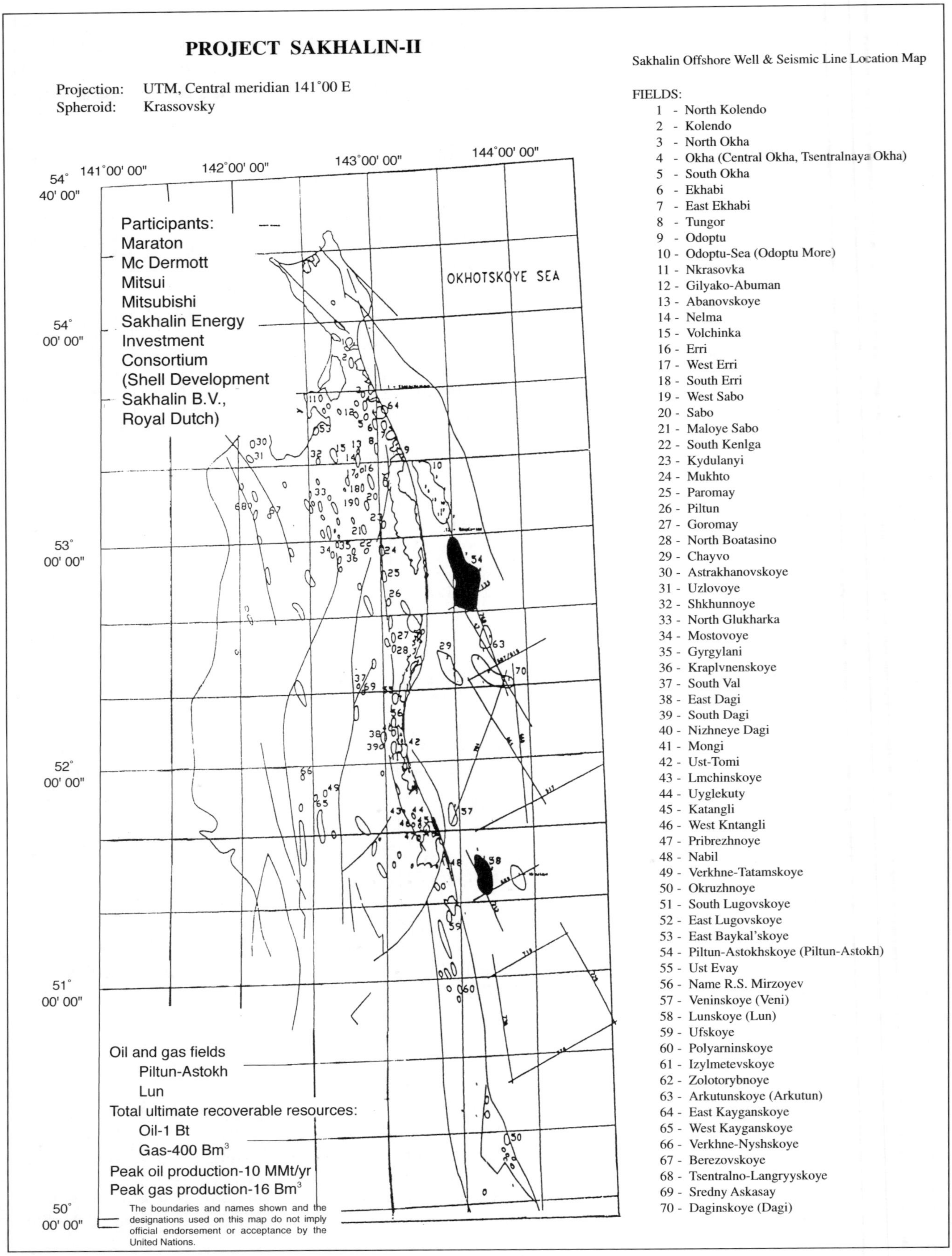

Figure 23. The Pitun-Astokh and Lun oil and gas fields-Project Sakhalin-II

PROJECT SAKHALIN-III

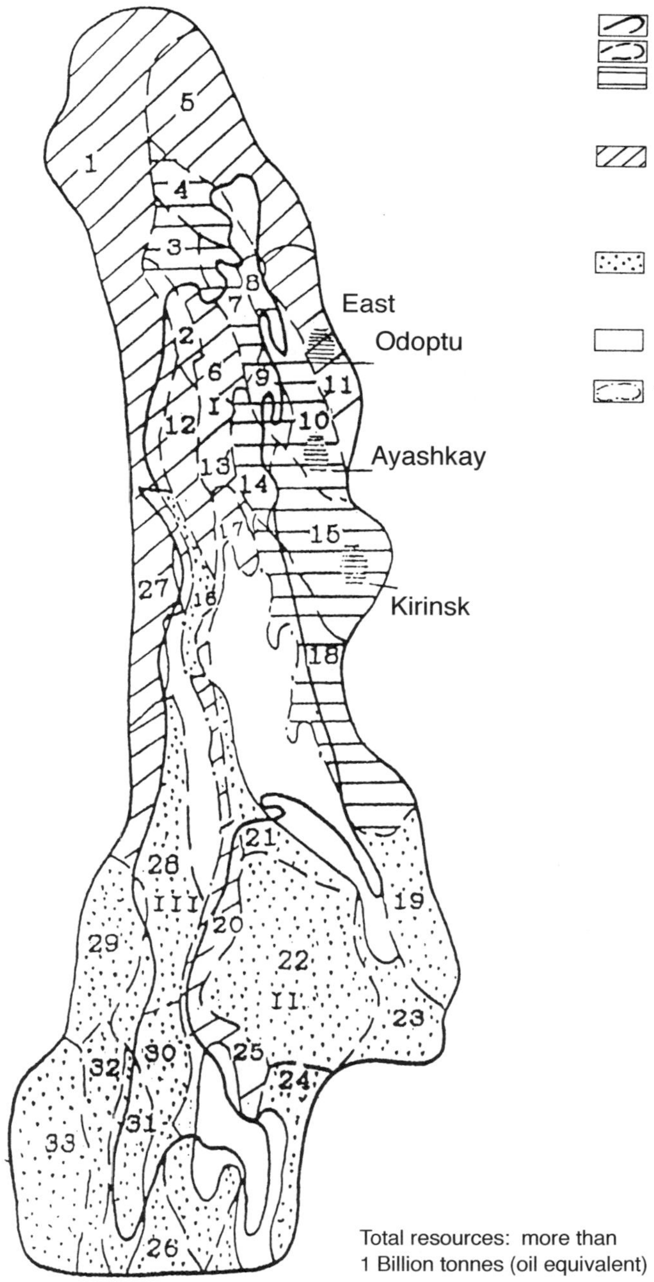

1. Political boundary
2. Boundaries of petroliferous zones;
3. Areas with resource densities more Than 50 Mt/km, promising for large and mean-size discoveries;
4. Area with resource densities of 10-50 Mt/km, promising for mostly small and sometimes mean (depending on the trap size) or large discoveries;
5. Areas with resource densities less than 10 Mt/km, promising for mostly small, less commonly mean-size discoveries;
6. Onshore and offshore areas with low Prospects or non-prospective;
7. Boundaries of petroliferous regions: I-Northern, II-Central, III-Southwestern.

Petroliferous zones with proven HC Reserves:
2-Langry,
3-Astrakhanovka,
4-Espenberg,
6-Gyrgylanyi,
7-Volchinka-Sabo,
8-Okha-Ekhabi,
9-Paromay-Chayvo,
10-Odoptu,
13-West Dagi
14-East Dagi,
15-Nyyvo,
17-Nysh-Tym,
18-Lun-Pogranichnyy,
26-Aniva,
29-Krasnogorsk Potentially petroliferous zones:
1-Near-mainland,
5-Tri BRATA,
11-East Odoptu,
12-Engizpal,
16-West Tym,
19-Rymnikovo,
20-Onor-Makarov,
21-Pogranichnyy,
23-Vladimirovo,
24-Tonino-Aniva,
25-Susunay,
27-Arkovo,
28-Lopatin,
30-Tomari,
31-Krilyon,
32-Kusnetsovsk,
33-Moneron.

Total resources: more than 1 Billion tonnes (oil equivalent)

The boundaries and names shown and the designations used on this map do not imply official endorsement or acceptance by the United Nations.

Figure 24. Petroliferous regions and zones of the Sakhalin Oblast, Project Sakhalin-III

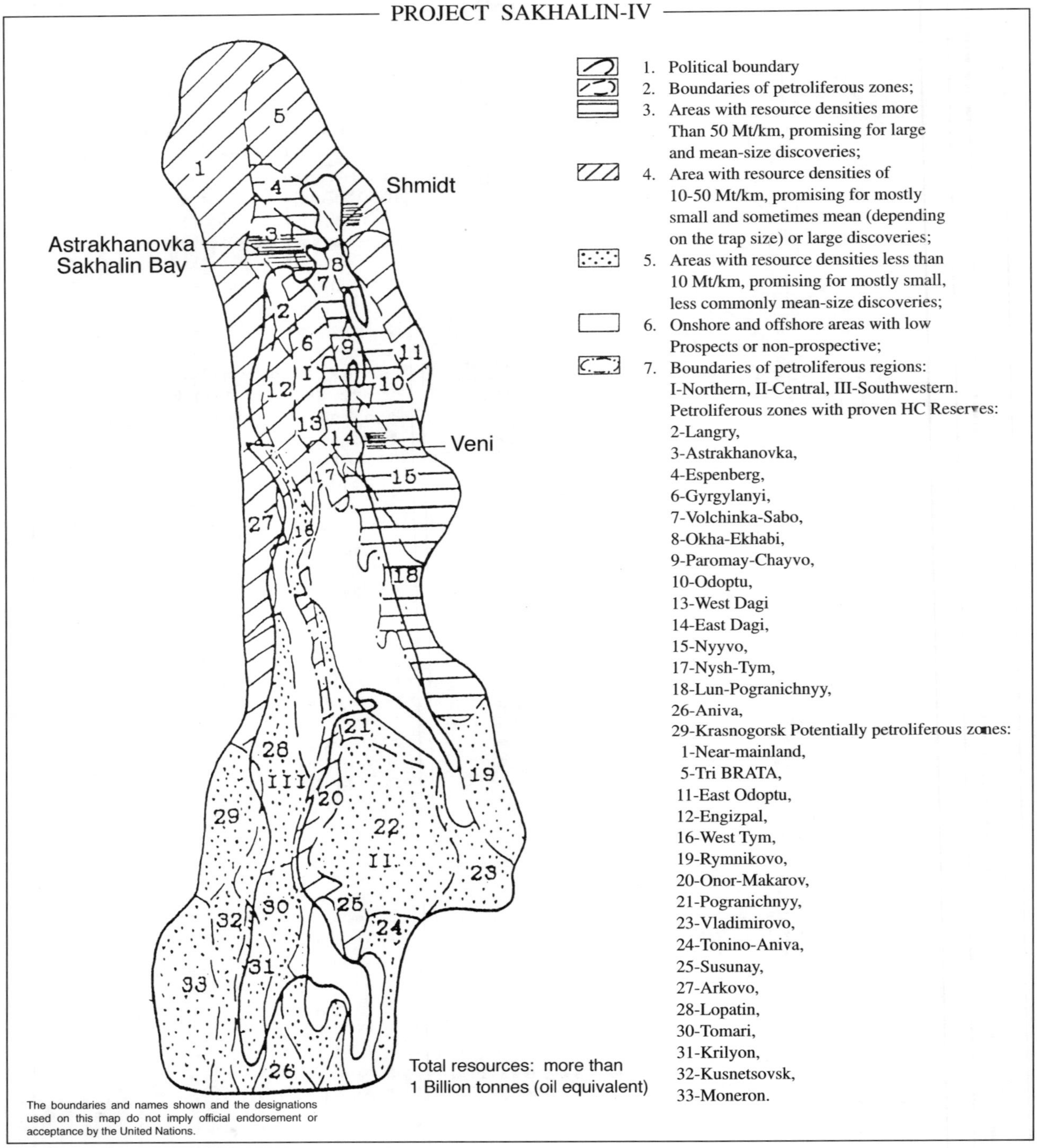

Figure 25. Petroliferous regions and zones of the Sakhalin Oblast, Project Sakhalin-IV

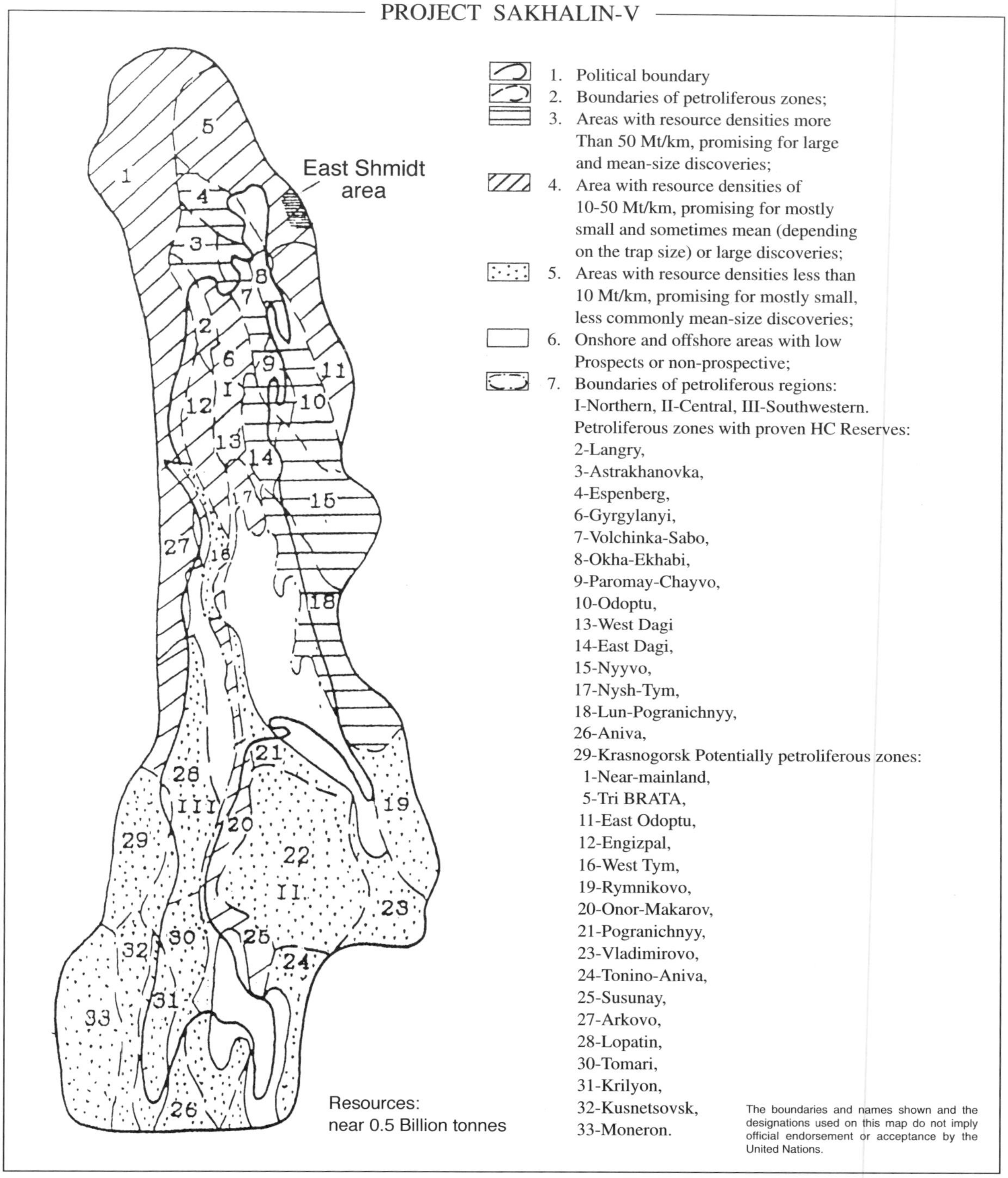

Figure 26. Petroliferous regions and zones of the Sakhalin Oblast, Project Sakhalin-V

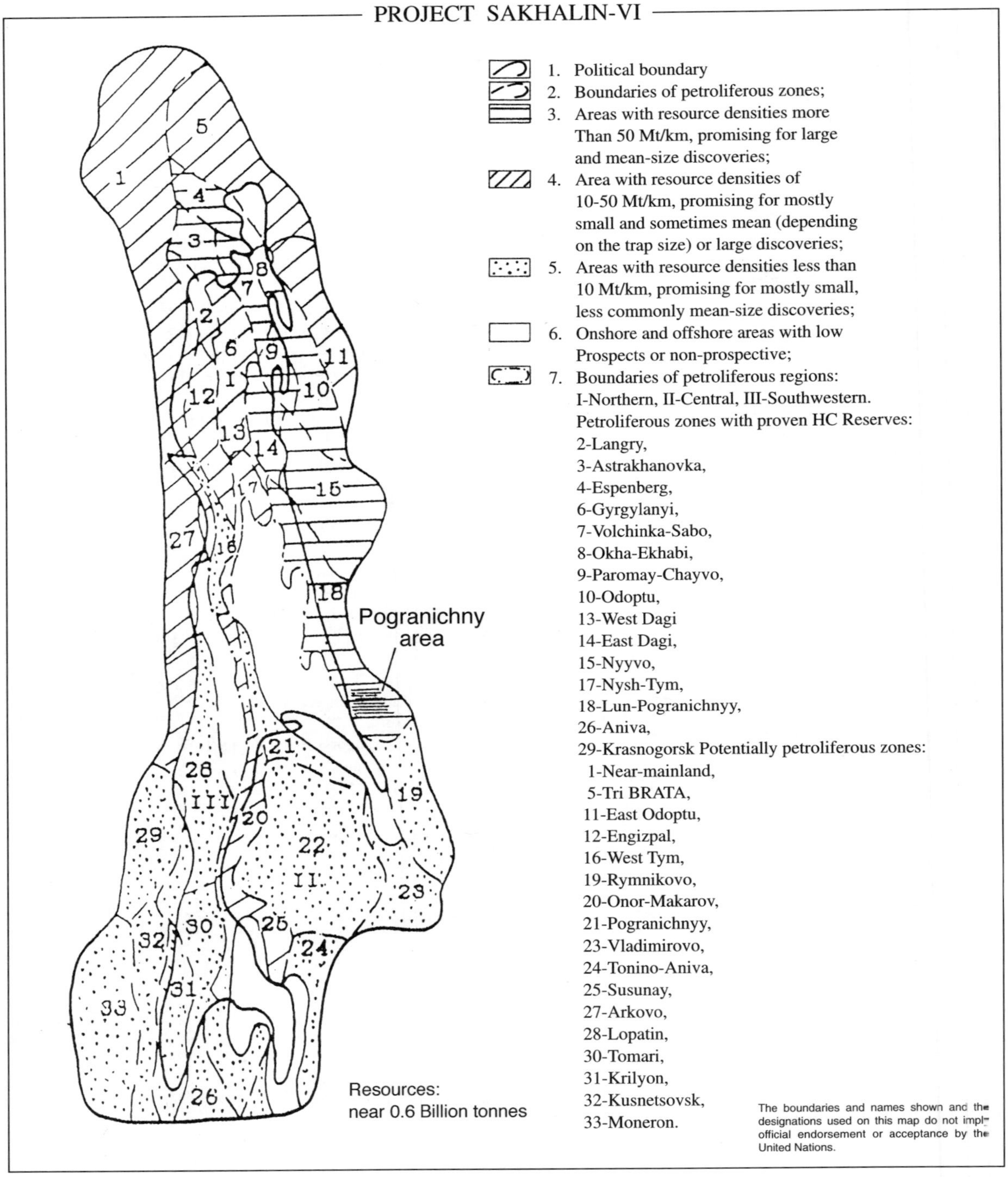

Figure 27. Petroliferous regions and zones of the Sakhalin Oblast, Project Sakhalin-VI

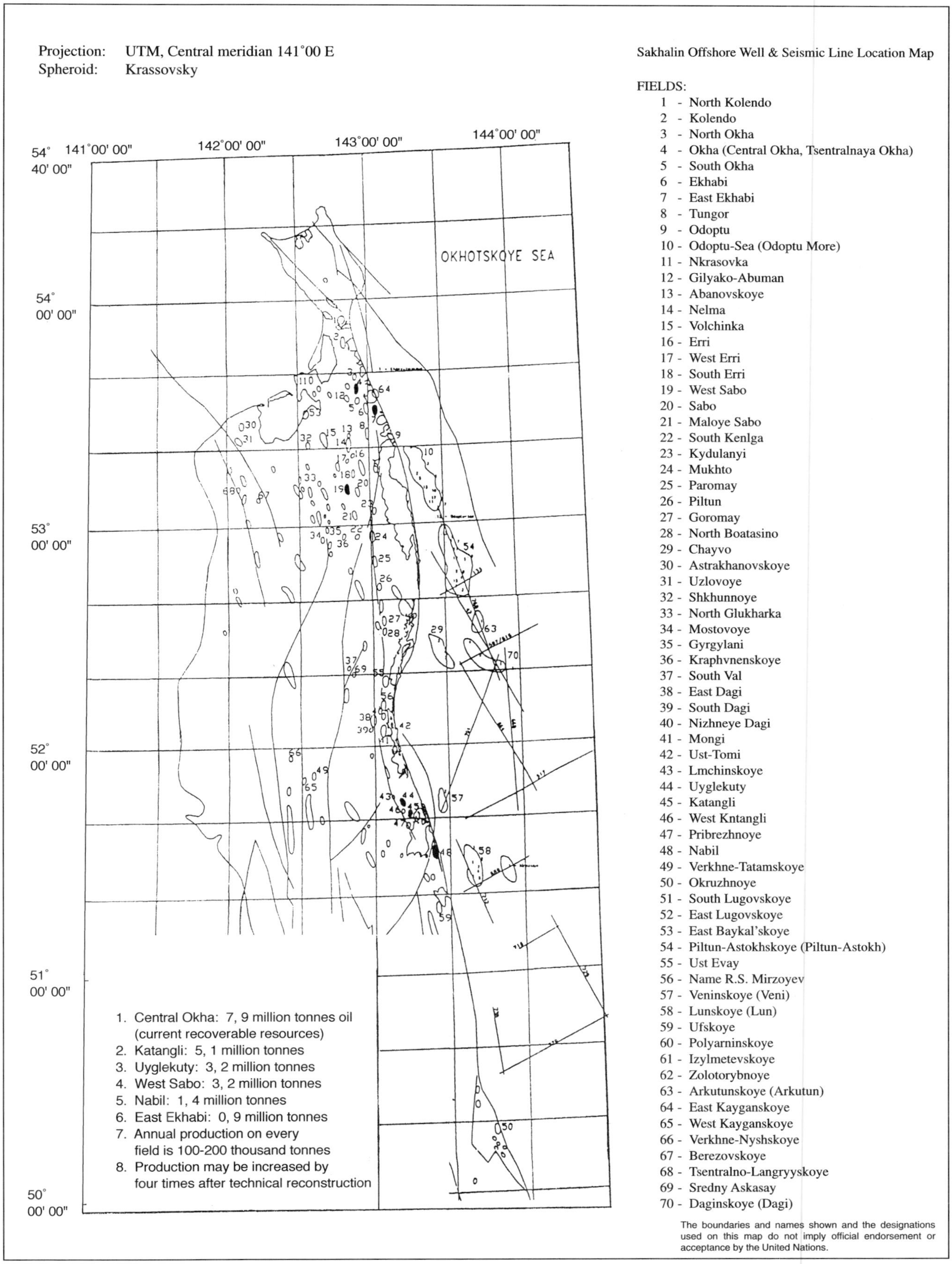

Figure 28. Development of oil fields in Sakhalin island in 1995

The same decree submitted the state federal law to the amendments and additions to Article 2 of the Federal Law on Production Sharing Agreements. This law specifies the amendments to the order of compilation and official approval of the list of subsurface areas permitted for bidding on production sharing terms, and the criteria for the selection of the areas to be included in the list.

REFERENCES

1. Alferov, S.Ye., Genkin, B.M., 1993. K. probleme osvoyeniya neftegazovykh resursov nedr Rossiyskoy Federatsii (Development of petroleum resources of the Russian Federation). VNIIOENG, Geologiya, Geofizika, i Razrabotka Neftyanykh Mestorozhdeniy, No. 6, pp. 11-14.

2. Bakun, N.N., Bobylev, V.V., Bushkova, O.P., Ivanov, S.V., 1994. Usloviya obrazovaniya gazonosnykh verkhnepaleotsen-nizhneoligotsenovykh otlozheniy Kolpakovskogo progiba (Zapadnaya Kamchatka) (Genesis of gas-bearing upper Paleocene-lower Oligocene deposits of the Kolpakovo Trough, West Kamchatka). Geologiya Nefti i Gaza, No. 2, pp. 30-34.

3. Bogdanchikov, S.M., Koblov, E.C., 1995. Syryevaya baza uglevodorodov Sakhalinskoy neflegbazonosnoy oblasti (Hydrocarbon resource base of the Sakhalin petroliferous region). Mineralnyye Resursy Rossii, 1995, No. 1, pp. 19-28.

4. Bulletin Pravovoy I Kommercheskoy Informatsii, Issue 38, August 15, 1994.

5. Geological Atlas of Japan. Geological Survey of Japan, 1982.

6. Gnibidenko, G.S., 1979. Tektonika dna okrainnykh morey Dalnego Vostoka (Sea floor tectonics of the Far East marginal seas). Nauka, Moscow, 161 pp.

7. Gololobov, Y.N., 1993. Kinematic analysis of the Sakhalin fault zones for oil and gas prospecting. AAPG Bull., v. 77, No. 9, p. 1627.

8. Khartukov, E.M., 1994. Potential of Pacific frontiers of former Soviet oil, gas empire. Oil and Gas Journal, v. 92, No. 17, pp. 69-73.

9. Khvedchuk, J., 1993. The Petroleum Basins of the Sea of Okhotsk. AAPG Bull., v. 77, No. 9, p. 1637.

10. Konoplyanik, A.A., Oganesyan, S.A., Retyunin, A.P., 1992. Neft i gaz Sakhalinskogo shelfa: Konkurs inofirm (Oil and gas of Sakhalin shelf: bidding round for foreign companies). Ekon. i Khozraschet Predpriyatiy Toplvno-Energeticheskogo Kompleksa: Meshotraslevoy Informatsionnyy Byulleten, No. 3, pp. 7-18.

11. Mavrinski Y., Koblov E., 1993. Petroleum basins of Sakhalin and adjacent shelf. AAPG Bull., v. 77, No. 9, p. 67.

12. MINI, RNB, 1993, No. 23, p. 12.

13. PIOL, 1994, v. 10, No. 32, p. 5.

14. PIOL, 1995, v. 11, No. 8, p. 3.

15. Avdeyko, G.A., Antonov, A.Yu., Volynets, O.N., et al., 1992. Podvodnyy vulkanism i zonalnost Kurilskoy ostrovnoy dugi (Submarine volcanism and zoning of the Kuril island arc). Editor-in-chief Yu. M. Pushcharovskiy, Nauka, Moscow, 528 pp.

16. Ravdonikas, O.V., Ivanov, G.I., Tuysk, Kh.A., 1987. Prirodnyye gazy Yuzhnogo Sakhalina (Natural gases of South Sakhalin). Tikhookeanskaya Geologiya, No. 6, pp. 50-61.

17. Rising gas and oil production in the Russian Far East will recast Asia-Pacific energy markets. Petr. Econ., 1994, v. 61, No. 9, pp. 43-47.

18. RNB, 1994, No. 43 (16. X), pp. 1-2.

19. Savostin, L., Kuznetsov, N., 1993. Oil and gas potential of the West Kamchatka trough. AAPG Bull., v. 77, No. 9, p. 1661.

20. Sergevey, K.F., 1976. Tektonika Kurilskoy ostrovonoy sistemy (Tectonics of the Kuril island system). Nauka, Moscow.

21. Bikkenina, S.K., Anosov, G.I., Argebtov, V.V., Sergeyev, K.F., 1987. Stroyeniye zemnoy kory yuzhnoy chasti Okhotskogo morya po seysmicheskim dannym (Crustal structure of the southern Sea of Okhotsk from seismic data). Editor-in-chief Krasnyy M.L., Moscow, Nauka, 87 pp.

22. Structure and dynamics of the lithosphere and Astenosphere of the Okhotsk Sea region. Results of research on the International Geophysical projects. Ed. Rodnikov A.G., Tuesov I.K., Charachinov B.V. Russian Academy of Sciences National Geophysical Committee. Moscow 1996.

23. Tronov, Yu.A., Kharakhinov, V.V., Kononov, V.E., Pudikov, E.G., 1987. Severo-Tatarskiy neftegbazonosnyy basseyn (North Tatar petroliferous basin). Tikhookeeanskaya Geologiya, No. 6, pp. 45-47.

24. Tuyezov I.K., 1988. Karta teplovogo potoka Tikhogo okeana I prilegayuschikh kontinentov (poyasnitelnaya zapiska), Khabarovsk, 33 pp.

25. Tyutrin, I.I., Volgin, P.F., Tadyush, V.M., Motovilov, Yu.V., 1986. Glubinnyye razlomy kak odin iz opredelyayushchikh faktorov razmeshcheniya neftegazovykh mestorozheniy na Sakhaline (Deep-seated faults as a determining factor of pertroleum fields distribution on Sakhalin). Tikhookeanskaya Geologiya, No. 6, pp. 53-57.

26. Vasilkovskiy, N.P., Yevlanaov, Yu.B., 1981. Geologiya Yaponskogo morya s pozitsiy yeago epiokeanicheskoy prirody (Geology of the Sea of Japan in terms of its epioceanic origin) in: Geologiya Dalnevostochnoy Okrainy Azii, Vladivostok: DVNTS AN SSSR, 148 p.

PART TWO

SUBREGIONAL PROSPECTS IN MINERAL RESOURCES DEVELOPMENT AND SCOPE FOR COOPERATION IN THE NORTH-EAST ASIA

CONTENTS

1. WORLD OIL AND GAS RESOURCES AND THE POTENTIAL OF THE NORTH-EAST ASIA[1]

1.0 INTRODUCTION

One of the most important consequences of the oil crisis, early in the 1970s was the recognition that hydrocarbon resources are far from unlimited. From this time, general interest in world oil and gas resource estimates has dramatically increased with the concern as to whether or not, and to what extent, the resources will be sufficient for future generations. In other words, how much time is left for humanity to find an appropriate alternative for oil and gas? What are the structure and dynamics of fossil fuel consumption?

At the beginning of this century, world energy consumption was not more than I Btoe. In the middle of the century, it amounted to 2.5 Btoe; in 1990, to 11 Btoe. In 2000, fossil fuel consumption is predicted to grow to I8 Btoe. This growth of fossil fuel consumption is taking place, in spite of active precautions to reduce the energy consumption of industrial processes, and stringent energy conservation.

Prior to the mid-1950s, more than half of all energy requirements were provided by coal, but as new oil and gas fields were discovered, with extremely low production costs (Middle East, North Africa), oil's contribution started to grow and reached 42 per cent, or 65 per cent jointly with gas, by the mid-1960s.

By the beginning of the 1990s, the world fossil fuel balance was as follows: oil accounted for 35.6 per cent; gas, 13.8 per cent; electric power, 12.4 per cent; and non-commercial fuels, 7.5 per cent (figure 29).

According to most of the forecasts, the oil and gas percentage will decrease and stabilize at 45 per cent (oil, 25 per cent, and gas, 20 per cent, respectively). Nevertheless, actual hydrocarbon fuel consumption will grow, even if at a lower rate.

Over the last 5 years, oil consumption grew at a rate of 2 per cent per annum on average, and amounted to 3.3 Bt in 1995. It is predicted to reach 3.8 Bt in 2000 and 5 Bt in 2020. To satisfy these requirements, it is necessary to recover about 100 Bt oil in the coming 25 years. Natural gas consumption develops at similar rates.

2.0 EVOLUTION OF WORLD HYDROCARBON RESOURCE ESTIMATES

As is obvious from figure 30, oil resource estimates have shown a distinct tendency to increase over the last 50 years. Most clearly, this tendency manifested itself in the period between the early 1940s and 1965, when the estimates grew from 60-80 Bt to 335 Bt (Hubbert, 1967). This was due both to improvements in resource evaluation techniques and the contribution of new large sedimentary basins where new discoveries had been recently made (Middle East, North Africa, North Sea, Western Siberia).

In the years that followed, in spite of the general tendency to grow, resource estimates remained largely around 300+ Bt. Note, however, that we are speaking about "traditional", or common oil recoverable by well known and widely applied primary and secondary recovery techniques.

Gas resource estimates underwent a similar history. Over the last decade, most of the estimates fell within 290-340 Tm^3.

3.0 OIL AND GAS RESOURCE ESTIMATES

VNIIZARUBEZHGEOLOGIA has been involved in world oil and gas resource evaluation over the last 30 years, and each 5 years the estimates are corrected with respect to new geological and geophysical data and the improvement of evaluation techniques.

1 V.I. Vyssotski, Petroleum Institute, JSC "VNIIZARUBEZHGEOLOGIA", Moscow, Russia

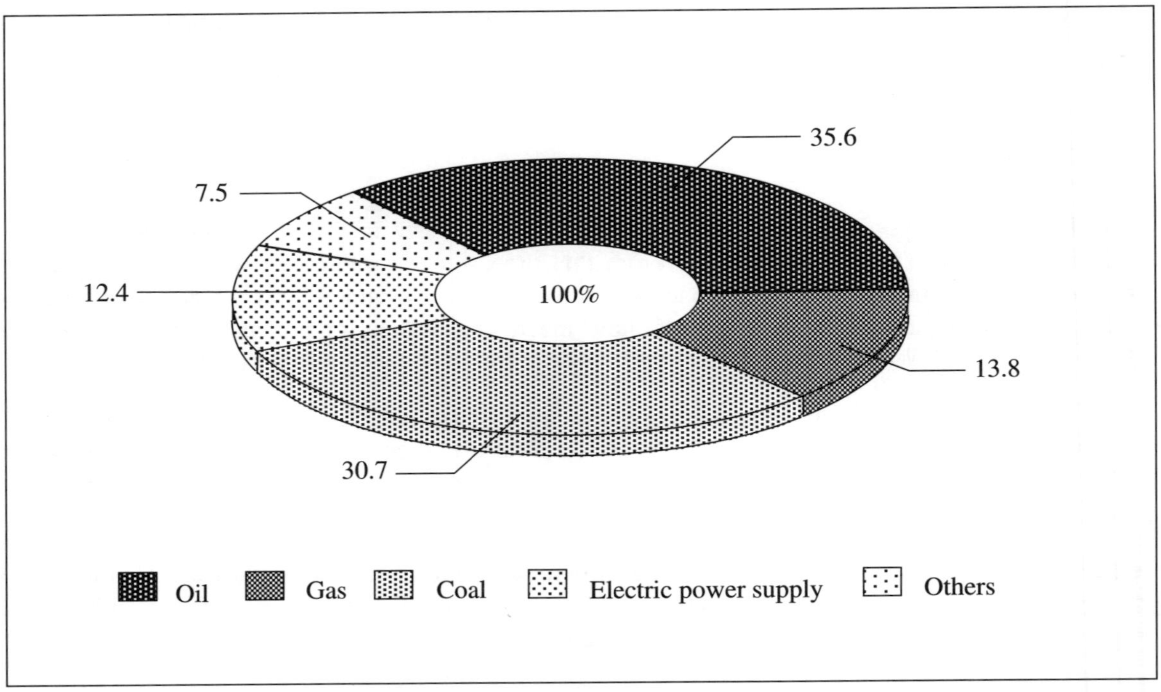

Figure 29. World fuel and energy distribution

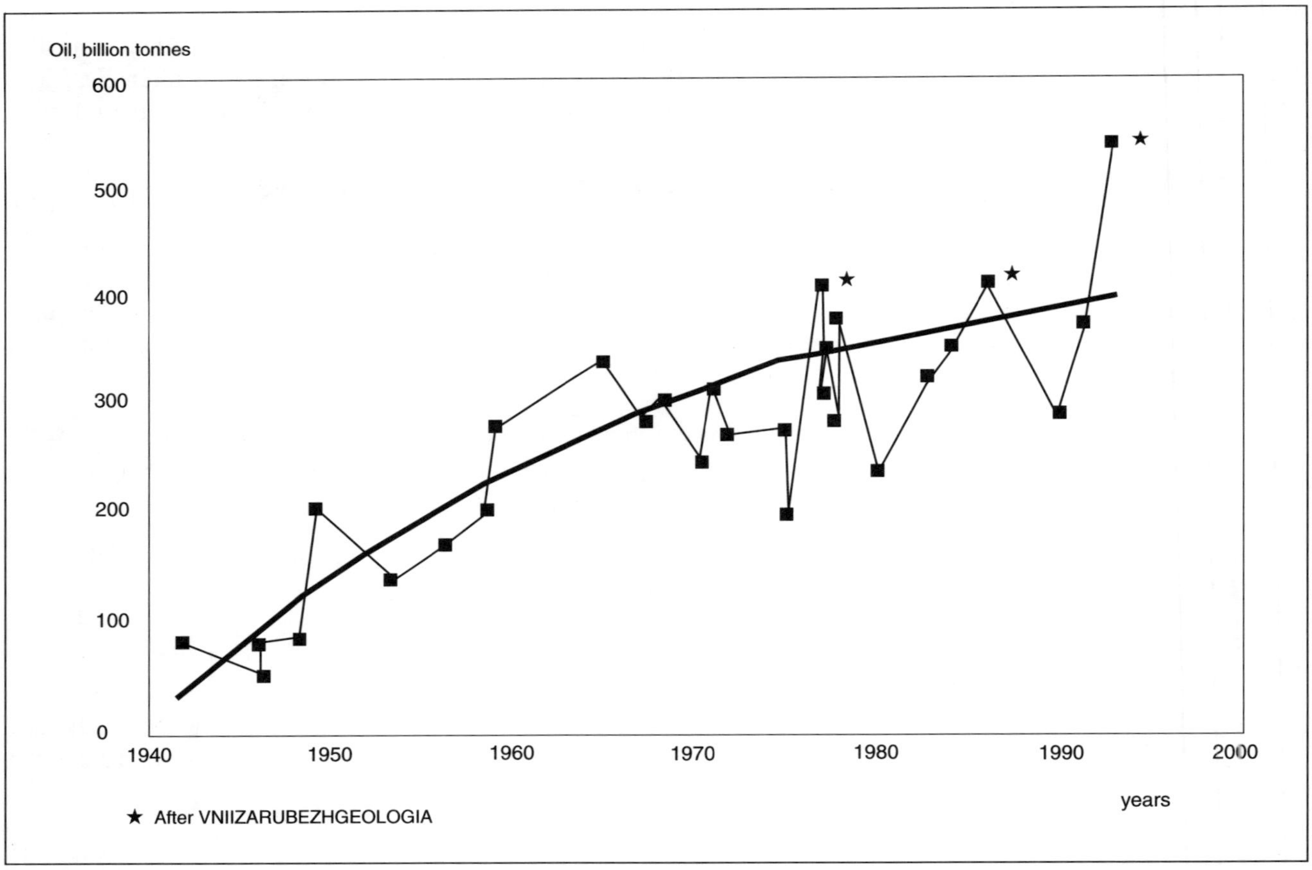

Figure 30. Evolution of world primary recoverable oil resources valuation

3.1 Petroliferous basin as the main object of estimation

Petroliferous basins (sedimentary basins) of various size, structure, and geologic history, are considered as the main object of hydrocarbon (HC) resource evaluation. These basins are morphologically expressed in the modern crustal structure and filled with sediments capable of HC generation, accumulation, and retention. Based on this concept, 511 basins are recognized throughout the world including Antarctica, of which 226 contain proven commercial HC accumulations, and are therefore classified as petroliferous basins (PBs). In the basins the presence of oil and gas pools is inferred, and they are termed as possibly petroliferous (PPBs).

The PBs and PPBs were studied on the basis of a single concept. Numerous and various data on basin structure and evolution, depositional environment, and formation of HC pools in different phase states were analyzed. Subject to particularly detailed analysis were the data on sediment lithology, structure and origin of the main structural interfaces, a real and vertical distribution of rock thickness and facies within productive and prospective sequences. Particular emphasis was placed on the studies of reservoirs, source rocks, seals, and spatial distribution and structure of oil and gas fields.

Generalization of the data on geologic structure, evolution, and petroleum potential of the sedimentary basins of the world enabled us to create an original classification of PBs and PPBs based on the present-day geodynamic setting (Vyssotsky et al., 1996).

On the basis of this classification, all basins of the world were systematized, and basin categories, groups and subgroups, types and subtypes were defined (figure 31).

According to crustal composition and position within the lithospheric plates, all basins are classified into three categories: continental, oceanic, and transitional. In terms of spatial relationships with the main tectonic elements inside the plates, basin categories are subdivided into groups. For example, the continental category is subdivided into two groups of basins (platform and orogenic belts); the transitional, into four groups (relict margins, continental and oceanic margins, and interplate). The oceanic group of basins is represented by one group, thalassocratonic.

Basin groups are subdivided into types according to geologic history and structure. In the former case, such types as cratonic, cratogenic, postplatform, collisional, subduction-related, etc., are recognized. Referred to as cratogenic basins are those with ancient Archean-Early Proterozoic basements, and as cratogenic, those basins with younger (Baikalian, Caledonian, Hercynian, and Mesozoic) basements. The basins located on microcontinents are interpreted as continental. Depending on geologic structure, they are classified as platform cratonic or cratogenic basins.

In the latter case (in terms of structure), the following types are recognized: synclinorium; rift; blocky; platform-folded; inner deep-water seas; European-type pericontinental backarc; Pacific-type backarc, forearc, and inter-arc; and marginal seas, etc. Some types, e.g. intrafolded, are subdivided into subtypes. Recognition of the platform-folded basins, the distinguishing feature of which is the presence of foredeeps, was conducted only for the orogenic belts formed during the Alpine tectonic cycle.

The basins of active oceanic margins – forearc, interarc, and backarc – are recognized only at the junctions of the oceanic sectors of the plates, for example, the Pacific and Indian-Australian plate; the Philippine Sea and North American plate; the Caribbean with North American and South American plates. At the junctions of oceanic and continental plates, active continental margins are recognized without a further subdivision into forearc, interarc, and backarc basins.

The classification of the basins is supplemented by geologic models of all recognized basin types (figures 32 and 33). Also, correlations are made with the classifications most common in the western countries (Bally and Snelson, 1987; Klemme, 1984).

Detailed investigation of basin structures and the spatial distribution of oil and gas fields, along with the accomplished classification, enabled us to solve the main problem – a quantitative estimate of the total HC resources in all petroleum prospective regions of the world.

CLASSIFICATION OF PETROLIFEROUS BASINS

Compiled by Yu.G. Namestnikov

Table 1

Category: By the underlying crust composition and location within lithospheric plates	Group and subgroup: By confinement to major tectonic elements of plates		Type and subtype: By burial geological history	Type and subtype: By geological structure		Principal structural and morphological forms	Type model of petroliferous basin	Type of the basin: After A.W. Bally (1980)	Type of the basin: After H.D. Klemma (1984)	Examples of typical petroliferous basins and their numbers in the map
Continental	of platforms		Cratonic (ancient)	synclinore / rift		syneclise / rift sviakogene		121 / 222	1.11a / HIA	Middle Amazones (137), Ordoz (351) / Ohed (241), Onepr-pripyat (184)
Continental	of platforms		cratogenic (young)	synclinore / rift		depres-sion / rift sviakogene		1211, 1212 / 1211	IIB / KLA	west Siberion (271), England-Paris 199) / Songred
Continental	of platforms		post-platform	blocky		Intermontane depression		222	HA	Uints-Piacance (65), Green River (61)
Continental	of orogenic belts		convergent	collision	platform-folded	foredeep		41, 22, 221	HCa	Persian Gulf (317), North Pre-Carpathian (203)
Continental	of orogenic belts		convergent	collision	on median massas; on orogenic belts	Intermontane depression		321, 322	HIIBc	Pannonian (206), Vienna-Moravian (205), Greet Valley (74)
Continental	of orogenic belts		convergent	collision	nappefolded	thrust, nappe		41.42	HCb	Carpathian (204)
Transition	of relict margine	continental passive	paleo-divergent	of marginal seas		onshore-shelf-continental slope-piedmant		33	HIBb	Coastal Columbian (121), Bonaire-Caniacc (124)
Transition	of relict margine	continental passive and oceanic active	paleo-divergent-conver-gent	of inner deep-water seas		onshore-shelf continental slope-deep-water depression		1143	HCc, IV	Gulf Mexico (87), South Caspian (289)
Transition	of continental margine	passive	divergent	pericon-tinental		pericontinental depression, shelf-continental slope-piedmant delta		114	HIc, IV	Campoe (158), Gulf of Guinea (248), Beaufort (15)
Transition	of continental margine	active	convergent	of Pacific type	back-arc pericon-tinental	depression, syneclise		3132	HIBa	North Sumatra (435), North Jawa (441)
Transition	of continental margine	active	convergent	of Pacific type	fore-arc pericon-tinental	depression, deep-water tranch		1141	HIC, HB	Bonaperte Gulf (449)
Transition	of continental margine	active	convergent	of Andean type		depression-party deep water tranch		21- OC	–	Lebu Arauko (161)
Transition	of continental margine	active	convergent	back-arc of European type		depression		3122	HIBc	Crete (210)
Transition	of oceanic margine	active	convergent	of Pacific type	back-arc	depression, trough		32	HIb	Uetsu (372)
Transition	of oceanic margine	active	convergent	of Pacific type	tore-arc	terrace-trough		311.42	V	Kuriles (369), Barbados-Tobago (116), Benkiden (440)
Transition	of oceanic margine	active	convergent	of Pacific type	interarc	depression		332	HIB	Kagayan (418)
Transition	of oceanic margine	active	convergent	of marginal seas		depressions, deep-water basins		3121- OC	–	Okhatsk (288)
Transition	Interplate			rift		rift, graben		111	HIA/XIC	Red Sea (237), Gulf of Aden (322), Gulf of California (91)
Oceanic	thaiassocratonic			of synclinore / of rifts		deep-water basins / rifts, axial rift valley		113 – OC	–	Gulf of Bengal (403)

Note: Vertical section of the sedimentary cover may contain a Combination of various types of the basins.

Legend

- sedimentary rock
- continental crust
- oceanic crust
- dislocated sedimentary rocks partly metamorphosed
- volcanic ridges
- nappes, overthrowsts sedimentary rocks

Figure 31. Classification of petroliferous basins

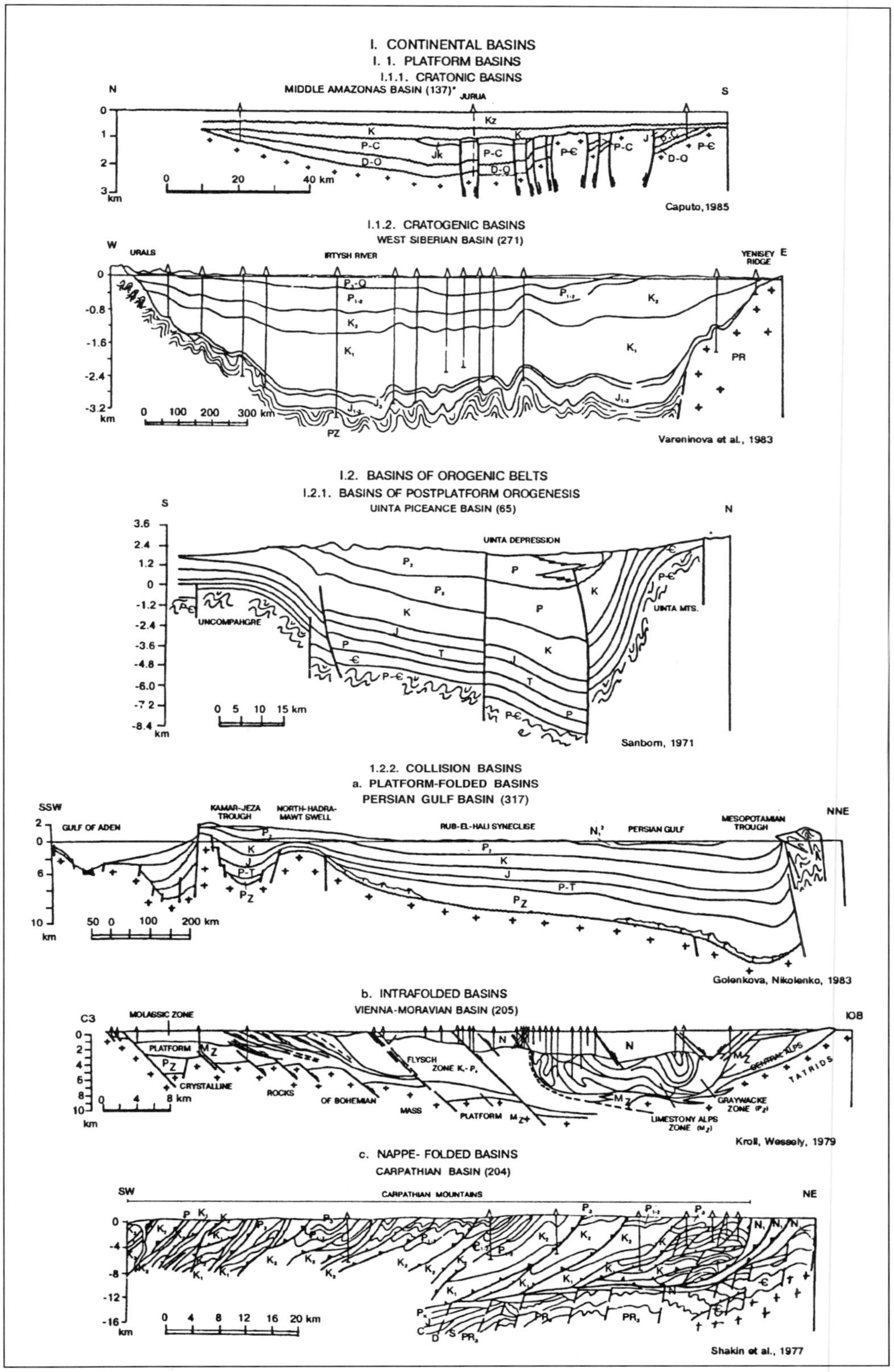

* - Numbers of basins in the map.

Figure 32. Geological sections of petroliferous continental basins

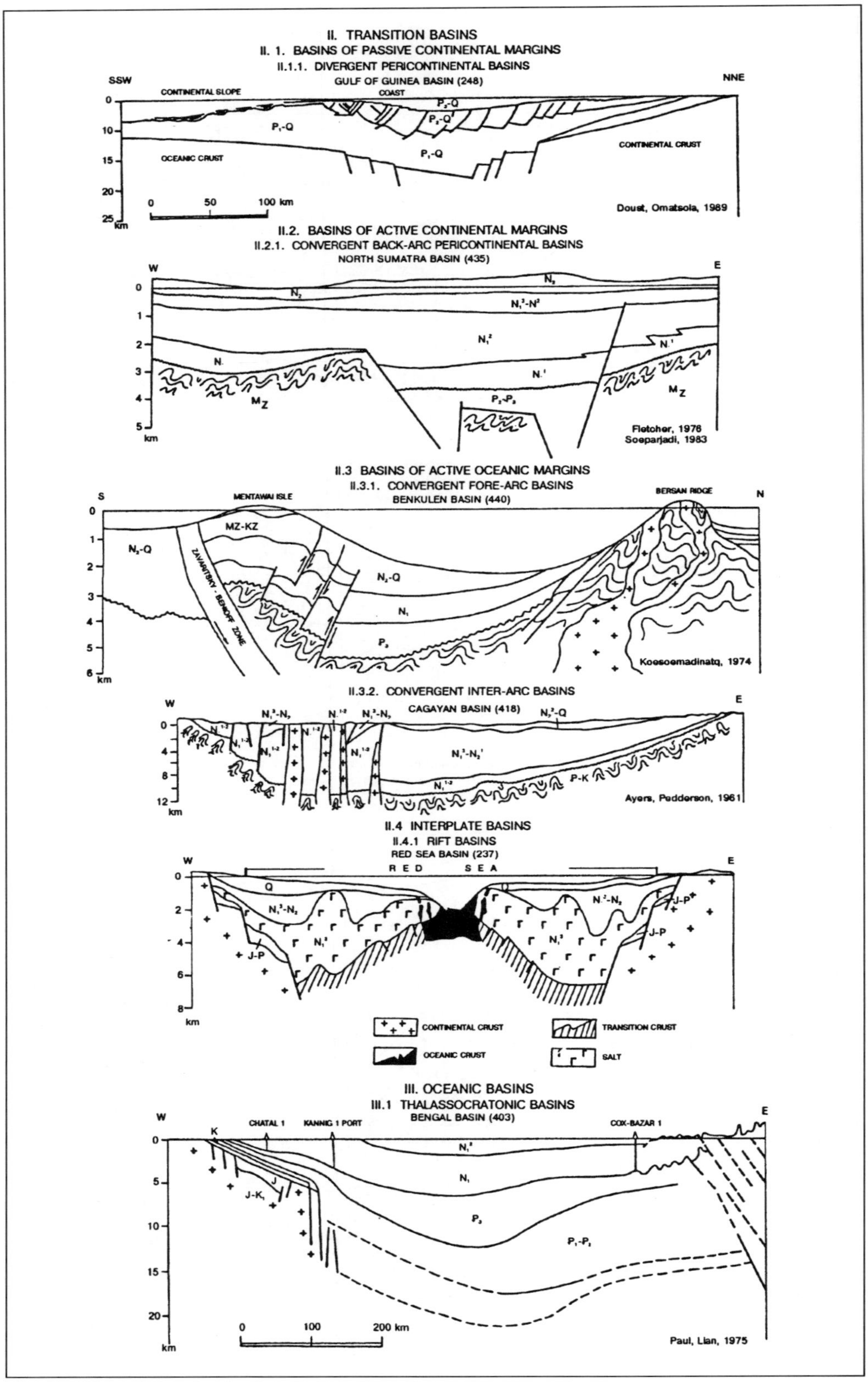

Figure 33. Geological sections of petroliferous transitional and oceanic basins

3.2 Methods of resource estimation

Hydrocarbon resource estimates for most basins of the world are made using two main methods widely applied in the former USSR (Instruction on the Quantitative Estimate of Forecast Oil, Gas, and Condensate Resources, 1983).

The *volume-statistical* method of oil and gas resource evaluation and the versions of this method, are based on the concept that the initial total in place HC resources ($ITIPR_{HC}$) are genetically and spatially related to sedimentary deposits. Generally, if a sedimentary sequence is the source of hydrocarbons and at the same time hosts HC accumulations, then the $ITIPR_{HC}$ volume of the object under consideration is to a certain extent functionally dependent on the sedimentary fill volume. Another version of the volume-statistical method suggests an empirically defined relationship between the variations of the volume density of resources (i.e. oil and gas reserves per unit volume of the sedimentary fill, ρv) and one of the parameters characteristic of the basin's sedimentary fill – the proportion of natural reservoirs in the total sedimentary fill volume (efficient holding capacity coefficient, Kec).

The graphic expression of the above mentioned versions of the volume-statistical method, reflecting the empirically determined relationship,

$$ITIPR_{HC} = P(V) \text{ and } \rho v = f(Kec)$$

is shown in figures 34 and 35.

According to the first version, the objects with sedimentary fill volumes between 10,000 and 3,000 km^3 show a practically linear dependence, corresponding to the following equation:

$$\ln ITIPR_{HC} = 1.19 \ln V - 6.47$$

For larger objects, it is described by the following equation:

$$ITIPR_{HC} = ({}_{e}1.3 \times 10^{-4v} - 1)$$

In the second version, the following formula was applied:

$$\rho V = e^{4.69} \times e^{2.67Kec} \times K_{ec}\, 1.12$$

where $ITIPR_{HC}$ are in MMtoe (1 t oil = 1,000 m^3 gas); V, in thou km^3; K_{ec}, as a fraction.

The volume-statistical method enables us to give a first approximation estimate of HC resources in frontier sedimentary basins or their larger parts, for which the theoretical possibility of petroleum generation and accumulation can be assumed, and the sedimentary cover volume and the qualitative characteristics of petroleum potential can be estimated according to the available geologic and geophysical data. In this case the first version of calculations is applicable (figure 34). In the cases when an additional opportunity arises to analyze the sedimentary sequence of the prospect and estimate the Kec value, if even in general, the second version of calculations is applied (figure 35).

Not all the reservoir rocks are referred to as natural reservoirs. Excluded from the "effective porosity", for example, are the permeable seams lacking competent seals above them; thick reservoir units without any intraformational permeability barriers, because they are easily drained by formation water, etc. The proportion of natural reservoirs in the sequence or in the basin fill is characterized by effective porosity coefficient, and depends on the lithology of sedimentary sequences and complexes. This coefficient widely varies from 0.001 to 0.3-0.45. The lowest values are, as a rule, characteristic of coarse-grained deposits of orogenic complexes; the highest, of biogenic and bioclastic limestones (reefs in particular).

The objects of estimate in the *volume-balance method* of HC resource evaluation are large self-regulated equilibrium petroleum-hydrodynamic systems, laterally coinciding either with the whole basin or with the constituent regions, sags, petroleum accumulation zones, and vertically confined to certain lithostratigraphic units. The petroleum-hydrodynamic system (PHDS) volume encompasses the total volume of natural reservoirs present in

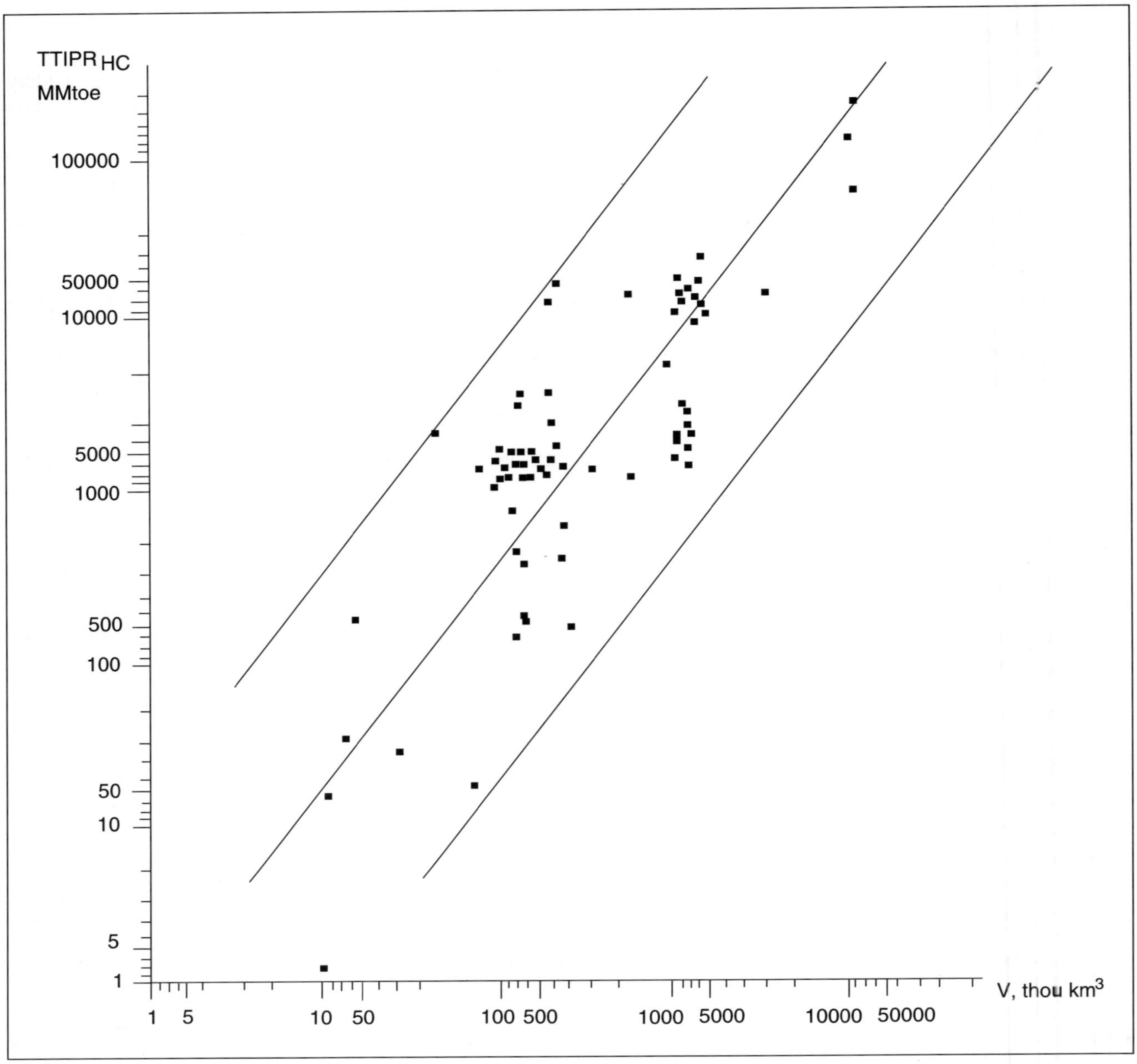

Figure 34. Initial total in place hydrocarbon resources (HC) (TTIPRhc) vs basin fill volume

the object under consideration and filled with three interrelated components, oil, gas and formation water. The ratio between the volume in place of oil and gas saturated natural reservoirs (pools) and the total PHDS volume is the concentration coefficient (respectively oil, gas, or total HC concentration coefficient).

In the absence of the data necessary to calculate the oil (gas, total HC) concentration coefficient for a certain region, it is expedient to use the average values of that parameter and correction coefficients, depending on the qualitative estimate of petroleum potential.

Large systems with volumes in excess of 8,000 km^3 for cratonic depressions; 1,000 km^3, for foredeeps; and 300 km^3, for intermontane ("intrafolded") depressions are characterized by average oil concentration coefficients of about 15 x 10^{-5}; free gas, 35 x 10^{-5}; total HC, 30 x 10^{-5}. In the cases when natural reservoir volumes are less than the above values, or the qualitative estimate of the petroleum potential is different from the average, the correction coefficients increasing or decreasing the average concentration coefficients by 2-5 times are introduced.

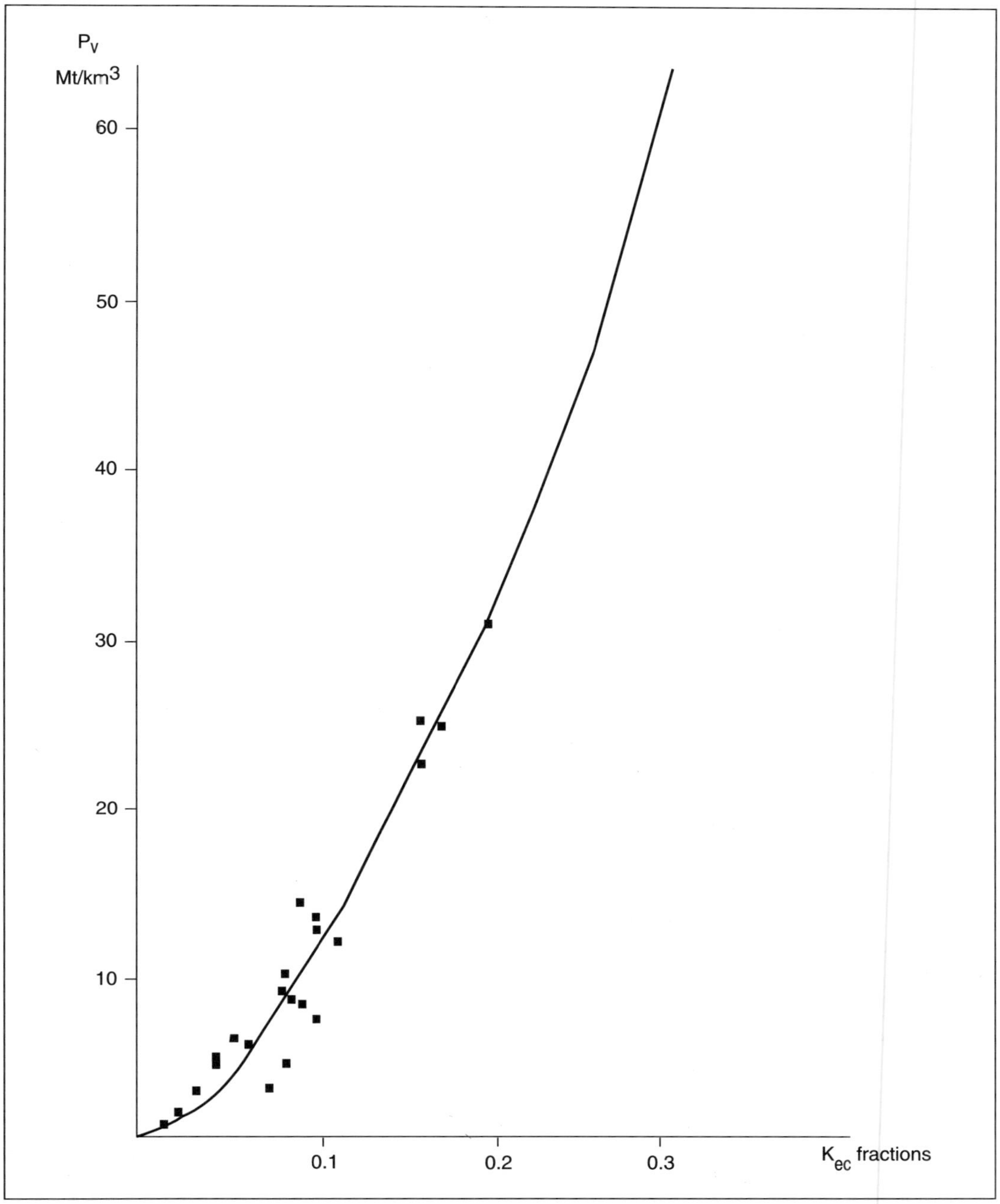

Figure 35. Volume density of initial HC resources (pv) vs effective capacity coefficient of the basin fill (Kec)

Allowances for the changing conditions of the qualitative estimate of petroleum potential are usually made as expert estimates.

Natural reservoir volumes of the objects of estimate are calculated by multiplication of sedimentary cover volumes by the effective holding capacity coefficient of these deposits (Kc), which characterizes the proportion of natural reservoirs in the total thickness of the sequence.

Based on a generalized modified version of the volume-balance method, in the absence of any data on temperature/pressure, porosity, and other in-place conditions of the pay zones, the original total in-place HC resources are calculated using the following formula:

$ITIPR_{HC} = V_{nr} \times (\varphi_{HC} \times \gamma \times 103$ (MMtoe),

where V_{nr}, is the volume of natural reservoirs, km^3 ;
φ_{HC} – coefficient of total HC concentration, fractions;
γ – average density of total HC, g/cm^3

It should be noted that both of the above methods were applied for the evaluation of frontier basins or, in the cases when we did not have access to sufficient initial data, to apply more accurate calculations.

The latter comprise comparative geologic and volume-genetic methods of resource evaluation. The comparative geologic methods are based on the selection of a standard mature basin and its comparison in terms of geologic parameters with the basin under evaluation. In doing so, correction coefficients are introduced to make up for the difference in the compared sediment thicknesses, reservoirs, seals, etc.

We have applied a modified version of these methods (standard scanning method, worked out in VNIIZARUBEZHGEOLOGIA) for the resource evaluation in the basins related to continental margins. A number of basins for which the data on source organic matter composition and maturity were available were evaluated using the volume-genetic method. The method, essentially, comprises calculation of the probable volume of hydrocarbons which could be generated in the source rocks; then evaluation of hydrocarbon migration from these rocks; and then, taking into account migration losses, the amount of oil and gas which could accumulate in the pools is determined. It should be noted that this method is one of the most widespread in many countries. In recent years it was considerably modified in terms of petroleum systems concepts (The Petroleum System – from Source to Trap, 1994).

The quantity of hydrocarbons, which a certain petroleum system is capable of generating, is calculated on the basis of the estimated difference between the hydrogen indices of the kerogens in immature and mature source rocks. The difference is interpreted as the additional generative potential of the organic matter, and the total amount of the generated hydrocarbons is estimated on the basis of the total source rock volume. The latter, in its turn, is calculated by multiplication of the average net thickness of each source rock unit by the area of its occurrence. To sumarize, the quantitative estimate of a petroleum system requires a number of geologic and geochemical parameters determined both by traditional regional geologic survey methods (for the source rock characteristics, the occurrence area and thickness) and specific geochemical methods, particularly Rock Eval pyrolysis enabling us to determine parameters such as total OM content, maturity, and hydrogen index (HC content in organic matter). So, the first phase of the quantitative estimate of the petroleum system implies determining of the total amount of hydrocarbons generated (HCG). The second phase implies evaluation of the petroleum system efficiency and the amount of hydrocarbons capable of accumulation in the pools (HCA). The efficiency of the system is expressed by the accumulation coefficient (GAE). Petroleum system efficiency is controlled mainly by HC pool development conditions, or the conditions of HC accumulation within a trap. The main part in this process belongs to the size and capacity of the traps, reservoir properties of the pay zones, presence of competent seals, and migration pathways of the generated hydrocarbons to the possible traps, i.e. the spatial distribution of source rocks and reservoir horizons. Apart from the spatial relationship between the source rocks and reservoirs, also important is the timing of maturation and trap formation. The minimum losses of the generated hydrocarbons are characteristic of the systems where most of the traps had been formed the moment the organic matter entered the oil window. Evaluation of the generated and accumulated hydrocarbons were made using the following formulae, (Schmoker, 1994; Magoon & Valin, 1994).

M (mg TOC) = TOC (wt%) x d (g/cm^3) x V (cm^3)
R (mg HC/TOC) = HIo (mg HC/g TOC) – HIp (mg HC/g TOC)
HCG (kg HC) = R (mg HC/g TOC) x M (g TOC) x 10^{-6} (kg/mg)
HCA = HCG x GAE

M – total organic matter content in the rock, wt%;
R – volume of the hydrocarbons generated per unit mass of organic carbon, mg HC/g TOC;
HCG – total volume of hydrocarbons generated, kg;
HCA – total volume of hydrocarbons accumulated, kg;
GAE – system efficiency coefficient;

TOC – total organic carbon content in the rock, wt%;
V – source rock volume, cm^3;
d – source rock density, g/cm^3;
HIo – hydrogen index of immature rock, g HC/kg TOC;
HIp – hydrogen index of mature rock, g HC/kg TOC.

Note that we have applied this method for resource calculation, particularly in a number of basins in North-East Asia. The results of oil and gas resource evaluation for the whole world and various regions and countries were compared with the estimates obtained by other organizations and researchers, the US Geological Survey in particular, the results of which were demonstrated at all latest World Oil Congresses (Masters et al., 1991, 1994).

4.0 OIL AND GAS RESOURCES AND THEIR GEOGRAPHICAL DISTRIBUTION

According to our calculations, the total initial recoverable oil resources of the world are 540 Bt; gas, 546 Tm^3 (figures 36 and 37). Let us compare these with the latest world petroleum potential estimates performed by the US Geological Survey, which are 372 Bt oil and 358 Tm^3 gas (Masters et al., 1994). Without going into the details of differences in calculations, two principals are noteworthy in our opinion, and the following reasons account for these differences. The American estimates are based on considerably underestimated oil and gas resources of the basins in the CIS countries. Indeed, Masters et al. (1994) estimate gas resources in the CIS at 107 Tm^3, whereas the official GAZPROM estimate (prepared jointly by many scientific organizations in the former USSR) is 260 Tm^3. There are similar differences in the CIS oil resource estimate. In addition, the American estimates did not take into account the deep sea basin resources (outside the continental shelf). According to our estimate, the latter are 36 Bt oil and 63 Tm^3 gas.

The richest oil resources are found in the countries of the Middle East (figure 38). Our estimate of these resources is 215 Bt, or 40 per cent of the total world oil resources. Of that, 97 Bt are undiscovered resources. The second richest are the CIS countries, accounting for 22 per cent of world oil resources (118 Bt). Of that, 80 Bt are unexplored.

Next come Latin America and North America (11 per cent and 10 per cent, respectively); then Africa and Central Asia and the Far East (6.5 per cent and 5 per cent, respectively). The rest of the regions (SE Asia, West Europe, Australia and Oceania) jointly account for a trifle over 5 per cent of the world's oil resources.

Geographical distribution of gas resources (figure 39) shows a different pattern. In this case the leading position belongs to the CIS countries, the resources of which are estimated at 260 Tm^3, or 46 per cent of the world gas potential. Of that 182 Tm^3, or 70 per cent, are undiscovered. Next come the Middle East countries (18 per cent). The proportion of each of the other regions of the world is not more than 10 per cent of the total world gas resources.

Noteworthy are the region by region variations of the reserves to resources ratio, indicating the proportion of identified oil and gas reserves in the initial total resource volume (figures 40 and 41). In most regions of the world, this ratio for oil ranges from 45 per cent (South and South-East Asia) to 58 per cent (West Europe). Only in the CIS countries, Central Asia and Far East is it just 33 per cent and 25 per cent, respectively.

Gas reserves to resources ratio is the highest in West Europe (51 per cent), Middle East (45 per cent), and North America (45 per cent). In the other regions, it ranges between 30 per cent and 38 per cent. The exceptions are Central Asia and the Far East, where gas reserves to resources ratio is just 12 per cent.

In summary, oil and gas reserves to resources ratios in Central Asia, and in the Far east, are the lowest among the other large geographical regions of the world.

5.0 PETROLEUM POTENTIAL OF THE NORTH-EAST ASIAN SEDIMENTARY BASINS

North-East Asia includes China, Mongolia, the People's Democratic Republic of Korea, the Republic of Korea, Japan, and the Okhotsk region of Russia. The total oil resources within the area, according to our estimate, are 27.7 Bt; gas resources, 21.6 Tm^3.

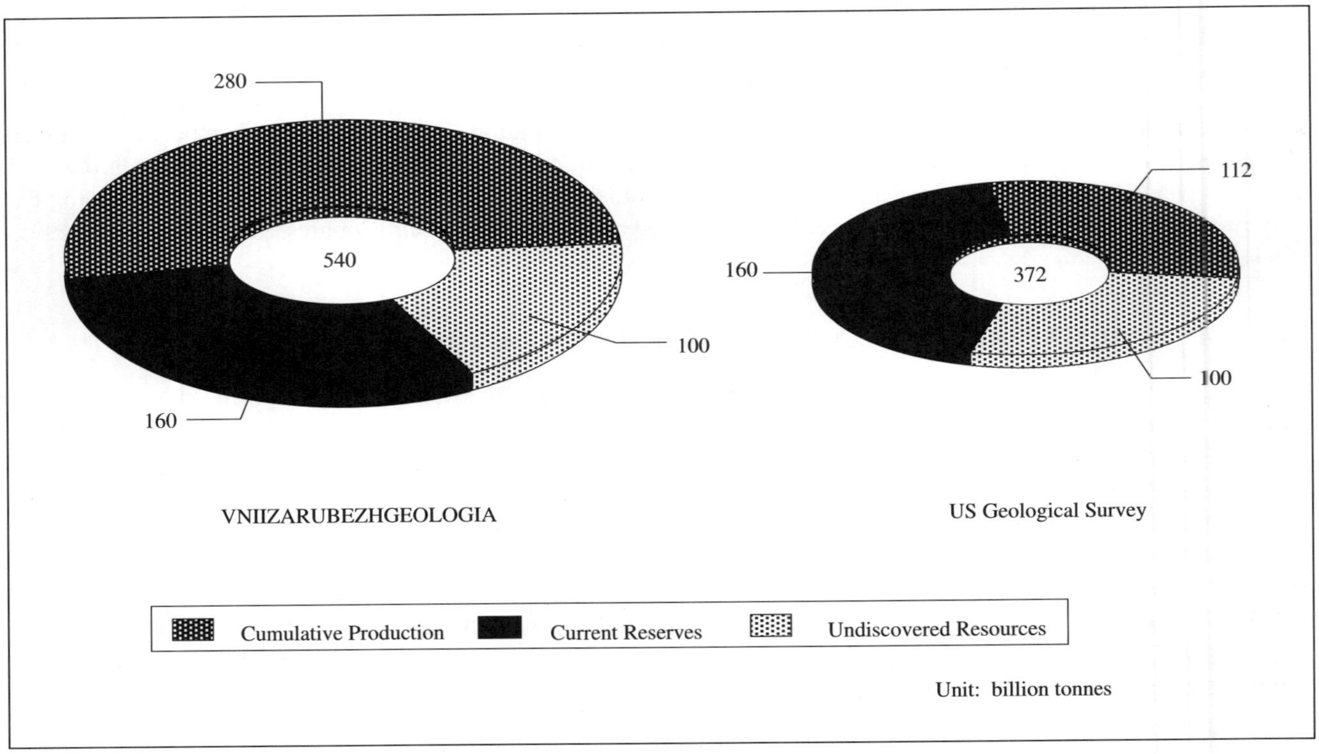

Figure 36. World oil resources estimation

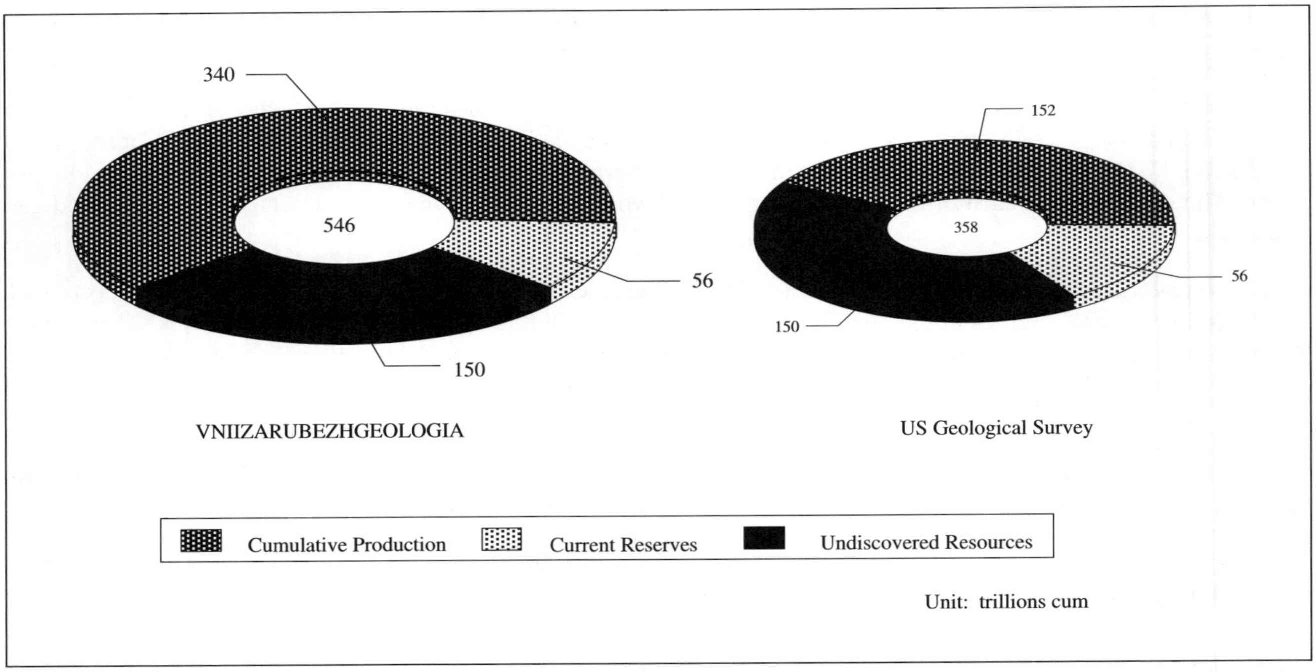

Figure 37. World gas resources estimation

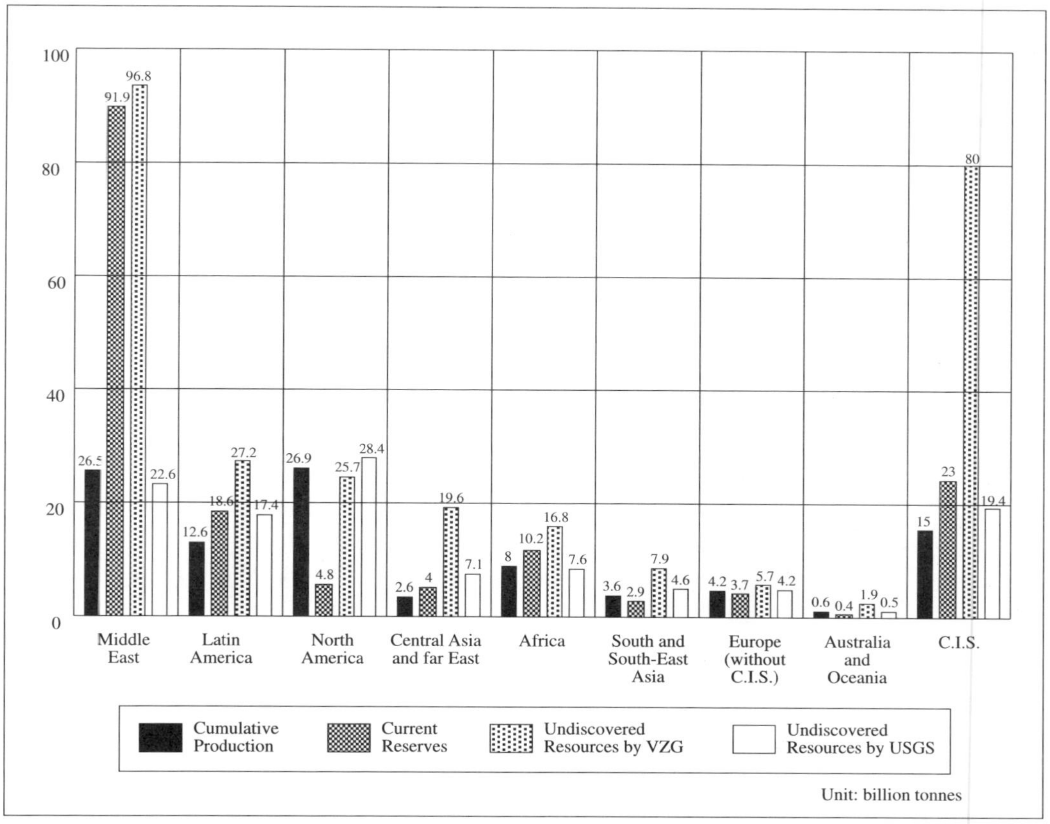

Figure 38. Geographic distribution of oil resources

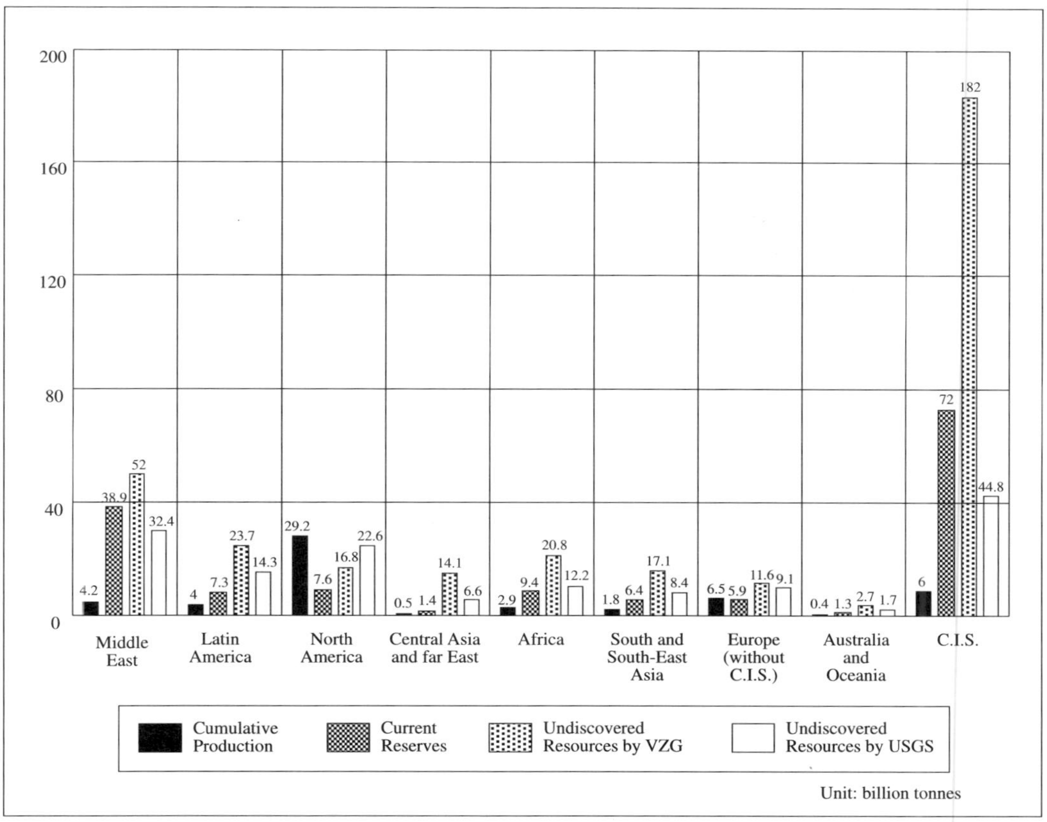

Figure 39. Geographic distribution of gas resources

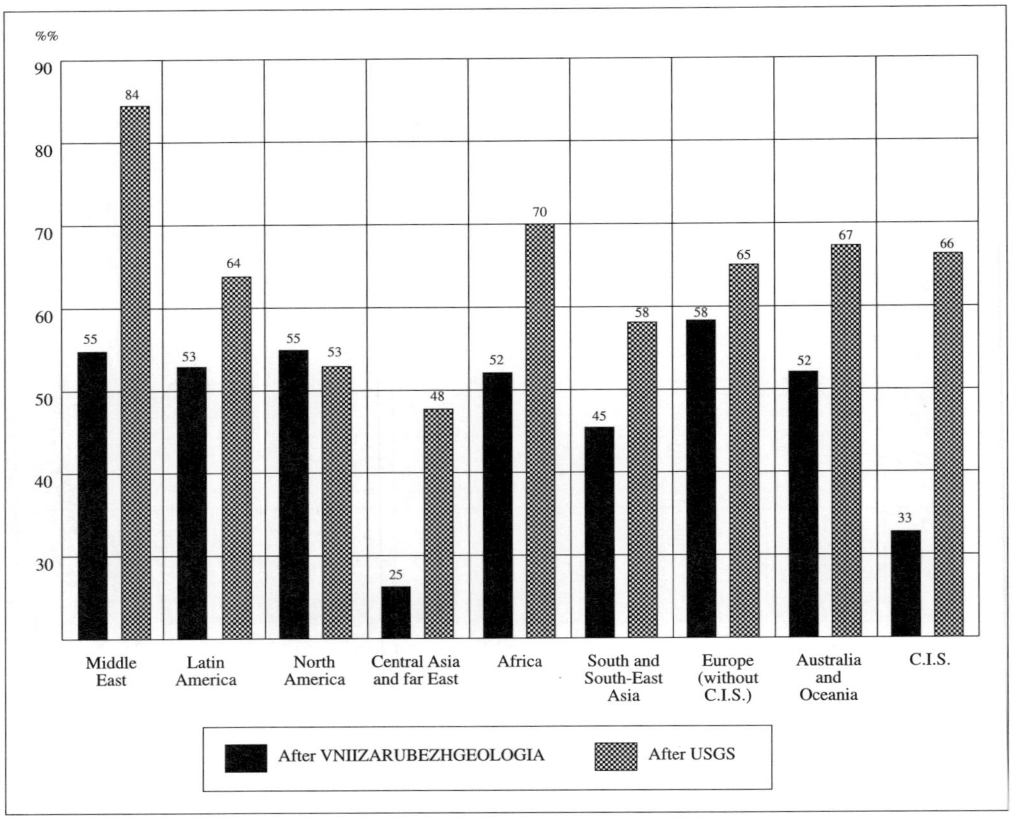

Figure 40. Regional exploration degree of the world oil resoruces

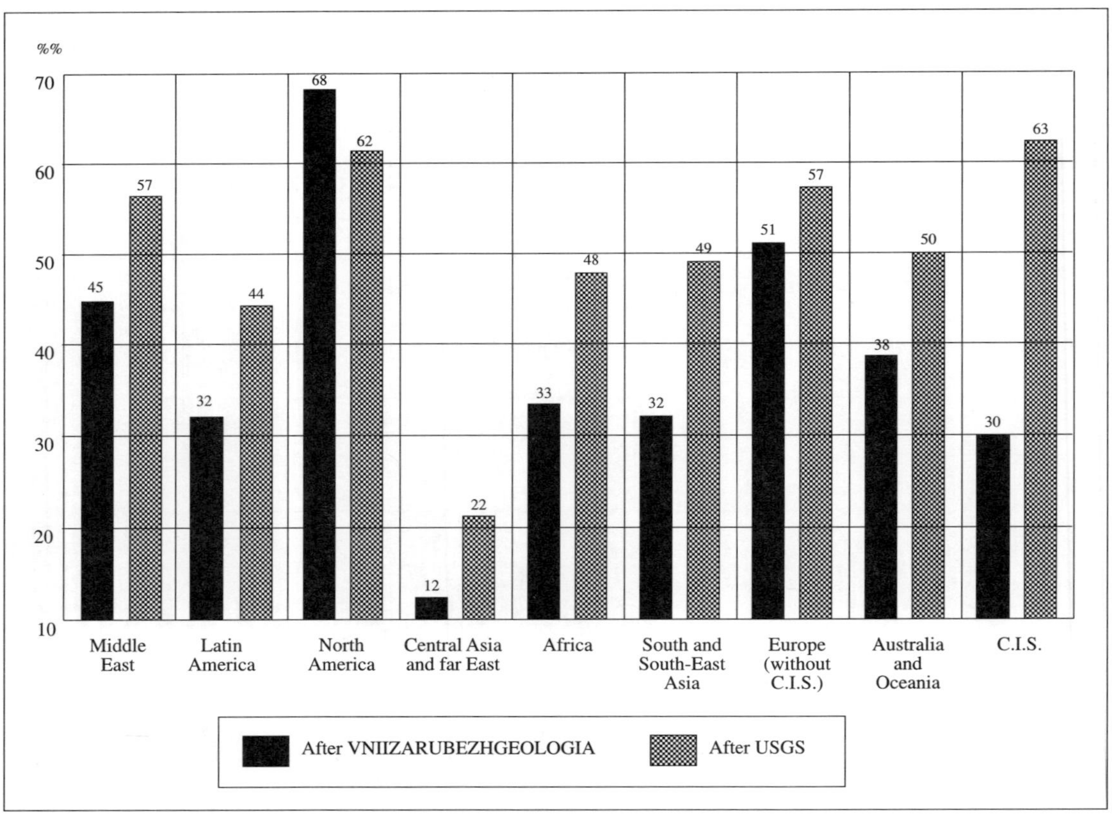

Figure 41. Regional exploration degree of the world gas resources

The bulk of oil resources lies in the sedimentary basins of China (88.8 per cent); gas, in the basins of China (62 per cent) and Russia's Far East (25.9 per cent). Japan accounts for a trifle more than 4 per cent oil and 8 per cent gas. The rest of the countries do not play significant parts in the total North-East Asian resource potential (figure 42).

In China, 39 sedimentary basins with a total area of 3,900,000 sqare kilometres have been recognized (Salmanov, Vyssotski, 1991). In 21 of them, oil and/or gas fields have been discovered. The initial recoverable oil resources, according to our estimates, are 24.6 Bt; gas, 13.4 Tm^3. Spatial distribution of hydrocarbon resources is extremely uneven. Two basins, Tarim and North China, account for 34 per cent of the country's total oil resources; and if three more basins are added – Sungliao, East China Sea, and Sichuan – the five basins will account for almost 60 per cent of the total oil resources.

Still more uneven is the spatial distribution of the country's gas resources. Two basins, Tarim and Sichuan, account for 57 per cent of China's gas potential. The bulk of undiscovered gas and oil resources is concentrated in the Tarim basin in western China. According to our estimate, its potential resources are 4.4 Bt oil and 4.8 Tm^3 gas. According to Chinese workers, oil resources of the Tarim Basin are somewhat less – 3.7 Bt – and gas resources are much more, 6.7 Tm^3 (Desheng et al., 1996).

The second richest in petroleum potential among the North-East Asian countries is the Russia's Far East. It comprises two vast megabasins: the Okhotsk megabasin, consisting of 12 basins, and the Japan megabasin, including two sedimentary basins.

The total area of the basins is 1.85 million square kilometres. Total potential resources in all sedimentary basins of the Russian Federation's Far East are estimated at 1.5 Bt oil and 5.6 Tm^3 gas. Most of them (60 per cent oil and 69 per cent gas) are concentrated in the North Sakhalin basin. The second important basin of the region, Ishikai-West Sakhalin, accounts for 20 per cent oil and 12.5 per cent gas resources of the Russia's Far East.

In Japan, the main oil and gas potential is confined to three basins: Tsushima, Semanta, and Abakuma. These basins account for 50 per cent of the country's oil resources and 64 per cent of its gas resources.

In the Republic of Korea, 10 sedimentary basins with a total area of 311,000 sqare kilometres have been recognized. According to our estimate, total recoverable resources of these basins are 350 MMt oil and 800 Bm^3 gas. Most of the resources (63 per cent) lie in three basins, Ulleung, Yellow Sea, and Socotra.

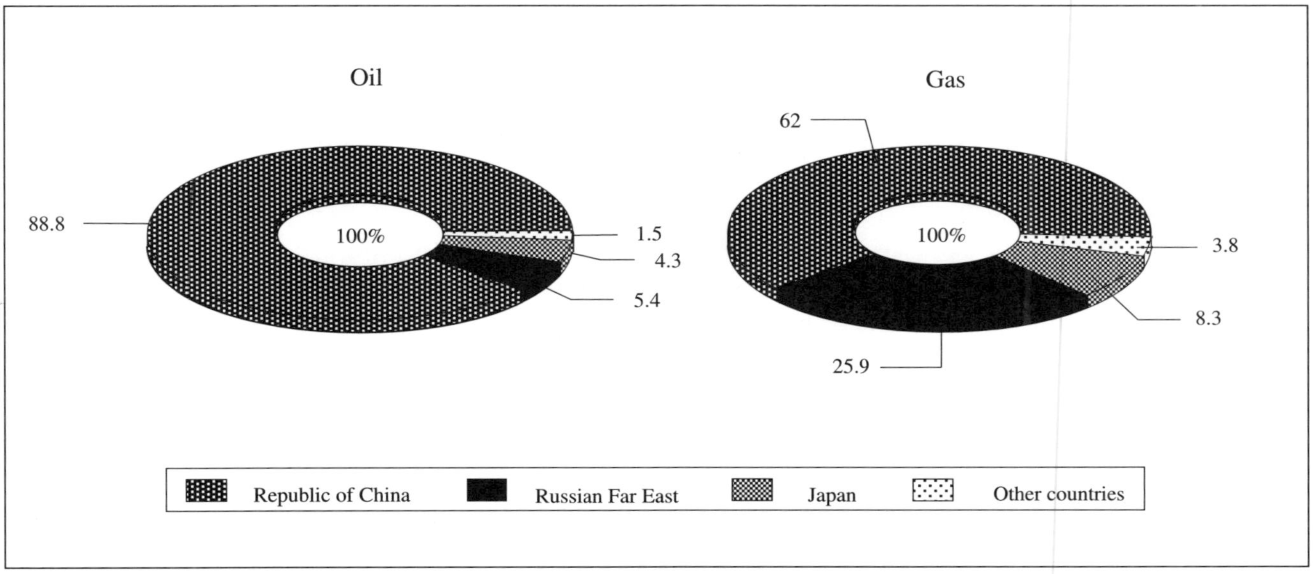

Figure 42. Distribution of oil and gas resources in North-East Asia

Oil and gas resources of Mongolia and the Democratic People's Republic of Korea are insignificant, a few tens of MMtoe.

Of course we are aware that our estimates cannot always be sufficiently reliable. Naturally, they will be improved as new geological data are acquired in the course of sedimentary basin exploration. Nevertheless, they provide a benchmark on the basis of which possible oil and gas production rates in Northeastern Asia can be predicted. According to the analysis of the mature basins of the world, peak oil production within them may amount to 1.5-2 per cent of the initial recoverable HC resources (Kalinin, 1979; Modelevsky, 1983; Baturin, 1996). It will be noted, however, that only 50 per cent of the total resources in place are efficient for exploration, or active resources. The rest of the resources, poorly efficient or passive, are concentrated in small and extra small fields and cannot be economically recovered under the present or close to the present oil and gas prices. Even if the prices are twice as high as at present, only 50 per cent of the passive resources may become active, i.e. commercially exploitable.

Therefore, the forecasts of peak production rates in conditions of present-day or close to present oil prices should be based on the active part of resources.

From the above reasoning it may be suggested that peak annual oil production in North-East Asia should be expected at about 205-270 MMt oil and 160-215 Bm^3 gas. This level will be determined by the tempo of oil and gas resource exploration in China and the Russia's Far East.

6.0 CONCLUSIONS

According to our estimates, the recoverable oil resources of the world are 540 Bt oil and 546 Tm^3 gas. These resources are quite sufficient to fulfill mankind's demand for this type of energy source during the next century. The least explored among the largest geographical regions of the world are the hydrocarbon resources of North-East Asian countries. Reserves to resources ratio within them is just 12 per cent. Exploration of the active part of these resources enables us to predict possible peak production at 205-270 MMt oil and 160-215 Bm^3 gas. The achievement of this level will depend mostly on the oil and gas exploration tempo in China and Russia's Far East.

REFERENCES

1. Bally A.W. and Snelson S., 1987. Realms of Subsidence Geologic Basins 1. Classification, Modelling, and Predictive Statigraphy. Compiled by Norman H. Foster and Edward A. Beaumont: AAPG, p. 1-86.

2. Baturin IO.N., 1996. Geologo-ekonomicheskaia structura NSR nefti osnovnih neftedobuveiustchih regionov Rossii: Geologia nefti i gaza, No. 10, p. 14-20.

3. Hubbert, M.K., 1967. Degree of advancement of Petroleum Exploration in USA: Bull. Amer. Assoc. Petrol. Geol., 51 No. 11, p. 2207-2227.

4. Kalinin N.A. ed., 1983. Kadastr zarubezhnyh stran, obladaiuschih prirodnymi resursami nefti i gaza, tom I,: Nedra, 333 p.

5. Klemme H. Douglas, 1984. Types of petroliferous basins Geologic Basins 1. Classification, Modelling, and Predictive Stratigraphy. Compiled by Norman H. Foster and Edward A. Beaumont: AAPG, p. 87-101.

6. Magoon, L.B. and Dow, W.G., 1994. The Petroleum System – From Source to Trap,: in Leslie B. Magoon and Wallace G. Dow, eds., The Petroleuim System – From Source to Trap: AAPG Memoir 60, p. 3-24.

7. Li Desheng, Liang Digang, Jia Chengzao, Wang Gang, Wu Qizhi, and He Dengfa, 1996. Hydrocarbon Accumulation in the Tarim Basin, China: AAPG Bulletin, No. 10, p. 1587-1603.

8. Masters Cli., Attanasi K, Root D., 1991. World Resources of Crude Oil and Natural Gas; Preprint of the 13th World Petroleum Congress, Buenos Aires, p. 1-16.

9. Masters Ch., Attanasi K, Root D., 1994. World Petroleum Assessment and Analysis: Preprint Proceedings of the 14th World Pertroleum Congress, Stavenger, P. 146-152.

10. Metodichskiye ukazania po colichestvenoyi ocenke prognoznyih resursov nefti, gaza i condensata, 1983: Moscow, 214 p.

11. Salmanov F.K., Vysotsky V.I., 1990, Kitai glazami geologa, 1991: Nedra, 96 p.

12. Modelevsky M.S., 1979. Sovremennye predstavlenia o resursah nefti i gaza i metody ih ozenki: Itogi nauki i techniki, seria Geologicheskie i geohimacheskie metody, 150 p.

13. Schmoker J.W. 1994. Volumetric Calculation of Hydrocarbons Generated: in Leslie B. Magoon and Wallace G. Dow, eds., The Petroleum System – From Source to Trap: AAPG Memoir 60, p. 323-326.

14. Vysotsky V.I., Isaev E.N., Klestchev K.A., Miletenco N.v., Namestnicov J.G., and Fedorov D.L., 1996. Neftegasovyi potencial osadochnyih basseynov mira; Otechestvenaya geologia, No. 2, p. 6-14.

CONTENTS

CONTENTS *(continued)*

2. REVIEW OF MINERAL RESOURCE POTENTIAL OF THE NORTH-EAST ASIA IN SUPPORT OF SUSTAINABLE DEVELOPMENT OF COASTAL AND OFFSHORE AREAS[1]

1.0 INTRODUCTION

1.1 Definition and characteristics of the area

North-East Asia, as defined for this study, includes the Democratic People's Republic of Korea, Japan, Mongolia, the northeastern portion of the People's Republic of China (Heilongjiang, Jilin and Liaoning Provinces) and the Russian Far East. The countries of North-East Asia include one of the least developed areas of the world (Russian Far East), two of the fastest growing economies (People's Republic of China and the Republic of Korea), two recently emerging countries (Mongolia and the Democratic People's Republic of Korea) and one member of the OECD community (Japan).

The development of the mineral and energy resources of the above defined broader North-East Asia area can, and should, be viewed in the context of the overall development of the Asia-Pacific region. However, within the North-East Asia region itself there has been a recent focus on the development of mineral and energy resources for local internal integrated development of the region, as outlined in the UNDP sponsored Tumen River Development Programme, through the "Tumen River Economic Development Area" (TREDA). TREDA is a smaller area consisting of the local administrative areas adjacent to the Tumen River that forms part of the boundary between the Democratic People's Republic of Korea (DPRK), the Peoples Republic of China (PRC) and the Russian Federation (RF). These local administrative areas comprise North Hamgyong Province in the DPRK, the Yanbian Autonomous Prefecture in Jilin Province in the PRC and the southern and central part of the Primorsky Territory (Krai) in the RF. The development of TREDA will be closely linked and an integral part of the broader development of North-East Asia. For purposes of the present study the primary focus will be on the broader area of North-East Asia.

The North-East Asia area also centres on three great marine areas i.e. the Yellow Sea, the Sea of Japan and the southern portion of the Sea of Okhotsk which all share the unique characteristic of being "semi-enclosed" seas.

1.2 General geography of the area

The area of North-East Asia is characterized by a number of diverse and environmentally significant marine and terrestrial ecosystems; many of which are both poorly defined and understood, but which are known to have global importance. In the majority of the region, detailed environmental studies have only recently been initiated and as a result the information level in many areas is often poor to non-existent. The majority of the area of North-East Asia lies within the Amur/Shakalin/Manchurian bio-region, whereas, the marine environments are largely contained within areas of the above mentioned Sea of Japan, Sea of Okhotsk and the Yellow Sea.

The Amur/Shaklin/Manchurian bio-region is characterized by (a) mixed coniferous and broad-leaved forests of the Amur region, northern Mongolia and northeast China (b) the desert areas of central and southern Mongolia, the high mountains of northern Democratic People's Republic of Korea and adjacent areas of the PRC and the wetlands associated with the Amur and Tumen river drainages.

Conversely, the marine environments are within and adjacent to the semi-enclosed Sea of Japan, the Sea of Okhotsk and the Yellow Sea. The semi-enclosed nature of the seas is of critical importance when considering the environmental impacts of ongoing and potential mineral and energy developments, largely because of limiting ocean currents and the free flow of waters in and out of the seas. As a result pollutants are essentially "trapped" and concentrated within the sea basins. Regional cooperation among the nations of North-East Asia is essential if the marine environment is to be protected and available for sustainable development. Potential mineral developments, nearshore and offshore, within the three seas are discussed in some detail in this paper.

---

[1] Dr. Allen L. Clark, Programme on Resources, Honolulu, Hawaii

2.0 STATUS OF MINERAL RESOURCE DEVELOPMENT

The economic diversity of the region can be seen by evaluating some of the key socio-economic indicators of the various nations/regions of North-East Asia (table 25.)

Table 25. Socio-economic indicators of North-East Asia

Country/Area	*Population (Millions)*	*Area (000 Km²)*	*GNP (Mil. US$)*	*Total Mineral Ex. (M US$)*	*Total Energy Ex. (M US$)*
Democratic People's Repulic Korea	22.5	121.8	21,300	296.6	58.4
Japan	125.5	377,835	2,527,000	21,387.8	1,697.4
Republic of Mongolia	2.5	1,565	4,400	316.0 (est)	N A.
Northeast China	106.9	6,216.0	71,530	8,896	N A.
Republic of Korea	44.1	99.2	580,000	661.6	686.6
Russian Far East	8.00	377.7	13,054 (est)	24.38	230.9

An analysis of table 25 reveals several factors that are importance in terms of assessing the overall development potential of North-East Asia and in particular the following factors are noteworthy.

First, an evaluation of the relationship between population and geographic area clearly shows the relatively sparse population base of Mongolia (1.59 individuals/km^2), the intermediate population density of the Russian Far East (21.18 individuals/km^2) and northeast China (17.2 individuals/km ^{2}) and the high population densities of the Democratic People' Republic of Korea (184.7 individuals/km ^{2}) and the highest population densities of the region in the Republic of Korea (444.5 individuals/km ^{2}) and Japan (332.2 individuals/km ^{2}). An evaluation of the population densities of the individual nations/areas is particularly important when considering (a) the potential workforce available for resource development, (b) the impact of resource development on economic growth and per capita income, and the potential impact of resource development on environmental quality.

Second, is the level of per capita income between the individual nations/areas of the region which range from highs of US$20,200 in Japan and US$13,152 in the Republic of Korea to a low of US$669 for northeast China. The per capita incomes of the Russian Far East, Mongolia and the Democratic People's Republic of Korea are US$1,760, US$1,631 and US$946, respectively. For the majority of the countries/areas, excluding Japan and the Republic of Korea, the development and utilization (export or domestic consumption) of domestic mineral and energy resources will be key to national development and to raising the levels of per capita income. In the case of Japan and the Republic of Korea, the importation rather than the development of domestic resources, will be key to overall economic growth and the increase in per capita income.

Third, although table 25 shows that all of the countries of the region are, to a greater or lesser degree, exporters of both mineral and energy resources. It should be emphasised that only the Russian Far East is a net energy exporter, primarily in unprocessed crude oil and natural gas, whereas the exports of Japan and the Republic of Korea are both small in quantity and are primarily processed fuels – in both cases they are by far net energy importers rather than exporters. Both Mongolia and the Democratic People's Republic of Korea are virtually totally import dependent for oil and gas and must also import coal, although there are domestic resources presently being extracted and utilized.

Just as the levels and forms of economic development are highly variable between the individual countries, so are the respective mineral and energy resource endowments of the individual Nations and their respective ecosystems. The resource rich areas of Mongolia and the onshore and offshore areas of the Russian Far East perhaps represent the most resource rich and ecologically diverse portions of the region, whereas, because of a more mature, developed and exploited mineral and energy sector, the mineral resource potential of Japan and the Republic of Korea is perhaps the lowest of the region and both countries have ecosystems which have been heavily impacted environmentally. In between lies the relatively unknown resource potential, and ecological diversity of northeast China and that of the Democratic People's Republic of Korea. In both of these areas, the importance of resource development is expected to be large and the environmental impact significant – indeed in many areas past resource development has left a highly impacted environment.

As can be seen from the above the nations that constitute North-East Asia, they differ greatly in a number of areas and it is this high degree of variability which presents both the great challenges and opportunities for environmentally sound and sustainable mineral and energy development in the region.

2.1 Proposed plan/rationale for integrated development

In the last century, North-East Asia has been characterized by changing patterns of "amity and enmity, frequent tension between the strongest powers in the region, and resultant attempts to forge alliances with the lesser powers (Valencia and Dorian, 1996). For many years, the interest of the world in this region was focused on the political stability of each of the countries and the region as a whole, and from a resources perspective, what resources were available to support the largely political or military efforts of each of the countries, especially Russia, China, and the Democratic People's Republic of Korea.

However, more recently, with the breakup and restructuring of the former USSR, the resulting liberation and independence of Mongolia, the far-reaching economic development of China, the opening up of the Democratic People's Republic of Korea and the further growth of economic power of the Republic of Korea and Japan and trade with international partners, each country has proceeded to find their niche in the global economic world and moved to normalize relations with their neighbours and trading partners. Some observers have characterized this economic interaction across ideological and political boundaries within North-East Asia as creating a "soft" regionalism in North-East Asia, a regionalism that lacks organizational structure, or regional institutions, but which is accepted by and loosely encouraged by the respective governments (Valencia and Dorian, 1996).

With the development of this "soft" regionalism, there is an increasing interest by the world's resources community to the region's resource potential, both onshore and offshore, of the region and the potential of the region to successfully exploit and utilize those resources.

However, even with the development of "soft" regionalism, there remain significant economic, political, technical and environmental problems associated with any resource development. The most significant political problems are exemplified by the border disputes between Japan and Russia concerning the Kuril Islands/Northern Territories. This will likely limit Japan's investment in Russian resource development. Moreover, concerns with respect to political and economic stability in Russia and the Democratic People's Republic of Korea is, and will continue, to inhibit foreign investment in these areas – particularly with respect to minerals and to a lesser extent with respect to oil and gas. Within the North-East Asia region, the Democratic People's Republic of Korea (despite recent talk of opening some sectors to foreign investment and the creation of an Special Economic Zone (SEZ)) remains one of the world's most underdeveloped and stagnant economies. Moreover, it is anticipated by many potential investors (Clark, 1997, Valencia and Dorian, 1996) that inter-regional transportation from resource points to end-users will require crossing many provinces and/or countries, each of which will require permission, likely charges and mitigation of environmental effects.

Nevertheless, regional development under the proposed TREDA programme is proceeding with the recognition that there is both an inter-dependency within North-East Asia for mutual economic and social development and that the development of the region's mineral and energy resources is key to achieving this development. This regional development of mineral and energy resources will be greatly assisted by other activities within TREDA such as the development of a regional gas pipeline system, regional infrastructure (roads and telecommunications and Special Economic Zones (SEZs). Indeed, the creation of new SEZs within the Democratic People's Republic of Korea is likely to have a great impact in opening up the region and fostering cooperation and development. Similarly, the port of Pongyang, and Sinuiju, on the northwest Chinese border, are mentioned as possible candidates for new SEZs. However, it appears that China, the Republic of Korea and Japan may, because of the access it gives them to previously inaccessible areas, will become the major foci and drivers of the regions development.

2.2 Role of mineral and energy resources in regional development

Minerals and energy developments in North-East Asia must be viewed in the broader context of overall mineral and energy development within the Asia-Pacific region specifically and to a lesser extent globally as future developments will be driven both by domestic needs and for the development of resources for export to the region.

Energy Demand in Asia-Pacific – There is a strong linkage between economic growth and increasing demand for energy, both historically and at present; which represents a trend which is expected to continue into the future with far-reaching national, regional and global resource and environmental implications. An analysis of the structure of energy demand for the Asia-Pacific region, based on projections to 2005, is shown in table 26.

Table 26. Asia-Pacific energy consumption by fuel source[1]

Source	*1990 (per cent)*	*2005 (per cent)*
Oil	48	42
Gas	10	16
Coal	31	26
Hydro	6	7
Nuclear	6	9

Source: Fesharaki, et. al., 1995.

[1] Excludes China due to high coal consumption and dominant role in the region.

An analysis of table 26 reveals several important aspects of the energy demand structure for the year 2005 which has dramatic implications for long term energy development prospects of North-East Asia. In particular:

- The share of oil in the energy mix will decrease from 47 per cent to as low as 37 per cent depending on price;
- The major increase in energy share with an increase in utilization of gas in the energy mix; and
- Coal utilization with remain fairly constant or decline slightly.

Although table 26 shows that the energy demand structure will change significantly by the year 2005 it must be stressed that *absolute energy demand will increase significantly* and significant supplies may need to come from areas of North-East Asia. This will be particularly true with respect to gas development because of the increasing percentage of gas in the energy mix. Overall, this has several significant implications for both economic development and the environment in the Asia-Pacific region overall and the countries of North-East Asia specifically. Among the most significant are the following:

Firstly, the declining share of oil in the energy mix is largely attributable to declining reserves and production in the region, much of which is a low sulfur crude. Increasing reliance on imported oil in the region will be for high sulfur crudes from the Middle East. Refining and processing of higher sulfur stocks may well increase harmful emission – in particular SO_x. Similarly, such crudes will produce larger quantities of solid wastes.

Secondly, the increased use of natural gas in the energy mix is perhaps a mixed blessing for the environment and the economy. Being in relatively abundant supply and at a reasonable price, natural gas utilization can increase with marginal impact upon economic growth: although some large costs for conversion and new systems will be required. Similarly, processing emissions will be significantly lower in SO_x and CO_2 than comparable processing of oil, as natural gas releases approximately 14 kilograms of carbon dioxide per billion joules while oil releases approximately 20 kg and coal almost twice as much with 24 kg per billion joules.

Thirdly, although the percentage of coal in the energy mix is expected to decline slightly it must be remembered that the absolute demand for coal will rise significantly. The result will be a continued increase in the amount of coal required within the region and the countries of North-East Asia in particular. A major environmental impact of the energy growth and utilization will be in the production of the greenhouse gases with well known effects, effects that need not be recounted herein, in terms of global warming, acid rain, and the global hydrologic cycle.

Minerals Demand in Asia-Pacific – As with energy, the demand for most metals in the Asia-Pacific region is increasing rapidly, driven not only by increased use within individual countries but also because both within the region and internationally, many countries of North-East Asia are also major producers for the world market. It is not the purpose of this paper to address the issues of energy and mineral production for the world market, but it must be noted, that in both cases the attendant environmental costs of their production and processing are large.

Within the rapidly developing nations of North-East Asia, metals demand for selected commodities will increase on average by 2 to 3 times by the year 2010. To meet this rapidly increasing demand for metals in North-East Asia and the Asia-Pacific region overall, over 80 new mineral developments and/or expansion of facilities have been planned, proposed or are presently under way, with a total capital investment in excess of US$20 billion. However, to date very little of this planned development is in the North-East Asian area and the potential for future development within the mineral and energy sectors is the subject of the following analysis.

Minerals Issues in North-East Asia – As with many lesser developed areas, the availability of mineral and energy resources, has directly influenced the levels of development, industrialization and trade and for many countries of North-East Asia has required, and will require, high levels of both private sector and development assistance if the region is to fully realize the resource potential of North-East Asia.

The countries included in North-East Asia can in general be designated as resource poor or resource rich, based on the potential of domestic resources to meet a countries or regions overall development objectives. With this as a definition, the Democratic Peoples Republic of Korea, the Republic of Korea, Japan and northeast China would be termed resource poor, whereas, Mongolia and the Russian Far East would both be considered resource rich. In the case of Japan and the Republic of Korea, both are considered resource poor as they have very few domestic reserves of most metals and both are major importers of minerals. Conversely, both the Democratic Peoples Republic of Korea and China have considerable mineral and resource potential, however, both are also major importers of key minerals because high domestic consumption exceeds domestic supply; therefore, they are considered resource poor as well.

The same is true for the oil and gas industry where both China and Russia have large oil and gas resources and reserves. However, China's large internal demand for oil and gas has recently made them a net importer of oil along with Japan and the Republic of Korea. Overall, the nations of North-East Asia are primarily mineral and energy importers which serves as an incentive for the exploration, discovery and utilization of the region's untapped mineral and energy resources: particularly in the Russian Far East and Mongolia.

Importantly, it should be noted that the major areas of mineral and resource potential (Mongolia, the Russian Far East and to some extent areas of the Democratic People's Republic of Korea) are also the areas where: (a) least is known with respect to mineral and energy potential, (b) with poorly developed or non-existent infrastructure and (c) where mineral and energy development laws and regulations are poorly understood and/or enforced. In essence, they are areas of lesser priority, at the present time, for mineral and energy exploration and exploitation. The degree to which the above problems can be resolved or alleviated will to a large extent determine both the pace and the scope of mineral and energy development.

3.0 MINERAL AND ENERGY RESOURCES OF NORTH-EAST ASIA

North-East Asia is one of the last frontiers for the development of energy and mineral resources. The area already produces large quantities of coal (Democratic People's Republic of Korea, northeast China), crude oil (northeast China), cement (Japan) and gold (Russian Far East, northeastern China) (Dorian et al., 1993) and while exploration is increasing, exploration efforts as compared to other places worldwide is still sparse. Nevertheless, with the exception of Japan and the Republic of Korea, the potential for new mineral and energy developments within North-East Asia, resulting from new discoveries and/or extensions of known deposits/areas, is considered very good. Within Japan and the Republic of Korea the potential is considerably less, however, the recent discovery of the Hishikari gold mine in Japan indicates that a limited number of deposits will continue to be found even in these maturely explored and developed areas. Nevertheless, as stated previously, the mineral and energy potential of Mongolia and the Russian Far East appears to be greater than for other areas of North-East Asia and to a large extent regional economic development will be at least in part, be linked to the pace of exploration and development of mineral and energy resources in these two areas.

In the following brief review of known and potential mineral and energy resources of North-East Asia, particular emphasis is placed on Mongolia and the Russian Far East because of their anticipated larger resource potential. In addition, a major emphasis is placed on the mineral potential of the ocean and near shore areas of the Bo Hai/Yellow Sea, the Sea of Japan and the southern portion of the Sea of Okhotsk. The mineral potential of the individual countries is also briefly reviewed, including an overview of the existing minerals industry and the more important mineral resources of the region and concludes with a short analysis of major development issues.

The long term resource potential for deeper ocean oil and gas, manganese nodules, cobalt crusts and polymetallic sulfides are not included. For information on these resources and their potential, the reader is referred to Clark, et al., 1988; Clark, 1990; Clark and Li, 1991a, 1991b.

The energy potential of North-East Asia (oil and gas) is dealt with separately.

3.1 Bo Hai Sea/Yellow Sea

The mineral resource potential of the Bo Hai Sea/Yellow Sea has been discussed in detail by Clark and Li, 1991a, particularly in terms of the regions gold and heavy mineral placer potential and the following information is taken from that analysis and reproduced in part below.

The Bo Hai Sea area of northeast China is adjacent to onshore occurrences of gold lodes in Shandong and Liaoning which have made the region the largest gold-producing area in China. Additionally, the onshore area has undergone prolonged periods of erosion during which the onshore deposits have contributed an unknown, but undoubtedly large, quantity of gold to the Bo Hai Sea and surrounding environs. As a result, the Bo Hai Sea is considered to be highly prospective for the occurrence of several types of placer gold occurrences: some occurrence modes have been identified from existing occurrences, whilst other modes can be inferred from placer occurrences elsewhere in the world.

This paper focuses primarily on the placer gold potential of the Bo Hai Sea area, however, it is recognized that the potential for placer deposits of titaniferous magnetite, zircon and monazite exists within the region. As heavy minerals would basically follow the same distribution and concentration patterns as placer gold, the general conclusion, with respect to placer gold occurrences, would also apply to deposits of heavy minerals.

Because of the high potential for gold placers in the Bo Hai Sea area, the importance of gold to the economic development of China and the importance of defining specific exploration targets for gold placers, this paper provides a synthesis of existing knowledge on the geology and mineral resources of the Bo Hai Sea area (onshore and offshore) and postulates areas for future exploration in these offshore environs.

General Geology – The Bo Hai Sea, also called the Gulf of Bo Hai, is a shallow epicontinental sea, located adjacent to northeast mainland China and the west Korea peninsula, occupying an area of about 220,000 km^2. The Bo Hai Sea is bounded by the Liaodong peninsula to the north; the Shangdong peninsula and Yellow Sea to the south; the east China coast to the west and the Korea peninsula to the east (figure 43).

The Bo Hai Sea occurs as a sedimentary basin, mainly formed by Tertiary regional block-faulting, on a portion of the China-Korea Platform. Within the basin, the thickness of the Tertiary ranges from 4 to 11 km and is overlaid on a pre-Tertiary basement (Hu et al., 1989).

The land areas surrounding the Bo Hai Sea, based on geologic and geomorphic characteristics, can be divided into five major coastal geological regions:

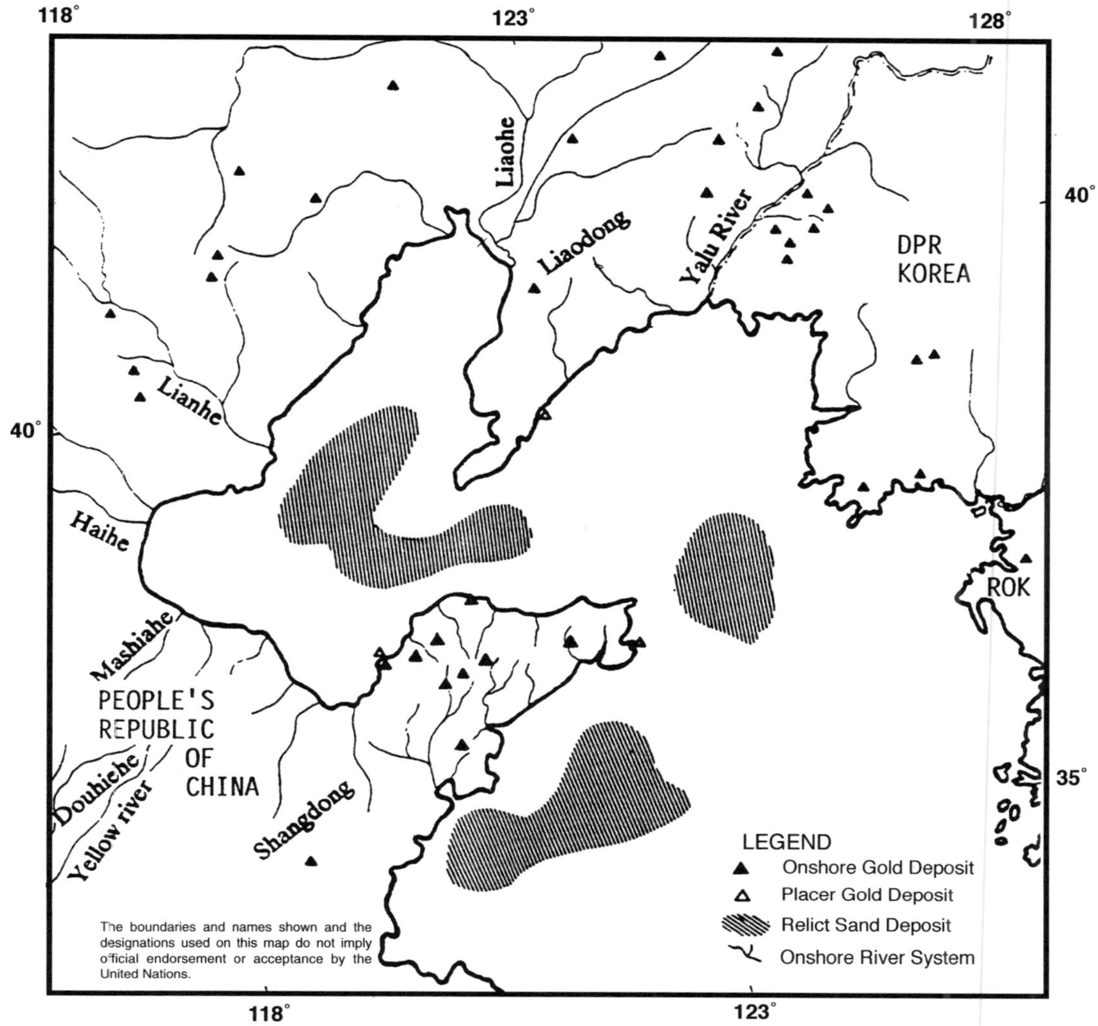

Source: Gold deposits of the People's Republic of China, 1979.

Figure 43. Distribution of the drainage systems and gold deposits in the Bo Hai Gulf and its adjacent land areas

1. The uplifted Liaodong-Korea peninsula – comprised mainly of metamorphic and igneous rocks with a high content of zircon and titaniferous magnetite, ranging 32-666 g/t and 3-217 g/t respectively (Tan and others, 1988). Additionally, most of the epithermal gold deposits found in the region occur within fracture zones of the metamorphic rocks.
2. The Liaohe Quaternary depression – located in the drainage areas of the Liao River covers an area of about 10,000 km^2 and is composed of unsolidified Quaternary sedimentary cover and an underlying Pre-Quaternary basement comprised of Tertiary oil-bearing sedimentary rocks with a thickness up to 10 km (Hu, et al., 1989).
3. The north Hebie uplift – located in the northwest of the Bo Hai Sea is represented by a NE trending low mountain range consisting of intrusive (with minor Mo), sedimentary and minor metamorphic rocks.
4. The north Shangdong Quaternary depression – situated adjacent to the western margin of the Bo Hai Sea, is filled with unconsolidated sediments of probable early Quaternary age.
5. The uplifted Shangdong peninsula region – located adjacent to the southern margin of the Bo Hai Sea, is mainly comprised of metamorphic and igneous rocks. The metamorphic rocks are considered as the major source of gold for the almost 200 epithermal gold deposits in the Shangdong peninsula.

Onshore Mineral Deposits – Onshore geological exploration for gold has discovered many epithermal lode and Quaternary placer gold deposits. As shown in figure 43, most of the major gold deposits found on land are located in the Liaodong-Korea and Shangdong peninsula areas. Although the gold deposits of the area vary greatly in terms of size and tenor they can be generally classified as belonging to the low-sulfide Au-Quartz vein type deposit, characterized by the mother-lode deposits of the United States, the Ballarat goldfield of Australia and the goldfield of Nova Scotia. A summary of the geological attributes of the major gold deposit types in the Liaodong-Korea and Shangdong peninsula areas is given in table 27.

Gold deposits within the Liaodong-Korea and Shangdong peninsula areas range in size from small single vein deposits to the Zhiao-Yi gold belt of the Shangdong peninsula which encompasses a gold producing area approximately 60 km x 10 km within which over 100 gold vein deposits occur. The gold grade of these deposits ranges from 1-52 g/t and range in Au values 2-10 g/t (Yu, 1991).

Based on this data, two conclusions seem appropriate, i.e., (1) future exploration onshore may lead to additional gold placer deposits associated with present and paleo-drainage systems, and (2) that large quantities of placer materials have been transported to the Bo Hai Sea and additional placer deposits may be discovered offshore as well.

Table 27. General attributes of Liaodong-Korea/Shangdong gold deposits

Major Deposit Types	*Mode of Occurrence*	*Major Ore Type*	Major *Guange Minerals*
Individual gold-quartz veins	Single or complex veins	Gold-polymetallic sulfide wolframites	Quartz, carbonate, sericite, chlorite
Composite gold-quartz veins	Steeply dipping	Gold-molybdenite	
Gold-quartz vein networks with disseminated ore	Cross-cut gneissoity Fracture zone control	Gold	
Zonation	*Mineralology of Gold-Polymetallic-Wolframite Ores*		
Core-gold-polymetallic sulfides-wolframite	Ore minerals	Major sulfides	Minor sulfides/oxides Sphalerite
Margin-gold polymetallic sulfides	Electrum	Pyrite	Molybdenite
Vertical-Pb-Zn-Cu increase with depth	Au/Ag tellurides	Galena	Scheelite
		Chalcopyrite	Wolframite
	Ore Characteristics	*Wall-Rock Alteration*	*Accessory/Trace Minerals*
Silicification-sericitization-pyritization-carbonatization-chloritization-k-feldspathization-albitization	Au-Ag ratio: 1:2-2:1 Gold fineness: moderate-high average	Native gold Electrum-gold/silver tellurides	Pyrrhotite Bismuthinite

Sources: Ikonnikov (1975); Wang et al. (1984) and Werner (1988).

Offshore Mineral Deposits – To date only two small placer gold deposits have been found in the offshore area of Bo Hai. The first deposit found, called the Zhinzonwan placer (figure 44) gold deposit, is located on the east coast of the Liaodong peninsula and the second, the east Liazhouwan placer gold deposit, occurs on the northwest coast of the Shangdong peninsula (figure 44). These placer deposits generally extend parallel to the coastline and encompass an area several kilometres long, several hundred metres wide and 1-3.5 m in thickness (table 27). The known gold placers occur within beach, sand bar and wave-cut terrace environments.

According to Tan and others (1988), the Zhinzonwan placer gold deposit encompasses an area of 8 km^2. The placer gold is primarily concentrated in littoral beach, ancient bay, longshore sand bars and wave-cut terrace environments. The sediments of the littoral beach and ancient bay deposits can be divided into two units, an upper mud-sand layer containing only a small quantity of heavy minerals and a lower sand-pebble layer (0.05-3.0 m) enriched in gold and zircon, magnetite, monazite and rutile. The average gold grade in the sediments is approximately 0.15 g/m^3, however, the grade may be as high as 0.38 g/m^3. Within some localized areas, such as areas surrounding reef environments, the gold placer grade may be as high as 5.2 g/m^3. The longshore sand bar deposits normally consists of four units, differentiated on the basis of different color and grain size. From top to bottom, the units are: a 0.5 m thick black sand; a 1 m thick white sand; a 0.5-5 m thick sandy pebble unit; and a 4 m pebble sand unit. The four units from top to bottom contain gold grades of 0.1-0.62, 0.09-0.25, 0.015-0.1 and 0.25-0.38 g/m^3 respectively. The wave-cut terraces situated 5-10 m below the sea level commonly have a gold grade of 0.075 g/m^3. As a whole, the grade of the placer gold tends to increase seaward. In the east Liazhouwan, the placer gold is mainly concentrated in longshore sand bars. The average gold grade in the sand bar is approximately 0.025 g/t (table 28).

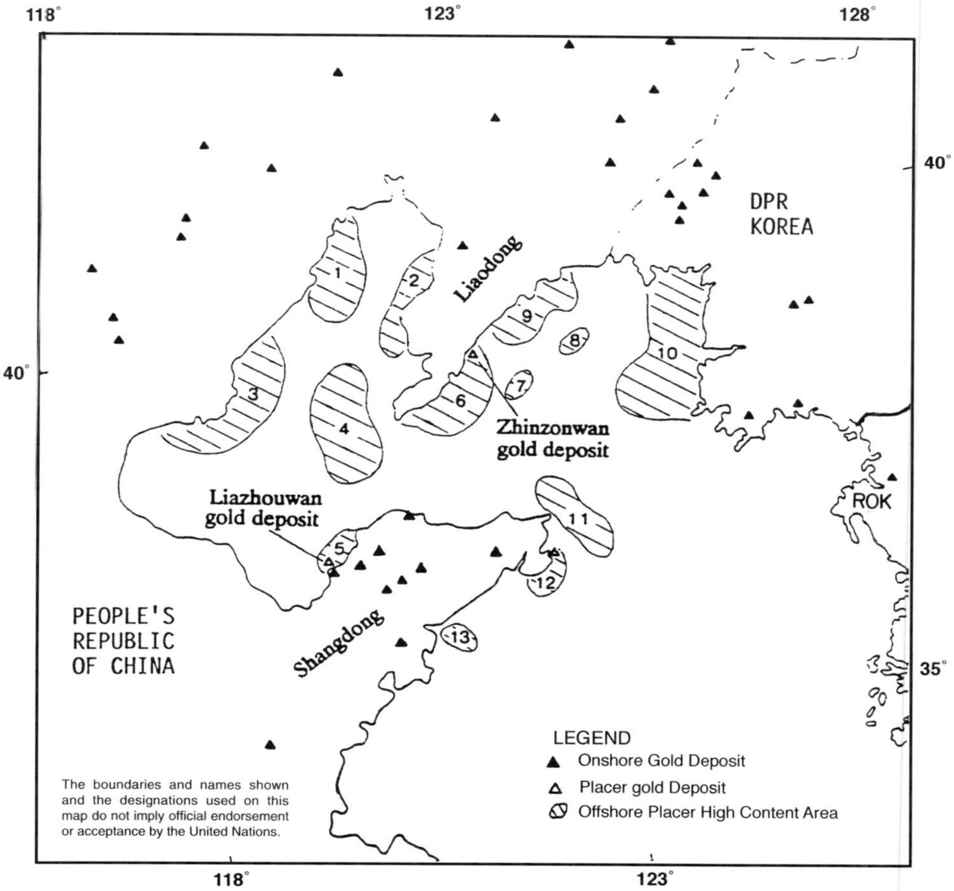

Source: Teng, 1988.

Figure 44. Sketch map showing the distribution of gold placer high-content areas in the Bo Hai Gulf

Table 28. Littoral mineral placers in the Bo Hai area

No.	*Location*	*Mineral Type*	*Concentrate Grade*	*Morphology*
1	Luguhe Mouth	Titaniferous magnetite	20-30%	Sand bar
2	East Liaohe Bay	Zircon, titaniferous magnetite	5%, 20%	Sand bar
3	Liaohe Mouth	Titaniferous magnetite, zircon	5%, 5%, 20-30%	Submerged river
4	Liaodong Shoal	Titaniferous magnetite, zircon	20-30%, 5%	Submerged beach
5	East Liazhouwan	Gold	>0.025 g/t	Beach sand bar
6	Zhiuzonwan	Gold	0.15-5.2 g/m^3	Beach sand bar
7	Bo Hai Strait	Titaniferous magnetite	>10%	Submerged terrace
8	East Bo Hai Strait	Zircon	>1%	Submerged terrace
9	Liaonan Zhanhe	Zircon	>1%	Beach
10	South Zhon Island	Garnet	>20%	Beach
11	East Haiyan Island	Titaniferous magnetite	>10%	Island sand bar
12	Zenzhanzhou	Zircon	>20%	Submerged river
13	Shi Dao	Zircon	0.0169-0.0597%	Beach
14	Dingzhiguang	Zircon	0.0628-0.2724%	River mouth

Sources: Institute of Oceanology (1964) and Bureau of Oceanic Geological Investigation (1983).
Note: The grades for No. 1-11 are measured after sampled grains are sorted in 0.1-0.05 mm size.

Besides the known placer gold deposits, the Institute of Oceanology, Academia Sinica and Bureau of Ocean Geological Investigation – Ministry of Geology, have identified at least 12 additional areas in the Bo Hai Sea which are considered prospective for further placer discoveries. The placer types, other than gold, include titaniferous magnetite, zircon, garnet and magnetite. Their characteristics and distribution are shown in table 28 and figure 44.

Mineral Development and the Environment – In the Bo Hai Sea/Yellow Sea portion of North-East Asia it is anticipated that onshore gold mining will continue and expand and that known occurrences of gold and heavy mineral deposits in the near shore areas will continue to be developed and exploited. The presence of increasing levels of onshore and offshore gold mining raises three major environmental problems which need to be addressed, both at the present time and in the future. These are: tailings discharges into the rivers and sea environment, toxic chemical discharges (associated with gold recovery) into the rivers and the sea, and mine rehabilitation and reclamation – particularly with regard to nearshore placer and mining operations. Virtually all of these problems will be exacerbated as mining moves to lower grade material and produces larger quantities of waste rock and tailings.

Nearshore mineral sand and placer gold deposits are largely undeveloped along the coastline of the Democratic People's Republic of Korea, but it is anticipated that these occurrences will be further explored and developed in the future. If this is the case the same environmental problems that occur adjacent to the People's Republic of China will also be encountered and will need to be dealt with.

The semi-enclosed nature of the Yellow Sea is further exacerbated by the even more restricted nature of the Bo Hai Sea and, therefore, particular attention should be paid to the prevention of mining related pollution being carried into either the Bo Hai or the Yellow Sea.

3.2 Sea of Japan

According to Morgan and Valencia (1992) five straits (the Korea, Shimonoseki, Tsugaru, La Perouse and Tartar) separate the waters of the Sea of Japan from the East China Sea to the south, the Sea of Okhotsk to the north-northeast and the Pacific Ocean to the east. Unlike the Bo Hai Sea/Yellow Sea area which is both shallow and relatively uniform, the seafloor of the Sea of Japan consists of numerous troughs, basins, abyssal plains, various ridges and an overall greater depth (1,000 + metres).

Unlike the Bo Hai Sea/Yellow Sea area, where there is a clear association between onshore gold deposits and heavy mineral bearing rock units which provide a variety of residual minerals for placer deposits, there has been few gold and/or heavy mineral deposits developed along the margins of the Sea of Japan. An exception to this generalization is the occurrence of three known placer deposits of titaniferous magnetite which have been reported on the southwest coast of Japan. This lack of nearshore mineral deposits is attributed to two features of the Sea of Japan which differentiate it from the Bo Hai Sea/Yellow Sea region:

- Firstly, the coastal and shelf areas of the Sea of Japan descend steeply (< 10°), in particular on the western margin of the Sea, into the sea basin. This results in there being less area and less permissive hydrology for the formation of near shore deposits.

- Secondly, the older metamorphic and igneous rocks, which are the source for the main deposits in the Bo Hai Sea/Yellow Sea, only occur nearshore in the vicinity of the Democratic People's Republic of Korea (where one ilmenite, zircon, monazite placer deposit is known to occur). Japan's coastline is composed primarily of much younger Mesozoic and Tertiary rock types which are less conducive to forming nearshore placer deposits.

As a result, it is considered unlikely that there will be any significant nearshore mineral development in the margins of the Japan Sea.

3.3 Sea of Okhotsk

Overall the Sea of Okhotsk can best be defined as a combination of the attributes of the both the Bo Hai/Yellow Sea and the Sea of Japan in that the western margin consists of a relatively shallow and gently dipping shelf (similar to the Bo Hai/Yellow Sea), whereas the central and eastern portions are characterized by steeply dipping slopes and deep basins such as the Kuril Basin in the southern portion of the Sea of Okhotsk.

In terms of known nearshore mineral occurrences, the Sea of Okhotsk, would appear to be equally deficient in nearshore placer deposits as the Sea of Japan. However, because of the large number of known mineral occurrences of onshore gold deposits, it is suspected that placer gold deposits may occur on the mainland (Sakhalin) and eastern (Kamchatka) margins of the Sea of Okhotsk. The lack of discoveries is likely attributable to both (a) a lack of exploration for placer and nearshore deposits, (b) a focus of exploration on onland deposits, and (c) difficult weather and ocean conditions.

3.4 Environmental impact of mineral development

In the case of both the Sea of Japan and the Sea of Okhotsk, any developments on the western margins of these Seas will occur in relatively pristine areas, where both the onshore and offshore biota will be extremely sensitive to any nearshore or offshore development. Potential placer mining activities in particular can, and often do, lead to the generation of large volumes of very fine beach and nearshore materials into immediate offshore areas. In both the Sea of Japan and the Sea of Okhotsk such materials could have a deleterious effect on fisheries.

More extensive operations, particularly for the recovery of gold, can also result in potentially damaging impacts from the effects of mercury and cyanide chemicals that are sometimes used in the recovery of the fine gold from placer deposits.

4.0 MONGOLIA

Mongolia has significant geologic potential for economic mineral discoveries and further development of the country's mining industry.

The Mongolia's Government is undertaking intensive work and making great efforts to attract foreign investment to the Mongolian mining industry. At present there are approximately 200 mines or deposits being worked in Mongolia. The largest mines are copper, molybdenum, fluorspar, coal and gold, from which Mongolia is a major exporter of copper, molybdenum and fluorspar concentrates (figure 45).

Mongolia is the 17th largest country of the world and lies in the northern portion of Central Asian plateau between Russia and China. Although a vast country, the nation has poorly developed infrastructure which was, and is, perhaps the main drawback to the development of the countries' mining and energy sectors. At present, Mongolia has only two highways (the Ulaanbaatar-Suhbaatar, 360 km in length, and the Ulaanbaatar-Arvaiheer, 430 km in length) and two railway lines (the Trans-Mongolian from the northern border (Suhbaatar town) to the southern border (Zamin ud), 1,089 km in length, and the eastern Mongolian from Choibalsan to Ereen, 238 km in length).

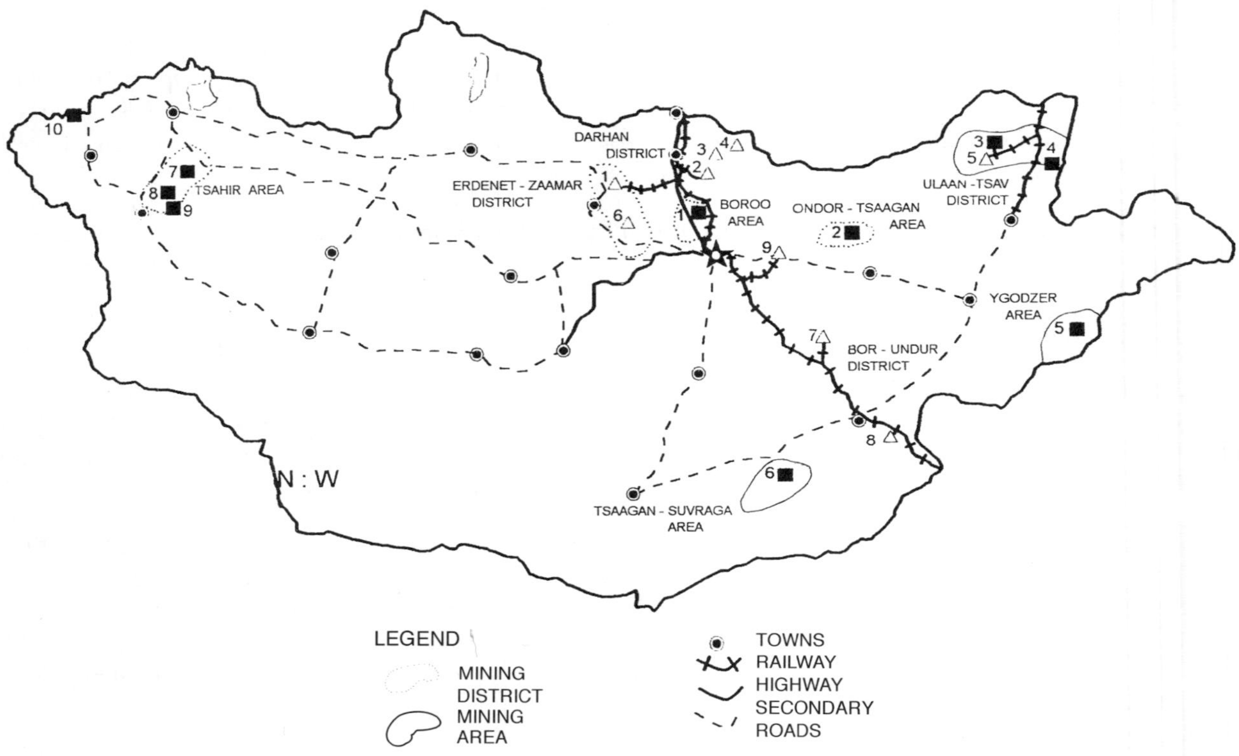

■ DEPOSIT
1. BOROO: Au
2. ONDOR-TSAGAAN: W
3. ULAAN: Pb, Zn, Ag
4. TSAV: Pb, Zn, Ag
5. YGODZER: W, Mo
6. TSAGAAN-SUVRAGA: Cu, Mo
7. ULAAN-TOLGOI: Ta, Nb, Zr, Ree
8. TSAHIR: Ta, Nb, Zr, Ree
9. HALZAN-BUREGTEI: Tz, Nb, Zr, Ree
10. ASGAT: Ag, Pb

△ MINES
1. ERDENET: Cu, Mo
2. SHAREENGOL: Coal
3. IKH-ALT: Au
4. TOLGOIT: Au
5. MARDAI: U
6. ZAAMAR: Au
7. BOR-UNDUR: Fl
8. URGEN: Fl
9. BAGA-NUUR: Coal

Figure 45. Major mining districts and areas of Mongolia

In 1992 the Mongolian government signed an agreement with China to use the Tenjin seaport (the distance between Ulaanbaatar and Tenjin is 840 miles) and also currently accesses the seaport Nakhodka in Russia (the distance from Ulaanbaatar is 1,260 miles).

4.1 Mining Industry

At present, Mongolia exports copper, molybdenum, fluorite and tin concentrates, fluorspar ore, and coal, accounting for nearly 20 per cent of the country's national income and 40 per cent of its export revenue. It produces 15 per cent of the total world supply of fluorite and, among former socialist countries, is a major producer and exporter of Cu-Mo concentrate.

The major production centres (figure 45) are: Erdenet for copper, molybdenum; Bor-Undor, Urgen, Har-Airak and Berkh for fluorite; Tolgoit, Zaamar and Bayanhongor for gold; and Baga-nuur, Shareen-gol and Aduunchuluu for Coal. The main mining complexes at Erdenet, Bor-Undur, Baga nuur, Hotol, and Tolgoit were established with Soviet aid. Two large Mongolian-Russian joint ventures, "Erdenet" (Cu-Mo) and "Mongolsovtsvetmet" (fluorite, gold) are responsible for the major production.

4.1.1 Erdenet-Zaamar mining district

The Erdenet-Zaamar mining district is centred on the Erdenetiin ovoo (Erdenet) copper-molybdenum deposit and the associated Zaamar gold field, consisting of both placer and primary gold occurrences. Erdenet, the Mongol-Russian JV "Erdenet" is the largest operating mine in Mongolia producing approximately 23,000 tonnes of 35 per cent Cu-grade concentrate per day. The reserves of Erdenet are estimated at 2.3 billion tonnes of ore, grading 0.5 per cent Cu, 0.014 per cent Mo, 1.81 g/t Ag.

Within the immediate mine area, several additional mineralized porphyry systems are known to occur, but have not yet been explored in detail.

Approximately 130 km southeast of Erdenet is the Zaamar gold district which consists of more than 15 individual gold placers and over 150 gold-quartz veins associated with Cambrian shales. Gold-placer reserves are estimated at 85 t of gold, with an average grade of 0.7-1.5 g/m^3, the placer deposits vary greatly in size and origin. The largest placer deposit contains about 52 t of gold, whereas the smallest approximately 100 kg. Some placers have extremely high grades of 8-13 g/m^3, (up to 100 g/m^3 within the enriched placers).

Studies of primary gold mineralization have identified one large Au bearing vein, with a length of 800 m, a width 1-6 m, and a down dip extension of 250 m which has an average grade of 10 g/t. In total this single vein is estimated to contain approximately 10 T of gold.

4.1.2 Bor-Undur mining district

The Bor-Undur district encompasses an area approximately 34,000 sq km and is centered on the Bor-Undur fluorite mine, which is jointly managed by Mongolia and Russia through the JV "Mongolsovtsvetmet." The Bor-Undur district contains more than 14 individual fluorite deposits of which 7 are presently being mined. The biggest deposit is Bor-Undur, containing 12.0 million tonnes of ore, grading 46.5 per cent CaF_2 and producing 130 Kt per year of fluorite concentrate. All the products (i.e. fluorite concentrate and fluorite ores) are shipped to Russia.

4.1.3 Darhan mining district

The Darhan mining district produces a variety of minerals including coal (Shareen gol open-pit) and gold (Tolgoit and Ikh-Alt placer mines both belong to JV "Mongolsovtsvetmet"). Also within the district there are several iron-skarn deposits and rare-metal occurrences. At least 10 placers are ready for mining and the total gold reserves of the district have been estimated at 20 t.

4.1.4 Ulaan-Tsav mining district

The Ulaan-Tsav mining district consists primarily of two deposits, i.e., the Ulaan and the Tsav. The Ulaan deposit has total reserves of 68.1 Mt of ore, of which 38.8 Mt of ore, grading 1.2 per cent Pb; 2.0 per cent Zn; 53 g/t Ag; and 0.21 g/t Au, is considered amenable to open-pit mining. The Tsav deposit located near an existing railway, and has reserves of 7 MT which average 6 per cent Pb; 3-4 per cent Zn; 232 g/t Ag, and 0.8 g/t Au.

Three additional unexplored areas near Tsav are: the Bayan-Uul (Pb, Zn, Au-Ag), Altantolgoi (Pb-Zn-Ag) and Salhit (Pb-Zn-Ag) which may have considerable potential. In addition to the lead-zinc deposits of the district, there are several fluorite deposits, and three uranium deposits.

4.2 Prospective areas

Boroo Area (gold) – Located approximately 150 km north of Ulaanbaatar (figure 45), the Boroo gold deposit was previously investigated by a Mongol-Germany JV. It has proven reserves of about 30 t of gold, with average grades of 4-5 g/t. Associated gold placer deposits in adjacent valleys are estimated to contain approximately 2.0 t of gold. In the area surrounding the Boroo deposit, there are several gold-quartz veins and quartz-vein systems (Sujegtei and Naran Tolgoi occurrences).

Ondor-Tsagaan Area (tungsten, molybdenum, lead, zinc) – Located approximately 270 km east of Ulaanbaatar, the Ondor-Tsagaan tungsten-molybdenum deposit has been jointly explored by Mongolia and previously by "CMEA International Geological Expedition." The deposit is a ribbon-like zone approximately 1,800 m in length, 600-800 m wide and 10-800 m in depth, with reserves estimated at approximately 186 Mt. The average ore grade is 0.17 per cent W_2O_3, 0.078 per cent Mo with associated Bi (0.08 per cent) and BeO (0.031 per cent).

Located nearby there also occurs the Mungun-Undur lead-zinc deposit with reserves of 13 Mt, containing Pb 1.26 per cent, Zn 0.97 per cent and Ag 95 g/t.

Ygodzer Area – Located in southeastern Mongolia near the China border, the Ygodzer area contains several occurrences of tungsten-molybdenum mineralization, primarily as vein and greisen type ore bodies. The largest area is the Ygodzer deposit consisting of over a dozen veins and an elongate orebody. In total, the Ygodzer mineralization encompasses more than 2.0 sq km (1,700 m x 100-120 m) with an average grade of 0.2 per cent W_2O_3 and 0.56 per cent Mo and reserves of 21.6 Mt. Numerous prospects occur throughout the area indicating a strong potential for future discoveries.

Tsagaan-Suvraga Area (copper, molybdenum) – Located in southeastern part of Mongolia in the Gobi Desert. Within the area there occur several granite porphyries, and the granosyenites are known to contain copper and molybdenum mineralization. Two ore bodies, with total reserves of 222 Mt of ore grading 0 54 Cu; 0.019 per cent Mo; 2.64 g/t Ag, and 0.39 per cent Re have been identified. Several additional copper occurrences are located adjacent to the Tsagaan-Suvraga deposit.

Tsahir Area – Located in western Mongolia between Lake Hyargas and Lake Harus, approximately 1,000 km west of Ulaanbaatar. The area was discovered in 1989-1990, and contains several rich occurrences of rare earths, zirconium, tantalum and niobium mineralization occurs in altered alkalic arfedsonite granites, nordmarkites and syenites. The occurrences are situated in large altered zones with typical zones being 300-1,200 m in length, 50-400 m in width and extending 100-250 m in depth. Reserves are estimated at 500-1,100 Mt with an average grade of 1.27 per cent Zr, 0.2-0.5 Nb, 0.01-0.1 per cent Ta and 0.08-0.5 per cent Y. The largest areas are Halzan-Buregtei, Tsahir and Ulaan Tolgoi.

4.3 Gold

Known gold mineralization in Mongolia are represented by three economic types: veins, mineralized zones and placers. Although vein occurrences are the most widely developed of all gold occurrences, the bulk of proven gold reserves is associated with placers and mineralized zones. Table 29 provides a summary of the gold occurrences in Mongolia. On the basis of geological data (tectonic structure, mineral occurrences, age of

Table 29. The characteristics of mineral types of gold deposits

Type	*Mineral type, genetic type of placers*	*Morphology, Parameters*	*Mineralogy/alterations*	*Content g/t, g/m³*	**Reserves*	***Age of Mineralization*
Vein	Gold-sulfide-quartz	Irregular veins, subparallel series of veins. Length 200-2,200 m, thickness 1-5 m, downdip 50-400 m	Native gold, silver, chalcopyrite, covellite, pyrite, galena, sphalerite, molybdenite, argentite. Beresitization, K-feldspatization pyritization	Au: 4-45 g/t Ag: 30-90 g/t Cu: 0.8-1.5%	Au: 1-30 t Ag: 2-25 t	PZ-MZ
	Gold-quartz	Irregular veins, series of veins. Length 200-1,000 m, thickness 1-5 m, downdip 100-350 m	Native gold, silver, pyrite, arsenopyrite/Beresitization, albitization	Au: 2-30 g/t Ag: 5-15 g/t	Au: 1-15 t Ag: 1-10 t	PZ_2-MZ_1
Minera-lized zone	Gold-Sulfide-Quartz	Horizontal Lode-like metasomatic zones. Length 20-70 m, thickness 401- m, downdip 150-300 m	Native gold, arsenopyrite, pyrite, sphalerite, chalcopyrite, galena, argentite, native silver/ Chloritization, albitization, beresitization	Au: 4-6 g/t	Au: 10-30 t	PZ_3-MZ_1
Placers	Alluvial stream, terrace. Buried.	Ribbon-like placers. Length 1-15 m, width 100-850 m, thickness 1-2.5 m	Panned gold	Au: 0.8-14 g/m³	Au: 1-50 t	N-P

**Age subscription: 1 = early, 2 = middle, 3 = late

mineralization and gold content) ten main gold metallogenic provinces (table 30) within Mongolia have been defined. However, it should be emphasized that because of a scarcity of geological knowledge concerning the western and southern portion of Mongolia, and an emphasis on only placers and/or quartz vein mineralization in the areas studied, the true gold potential of these regions is largely conjectural. Nevertheless, the number of occurrences suggests a high potential for further gold discoveries in the region.

Additionally, geologic data indicates that occurrences of (a) gold-bearing conglomerates of Cretaceous and Tertiary age can be expected in southern and western Mongolia, (b) gold mineralization associated with ancient volcano-sedimentary areas in Central Mongolia and (c) ancient epithermal gold occurrences (island arc associations) may be expected in the Kharkhira, Ulzeet, and southern Mongolian gold provinces.

4.4 Base metals (Cu, Mo, Pb, Zn)

Copper-molybdenum (Cu-Mo) occurrences in Mongolia are predominantly of three deposit types, i.e., late Paleozoic-early Mesozoic Cu-Mo porphyries, Cu-Ni magmatic segregations associated with grabbo of unknown age and Paleozoic-Mesozoic stratabound Cu. The basic characteristics of each deposit type are given in table 31.

The Cu-Mo deposits occur within four main base-metal provinces in Mongolia, i.e., the Altai, north Mongolian, central Mongolian and south Mongolian provinces.

Mongolia also has a large number of middle to late Mesozoic Pb-Zn deposits of various deposit types, largely in the eastern part of the country (central Mongolia and Altai Province). The predominant deposit types are: skarn, mineralized explosive pipes, vein, and mineralized zones. Their characteristics are described in table 32. The known base metal deposits were mainly formed in the middle-late Mesozoic are located in the eastern part of country.

The northern, central and southern metallogenic provinces of Mongolia have a high potential for the discovery of new Cu, Mo, Pb, and Zn deposits. In the north and south Mongolian provinces, there is greater potential for Cu-Mo porphyries, and in central Mongolia there is potential for Pb, Zn and Cu-Mo porphyry deposits.

Potential discovery of large scale copper and/or polymetallic mineralization in the Altai province in south Mongolia is high, where there is an intersection of three regional geologic structures (Altai metallogenic province, Beitianshan tectonostratigraphic terrane and a complex zone of faults).

Regionally, the Altai metallogenic province includes areas of Russia (Gornii Altai), China (Altai), and Mongolia (Mongolian Altai). In the Gornii Altai of Russia large deposits of lead and zinc have been discovered, whereas on the Russia-Mongolian border area, hydrothermal Pb-Zn deposits, gabbro hosted Cu-Ni deposits and volcano-clastic Fe mineralization have been identified.

In the Altai zone of China, in mid-Paleozoic sandstones and shales, volcanoclastic deposits of synorogenic-synvolcanic massive Cu-Ni sulfides occur. Estimates by Clark, et al. (1989) indicate that at least two major synorogenic massive sulfide deposits, in addition to Kuroko-type massive sulfide deposits will be discovered. Based on the same criteria as used for the China Altai region, it is estimated that similar deposits may occur in the adjacent Mongolian Altai.

4.5 Fluorite

Fluorite mineralization is widespread throughout Mongolia, occuring in rocks of Late Paleozoic to Late Mesozoic age. Economic mineralization occurred during late Mesozoic/late Jurassic-early Cretaceous epochs. The fluorite mineralization occurs as two deposit types: epithermal vein and metasomatic ore bodies.

Late Mesozoic fluorite mineralization is associated with rare earth and lead-zinc mineralization but not in economical quantities. Three major fluorite provinces are defined in Mongolia, i.e., the north Mongolian, trans-Mongolian and south Mongolian with the trans-Mongolian province having the largest reserves and being the most actively exploited.

Table 30. The characteristics of gold metallogenic provinces

No.	*Zone*	*Parameters, Location*	*Ore Hosted Formations, age*	*Mineral type*
1	Mongol-Altai (is SE part of Talitsko-Mongol-Altain zone, which involves territory of Russia, China, Mongolia)	Length-450-470 km; width: 30-6 km Southeast direction. In Western part of Mongolia Mongolian Altai Range	Late Devonian-late Paleozoic granites, intruding Paleozoic (E_1-0) volcano-terrigenous rocks	Placers. Veins,
2	Kharkhira (united with Mongol-Altai zone)	Length: 900 km Width: 60-100 km southeast direction from L. Achit up to south edge of Mongolian Altai	E-O terrigenous, O-S volcano terrigenous deposits, O-S granites	Veins (gold-sulfide-quartz)
3	Djida	Length: 150 km Width: 30-150 km Northeast trending. Northern Mongolia, east from Lake Hobsgol	V-E, volcano-terrigenous metamorphized rocks, PZ_{2-1} granitoids, P-T effusive rocks	Veins (gold-sulfide-quartz, gold-quartz)
4	North-Hentii (main gold district)	Northeast trending in northern Mongolia. Length: 400 km Width: 50-750 km.	R-E_1 gneisses, sandstones, PZ_1-shales, sandstones, D, P, T and J-granitoids. Cretaceous conglomerates	Veins (gold-sulfide-quartz, gold-quartz), mineralized zones, placers
5	Onon-Ulz	Length: 460 km Width: 50-150 km Northeast trending in the interfluve of rivers Onon and Ulz	R_{1-2}, V-E_1 metamorphic shales, schists, sandstones, PZ_{2-3} effusive rocks, PZ_3-MZ intrusive rocks	Veins (gold-sulfide-quatz, gold-quartz), placers
6	Dundgovi	Length: 560-600 km; Width 30-100 km. Northeast trending in eastern Mongolia	PZ_1-granitoids, PZ_{2-3}, MZ volcanic rocks,	Veins (gold-quartz) rarely R_{1-2} green shales black shales
7	Southwest Hentii	Length: 200 km; width 50-90 km northeast trending	Riphean PZ_{2-3} shales, sandstones, MZ-volcanics, P, T, J granitoids, alkalic granites	Veins (gold-quartz, gold-sulfide-quartz)
8	Bayanhongor	Length 350 km; width 60-90 km northwest trending in the southern spurs of Hangai mountains	Precambrian gneiss, migmatite, PZ_1-shales, schist, ophiolite, granite	Veins (gold-sulfide-quartz, gold-quartz), placers
9	Ulzeet	Length 800 km; width 50-100 km. Direction strike in south Mongolia	PZ_1-shales, sandstones, mudstones, PZ_{2-3} diabase, gabbro, granites	Veins (gold-quartz, gold-quartz-sulfide), stockwork type
10	South Mongolian	Length 600 km; width 90-120 km, sublatitude direction south part of Gobi desert)	PZ_1-shales, sandstones, P_3, C_1 intrusive rocks, alkalic granites, syenites, ophiolotes	Veins (gold-quartz), placers

Table 31. The characteristics of base metal metallogenic provinces

No.	*Zone*	*Parameters, Location*	*Host rocks (age)*	*Type*
1	Altai Regional basemetal belt involving areas of Russia, China, Mongolia	Length 1,300 km Width 60-200 km direction northwest Whole Mongol-Altai range	Various volcano-sedimentary, metamorphic, effusive rocks $AR\text{-}PR_1\text{-}K_1$. Different age granitoids, gabbro, alkalic rocks, ophiolites.	Disseminated porphyry stratabound, magmatic segregation Cu-Ni with gabbroids, Pb-Zn stratabound, sakrn
2	North Mongolia (includes the Erdentinovoo deposit)	Length 680 km Width 150-250 km Direction ENE	P_1-trachyandesite, trachybasalt, dacite, basalt-rhyolite association. $P_{1\text{-}2}$ granitoids, porphyry granodiorites, granites	Porphyry
3	Central Mongolian	Length 1,500 km Width 60-300 km. Direction north-east from Bayanhongor town to north-east boundary	PZ_3-MZ dacite, rhyolites, andesite, basalts, PZ_2-granites, loeucogranites.	Disseminated porphyry, Cu-Mo, skarn, vein, volcanic Pb-Zn
4	South Mongolian	Length 650 km Width 60-200 km Direction WE Southeast portion of Gobi desert	PZ_2 volcano-terrigenous rocks, $P_{2\text{-}3}$ sublakalic granitoid, syenites, granosyenites, porphyry-granites	Porphyry, Cu-Mo, Pb-Zn skarn

Table 32. The characteristics of mineral types of lead-zinc deposits

Type	*Mineral type*	*Morphology/ Parameters*	*Minerals, alteration*	*Content*	*Reserves*	*Age of mineralization*
Skarn	Magnetite-sphalerite sulfide	Lode and lenses, gentle to steeply dipping. Length 50-200 m Width 10-50 m Downdip 50-450 m	Sphalerite, galena, magnetite, chalcopyrite, pyroxene/ skarnization	Zn 10-15% Fe 23.0%	Zn 0.9 MT Fe 1.5-2 m	MZ_2
Mineralized breccia pipes	Sphalerite-sulfide fluorite	Pipelike bodies. Length 100-150 m Width 40-60 m Downdip 900 m	Sphalerite, galena, chalcopyrite, pyrite, fluorite, siderite, epidote/ epidotization, amphibolitization	Zn 1.5% Pb 1.25% Ag 46 2/t Au 0.2 g/t	Zn 0.74 mln t Pb 0.42 mln t Ag 2.0 t Au 7 t	K_1
Vein	Galena-sulfide	En-echelon steep veins. Length 400-3,000 m Width 7-3.0 m Downdip 600 m and more	Ganelna, sphalerite, chalcopyrite, pyrite, siderite. Argilliation, silicification	Pg 6-12 % Zn 3-4% Ag 200-400 g/t Au 0.8-1 g/t	Pb 60 ths t Zn 31 ths t Ag 218 t Ag 1 t	K_1
Mineralized zones	Galena-sphalerite sericite	Steeply dipping mineralized zones. Length 200-1,300 m Width 1-70 m Downdip 300 m	Galena sphalerite, pyrite, chalcopyrite/ sericitization, choliritization	Pb 1-1.5% Zn 1% Ag 100-110 g/t	Pb 0.1 t Zn 0.08 t Ag 1,000 t	MZ_1

4.6 Rare-earth metals (REM)

Genetically and spatially REMs (table 33) in Mongolia are primarily associated with felsic and alkalic rocks of Late Mesozoic, Early Mesozoic and Late Paleozoic age, with mineralization occurring as (a) stockwork mineralization in granitoids, (b) vein and skarn mineralization in granites, and (c) on the contact of granites and carbonate rocks.

Within Mongolia, six REM provinces have been identified; the Altai, north Mongolia, Hentii, Hangai, southeast Mongolia, and south Mongolia.

Table 33. The characteristics of REE and REM metallogenic provinces

No.	*Zone*	*Parameters/location*	*Age and Host Rocks*	*Type*
1	Altai	Length 600-km, width 100-300 km. In western Mongolia. Mongolian Altai range and Great Depression Lakes	Ordovician and Devonian granites, granodiorites, granosyenites, Li-F leucogranites	Ta, Nb, Zr with alkalic granites. W, Sn vein
2	North-Mongolian (in adjoining area of Russia big deposits of REM)	Length 840 km, width 60-300 km, Northern Mongolia. Hobsgol mountains	Cambrian, Ordovician and Devonian granites, granodiorites, alkalic granitoids	Ta, Nb with alkalic granites. W, Sn stockwork
3	Hentii	Length 800 km, width 200-360 km, Hentii mountain, basins of Herlen, Ulz rivers	Mid-Late Mesozoic granites, granodiorites, rhyolites, dacites. Mid-Paleozoic granites, Li-F leucogranites	W, Sn-stockwork, W, Sn-veins Sn-placers Sn-skarn
4	Hangai	Length 540-km Width 120 km Hangai mountain	Mid-Paleozoic granites, granodiorites, alkalic granites, rhyolites, dacites	W, Sn-veins S, Wn-stockworks
5	Southeast Mongolian	Length 1,000 km Width 60-120 km	Triassic, Jurassic granites, rhyolites, Devonian alkalic granites, granosyenites	W, Mo greisen bodies, veins. REE with volcano – plutonic complexes
6	South Mongolian	Length 540 km Width 100-180 km In Southern Mongolia, surround city Dalanzadgad	Mostly Late Paleozoic granites, alkalic granites. Late-Mesozoic trachytes, rhyolites, dacites	Ta, Nb-with alkalic granites. REE with volcano-plutonic complexes

4.7 Coal

Tavan-Tolgoi – The Tavan-Tolgoi coal deposit (figure 45) is unique within the country both in terms of its size, covering 90 sq km within the Ulannuur coal-bearing depression, and the high quality coal. The deposit is located 560 km south-southeast of Ulaanbaatar and 90 km east of Dalanzadgad, the administrative centre of the Omnogov aimak. The Late Permian coal-bearing sequence is 965 m thick in the central part of the syncline and occurs within units of intercalated sandstone, siltstone, claystone, and conglomerates. Overall, the coal-bearing sequence includes 16 coal seams ranging in thickness from 2 to 72 m. The aggregate thickness of coal beds, in the central part of the structure, reaches 163 m, with the maximum depth of the base of the lowermost seams occurring at 945 m.

Chemical analyses made by the Eastern Coal-Chemical Institute (Sverdlovsk, Russia) indicate that coals from the Tavan-Tolgoi deposits is similar in quality to the main coal types exported by the U.S., Canada and Australia.

Coal resources in the Ulannuur basin, including the Tavan-Tolgoi, Ukha-Khudag, Bortolgoi and eastern areas (to the maximum coal seam depth of 940 m) are estimated to be 7.0 Bt, of which 2.8 Bt is amenable to open pit mining to a depth of 340 m, and which includes 4.0 Bt of coking coals of varying quality.

Baga-Nuur – The Baga-nuur deposit is located 110 km east of Ulaanbaatar in the central part of the Tabunsubatuin depression and has been operating since 1978. The 480-500 m thick coal-bearing sequence, comprised of intercalated sandstones, claystones and siltstones, occur within a 5 km wide and 14 km long depression. Two coal-bearing horizons, containing 24 coal seams and interbeds are recognized within the coal-bearing sequence and contain approximately 600 Mt of coal amenable to open-pit mining.

5.0 RUSSIAN FAR EAST

5.1 Introduction

The mineral resource sector of the former Soviet Union was the world's largest with an annual commodity production value of US$180,000 million in 1991. Not only did the former Soviet Union have the world's largest minerals industry, it also ranked number one in the world in terms of proven reserves of barite, cobalt, fluorspar, iron ore, nickel, tin, titanium, tungsten and in brown coal, gas and oil.

This vast mineral potential includes over 18,000 individual deposits which have been recorded in the State Balance of National Wealth records. Approximately 5,000 of the recorded deposits are under industrial exploitation, 4,700 are reserved for future mining and nearly 9,000 are planned to be mined in the near future. According to Borisovich (1991), the proportion of proven reserves set aside for future mining is relatively large and for certain minerals, it may comprise as much as 25-40 per cent of total proven reserves. The State Balance also includes a large number of deposits not scheduled for mining until after 2015. These deposits include 30 per cent of proven reserves of iron ore and copper, 14 per cent of lead-zinc, 39 per cent of bauxite, 24 per cent of titanium, 32 per cent of phosphate, 16 per cent of fluorspar and 45 per cent of brown coal and are mostly located in the Russian Far East.

The Russian Far East also enjoys a comparative advantage in that over 120 different types of metallic, non-metallic and energy resource occurrences, and is endowed with large reserves of antimony, boron, diamonds, gold, platinum, silver, tin, and tungsten. In particular, the area contains the largest reserves of gold and diamonds, which presents opportunities for the region to attract foreign investment.

5.2 Geology and mineral resources

The Russian Far East constitutes an area roughly the size of the United States, comprised of the Siberian Platform on the west, complex fold belts to the north and south, volcanogenic zones to the east as well as the shelf areas of the Sea of Okhotsk and the Pacific and Arctic Oceans (figure 46). Any discussion of the detailed geology of the Russian Far East and its mineral deposits is beyond this paper. However, the discussion will focus on the general characteristics of the major geologic zones of the Russian Far East and their associated mineral deposits.

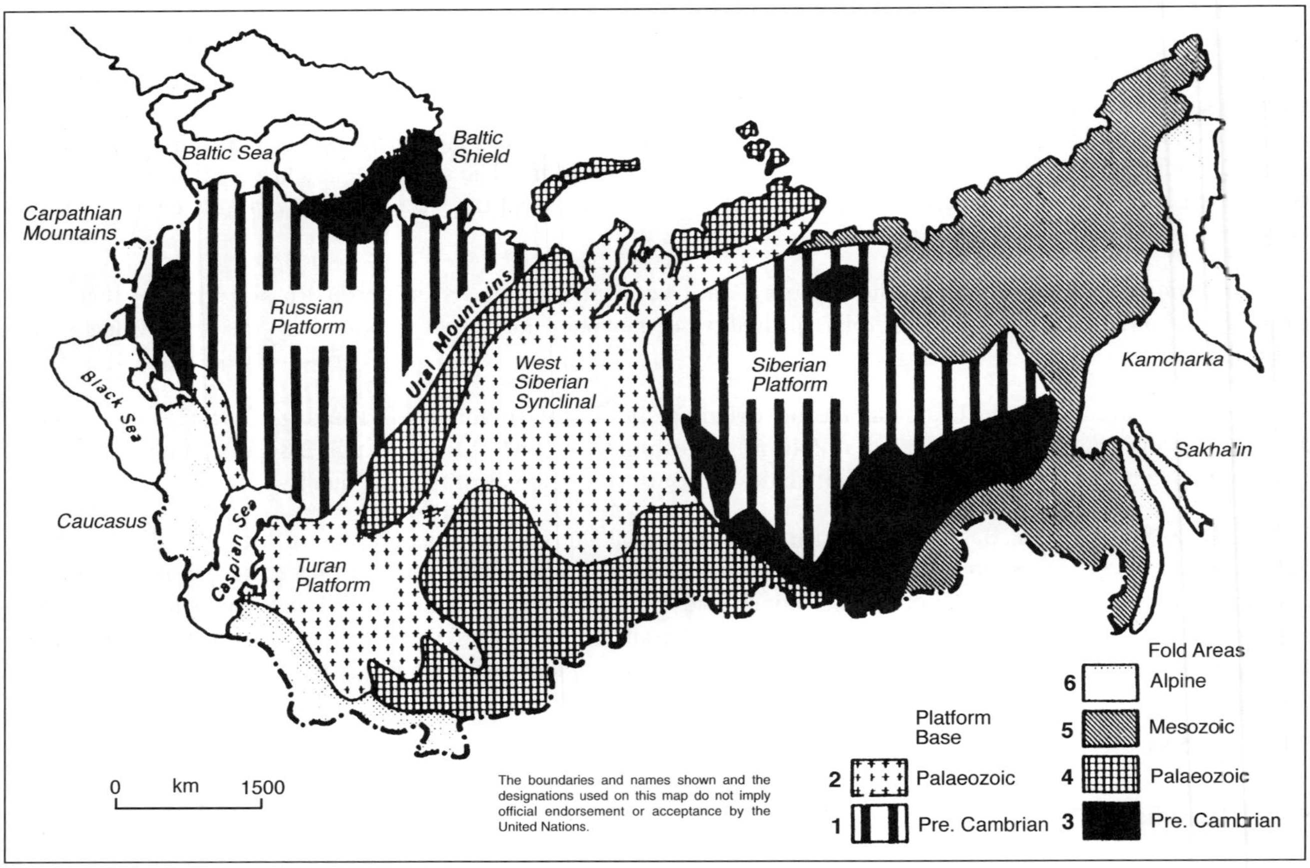

Figure 46. Major geological provinces of the Russian Far East and adjacent areas

Siberian Platform – The Siberian Platform is one of the world's most extensive platforms formed upon a basement of gneisses and crystalline schists of early Precambrian (Archean) age, with elevated massif areas separating large depressions. The large depressions serve as the main basins for rivers developed on the Platform. The main structural elements defined by the massifs and depressions are the elevated Aldan Shield and Lena Yenisei platform area, the eastern portion of which forms the Viljui depression (syncline) and the southern portion of the Angora-Lena depression. The Angora-Lena depression is in turn separated from the Yenisei depression, located on the western margin of the platform, by the Pelelui uplift. These sequences of massifs and depressions and their associated rivers are hosts for several occurrences of the richest and largest gold placers in the Far East. Mineral deposits associated with the Siberian Platform include: the world famous Norlisk Cu-Ni-PGE deposits; extensive siderite iron ores; copper-bearing red bed sandstone deposits; and extensive gold and platinum placer deposits associated with the massif and "trap" areas respectively; diamond placer deposits associated with kimberlite pipes; and bauxite deposits occurring both as lateritic deposits and as metamorphosed diasporic deposits intercalated with Lower Cambrian algal reef limestones.

Additional mineral occurrences include deposits of mica, phlogophite, graphite and rock salt.

Mongolian-Okhotsk Fold Area – Adjacent and south-southeast of the Siberian Platform occurs the Mongolian-Okhotsk Fold area which is inferred to be separated from the main Siberian Platform area by a major geosuture zone. The Mongolian-Okhotsk fold area is in fact composed of three individual fold zones, i.e., the eastern-Zabail al, Upper Amur and Priokhotsk zones. The Mongolian-Okhotsk fold area in the Far East is of particular importance. It is the northernmost extension of the Angora geosyncline area which extends into southern Kazakhstan where it is the host province for the world famous Dzhezkazgan porphyry copper deposit.

Within the Russian Far East, the Mongolian-Okhotsk fold area is known for occurrences of tin, tungsten, molybdenum, arsenic, antimony and diverse polymetallic (Cu-Pb-Zn) deposits. Additionally, a broad zone of placer and smaller lode gold deposits are concentrated just south of the area in association with the Archean and Lower Proterozoic gneisses in the Bureja massif. Large Jasperlite deposits of iron also occur in association with the Riphean strata of the area, but to date are not considered economic.

Pacific Fold Belt – The Pacific fold belt lies east of the Siberian Platform and although it contains rock units of all geologic eras, the most characteristic feature of the region is the predominance of Mesozoic and Cenozoic rock units of diverse origin, including marine limestone, shale and sandstone sediments; terrestrial coal-bearing units; extensive extrusive rocks (largely quartz porphyries) and complex sequences of volcanic rocks.

The major mineral deposits associated with the Pacific Fold Belt are lode and placer gold deposits associated with both Cretaceous and Neogene granitic porphyries; tin-bearing granites and placers; polymetallic lead, zinc, copper and silver vein and disseminated deposits; molybdenum-tungsten greissen and disseminated deposits; as well as minor deposits of antimony, mercury, arsenic, jasperlite iron deposits and sedimentary manganese deposits. Graphite, fluorite, and native sulphur (associated with extinct volcanoes in Kamhuatka) also occur.

5.3 Mineral deposits of the Far East

The mineral deposits of the Far East are numerous and diverse and given that the area has only been prospected by reconnaissance methods, the number and types of deposits discovered in future is expected to increase.

Geological-Structural-Metallogenic Zones – Although the geology and structure of the Far East is rather easy to classify in terms of broad geologic areas, the details of the region's geology, structure and mineral deposits is both complex and in many instances less well understood. Nevertheless, an evaluation of the geologic and structural divisions (figure 47) and the metallogenic zones (figure 48) of the Far East, both based on the works of Kunaev (1984), demonstrate a clear association between individual metallogenic zones, individual geological-structural zones and specific types of mineral deposits. Based on these associations, Clark and Sekisov (1993) evaluated the present and future development potential of individual deposit types, both for the present and the future (2000) for the Russian Far East, this evaluation is given in table 34.

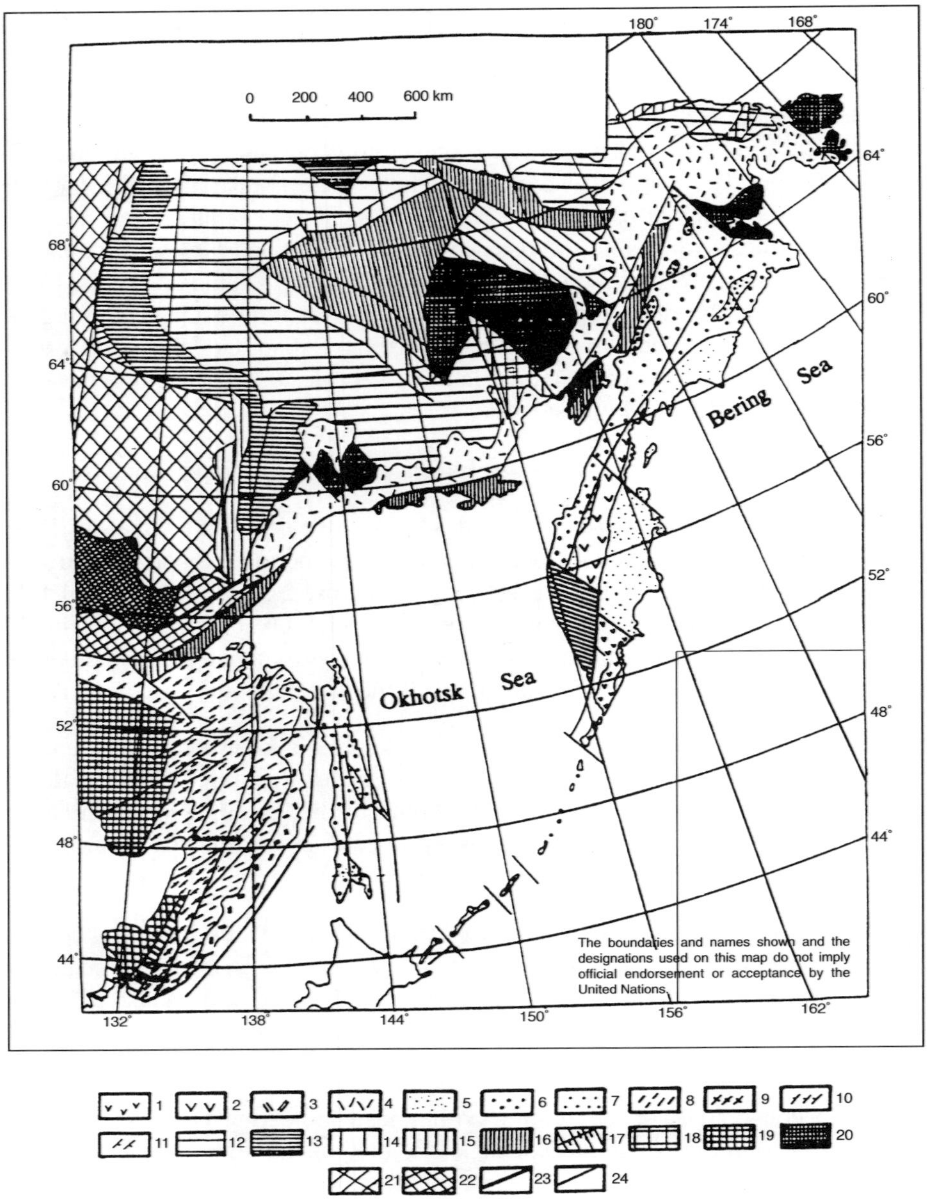

Legend for Figure 47

Geological and Structural Zones of the Russian Far East (After Kunaev, I.V., 1984)

No. 1-4	Volcanogenic belts representing marginal oceanic "Island Arc Andesites" (1) and marginal continental "rhyolite rich" sequences (2), (3), (4). Ages K_1-N
No. 5-7	Geosynchinal folded systems an zones consisting of volcanogenic and silicious-terrigenous complexes with zones of ophiolite complexes. Ages J_3-N
No. 8-10	Geosynclinal volcanogenic and silicious-terrigenous complexes without ophiolites. Ages PZ_2-K
No. 11-12	Geosynclinal terrigenous complexes. Ages PZ-J_3
No. 13-14	Pericratonal aulocogenic terrigenous-silicious complexes. Ages PR_3-PZ
No. 15	Undivided geosynclinal complexes, volcanogenic-silicious-terrigenous and volcanogenic-terrigenous complexes. Age PZ-MZ
No. 16	Massif complexes near geosynclinal basins and foredeeps. Age MZ-PZ
No. 17-19	Basement and cover complexes. Age PZ-PZ_2 and PR_2, respectively.
No. 20	Platform of terrigenous-silicious complexes. Age PR_3-PZ_2
No. 21-22	Basement complexes. Age AR_2
No. 23-24	Faults.

Source: Kunaev, 1984.

Figure 47. Geological and structural zones of the Russian Far East

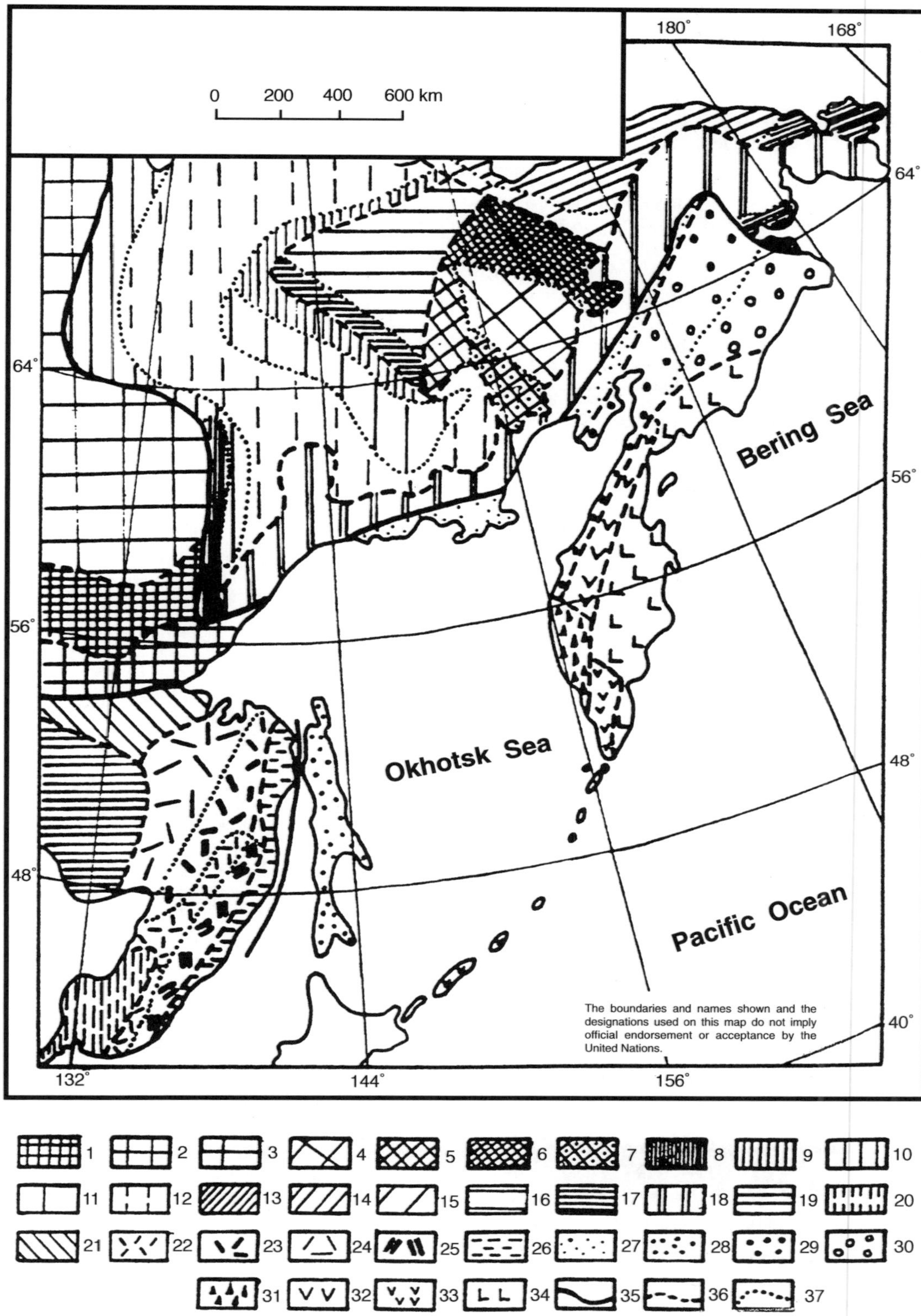

Source: Kunaev, 1984.

Figure 48. Metallogenic regions, provinces and sub-provinces of the Russian Far East

Legend for Figure 48

		Geologic-Structural Unit[2]	
No.	*Metallogenic Regions/Provinces/Subprovinces*[1]	*Unit No.*	*Age*
	Eastern Siberian Region:		
1.	Aldano-Amginshaya Province: F-AR; PK-PR_3; Au(Mo, Pb, Zn)-J_3-K_1;	20	PR_3-PZ_2
2.	Stanovaya Province: Au, Ti, Fe-AR_2; PK-PR_3; Mo; Pb; Zn-J_3-K_1;	22	AR_1
	Au, Ag, Cu, (Pb, Zn, Sb)-K_{1-2};		
3.	Leno-Viljuiskaya Province	22	AR_2
	Verkhoyano-Chukotskaya Region:		
4.	Omolonshaya Subprovince: Fe-AR, Au, Ag, (Cu)-D_{2-3};	20	PR_3-PZ_2
5.	Prikolymskaya Subprovince: Cu, Pb, Zn, Fe-PR_3-PZ_2;	20	PR_3-PZ_2
6.	Omolonskaya Subprovince: Cu, Mo, Au, Ag (Pb, Zn)-K;	17	up to KZ
7.	Sugoyskaya Subprovince: Mo, Pb, Zn-K_2;	18	up to PZ_2
	Yano-Kolymskaya Province:		
8.	Sette-Dabanskkaya Subprovince: PK-PRc; Pb, Zn, Cu, Fe-PR_3-PZ_1;	15	PZ-MZ
9.	Omulevo-Polousnenskaya Subprovince: Cu, Pb, Zn, Sb, Hg-PZ_2; Mo, Au-J_3-K_1;	14	PR_3-PZ
10.	Verkhoyanskaya Subprovince: Au-J_2-K_1;	13	PZ_2
11.	Indigiro-Kolymskaya Subprovince: Au, Sb, (W, Mo)-J_3-K_1;	12	PZ_3
12.	Yano-Balygychanskaya Subprovince: Sn, W, Pb, Zn, (Sb)-K_{1-2}; Hg(Sb)-K_2-P_1;	12	PZ_3
	Chukotskaya Province:		
13.	Chukotsko-Anadyrskaya Subprovince: Au-J_3-K_1 ( );	20	PR_3-PZ_2
14.	Anjuisko-Chukotskaya Subprovince: Au(Sb)-K_1; Sn, W-K_2; Hg(Sb)-K_2-P_1;	12	PZ_3
15.	Yuzhno-Anjuiskaya Subprovince:	16	PZ_3-MZ
	Alazeyskaya Province:		
16.	Alazeyskaya Subprovince:	16	PZ_3-MZ
17.	Ilyn-tasskaya Subprovince: Pb, Zn, Ag	16	PZ_3-MZ
18.	**Okhotsko-Chukotskaya Province**: Au, Ag, Cu, Hg, (Pb, Zn, Sb)-K_2-P_1	4	K_1-R_1
	Amurskaya Region:		
	Hank-Burejinskaya Privince:		
19.	Burejinskaya Subprovince: Fe, Bp-PZ_2; Mo-PZ_3; Sn, Sb-K_2; Hg-P	19	up to PR_2
20.	Hankajskaya Subprovince: Sn, -PZ;	19	up to PR_2
21.	**Amursko-Okhotskaya Province**: Mn, Fe-Pz_2; Au(W)-PZ_3, MZ_1; Hg-P;	15, 9, 10	PZ-MZ
	Sikhote-Alynskaya Province:		
22.	Zentralno-Sikhote-Alynskaya Subprovince: W(Au)-K_2;	8	K
23.	Nizhne-Amurskaya Subprovince: Au(W, Mo, Sb)-K_2;	8	K
24.	Priamurskaya Subprovince: Sn(W-Cu)-K_2; Hg-P;	9	MZ_1-Z
	Vostochno-Sikhote-Alynskaya Subprovince: Sn, Pb, Zn-K_2P;	8	K
25.	**Pribrezhnaya Province**: Au, Ag, Hg, Sb, Au-P;	3	K_2-P
	Near the Pacific region:		
26.	**Taigonossko-Murgalskaya Province**: Mo-K_1, Au, Ag, Cu-K_2;	16	PZ_3-MZ
27.	**Sakhalinshaya Province**: Cr-K_1; Hg(Sb, W)-N;	6	J_3-K_3
	Zapadno-Kamchatsko-Koryakskaya Province:		
28.	Zapadno-Koryakskaya Subprovince: Cr-K_1; Sn(Ag)-P_3-N_1; Hg-N_1;	6	J_3-K_2
29.	Khatyrskay Subprovince: Cr-K_1; Hg, W-N;	6	J_3-K_2
30.	**Zentralno-Okhotskaya Province**: Cu, Ni-K_2; Cu(Mo)-N;	2	N
	Kurilo-Kamchatskaya Province:		
31.	Zenralno-Kamchatskaya Subprovince: Au, Ag, Pb, Zn, Hg, C-N;	2	N
32.	Kurilo-Yuzhno-Kamchatskaya Subprovince: Au, Ag, Cu, Pb, Zn, Hg, C-Q;	1	N-Q
33.	**Vostochno-Kamchatskaya Subprovince**: Cr-K_2; Hg-N;	5	K_2-P

1 Metallogenic regions/Provinces/Subprovinces of the Far East (according to Kunaev, I.V., 1984; Conventional signs to V. Ya Shesternev) index numbers key to map patterns and zones shown in figure 48.

2 Gelogical-Structural Units of the Far East (according to Kunaev, I.V., 1984; conventional signs to V. Ya Shesternev) unit numbers and associated age designations key to map patterns and zones shown in figure 47.

Table 34. Far East mineral deposit types and estimated present and future development potential

	Present Development Potential			*Future Development Potential*		
	High	*Medium*	*Low*	*High*	*Medium*	*Low*
Noril'sk Cu-Ni-PGE	X			X		
Alaskan PGE	X			X		
Placer PGE-Au	X			X		
Diamond Pipes	X			X		
Low-Sulfide Au-Quartz Veins	X			X		
Porphyry Cu		X		X		
Cu Skarn Deposits			X		X	
Porphyry Cu-Mo		X		X		
Sediment-Hosted Cu	X			X		
Volcanic-Hosted Cu-As-Sb			X			X
Algoma Fe			X		X	
Superior Fe			X			X
Fe Skarn Deposits			X		X	
Sn Veins		X			X	
Sn Skarn Deposits		X				X
Tungsten Skarn Deposits		X				X
Polymetallic Veins			X		X	
Sedimentary Exhalative Zn-Pb		X		X		
Southeast Missouri Pb-Zn			X	X		
Sandstone-Hosted Pb-Zn			X			X
Polymetallic Replacement Deposits			X		X	
Zn-Pb Skarn Deposits		X				X
Simple Sb Deposits			X			X
Serpentine-Hosted Asbestos			X			X
Volcanogenic Mn			X		X	
Sedimentary Mn			X		X	
Upwelling type Phosphate Deposits			X			X
Laterite Type Bauxite Deposits		X			X	
Lateritic Ni			X			X
Carbonatite Deposits			X		X	
Shoreline Placer Ti		X		X		

5.3.1 Mineral commodities

Iron ore – The Far East ranks first in known reserves of iron ore with more than 90 per cent of Russia's total reserves concentrated in over 180 placer and bedrock occurrences. Within the Far East, Yakutia has approximately 42 per cent of total reserves, with the Priamurye (24 per cent) and Magadan (21 per cent) areas in second and third place, respectively.

Tin – Within the region, the Pyrkakajsky, Pravourmijsky and Tigrinsky occurrences have the highest potential for future development as bedrock tin resources. Nevertheless, the largest current tin and future development targets of the Far East are placer deposits. The tin placers of the Vanka inlet, averaging 2.5 m in thickness and 600 gm/m^3, and associated with ilmenite, zircon, rutile and monazite are a primary development project. Additionally, large tin placers are being explored for the Bureja, Niman, Selemja and Amur Rivers basins and in Chukotka, where a new tin area in the Korgak upland is being actively explored.

Tungsten – The Far East contains approximately 10 per cent of Russia's tungsten reserves which are concentrated in approximately 40 placer and bedrock deposits. As tungsten is widely associated with tin, it occurs primarily in the same territories but with a slightly different resource distribution with 43 per cent of reserves in Primorye, 34 per cent in Yakutia and 15 per cent in Magadan. At present, large tin and tungsten deposits, largely placers, are being delineated in Budjal and particular interest is being shown in the large molybdenum-tungsten stockwork deposit at Burdainskoe.

Copper – Although the Far East is not a major copper producer nor does it have significant reserves, recent discoveries and work on the stratiform copper-sandstone deposits of Yakutia (the Udokan orebody with proven reserves of 700 MT of ore grading 1.5 per cent Cu) are primary development targets. Similarly, recently discovered Cu porphyry and Cu-Mo porphyry systems in the Khabarovsk Territory, Madang Province and the Yakutia Autonomous Region are also target areas.

Precious Metals – The Far East produces approximately 25 per cent of the CIS's total gold production and is first within the CIS in gold reserves with Yakutia and Magadan having the largest share of the reserves. At present, most gold is produced from placer deposits, particularly from the tributaries and main streams of the Amur, Indigirka and Kolyma Rivers. To a large extent, the gold potential of the Far East, particularly with respect to lode gold deposits, is relatively unknown. The known association of gold mineralization with older massif structures and acidic intrusives within the Pacific Fold Belt suggests potential new discoveries are possible.

Gold – Large gold placers have been discovered in several places within the Khabarovsk Territory, in Chukotka and on the shelves of the Japan and the Okhotsk Seas. More than 80 prospective areas have been discovered in the southern portion of the Primorye Territory, in the Shantar Islands, in the Okhotsk coast of the Lower Amur, in the Western coast of Kamchatka, by the shore of Eastern Chukotka, in the Chaun Inlet and the Vanka Inlet. The gold content in these placers varies between 250-600 mg/m^3, with some containing up to 22 g/m^3. Placers are tracable for up to 400 m and deep out into the sea. The thickness of mixed river and marine sands is up to 6 m.

In assessing the gold potential of the Far East, particularly for placer gold, it must be emphasized that many gold placers also produce substantial quantities of platinum metals, cassiterite, scheelite, ilmenite, magnetite, zircon, monazite, and rutile. Thus, a rare earth extraction industry may be viable in the future.

Silver – The Far East is a leader both within Russia and worldwide for silver reserves (some tens of thousand of tonnes). Magadan Province, Khabarovsk Territory, and Yakutia are respectively one, two and third in their silver reserves. Discoveries of the Serra-Potozy type are anticipated, owing to the discovery of silver concentrations in phosphorite deposits and in oxidation zones of polymetallic deposits.

Platinum – The Far East is rapidly expanding its platinum mining, particularly placer deposits associated with ultramafic massifs (Khanotor, Inogli and Chad in the Aldan Shield; the Uzbelsky massif in Chukotka and the Kuril massif in Kamchatka). In addition the bedrock sources of platinum associated with the complex copper porphyry ores of Chukotka and Yakutia, and the massive copper-nickel ores of Shanuch in Kamchatka provide targets for the present but indicate both areas and deposit types for future exploration and development.

Diamonds – The main diamond reserves (> 75 per cent) of Russia occur in the Far East and principally within four main kimberite provinces in the northwestern Yakutia, i.e. Malobotuobinsky (Mirninsky), Daldyno-Alakitsky (Ihalo-Udachnensky), Nizhnelensky and Anabarsky. Over 97 per cent of all diamond reserves are in hard rock deposits and only 3 per cent are concentrated in placer occurrences.

Overall, Russia's diamond mines hold reserves expected to last 25-40 years with reserves in open excavations lasting 15-17 years (Borisovich 1991). The Udachnoye deposit in Yakutia produces nearly 90 per cent of all of Russia's diamonds and contains proven reserves expected to last for 10-12 years. Recently, an extremely rich diamond deposit was discovered in the Arkhangelsk region.

Ferro-alloy Metals – The Far East is not a major producer of either ferro-alloy metals or of steel products, nevertheless the resource endowment of the region would appear to be favorable for future development of either a steel industry or for exploration and development of ferro-alloy ores for export. Iron ore deposits of the Far East are concentrated in four main districts, i.e., the Southern-Aldan (magnetite, martite, half-martite ores), Chara-

Tokinsky and Malo-Hingansky (ferruginous quartzite), and magnetite-hematite ores have also recently been discovered. The Malo-Hingan occurrences ores are characterized by a low phosphorus and sulphur content and are enriched (iron extraction is 68-87 per cent and its content in the concentrate is up to 65.8 per cent). Manganese ore in the Far East Territory has only been explored for in the Malo-Hingan district. Ore reserves in Southern Hingan occurrence are 8.6 million tonnes with an average manganese content of 21 per cent, 8.6 per cent tin, and 0.05 per cent phosphorus.

Chromite ore – occurrences remain undiscovered, however, chromite occurs in the Khonder platinum metals placer occurrence. Sand reserves of C_2 category are 36 million m^3 with a chromite content of up to 70 kilograms/m^3.

Titanium – resources of the Far East are poorly quantified, but exceed 1,890 million tonnes in the P_1 and P_2 categories. The majority of the resources are contained in placer deposits, however, the titaniferous ore of the Dzhgzhur ridge area is mined by open pit. Similar deposits occur in the Amur province.

Marine placers of titanium, mostly with associated iron, rare earth-bearing minerals, and locally gold, occur as coastal marine placers adjacent to Sakhalin, the Kuril Islands and the coast of the Primorye Territory. In the latter, marine placers of titaniferous magnetite have been traced for over 600 km. In the Kuril Islands, near Iturup Island, the Rucher occurrence has reserves of 10.5 million tonnes of TiO_2, of which only 15 per cent has been adequately explored. Resource estimates suggest a resource potential of approximately 3 billion tonnes.

Fertilizer mineral – Development in the Far East is small (approximately 15 per cent of domestic demand), nevertheless, the region has substantial reserves and resources of fertilizer minerals. For example, the apatite-titanium-magnetite ore occurrences of the Khalaro-Haninsky (Amur Province), Dzhugazhursky and Baladeksky (the Khabarovsk Territory) ore districts represent a potentially large source of both titanium and mineral fertilizer materials. The Moymakan, Gajuim, Dzhninsk apatite occurrences are prospective with geological reserves of 3.6 billion tonnes and P_2O_4 content of 6-9 per cent.

Building stone – The Far East possesses unlimited resources of various building stone. More than 1,400 occurrences of various building materials have been discovered. Within the Khabarovsk Territory alone there are more than 560 occurrences.

5.3.2 Future mineral development of the Far East

The mineral development of the Far East, as with the remainder of Russia, awaits the resolution of economic development issues before a clear vision of future development is possible. Nevertheless, several factors favor the further development of a minerals industry within the Far East. Among the more important are the following:

- The Far East already has an established minerals industry, based upon a large reserve base for several minerals.
- The Far East is richly endowed in precious metals (Au, Ag, PGE) and diamonds which are high value commodities requiring less infrastructure than other mineral commodities.
- The Far East has a large and diverse energy base upon which to plan future development in the region and to support further industrialization.
- The Far East potentially has access to the rapidly growing markets for minerals in the Asia-Pacific region and the large economies of Japan and the United States.
- Evidence indicates that the Far East may host large undiscovered deposits (gold, copper, diamonds) which could provide the catalyst for development in the region.
- The diverse mineral resource base of the region suggests the development of integrated mineral industries such as in steel, aluminum and base metals is possible.

Offsetting these positive factors are the well known and often recited problems with respect to (a) a harsh climate, (b) a lack of infrastructure, (c) inefficient and outmoded technology, (d) lack of capital, (e) lack of political stability and (f) a regressive minerals policy and legislation. All of these factors are true to a greater or lesser extent and clearly do now and for the immediate future remain as serious impediments to mineral development within the Far East. For the immediate future, it would be appropriate for the Far East to build upon its existing strengths. For example:

- Development of a mining-chemical industry such as the large Jaroslavsky Mining Enterprise which produces over 75 per cent of the CIS's fluorite from its Voznesonsky deposit in the Primorye Territory;
- Expand the production of major mineral deposits such as the Dukatskoye gold-silver vein deposit, while simultaneously increasing production in surrounding and nearby areas; and
- Improve the metallurgical processing capabilities of exiting facilities such as those at Pyrkakajsky, Pravourijsky and Tegrin where tin ore extraction levels are 40 per cent in Krasnorechnskaya, 50 per cent in Solnechnaya and 70 per cent in Hrustalnenskaya. Marginal improvements in recovery would substantially improve their profitability.

Similarly within the Primorsky Mining Enterprise, scheelite recovery is only 67 per cent and copper 50 per cent.

In the future, the Far East must look toward further development and expansion of its mining activities, but also toward developing more integrated industries, such as in steel and aluminum or optimise resource utilization by value adding. For example, through gold smelting and refining, jewelry production, and development of gem cutting capabilities.

Overall, futher development of the Far East mineral industry must be both coherent and steady with the objective of improving the well being of its peoples, whilst also demonstrating to the foreign investor that the Far East is a preferred place to invest, particularly with its enormous known and undiscovered resources.

6.0 DEMOCRATIC PEOPLE'S REPUBLIC OF KOREA

The Democratic People's Republic of Korea has relied heavily on its mineral endowment for much of its earlier economic development and has been a substantial producer of anthracite coal, copper, gold, graphite, iron ore, lead, magnesite, silver, tungsten and zinc. Although a major producer of metals and non-metals, mineral production to meet demand and to increase exports has been a chronic problem. In addition to coal, its primary resource, the Democratic People's Republic of Korea had, according to Trigubenko (1991), the following metallic and non-metallic reserves as of 1991 (table 35).

An analysis of the mineral reserves of the Democratic People's Republic of Korea clearly shows that, with the exception of magnesite, where its reserves are among the largest in the world, the mineral reserves of the Democratic People's Republic of Korea as presently known are perhaps only adequate for domestic use in iron ore and zinc. Gold and silver reserves are not large, but are ample for domestic consumption and/or export depending on government policy.

Table 35. Estimated mineral reserves of the Democratic People's Republic of Korea

Material	*Amount (Tonnes)*
Iron Ore	4.0-4.7 Billion
Zinc	4.0-4.5 Million ton
Lead Ore	3.0 Billion
Copper	0.8 Million
Gold	1 Thousand
Silver	5,000 Thousand
Magnesite	10 Billion

(After Trigubenko, 1991).

6.1 Major mineral deposits

Iron Ore – The Democratic People's Republic of Korea's Musan mine, near Chongjin, is estimated to have approximately 1.3 billion tonnes of iron ore and produces approximately 6.5 million tonnes of ore/yr and is scheduled to undergo further expansion to perhaps 15 million tonnes/yr by 2000. Major iron ore production is also reported from the Unryul, Songhung, Komdok, Tokhyong, Chaeyong and Sohaeri mines, which, together with the Musan mine, account for approximately 12 million tonnes/yr at the present time (EIU, 1996).

Zinc and Lead Ore – The Democratic People's Republic of Korea's zinc ore production was estimated to be 120,000 million tonnes per year in 1992 (EIU, 1996) with the majority of the ore coming from expansion of the Komdok mining complex mines. Lead ore production was estimated at 70,000 tonnes per year in 1992. At the present time new deposits are being investigated for development in Takgol and other southern districts.

Magnesite – The Democratic People's Republic of Korea has the largest and highest quality magnesite deposit in the world and presently produces approximately 1.5 million tonnes of dead burnt magnesite per year. This mine is the Yongyang mine (Tanchon), located in South Hamgyong Province, with reserves of 3.6 billion tonnes of magnesite and annual production of approximately 1.3 million tonnes/year. Other magnesite deposits within the Democratic People's Republic of Korea include those at Paekam (Yanggang Province), Kimchaek (North Hamgyong Province) and the recently opened Taechung mine (Kanyo Province) which is apparently a high grade deposit of magnesite.

Rare Earths – According to Dorian et. al. (1993) in 1988, Japan's International Trading Corp. and the state-owned Korea Ryongaksan General Trading Co. set up a joint venture company, the International Chemical Joint Venture Corp., to extract and refine rare-earth metals. The company has built a US$12 million plant in the Democratic People's Republic of Korea, capable of processing 1,500 t/y of monazite and extracting 400 t of rare-earth metals and oxides.[2] This is one of the few mineral joint ventures operating in the country.

Graphite – A large scale graphite mine, producing approximately 100,000 tonnes/yr, has been developed in the Hungsang area of South Hwanghae Province, while numerous other sites and small mines for graphite are known.

Other Minerals/Metals – Little is known of the size and/or production of the metal mines of the Democratic People's Republic of Korea, however, estimates of production for precious metals for 1989 were 50 Kg of silver and 5,000 Kg of Au. The primary gold producing area is located north of the Bo Hai Sea.

The Democratic People's Republic of Korea also produces significant amounts of tungsten, barite, fluorspar, phosphate, salt, sulfur and talc.

Coal – In the Democratic People's Republic of Korea, anthracite coal is the largest mineral volume produced and there are currently more than 40 anthracite mines operating in 1979, with estimated reserves at 35 million tonnes. Brown coal and lignite production was estimated at 8.6 million tonnes during 1979. The target for coal production in 1993 was 120 million tonnes (EIU, 1996), however it is estimated (Far Eastern Economic Review, 1996) that production was approximately 40 million tonnes/yr or approximately 1/3 the targeted production.

The most important coal reserves exist in the Democratic People's Republic of Korea's coastal Anju mining area (located northwest of Pyongyang). While some of the reserves are on-land, half of the important coal resources located in the Democratic People's Republic of Korea's Anju mining area are located under the seabed. In the Anju mining areas close to the sea, miners must pump six tonnes of seawater per ton of coal mined, due to saltwater intrusion into the low-lying coal seams.

In addition, much of the coal of the Democratic People's Republic of Korea is of varying quality, with some having an ash content of up to 65 per cent and heating values as low as 1,000 kcal/kg. Such poor quality coal, if left untreated, has a low combustion efficiency and produces large volumes of bottom and fly ash, causing disposal and other environmental problems (von Hippel and Hayes, 1996).

---

2 Mining Journal, Mining Annual Review, 1991, 1991, p. 83-103.

7.0 REPUBLIC OF KOREA

The minerals industry of the Republic of Korea has undergone, and continues to undergo a major structural change necessitated by the Nation's low mineral resource endowment and high demand for both concentrate and refined metals. In essence, the Republic of Korea has moved from the position of a major mining country, to that of a major smelting and refining country. It has involved forming joint ventures internationally in the mining sector. This transition has led to a decrease in the number of operating mines and to a decrease in the quantity of minerals produced domestically.

Traditionally, the Republic of Korea has been a significant producer of antimony, copper, gold, graphite, iron ore, lead, mercury, molybdenum and zinc either as primary mine products or as bi-products of poly-metallic mining operations. To date however, the mining industry in the Republic of Korea is in a continuing state of decline and the Nation has begun to look overseas for additional mineral supplies both for domestic consumption and for processing and re-export. The transition of the mining sector from one of a producer to that of a processor and consumer is closely following that of Japan, where there is now only one mine still operating. For this reason, no detailed assessment of the Nation's mineral resources is provided. The reader is reminded, however, that in the discussion of the Bo Hai Sea area there is considered potential for nearshore and offshore placer deposits of gold and titanium to be discovered.

The magnitude of the change can be seen in a comparison of the export and import value of ores and metals, as a percentage of total export and import value, from the period 1970-1990 when ore and metal exports fell from 5.7 per cent of total exports to 0.8 per cent in 1990, while at the same time ore and metal imports rose staying essentially the same at 5.6-5.8 per cent during the same period.

8.0 NORTHEAST CHINA (HEILONGJIANG, JILIN AND LIAONING)

The area of northeast China (figure 49) encompasses the provinces of Heilongjiang, Jilin and Liaoning which, although relatively rich in mineral resources, derive geatest benefits from their coal, oil and gas – in particular, the Daquin oil field in Heilongjiang Province which is discussed in the oil and gas section of this paper. In terms of mineral potential, the individual provinces can be ranked from high to low; as Liaoning, Jilin and Heilongjiang. To a large extent this differentiation of resource potential is attributed to the large area of Precambrian metamorphic and intrusive rocks which occur in Liaoning Province, as opposed to the large area of Cretaceous sediments which comprise the bulk of eastern and central Heilongjiang Province. The approximately equal amounts of older and younger rocks in Jilin Province accounts for it's "intermediate" mineral potential.

8.1 Mineral resources

8.1.1 Metallic minerals

The mineral resource potential of northeast China is highly diverse with both metallic and non-metallic mineral resources as can be seen from the number of occurrence of individual metals/deposit types shown in the following table 36.

Gold – In table 36 it can be seen that the dominant deposit types of the region are gold placer and lode deposits which have been discussed previously for the Bo Hai/Yellow Sea. It is interesting to note, that the relatively large number of placer deposits in Heilongjiang Province, as defined in the Mineral Resources Maps of China (Geological Publishing House, 1992), are concentrated along the border areas with the Democratic People's Republic of Korea and Russia and are extensions of known gold occurrences which occur in these adjoining countries. It is reported (National Council for U.S. China Trade, 1982) that the Tuenjiegou region produces approximately 35,000 ozs/year from three mines and in Hume county approximately 10,000 ozs per year are produced from river sands. Therefore, it is anticipated that the northeast and northern portions of Heilongjiang Province have the potential for future gold discoveries.

Copper – The copper mines of northeast China represent both classic porphyry copper systems and massive copper-polymetallic sulfide deposits, of which the Duobaoshan deposit in Heilongjiang is perhaps the best

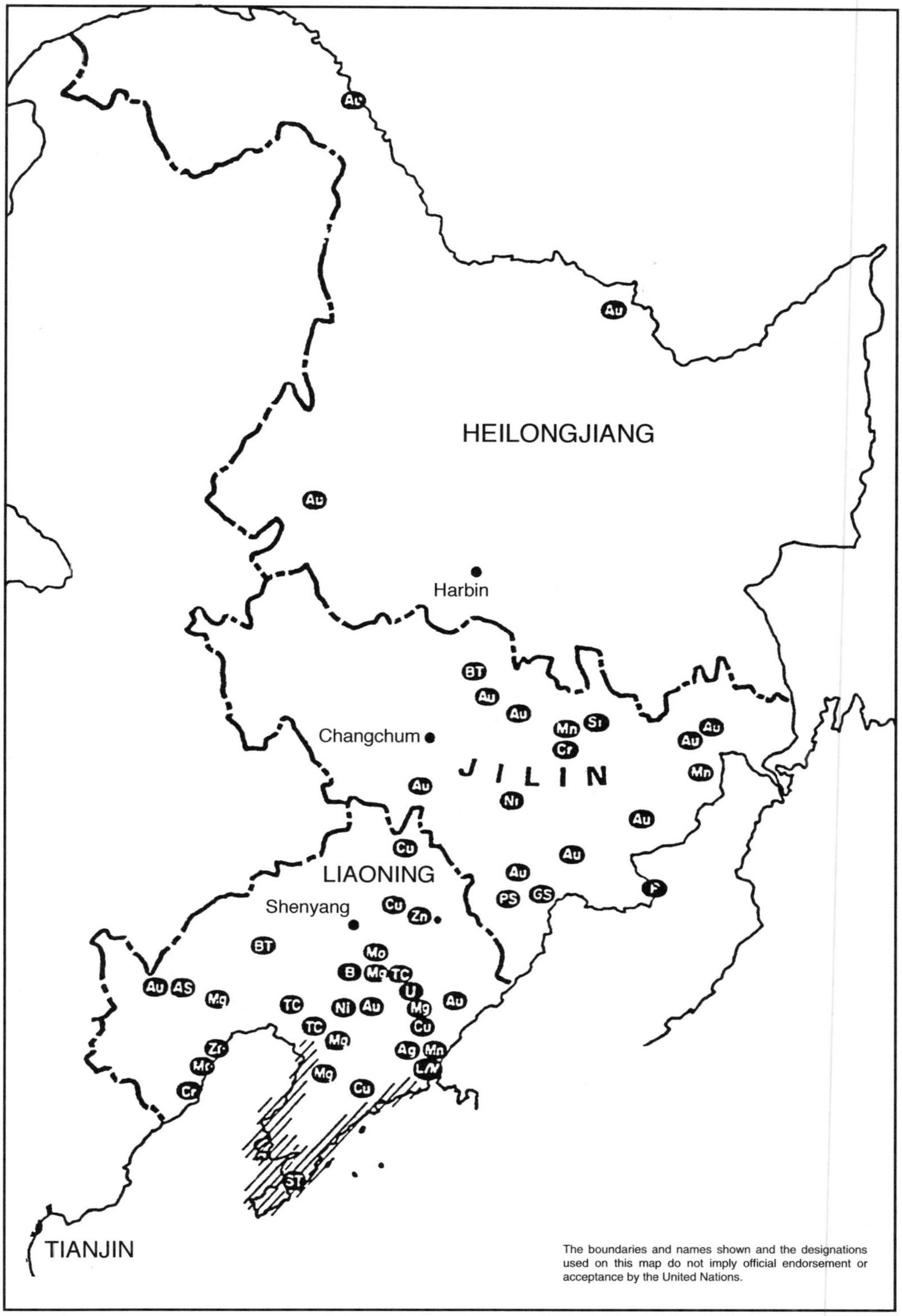

Figure 49. Mineral occurrences in the northeast of China

Table 36. Metallic and non-metallic mineral occurrences in northeast China

Metallic Mineral or Deposit Type	*Heilongjiang No. of Deposits*	*Liaoning No. of Deposits*	*Jilin No. of Deposits*
Metallic:			
Gold (Placers)	13	–	1
Gold (lode)	–	2	3
Silver			1
Copper	1	1	–
Copper-Zinc	1	2	–
Copper-Nickel	–	–	1
Molybdenum	1	4	1
Lead-Zinc	1	4	1
Pb-Zn-W-Mo	1	–	–
Iron	1	12	6
Manganese	–	1	–
Non-Metallic Mineral or Deposit Type:			
Asbestos	–	1	–
Bentonite	–	2	1
Diamond	–	3	–
Fire Clay	1	4	2
Fluorite	–	1	–
Glass Sand	1	1	1
Graphite	10	–	1
Gypsum	–	1	2
Kaolin	–	1	–
Magnesite	–	2	–
Phosphorous	-3	–	–
Pyrite	1	4	1

example, and porphyry skarns. It is anticipated that further exploration in northeast China will result in the discovery of other porphyry and massive sulfide systems.

In addition to the classic porphyry copper type deposit, massive sulfide and porphyry skarn deposits occur in conjunction with a wide range of other metals including zinc. These include Bailing and Gongpengzi in Heilongjiang Province; Tianbaoshan in Jilin Province; Chaihe and Bajiazi in Liaoning Province; and nickel at Chibiasong in Jilin Province. In general these types of occurrences are much smaller, with the exception of Chibiasong which is a world class deposit.

Molybdenum – The molybdenum deposits of northeast China occur primarily as porphyry systems, often in association with copper. The more significant of these deposits are Xiaojiayingzi, Lanjiagou and Yangjiazhangzi in Liaoning Province and Daheishan in Jiling Province. In addition to the traditional porphyry occurrences, skarn deposits are also quite common throughout the northeast area, however, these deposits are relatively modest in size.

Tungsten – China is one of the world's prominent producers of tungsten, mostly from simple to complex quartz vein type deposits (approximately 50 per cent of the Nation's reserves), from skarn deposits (approximately 25 per cent of the Nation's reserves) or more rarely from porphyry type deposits such as the Yangchuling deposit in Jiangxi Province. Rather surprisingly, few tungsten occurrences are known in northeast China.

Lead-Zinc – Although lead and zinc deposits occur throughout northeast China, occurring mostly as massive bedded and vein deposits within carbonate rocks, which may or may not contain appreciable amounts of volcanic material, the majority of the more significant deposits are concentrated in Liaoning Province where the Qingchengzi lead-zinc deposit is the largest and best example of this mineralization type. Other lead-zinc deposits of note in Liaoning Province are the Chaihe, Bajiazi and Wangbaoshan. In Heilongjiang Province the majority of lead and zinc occurances (in particular zinc), is associated with porphyry copper systems.

Iron Ore – Both Jilin and Liaoning Provinces have a substantial number of iron ore deposits which are primarily associated with highly metamorphosed Archean and Paleozoic volcanic sequences. The largest and best example of the Archean type deposit is the Anshan deposit in Liaoning Province which serves as a centre for iron and steel manufacturing in China. In Liaoning Province the major iron ore deposits, in addition to Anshan are Jiajiapuzi, Daheyan, Yanquianshan and Nanfen. In Jilin Province, in similar geologic environments, the iron ore deposits of Tadong and Banshigou are major producers, as is the Shuangyashan deposit in Heilongjiang Province.

Only the largest of the iron ore deposits in northeast China have been described, however, within this region, particularly in Liaoning Province, there are several hundred smaller scale iron ore mines operating.

8.1.2 Non-metallic minerals

Overall China posses a relatively abundant and complete spectrum of non-metallic commodities represented by 86 types located in over 6,700 localities (Geological Publishing House, 1992). In northeast China over approximately 55 major non-metallic mineral deposits occur in the three provinces, ranging from fire clay to diamonds. It is beyond the scope of this paper to attempt to deal with such a diverse spectrum of materials, however, some general comments are appropriate. In particular, it is noted from table 36 that:

- northeast China is a major producer of graphite (Heilongjiang Province), gypsum and pyrite (Jiling Province), Borax (Liaoning Province).
- Liaoning Province is a major producer of diamonds from the Laotiangou, Wafangdian and Toudogou deposits. In addition the Province also produces a significant quantity of famous Chinese jade from the Xiuyu Jade deposit.
- Liaoning Province is also a major producer of high quality magnesite which is particularly important regionally in that the Democratic People's Republic of Korea is the only other major producer of magnesite in North-East Asia.

Coal

China is the world's largest producer and consumer of coal, producing approximately 1.14 Bt of coal in 1993 (Johnson, 1994) and consuming 1.12 Bt. The development and utilization of it's coal resources is critical both to local and provincial development, as well as to National development.

The northeast area of China produced approximately 160 Mt of coal in 1995, and for the same year consumed approximately 170 Mt (Johnson, Personal communication), as a result northeast China, although host to substantial coal resources, is a net importer of coal in order to meet it's energy needs.

The majority of the coal resources in northeast China are of Jurassic-Cretaceous age and are hosted in structural basins. The vast majority of this coal is bituminous, with smaller amounts of lignite, particularly in Liaoning Province, and a limited amount of coking coal. Compared with other areas of China, northeast China is generally lacking in major coal basins and in high grade coal resources.

The environmental problems associated with coal mining in China are well known and extensive, and are due to past mining practices and lax government policy. This situation has been dramatically compounded by the large number of small scale coal mining operations throughout the country. These environmental problems are beginning to be addressed, but there is still a considerable need for existing mines to reach a reasonable environmental standard in their operations.

9.0 OIL AND GAS

The majority of nations in North-East Asia; the Democratic People's Republic of Korea, China, Japan, and Mongolia, can easily be classified as net oil and gas importers, whereas Russia is a net oil and gas exporter. China, on the other hand, has only recently changed from a net oil and gas exporter to a net importer of oil and gas and it's short to intermediate term role as a net importer or exporter is somewhat unclear. In the longer term, however, the

Nation will undoubtedly be a net importer. As the only net exporter of oil and gas Russia, and particularly the oil and gas potential of the Far East, is, therefore, a major factor in determining whether the region as a whole will have an adequate supply of oil and gas or if it will import dependence on the rest of the world.

To address the issue of potential oil and gas supply and developments (particularly pipelines) a brief review of the energy potential of northeast China and the Russian Far East is provided.

9.1 Oil and gas in northeast China

Northeast China accounts for approximately half of China's total oil output, which reached approximately 1,058 million barrels in 1993 (Fesharaki et al., 1995). Approximately 35 per cent of the total national oil production came from the Daqing field, China's biggest oil field, located in Heilongjiang which produces about 1.1 million barrels of oil per day. In addition to the massive production from the Daqing field, Liaoning province yields about 275,000 barrels per day and while Jilin produces approximately 70,000 barrels per day. North-East Asia produces approximately 40 per cent of China's total oil.

In assessing the long term oil and gas supply for China and the northeastern provinces in particular it must be noted that the Daqing field is a mature field and present production is mostly from secondary recovery. It is anticipated that production from the Daqing field will decline in the future, which will leave the northeast and eastern regions of China with a declining supply source.

The oil and gas potential of north central Heilongjiang has only been partially explored and much of the area is unexplored by modern seismic surveys. Therefore, it is possible that significant discoveries may be made in Heilongjiang Province. This may not, however, have a significant impact on the overall energy supply for China.

9.2 Oil and gas in the Russian Far East

At the present time, considerable world attention and in North-East Asia is focused on the potential for oil and gas developments within Russia and in the Far East (figure 50). The focus on the Far East is because of potential oil and gas developments within North-East Asia and because the developments are large and would provide a long term energy source. The development of Far East oil and gas resources is also important as they would provide supply diversity and security for importing countries of North-East Asia.

The remainig discussion on oil and gas potential and possible transport relies on the work of Dorian, J.P. and Khartukov, E.M., 1995 and Valencia, M.J., 1996, and the reader is referred to these three works for more details.

The Russian Far East (figure 50) is emerging as a new major source of hydrocarbon supplies for the rapidly growing Asia-Pacific energy market largely on the basis of its abundant but mainly undeveloped resources of potentially exportable oil and gas. As an exporter it is expected that beyond the year 2000, the now energy-deficient Russian Far East can substantially increase its self-sufficiency in crude oil, becoming a marginal exporter of light products (gasoline and jet fuel), and provide massive supplies of natural gas to neighboring Asian countries.

Like other parts of Russia, the energy future of the Far East and, in particular, the realization of its gas export targets, depends greatly upon the participation of foreign investors and the creation of joint ventures or joint-stock companies. Constructive international cooperation in developing this region's fuel resources will be critical for firmly integrating the Far East firmly integrated into the Asia-Pacific energy market. Foreign investment in the Russian oil and petroleum industry in 1995 is shown in table 37.

The Far East's hydrocarbon resource base is made up of oil and gas reserves discovered primarily on and off the northeast coast of Sakhalin Island in the Sea of Okhotsk, and in central and southwestern Yakutia, in the middle and lower reaches of the Vilyui River. There are also several minor oil and gas finds and a number of prospective areas in the Khabarovsk and Maritime Territories and the Magadan, Kamchatka, and Amur Regions. Still, these frontier areas are almost unexplored and cannot be expected to offer any significant commercial opportunities before the year 2000.

Figure 50. Oil and gas map of the Russian Far East

Table 37. Foreign oil and gas[a] ventures[b] in the former Soviet Union, 1995

Project Name/Location	*Known or Potential Partners*	*Total Investment[c]*	*Description*
Krasnodar region, Russia	ABB Lummus Crest Inc. (USA) and LUKoil	$1.5 billion	Plan to build a 180,000 b/d refinery in southern Russia. The refinery will use crude oil to produce gasoline, jet fuel, and diesel fuel, which will be transported along a planned 40 km pipeline to Novorossiysk.
Caspian Sea, Azerbaijan	Amoco (USA), Unocal (USA), McDermott (USA), Pennzoil (USA), British Petroleum, Statoil (Norwegian), Ramco (Scottish), TPAO (Turkish), Delta Nimir (Saudi-Arabian), SOCAR, LUKoil	$7 billion	Agreed to develop offshore Azeri field as well as Guneshli, Chirag, and Azeri fields. Offshore Azeri field estimated to hold 1.8 billion Bbls of oil and 2.5 trillion cubic feet of gas. Guneshli, Chirag, and Azeri fields estimated to have 4.4 billion bbls. of crude oil.
Kazakhstan and Russia	Bechtel (USA), Willbros (USA), and Oman Oil Co.- (Caspian Pipeline Consortium), Chevron and the Kazakhstan and Russian governments.	$1.25 billion	Tengiz field to Novorossiysk pipeline project; initial capacity expected to be 300,000 b/d, expandable to 1.5 million b/d. 3 year completion schedule has been adopted. Construction on Kazakh portion from Tengiz to Grozny underway.
Tengizchevroil-TCO (Joint Venture) Kazakhstan	Chevron (USA) and Kazakh Government	$1.5 billion	Agreement signed in 1993 to create 40-year 50/50 joint venture in order to develop Tengiz and Korolev oilfields. Currently producing 130,000 b/d.
Polar Lights (Joint Venture) Russia	Conoco (USA) and Arkhangelskgeologia (AAG)	$300 million	Developing Polyarnoye Siyaniye area. Completed 37-mile pipeline to connect with Komineft pipeline at Kharyage Field to southwest. Oil output currently at 23,000 b/d.
Sakhalin I Sakhalin Island, Russia	Exxon (USA), Sodeco (composed of Japan National Oil Corp., Cosmo Oil Co., and Itochu Corp.), and Russian Government	$15.2 billlion	Plan to develop three oilfields located northeast offshore of Sakhalin. Combined reserves estimated at 2.1 billion bbls. of oil and 15 trillion cubic feet of gas. Production will begin by 2003.
Sakhalin II Sakhalin Island, Russia	Marathon (USA), McDermot (USA), Shell (USA), Mitsui (Japan), Mitsubishi (Japan), Rosneftgaz, and Sakhalinmorneftegaz	$8-10 billion	Plan to dvelop two oil and gas fields (Pilton-Astokhskoye and Lumskoye) off of Sakhalin Island. Draft production agreement is nearing completion. If necessary laws are enacted, production will begin by 2000 with peak produciton of 175,000 b/d of crude oil and 1.5 bcf/d of gas.
Vanyoganneft (Joint Venture) West Siberia, Russia	Occidental (USA) and Chernogorneft	N.A.	Currently producing 45,000 b/d from 100 wells. Also developing 2 new fields nearby with total reserves expected to be more than 100 million barrels.
Kazakhstan	Power International (USA) and Kazakhstan government	$5-10 billion	Power International has signed a six-year agreement to develop and implement privatization programmes and to modernize and upgrade the country's power industry.
Timan Pechoria Co. Timan Pechoria, Russia	Texaco (USA), Exxon (USA), Amoco (USA), Norsk Hydro (Norwegian), Arkhangelskgeologia (AAG)	$50 billion	Plan to develop uninhabited oil-rich area in Siberia, where there is an estimated 2-5 billion barrels of crude oil. Agreement negotiations have hit a snag because of AAG's demands for a 50 per cent share.
LukAgip Western Siberia, Russia	Agip (Italy) and LUKoil	N.A.	Signed agreement to develop and produce oil in LUKoil's upstream concession in Siberia; Agip has become a partner in the already producing Siberian Vostochno Pridorozhnoye Field. LUKoil also gained share in Agip oil concessions in Tunisia.

Table 37. *(continued)*

Project Name/Location	*Known or Potential Partners*	*Total Investment*[c]	*Description*
Baku, Azerbaijan	Botas (Turkish) and SOCAR	$1.4 billion	Plan to build a 1,040 km pipeline to take Azerbaijani oil from Baku to Iran to Ceyhan, Turkey on the Mediterranean Sea. Construction begun in early 1994; estimated final capacity is 800,000 b/d.
Turkmenistan	Botas (Turkish)	$8.5 billion	Proposed gas pipeline from Ashkabad, Turkmenistan to Turkey via the Caspian Sea, Azerbaijan, and Armenia. Throughput capacity expected to be 40 billion cubic metres of gas a year.
Karachaganak Project Kazakhstan	British Gas, Agip (Italy), and Karachaganakgazprom	$6-20 billion	Plan to develop the Karachagansk oil and gas condensate field in Kazakhstan with reserves of 2 billion bbls of oil and condensate and 20 TCF of gas. Currently producing 75,000 b/d of condensate and 350 mcf/d of gas.
Catkoneft (Joint Venture) Tymen Region, Russia	Cat Oil (Germany) and LUKoil Association members: Kogalymneftegaz, Langepasneftegaz, and Krasnoleninskneftegaz	N.A.	Formed partnership to export oil from LUKoil fields. Currently producing approximately 22,000 b/d.
Volgodeminoil (Joint Venture) Volgograd Region, Russia	Deminex (Veba Oel, Rwedea, Wintershall) and Nizhnevolzhskneft	$8 billion	Formed a joint venture in order to explore the produce oil in the Volgograd region; reserves in main Povolzhye Oilfield estimated at 120-150 million tonnes. Seismic studies and exploration drilling have started and production could begin by mid-1995.
Yugansk-Fracmaster (Joint Venture) Tyumen Region, Russia	Fracmaster Ltd. (Canadian), Royal Dutch. Shell Group, and Yuganskneftegaz	N.A.	Fracmastser has agreed to buy out Royal Dutch and Shell Group shares in crude oil joint venture. Currently producing about 35,000 b/d; target at end of 20-year project is 100,000 b/d.
Komi Arctic Oil (Joint Venture) Komi, Russia	Gulf Canada Resources, British Gas, Komineft and Ukhtaneftegazgeologia	N.A.	Currently producing about 15,000 b/d. Project is running into trouble because Gulf Canada Resources has not been allowed any exports since October 1993 and is still having to pay export taxes. Until these problems are resolved, they have halted investment.
Atyrau Region, Kazakhstan	Hydrocarbon Engineering (French), Ronar Services Ltd. (British), Nichimen (Japan), Marubeni (Japan), Nissho Iwai Itochu (Japan), JGC (Japan), Chiyoda (Japan), Rerrostaal (German) and the Kazakhstan government	$1.2 billion	Feasibility study for reconstruction of Atyrau refinery completed in March 1994. Japanese/European consortium will reconstruct and improve refinery; when completed utilization rate expected to increase to 85 per cent and refinery capacity to 120,000 b/d. Production of premium unleaded gasoline and distillates will greatly increase.
Iran and Turkmenistan	Iran and Turkmenistan governments	$7 billion	Signed agreement to build a gas pipeline which will carry gas from Turkmenistan to Europe via Iran and Turkey. The length of the pipeline will be 4,000 km and will transport and estimate 15 billion cubic metres.
Yakutia, Russia	Korea Petroleum Development Corp., Daewoo (S. Korea), Samsung (S. Korea), Yukong Ltd. (S. Korea) and Russian government	$15-20 billion	Agreement signed in June, 1994 to jointly develop a natural gas field in Yakutia. Eventually project will involve construction of a pipeline connecting Yakutsk, Khabarovsk, and Wonsan (North Korea) to Seoul, South Korea.
Uzbekistan	Marubeni Corp. (Japan), Chiyoda Corp. (Japan), and Uzbekneftegaz	$1 billion	Plan to build a 100,000 b/d capacity crude oil refinery southeast of Bakhara. Feasibility study completed in August 1994.

Table 37. *(continued)*

Project Name/Location	*Known or Potential Partners*	*Total Investment*[c]	*Description*
Sakha Republic, Russia	Tokyo Boeki Development Ltd. (Japan) and Far East Energy Inc.	$10-14 billion	Developing reserves of 1 trillion cu. metres of gas and 2.25 billion bbls. of oil. In ten years project expected to yield 300,000 b/d of oil and 1.2 billion cu. metres of gas per day. Pipelines to Japan and Korea also projected.
Russia	Tragaz (Italy) and Gazprom	$1.9 billion	Signed contract for Russia to buy equipment which will modernize and reconstruct export gas pipelines. Feasibility study indicates that pipeline modernization will save 5 billion cu. metres of gas.

Sources: Dino Pappas, January 1995, U.S. Department of Energy Office of Oil and Natural Gas Policy, U.S. Energy Ventures in the Former Soviet Union and Eastern Europe; Foreign Energy Ventures in the Former Soviet Union and Eastern Europe, Washington, D.C.

N.A. – Not Available.

[a] Not necessarily a complete list. Intended only as a representation of the types of projects being established or discussed.

[b] Projects are either planned, proposed, or existing.

[c] Estimated. Subject to change. All dollars are U.S. dollars.

At present, nearly 60 fields containing crude oil are known in the Russian Far East: 11 fields in Yakutia, 3 in the Magadan Region, and 45 in the Sakhalin Region, on the shelf of the Sea of Okhotsk. Initial potential resources in the region are estimated at more than 65 billion barrels, which includes 0.7 billion barrels of cumulative oil production.

Almost 60 per cent of the Russian Far East's total explored recoverable reserves of crude oil (or 1.4 billion barrels) are located on and off Sakhalin Island, and 40 per cent (about 1.0 billion barrels) are in Yakutia. Magadan, with 25 million bbl, accounts for only 1 per cent.

To date, the greatest level of foreign investor interest has been on the most sizable accumulations of hydrocarbons discovered off the northeastern coast of Sakhalin (figure 50) after 1976: Four major oil/gas/condensate fields, Odoptu-More found in 1977, Chaivo in 1979, Piltun-Astokhskoye in 1987, and Arkutun-Daginskoye in 1989, as well as the largest gas/condensate field Lunskoye in 1984. Explored recoverable reserves of the four offshore oil/gas/condensate fields (930 million barrels) account for two-thirds of total crude oil reserves of in the Sakhalin region, whereas well flow rates tested in this part of the island's shelf are many times higher than in onshore fields. Likewise, explored gas reserves at the Lunskoye field (more than 10 Tscf) constitute more than 45 per cent of total initial explored reserves of natural gas in some 55 known fields on and off the island. Fairly large gas fields are also known in Yakutia, with three of them (Sredne-Botuobinskoye, Sredne-Vilyuiskoye, and Sredne-Tyungskoye) each having initial explored reserves in excess of 5 Tscf (7.2, 5.8, and 5.4 Tscf, respectively).

9.3 North-East Asia gas pipelines

The issue of ever increasing oil and gas demand within North-East Asia and the need to develop additional supply from both within and from without the area has led to a number of competing proposals for energy co-operation and transportation (specifically pipelines) i.e. linking the area into an "energy community (Valencia and Dorian, 1996) These include (a) Japan's Energy Community Plan, (b) the former Soviet Union's Vostok Plan, modified by South Korea's energy proposal; (c) China's Energy Silk Route Plan and (d) the Irkutsk Plan".

The proposed Irkutsk Plan (figure 51) centres on North-East Asia specifically, and is representative of the scope of the proposed pipeline projects to link North-East Asia with the oil and gas, in particular, gas resources of the Far East. As would be expected the proposed pipeline projects would be extremely high cost and could be expected to have a high environmental impact on both the onshore and offshore areas of North-East Asia.

Indeed, growing environmental concerns in Japan and the Republic of Korea regarding the use of coal, oil and nuclear power, including the problems of air quality, trans-boundary acid rain, CO_2 buildup and nuclear safety issues, are making the use of natural gas more desirable. According to Valencia and Dorian (1996), imports of LNG were 74 per cent of world LNG trade and it is projected that in 2010, natural gas demand for these countries will be 181 billion cubic metres per year (bcm/y). Overall natural gas will be increasingly popular in North-East Asia because of security of supply, minimal price volatility and lesser environmental affect compared to other traditional sources of energy.

10.0 ENVIRONMENTAL IMPACTS AND COSTS OF RESOURCE DEVELOPMENT

10.1 Environmental impacts of resource development

Increasing energy and mineral demand to sustain and support the economic growth globally, in the Asia-Pacific region and the individual nations of North-East Asia economy will bring with it increasing environmental impacts. A recent analysis by Holdren (1990) clearly demonstrated, in terms of the global atmospheric environment, a direct link between environmental impacts and increasing energy and minerals utilization, and that energy is by far the largest contributor to these environmental impactsn (including the fact that energy use contributes a significant amount of metals to the global environment). Holdren's study further showed that energy utilization is by far the major producer of greenhouse gases which impact the global climate. As such, energy and minerals production and their use are perhaps the two components of economic development which have the greatest capacity to alter the global environment. These are critical factors in the future development of mineral and energy resources in North-East Asia, considering that oil and gas development is considered essential to the overall development of the region.

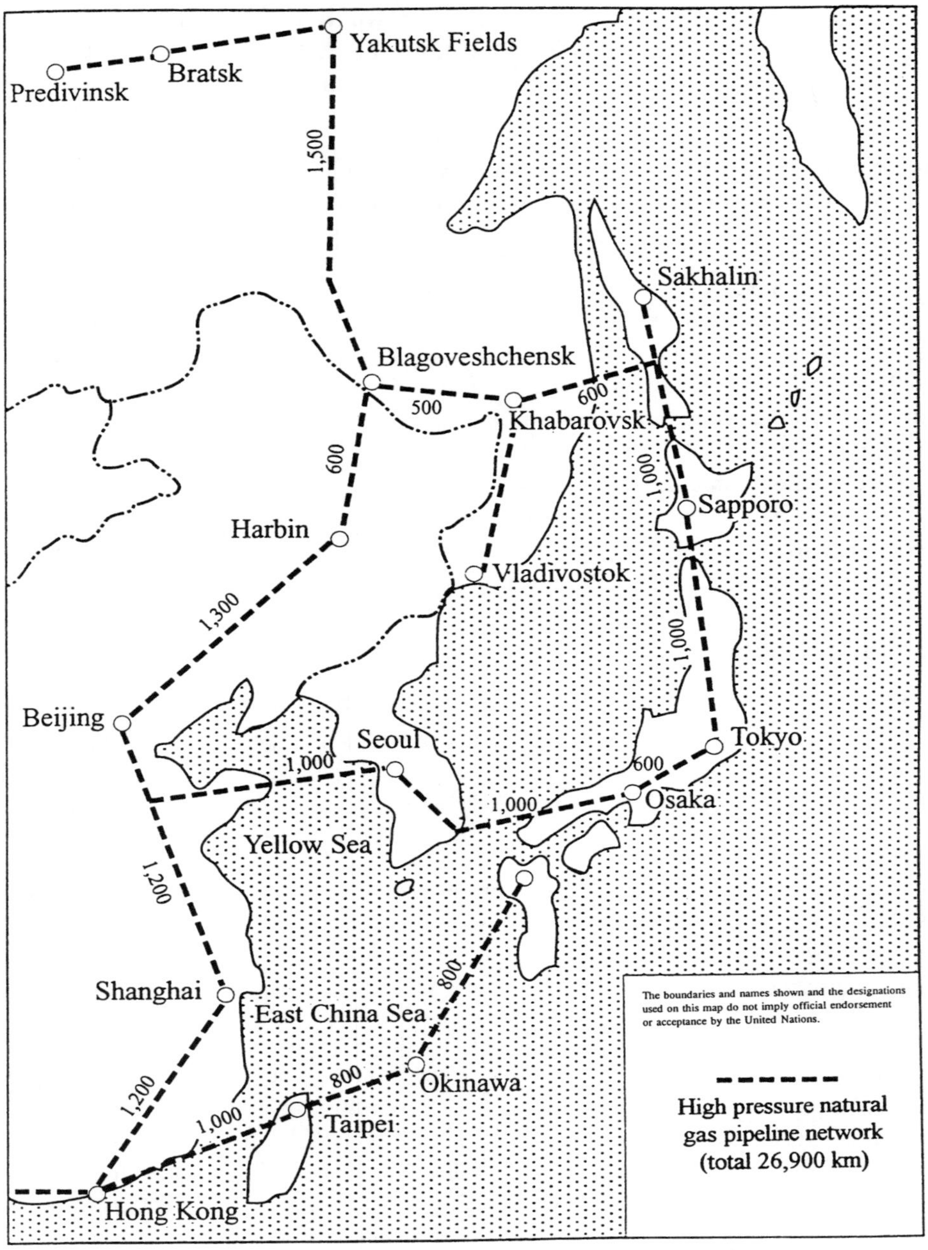

Figure 51. Gas pipeline grid for North-East Asia

Additionally, it must be stressed that the environmental degradation associated with resource development often increases as development matures. This occurs largely because with the depletion of presently known mineral and energy resources future increasing rates of demand will require that: (a) lower quality deposits be developed and utilized resulting in (b) the movement of larger and larger quantities of material, (c) increasing processing and (d) longer distances of transport: all of which will substantially add to environmental impacts. Additionally, it should not be overlooked that increasing demand, particularly in the conversion and use stage, is highly energy intensive, thus further increasing demand. Equally, is the need to recognize the growing magnitude of waste effluents from development activities need to be accommodated.

10.2 Environment and resource attributes of North-East Asia

When assessing the environmental impacts of energy and mineral development and their use in North-East Asia, there are a number of basic attributes of each nation which affect the scope and distribution of environmental impacts.

Firstly, most countries of the North-East Asia, with the exception of Mongolia, are either archipelagos (Japan) or have extensive shorelines. As such, many, if not most, environmental impacts will have not only an onland component but also an offshore effect. Indeed, this very archipelago/coastal nature also dictates that many activities (oil and gas and mineral sand developments) will be undertaken offshore. In both cases there will be a transfer of the environmental problem into the marine environment, often via rivers which flow into the ocean, where their effects may be distributed wider.

Secondly, is that for many countries of North-East Asia, new energy and mineral developments are taking place in two main areas, i.e., in remote onshore and offshore areas where the environment is virtually undisturbed *and/or* in and near densely populated urban centres where environmental problems are already large. In the former case, environmental impacts are highly visible, but hopefully, less damaging because of reduced pollution from newer facilities. In the latter case the environmental impacts may be less visible but are potentially more damaging as they are additional to an already stressed system.

Thirdly, the archipelago/coastal attributes, remoteness of mineral and energy developments and diverse *and/or* separate areas of energy and mineral developments lead to difficulties in communication and management with respect to environmental issues. In particular, their spatial distribution poses significant problems in monitoring and enforcing environmental regulations, necessitating a larger administrative and technical staff to cover the areas and activities. Such issues assume considerable importance when developing national environmental policies.

These factors have additional significance when considering energy and mineral resource development issues, such as the following:

1. Regionally and nationally, energy and mineral resources are usually not uniformly distributed in either quantity or quality, and as a result, development is concentrated in some areas, while absent in others; as also is the economic and environmental impacts of such developments.

2. Energy and mineral resources have distinct and variable modes of occurrence which dictates how they will be developed. For example, large deposits of coal, copper and phosphate are economically recoverable primarily by large-scale open-pit mining. Similarly, large deposits of offshore oil and gas usually require large-scale drilling and production facilities. In both cases, the environmental impacts of such developments are large, as are the economic benefits.

3. Energy and mineral exploitation itself requires large amounts of energy, therefore, the two industries are self-consuming; requiring their developments be juxtaposed or that each has access to the other's commodities. Clearly, there are energy requirements by the mining industry, and petrochemical complexes are major consumers of metal and minerals, e.g. rare earths.

4. Future discoveries and subsequent developments will continue to shift to the more prospective areas in North-East Asia with its emerging market economies. In particular, to areas which (a) are remote and inaccessible and with pristine environments, (b) require large infrastructure developments, and (c) which result in large projects which require significant expenditures to mitigate their environmental impacts.

Economic development which is based on increased energy and mineral development, carries with it high economic and environmental costs.

10.3 Assessing environmental costs

In the ever expanding field of environmental economics, one significant area of research is in defining direct (internalized) and indirect (externalized) environmental costs. In general, direct environmental costs can be divided into 5 main areas, they are:

1. **Assessment Costs** – Direct costs incurred in defining anticipated environmental impacts of an individual project or activity. Normally this includes base-line studies, environmental impact analyses and the preparation of an Environmental Impact Study (EIS).

2. **Prevention Costs** – Direct costs incurred during operations which prevent environmental impacts. Includes such items as mine tailings dams, and safety valves on offshore drilling platforms.

3. **Mitigation Costs** – Are those for mitigation of environmental impacts and are usually the largest direct cost. Unlike definition and prevention costs, primarily associated with new projects, mitigation costs are incurred by both new and existing facilities. Typically, energy and mineral developments have associated mitigation costs for their gaseous emission and effluent/discharge controls.

4. **Reclamation Cost** – Direct costs incurred in returning affected areas to pre-existing environmental conditions. These costs are often incremental through the life of the project or, may be an assessed cost at the end of the project.

5. **Compensation Costs** – Direct costs assessed as compensation to affected parties or costs assessed for irrecoverable damage to the environment. Such costs are often a result of civil or criminal litigation.

In dealing with direct environmental costs, direct costs 1-4 are generally assumed primarily as the cost of new and/or ongoing projects, However, direct costs 3-4 are often costs assumed by the state/province/national levels for environmental protection or restoration for areas of past development activities. The U.S. "Superfund" is one example of reclamation-compensation costs undertaken by government.

For countries of North-East Asia, direct environmental costs 1- 4 need to be included in developments in order to avoid the deferred reclamination and compensation costs.

Although considerable uncertainty exists concerning which technologies will be available in the future, these direct costs are quantifiable and their economic impacts evaluated. This is not the case with indirect environmental costs which in many instances are: (a) unknown, (b) difficult to measure, (c) difficult to value, and (d) difficult to integrate into economic planning.

What is clear, is that indirect environmental costs are not represented in standard measures of economic growth, such as GNP/GDP or per capita consumption. This accounting deficiency for indirect environmental costs is critical in the context of sustainable economic development when measured by GDP. An absence of indirect environmental cost accounting fails to recognize:

- The depletion of natural resources, in particular energy, minerals and the environmental components of water, air, soil and habitat.

- Environmental damage/degradation not being included in national accounts, whereas direct environmental costs such as reclamation are included, further distorting the national economic profile.

- Indirect environmental costs associated with a lack of conservation and efficient in resource utilization, requiring greater resource development and associated environmental degradation, is a hidden cost in many national economies.

A recent study by the World Resources Institute focused on the impact of indirect environmental costs by measuring an environmentally adjusted net domestic product (NDP) and net income (ENI). In an analysis of depreciation for oil, timber and top soil in Indonesia, it was found that the NDP growth rate for the period 1971-1984 was only 4 per cent, compared with an official GDP growth rate of 7.1 per cent. Although the actual NDP values might be calculated by several means, and undoubtedly would produce varying growth rates, the important issue is that adjusting economic growth indicators for indirect environmental costs produces a significantly reduced rate of growth. When viewed in the context of sustainable economic growth versus energy and mineral consumption, the conclusion seems inescapable: the efficiency of resource utilization in economic growth is much lower than previously predicted when indirect environmental costs are considered.

Measuring indirect environmental costs is at best difficult in most instances due primarily to (a) a lack of base line data against which to measure impacts, (b) the uncertainty surrounding the interactions within the environment in terms of cause and effect, and (c) the long time frames within which environmental impacts occur and total impact can be assessed. That indirect environmental costs are significant is easily substantiated, however, the actual costs are much more difficult to ascertain.

10.4 Direct environmental impacts and costs: energy and minerals

In any assessment of environmental impacts and costs, the rapidly increasing rate, scale and complexity of interactions associated with development, populations and consumerism presents an almost limitless number of effects. It is impossible in this overview to describe these environmental impacts and costs except in the most general terms: it is, however, sufficient to demonstrate that even at a summary level, both impacts and associated mitigation or prevention costs are high.

For both energy and minerals, the environmental impacts and associated costs occur throughout the cycle of exploration, development, extraction, beneficiation, processing and transportation. It can be said that the scale and complexity of the environmental impacts increases in the "downstream" industries of beneficiation and processing, as do costs of mitigation and prevention.

Energy development (oil and gas) within North-East Asia and worldwide occurs both onshore and offshore with somewhat differing environmental impacts, although environmental costs are similar for both types of operations (tables 38 and 39). Asia-Pacific countries are expected to invest between US$27-29 billion in refining and petrochemical processing during the 1990s. If this investment occurs, only 5-10 per cent of the total investment is for environmental controls, then a conservative estimate, of direct environmental costs would be approximately US$2 billion during the 1990s alone and only for the refining and processing industries.

Conversely, with the minerals industry, the environmental impacts are principally onland (table 40); although specific activities such as tin mining/dredging are offshore mining activities. Unlike the oil and gas industries, the very nature of minerals extraction, beneficiation and processing causes large and visible environmental impacts. For example, in the oil and gas industry, 95+ per cent of the extracted resource is utilized, while in the coal industry only 55-65 per cent of the extracted resource is utilized, and in the minerals industry, less than 10 per cent of the extracted resource is utilized. As a consequence, large quantities of waste rock, low-grade ores, tailings and slag must be stored and/or disposed of in an environmentally appropriate manner. Table 40 illustrates this need to handle and store large quantities of extracted material significantly increases the environmental impacts of mining.

As with energy, the anticipated expenditure for mineral development in the Asia-Pacific region is anticipated to be from US$20-22 billion during the 1990s. Based on estimates of direct environmental costs in other areas, ranging from 5-23 per cent of total mining project costs, the environmental costs incurred in the Asia-Pacific region (assuming a conservative 10 per cent of project costs) will be from US$2-2.2 billion for mineral developments alone during the 1990s. Clearly, direct environmental costs could be significantly higher within the region if existing environmental policies and regulations become more stringent.

As the scope and timing of energy and minerals development in the Asia-Pacific region in general, and North-East Asia countries specifically, will depend on numerous externalities (discoveries, economic growth, demand), the direct environmental impacts and costs are uncertain. What is not uncertain, however, is that direct environmental impacts and costs will occur and they will be substantial: sufficiently high that they will have an impact on economic development and the environmental quality of North-East Asia. If the costs economically and environmentally become too high, development may be retarded; conversely, if economic and environmental costs are balanced, then development may occur. There will need to be trade-offs, the developments must be sustainable both economically and ecologically.

Table 38. Environmental impacts and relative costs of offshore oil and gas development

	Relative[1] Impact	*Area[2] of Impact*	*Control[3] Costs*
Exploration			
o Small scale impact on sediments and sedimentation	M	VL	–
o Small scale disruption of benthic and pelagic organisms	M	VL	–
o Alteration of nearshore drainage	M	L	1
o Seismic disruption	M	L	–
o Local release of drilling fluids/muds	M	L	1
o Oil spillage (accidental or operational)	M	L	1
o Onshore staging area impacts	M	L	1
Development			
o Bottom and sub-bottom sediment disturbances	S	L	2
o Local release of drilling fluids/muds	S	L	2
o Oil spillage (accidental or operational)	S	P	2
o Gaseous emissions	M	P	2
o Increased disruption of benthic and pelagic organisms	S	L	2
o Modified current and sedimentation patterns	M	L	2-3
o Visual and aesthetic effects	S	L	L
Extraction			
o Extensive offshore or onshore disruption for storage and handling	S	L	2
o Oil spillage (accidental or operational)	H	P-R	3-5
o Micro-seismic events	S	P	2-3
o Groundwater/oceanic water pollution from recovery activities or natural effects	H	P-R	4-5
Processing			
o Surface disruption and modification	S	L	3
o Gaseous emissions (CO_x, SO_x, NO_x, particulates)	H	R	5
o Effluent emissions	H	P	4-5
o Residual wastes	S	L-P	4-5
o Spillage (accidental or operational)	S	L-P	4
o Surface and groundwater contamination	S	P	3
o Visual and aesthetic effects	S	L	3
Transport (Pre/Post Processing)			
o Surface disruption and modification (pipeline)	S	P-R	3
o Spillage/leakage	H	P-R	4-5

1 Qualitative estimate of the scope and permanency of the impact
M = Minor
S = Significant
H = High

2 Qualitative estimate of the areal extent of the impact
VL = Very local, i.e., confined largely to site of occurrence
L = Local, i.e., confined largely to immediate area of site of occurrence
P = Provincial, confined to broad surrounding area bounded by primary dispersal controls
R = Regional, broad distribution resulting from primary and secondary dispersal.

3 Associated costs of environmental impact, mitigation or prevention for medium to large-scale mining activities (does not include annual upkeep costs)

1. US$0 - 250 K
2. US$250 K - 1 M
3. US$1 - 10 M
4. US$10 - 50 M
5. US$50 or more

Table 39. Environmental impacts and relative costs of onshore oil and gas development

	Relative[1] Impact	*Area[2] of Impact*	*Control[3] Costs*
Exploration			
o Small scale vegetation and surface disruption (access)	M	L	1
o Minor erosion and sediment loading of streams and lakes	M	L	1
o Wildlife disturbance	S	L	1
o Visual and aesthetic impacts	M	L	1
o Seismic disruption	M	L	–
o Local release of drilling fluids/muds	M	VL	1
o Groundwater pollution	M	VL	1
Development			
o Extensive vegetation and surface disruption (access, infrastructure)	S	L	2
o Release of drilling fluids/muds	S	L	1
o Increased groundwater pollution	M	L	2
o Oil spillage (accidental or operational)	S	S	2
o Gaseous emissions	S	S	2
o Visual and aesthetic impacts	S	S	2
Extraction			
o Extensive vegetation and surface disruption (infrastructure)	S	S	3
o Oil spillage (accidental or operational)	S	H	4
o Micro-seismic events	S	H	4
o Modification and/or contamination of surface and groundwater	S-H	S	3
o Visual aesthetic impacts	S	H	3
Processing			
o Surface disruption and modification	S	L	3
o Gaseous emissions (CO_x, SO_x, NO_x, particulates)	H	R	5
o Effluent emissions	H	P	4-5
o Residual wastes	S	L-P	4-5
o Spillage (accidental or operational)	S	L-P	4
o Surface and groundwater contamination	S	P	3
o Visual and aesthetic effects	S	L	3
Transport (Pre/Post processing)			
o Surface disruption and modification (pipeline)	S	P-R	3
o Spillage/leakage (accidental or operational)	H	P-R	4-5

[1] Qualitative estimate of the scope and permanency of the impact
M = Minor
S = Significant
H = High

[2] Qualitative estimate of the areal extent of the impact
VL = Very local, i.e., confined largely to site of occurrence
L = Local, i.e., confined largely to immediate area of site of occurrence
P = Provincial, confined to broad surrounding area bounded by primary dispersal controls
R = Regional, broad distribution resulting from primary and secondary dispersal.

[3] Associated costs of environmental impact, mitigation or prevention for medium to large-scale mining activities (does not include annual upkeep costs)
1. US$0 - 250 K
2. US$250 K - 1 M
3. US$1 - 10 M
4. US$10 - 50 M
5. US$50 or more

Table 40. Environmental impacts and relative costs of mineral development (includes coal)

	Relative[1] Impact	*Area[2] of Impact*	*Control[3] Costs*
Exploration			
o Small-scale vegetation and surface disruption	M	VL	1
o Minor erosion and sediment loading of streams and lakes	M	VL	[illegible]
o Wildlife disturbance	S	L	2
o Visual and aesthetic impacts	M	VL	1
o Noise	M	L	1
o Local release of drilling fluids	M	VL	1
Development			
o Extensive vegetation and surface disruption (access, infrastructure and mine site)	S	P	2
o Increased erosion and sediment loading of streams and lakes	M	L	2
o Contamination of surface and ground water	M	L	2
o Waste soil/rock storage and/or redistribution	M	VL	3
o Wildlife disturbance	H	L	2
o Significant visual and aesthetic impacts	S	L	2
o Noise	S	L	2
o Reduced air quality (dust)	M	L	2
Extraction			
o Large-scale surface disruption and modification	H	L	3-4
o Erosion and sediment loading of streams and lakes	S	P	3
o Physical and chemical alteration of surface and groundwater quality and occurrence	S	L-P	3-4
o Mobilization and storage of low-grade and waste rock	S	VL	3-5
o Reduced air quality (dust, gases, equipment emissions)	S	L	2
o Natural leaching of low-grade and waste rock piles	M	L	2
o Visual and aesthetic impacts	H	L	3
o Land subsidence (underground mines)	M-S	VL	4
Beneficiation			
o Surface disruption and modifications	H	VL	3
o Chemical alteration of surface and groundwater	S	L-P	4
o Air quality (dust, chemical emissions)	S	P-R	5
o Tailings disposal (associated seepage)	S	L	5
o Hazardous waste (chemical effluents)	S	R	4
o Visual and aesthetic impacts	H	L	3
o Noise	H	L	3

Table 40. *(continued)*

	Relative[1] Impact	*Area[2] of Impact*	*Control[3] Costs*
Processing			
o Surface disruption and modification	H	L	3
o Tailings/slag disposal	H	L	4-5
o Waste water disposal	S	P	3-4
o Smelter emissions (SO_x, NO_x, particulates)	H	R	5
o Hazardous waste	H	P	4-5
o Chemical alteration of surface and ground water	S	R	4
o Visual and aesthetic effects	H	L	3
o Noise	S	L	3

[1] Qualitative estimate of the scope and permanency of the impact
M = Minor
S = Significant
H = High

[2] Qualitative estimate of the areal extent of the impact
VL = Very local, i.e., confined largely to site of occurrence
L = Local, i.e., confined largely to immediate area of site of occurrence
P = Provincial, confined to broad surrounding area bounded by primary dispersal controls
R = Regional, broad distribution resulting from primary and secondary dispersal.

[3] Associated costs of environmental impact, mitigation or prevention for medium to large-scale mining activities (does not include annual upkeep costs)

1. US$0 - 250 K
2. US$250 K - 1 M
3. US$1 - 10 M
4. US$10 - 50 M
5. US$50 or more

10.5 Non-conventional and alternative energy and minerals

The present analysis has focussed primarily upon conventional energy and minerals as the resource base for economic development within countries of North-East Asia. However, it is recognized that there are non-conventional or alternative sources of supply which will undoubtedly play an increasing role.

From an energy perspective, the alternatives range from nuclear power, through geothermal energy, solar, biomass, wind and tidal to hydropower. It is beyond the scope of this paper to discuss each of these alternatives, however, some general comments are made on the importance of these alternative energy sources within the region.

The development of hydro and other alternative energy sources within the region is anticipated to increase only marginally from 6 to 7 per cent by the year 2000. Notwithstanding the successes of hydropower in the region and geothermal and biomass energy in Indonesia and the Philippines, the role of alternative energy is anticipated to be small. This results from (a) the relative high costs of alternative energy generation by solar, and wind, (b) the limited availability of geothermal, (c) the limited efficiency of biomass for large-scale production of energy, and (d) the relative cost and difficulties with developing efficient tidal power. Although the major constraints at present are economic and technological, significant advances in reducing cost and developing the needed technologies are continuing to be made. With further advances these alternative energy sources may play an increasing role in the region.

Alternative energy sources are not without their environmental impacts and costs, although in most instances, they may be less than those associated with conventional energy sources. Conversely, the large areas impacted by hydro development, the land resources required for solar and wind generation, increased emissions of geothermal areas, and land and water impacts of biomass production all represent significant environmental impacts and costs, for energy efficiencies well below those of conventional sources.

For the minerals industries, the major trends have been with respect to (a) substitution of metals by plastics and ceramic products, (b) substitution of one metal for another, and (c) recycling. Each of the above acts differently in terms of reducing or transferring metals' demand and consumption, but each also has significant environmental impacts and costs.

Metals are also replacing non-metals, such as in the semiconductor chip industry where traditional silicon chips are being replaced by gallium arsenide chips. This is an example of substitution, which may result in higher environmental costs, but also increased metals usage.

Examples of metals substitution include aluminum for steel in beverage containers, aluminum for copper in electrical transmission lines, copper for gold as bonding wire on semiconductor chips and titanium for tungsten in tool and die applications. Overall, it may be argued that the substitution of one metal for another merely substitutes the environmental impacts of one for the other. However, in the case of aluminum as the substituting metal: in addition to its environmental impacts, one must include those associated with increased energy consumption. This arises as aluminum processing requires substantially more energy per unit of production than does copper or steel – the two metals it commonly replaces.

Recycling of metals is an ever increasing activity – particularly for many North-East Asian countries: and is one of the few activities which reduces the amount of metal needed to be produced. The recycling of metals, however, it a complex process requiring significant energy, often large quantities of chemicals and sometimes produces a high proportion of hazardous waste (5-35 per cent by volume). Impacts, costs and benefits associated with recycling must therefore, be balanced against the impacts and costs of mining.

With both energy and minerals, the technical opportunities for substituting non-conventional or alternative sources for conventional sources are significant.

The role of non-conventional or alternative energy and mineral sources within the Asia-Pacific region will be dependent on the economic and environmental trade-offs that each provides. At present, low cost conventional sources dominate in North-East Asia as they do globally.

BIBLIOGRAPHY

1. Clark, A.L. and Li, C., 1991a. "Marine Mineral Resources of the East China Sea: Scientific and Economic Opportunities," *Marine Mining,* vol. 10, pp. 117-144.

2. Clark, A.L. and Li, C., 1991b. "Gold Placer Deposits in the Bo Hai Sea," *Marine Mining*, vol. 10, pp. 195-214.

3. Clark, A.L. and Dorian, J.P., 1992. "Mineral Investment Conditions in Mongolia" in *Mineral Investment Conditions in Selected Countries of the Asia-Pacific Region,* United Nations, New York, pp. 151-179.

4. Clark, A.L., 1993. "Marine Mineral Resources of the South China Sea," *Marine Georesources and Geotechnology,* vol. 11, pp. 101-126.

5. Clark, A.L., 1993. "Southeast Asia Energy and Minerals Development: Environment and Economics," in *Minerals and Metals Trading,* compiled by Allen L. Clark, United Nations/Economic and Social Commission for Asia and the Pacific, New York, p. 134-158.

6. Clark, A.L. and Sekisov, G.V., 1993. "Geology, Mineral Resources and the Mineral Development Potential of the Russian Far East," in *CIS Energy and Mineral Development: Prospects, Problems and Opportunities for International Cooperation,* (eds.) Dorian, J.P., Minakir, P.A. and Borisovich, V.T., Kluwer Academic Publishers, pp. 127-156.

7. Clark, A.L., 1993. "Energy and Mineral Development: Environment and Economics" in *South-East Asia's Environmental Future: The Search for Sustainability,* ed., H. Brookfield and Y. Byron, Oxford University Press/UN University Press, pp. 75-102.

8. Dorian, J.P., Fridley, D. and Tressler, K., 1993. Multi-Lateral Resource Cooperation among North-East Asian Countries: Energy and Mineral Joint Venture Prospects, Journal of North-East Asian Studies, George Washington University, Washington, D.C., 43 p.

9. Dorian, J.P. and Khartukov, E.M., 1995. Joint Oil/Gas Ventures in the Commonwealth of Independent States: A Background Analysis, in European Energy Charter Treaty, Grahm and Trotman Limited, London, United Kingdom 28 p.

10. Economist Intelligence Unit, 1996. Country Profile South Korea and North Korea: 1995-1996. The Economist Intelligence Unit Limited, London, United Kingdom, 89 p.

11. Fesharaki, F. and Yamaguchi, N.D., 1991. Decade of Change in the Asia-Pacific Region: The Energy Outlook and Emerging Supply Demand Imbalance, Conference on Asia-Pacific Petrochemical and Refining Integration, East-West Center, Honolulu, pp. 29-55.

12. Fesharaki, F., Clark, A.L., and Intarapravich, D., 1995. Pacific Energy Outlook: Strategies and Policy Imperatives to 2010, Wast-West Center, Honolulu, Hawaii, 113 p.

13. Geological Publishing House, 1992. Mineral Resources Maps of China and Explanatory Notes, Chinese Institute of Geology and Mineral Resources Information, Institute of Mineral Deposits, Beijing, China, 48 pp. and 3 maps.

14. Johnson, C.J., 1994. Coal Suppliers From Asia for Asia, 7th World Coal Outlook Conference, MacQuerie University, Sydney Australia, 26 p.

15. Kunaev, I.V. (Compiler), 1984. Map of the Geological and Structural Zones of the Russian Far East and Map of the Metallogenic Regions, Provinces and Sub-Provinces of the Russian Far East, (In Russian), Khabarovsk, Russian Federation.

16. Morgan, J. and Valencia, M.J., 1992. Atlas for Marine Policy in East Asian Seas, University of California Press, Los Angeles, California, 152 pp.

17. National Council for US – China Trade, 1982. China's Minerals and Non-ferrous Metals Map, Washington, D.C., 1 map sheet with explanatory insets.

18. Noland, 1995. The North Korean Economy, Asia Pacific Economic Cooperation Working Paper Series 95-5, Inst. for Inter. Economics, Washington, D.C., 78 p.

19. U.S. Bureau of Mines, 1989. Mineral Industries of the Asia and the Pacific: 1989 International Review, U.S. Bureau of Mines, U.S. Gov. Printing Office, Washington, D.C., 383, pp.

20. Valencia, M.J., and Dorian, J.P., 1996. Multilateral Cooperation in North-East Asia's Energy Sector: Possibilities and Problems, Energy Workshop of North-East Asia Cooperation Dialogue V, Institute of Foreign Affairs and National Security, Seoul, Korea, September 11-12, 1996, 22 p.

21. Valencia, M.J., 1996. A Maritime Regime for North-East Asia, Oxford University Press, Hong Kong, 329 p.

CONTENTS

CONTENTS *(continued)*

3. MINERAL RESOURCES ASSESSMENT AND DEVELOPMENT IN THE NORTH-EAST ASIA[1]

INTRODUCTION

Although demand for some minerals has stabilized, and even declined in some developed countries, demand in the developing countries of South-East and North-East Asia, has been increasing. Demand for energy has however continued to increase in the developing as well as the developed countries and will continue to increase for several decades to come. For example, per capita consumption in the newly industrializing Republic of Korea has grown rapidly over the last 30 years to a level similar to that of the industrialized countries. Demand for minerals, both fuel and non-fuel, is likely to increase globally, and in Asia especially, mainly as a result of the continuing increase in population through most of the next century and the demand for improved living standards in developing countries.

On the supply-side however, minerals, fuel and non-fuel, are essentially non-renewable. Therefore, it is especially important that the rate of depletion, and the emphasis on recycling and economy of use be properly managed to ensure that the resources do not run out before acceptable substitutes are available. Sustainable consumption requires efficiency of extraction and use, recycling, new technologies and substitution, enabling reduction in the rate of increase of overall demand.

To address the issue of sustainable supply of mineral resources requires consideration of the whole mineral cycle, from mineral exploration and discovery through extraction, processing, manufacture, utilization to eventual recovery or disposal. The Oslo Round-table Conference on Sustainable Production and Consumption, 6-10 February 1995, underlined the importance of focussing on demand-side issues as a complement to the traditional supply-side approach. This study relates to sustainable supply and consumption patterns focusing on supply-side issues related to upstream mineral development activities.

Concerns about the environmental impact of mineral resources development and land use conflicts are also reducing the capacity to meet the demand for minerals. Therefore, the minerals supply issue should be addressed together with management of environment impacts, in the context of integrated land management.

The developing countries of the Asia Pacific region, including those countries in North-East Asia, have continually strived during the past decades to improve their capacities to capture the maximum economic and social benefits of their potential for mineral development. Open policies, new mining legislation and new fiscal regimes have been established in many developing countries to attract foreign direct investment and transfer clean technologies to develop their mining sectors.

During the past 10 to 15 years, needs for technical cooperation between countries have evolved from general information/data gathering through exploration to more sophisticated areas such as computerized information management, mineral policy formulation, investment promotion, environmental protection and the support of small-scale mining, mainly to assist countries to attract and benefit from foreign investment. In this regard, international organizations, including the United Nations system, have been requested to provide technical assistance in their efforts for mineral resource and policy development. Sub-regional economic and technical cooperation can play an important role in promoting the development of both developing and developed countries.

Chapter I of this study addresses the current level of development in North-East Asian countries, particularly with respect to both non-fuel minerals and fuel minerals. Under each of the mineral items, the presentation is organized country by country. The Chapter II of the study concentrates on the policies adopted by the North-East Asian countries in their efforts to develop their mineral resources. Chapter III presents the international cooperation activities related to development of minerals in the North-East Asia region.

---

[1] ESCAP Secretariat

Developmental issues, including the mineral resource base, resources assessment, development and trade, are presented as Part 3.1 of the Chapter III; the creation of a sound investment environment, including attraction of foreign direct investment (FDI) and establishment and improvement of legislation is presented in Part 3.2; The integration of environmental considerations into decision making for development of minerals, is addressed in Part 3.3; and a brief overview of some international cooperation programmes for the North-East Asia region, comprises Part 3.4.

Chapter IV, Conclusions and Recommendations, lists items that are considered important, and could serve as a reference in future considerations and discussions by governmental agencies and international institutions, in their efforts to develop the mineral resources of North-East Asia.

1.0 MINERAL RESOURCES ASSESSMENT AND DEVELOPMENT

1.1 Non-fuel minerals

1.1.1 Metallic minerals

1.1.1.1 China

Proven reserves of a number of mineral and energy resources in China are among the richest in the world, which includes antimony, barite (ranked 2, 1990), coal, fluorite (1), iron ore (2), rare earths (1, more than 80 per cent), talc (2, 1990), tin (20 per cent), tungsten, and vanadium. China is also a leading global mineral producer, with output of antimony ranked number 1 (accounting for 75 per cent in 1994), barite (1), raw coal (1, 29.5 per cent in 1995), fluorspar (1), gold (6), iron ore (3, 1993), lead and zinc (4, 1990) rare earths (1), salt (2), steel (2, 1993), tungsten ore (1, 73 per cent in 1992) and cement (18.1 per cent, 1991). Figures 52 and 53 show the administrative map of China and geographical distribution of China's metallic mineral resources respectively.

China is a large consumer of some mineral and energy products. China's consumption of coal ranks number 1, accounting for 29.0 per cent in 1995, gold (4), oil (3, 4.9 per cent in 1995), crude steel (3), rare earths (2).

China is also a big exporter of a few mineral ores such as tin (14 per cent, 1992), tungsten (1, 19 per cent in 1992), and a big importer for commodities such as unrefined copper (17 per cent in 1992), iron ore (4, 7 per cent in 1992).

Ferrous metals

According to statistics by the Chinese authorities in 1990, China had a total iron ore reserve of 50.1 billion tonnes, ranking second in the world. In 1993, China produced 87.3 Mt of pig iron and 88.7 Mt of crude steel and replaced the United States to become the second largest crude steel producer in the world after Japan.

Although widely distributed, China's iron ore reserves are relatively concentrated in 8 areas, of which the Anshan – benxi area in northeastern China is the largest, as shown in the table 41. Other important characteristics of China's iron ore reserves include: a) the iron content of the ores in China is only about 33 per cent on average; b) the rich ore reserve (greater than 45 per cent in iron content) in the country account for only 2.5 per cent of its total; c) magnetic-type ore is the most important to China in the size of its reserves (54 per cent of the country's total) and its utilization by the Chinese industries; and d) the iron ores in China are complicated in their composition, and associated with plenty of other components such as vanadium, titanium, copper, niobium and rare earths.

China now has a total steel output capacity of 100 Mt. Four mining districts each with an annual iron ore production of more than 10 Mt have been established. Most iron ore-rich areas in China have been developed except the Huoqiu area located in eastern China and adjacent to the Chang Jiang river, which is expected to be developed to form a new production base for iron ore in China in the 1990s.

Figure 52. Administrative map of China

The demand for steel in China is expected to reach 120 Mt by the year 2000 and the Ministry of Metallurgical Industry (MMI) plans to expand it to 135 Mt by the year 2000. MMI targets steel production from plants along the Chang Jiang river to be over 48 per cent, nearly 50 Mt, of the country's total output in the year 2000. By then, domestic steel companies are expected to supply 95 per cent of the steel products for the construction of railways, power stations and ships, and nearly 80 per cent of the oil pipes and automobile steel parts.

China's import of iron ore, mainly from Australia, Brazil, and India, has sharply increased since 1991, reaching 37.34 Mt in 1994. Although China also imports steel scrap, its ship breaking industry has become one of the major sources of the country's scrap steel, providing a total of 2.2 Mt scrap metal in 1992. In the mean time, China increased its import of steel bars and rods to about 10 Mt in 1994 from less than 100,000 tonnes in 1990/1991 (table 42). China is expected to remain in short supply of high-grade steel products with a deficit of 5.7 Mt by the year 2000.

China also exports iron and steel. At present, however, steel enterprises would sacrifice $17-18 for exporting each ton of rolled steel instead of selling them in China. Nevertheless, the government has encouraged them to gain a market share in the international arena even though it means sacrificing profits. Exports of steel products from China were expected to increase from the present level of 2 Mt to 4 Mt in the next several years.

Composed of officials from the ministries of Foreign Trade, Planning, Metallurgy Industry, and Internal Trade, as well as the State Administration for the Inspection of Import and Export Commodities, a steel

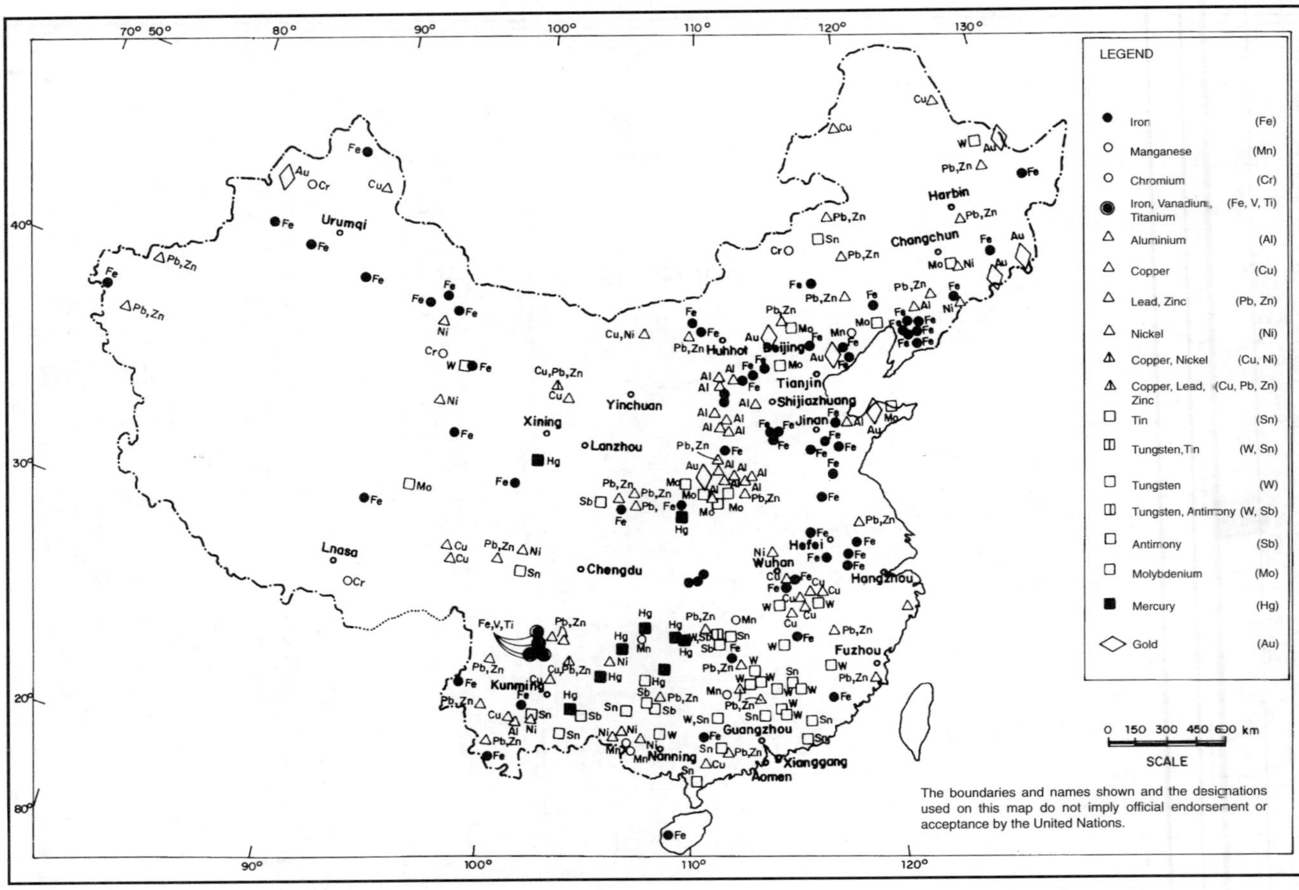

Source: Mineral Resources of China, 1992.

Figure 53. Distribution of metallic mineral resources in China

Table 41. Distribution of major iron ore reserves in China

	Iron ore reserve (Bt)	*Iron ore production (Mt)*	*Steel production (Mt), 1993*
Anshan – benxi area in Liaoning province	11.0	> 10	8.51 + 2.57
Qian'an (of Hebei Province) – Miyun (of Beijing) area	5.8	> 10	Shoudu 7.02
Panzhihua – Xichang area in Sichuan province	5.5	> 10	Panzhihua 2.42
Yichang – Enshi area in Hubei province	3.0		Wuhan 5.24
Wutai – Lanxian area in Shanxi province	2.6	3 - 5	Taiyuan 2.12
Nanjing (of Jiangsu province) – Ma'anshan (of Anhui province) – Lujiang area	2.3	> 10	Ma'anshan 2.13
Baotou – Bayan Obo area (of Inner Mongolia)	1.0	5 - 10	Baotou 3.08
Huoqiu area (of Anhui province)	1.0		Baoshan, Shanghai 6.98
Sub-total of the above	32.2 (64%)		40.07 (45%)
Others	17.9 (36%)		48.63 (55%)
Total China	50.1 (1990) world 2nd	235.7 (1993) world 3rd	88.70 (1993) world 2nd

Source: Mineral Resources of China, edited by Chinese Institute of Geology and Mineral Resources Information, Published by the China Building Materials Industry Press in 1992; and Mining Annual Review, 1994 and 1995, Published by Mining Journal, London.

Table 42. Production and trade of iron ore, pig iron, crude steel and steel products of China, 1988-1995

Year	*1988*	*1989*	*1990*	*1991*	*1992*	*1993*	*1994*	*1995*
Production (in '000 t)								
Iron ore	105,000	145,000	168,317	176,070	197,600	234,660	240,000	245,000
Pig iron	56,400	57,800	61,866	67,200	72,000	87,300	97,418	101,710
Steel, crude	59,000	61,200	66,038	70,570	80,000	88,680	92,610	92,970
Steel, rolled	47,000	48,700	51,209	56,380	65,300	75,900	79,390	79,390
Export (in t)					- SEP/0.7			
Ferrosilicon		82,592	244,386	319,997	270,000	340,000	300,000	
Pig/cast iron		530,000	380,000	620,000	580,000	340,000	1,540,000	
Steel:		880,000						
Bars and rods		279,163	800,963	1,115,033	820,000	220,000	580,000	
Shapes & sections		108,418	296,909	401,597	570,000	120,000	200,000	
Sheets and plates		286,238	515,007	456,775	400,000	310,000	430,000	
Tube and pipe		110,344	175,009	182,022		180,000	210,000	
Other			202,112	231,439				
Import (in t)					- SEP/0.7			
Iron ore	10,540,959	12,592,169	14,342,625	18,549,433	23,870,000	33,020,000	37,340,000	41,000,000
Pig/cast iron	869,911	690,000	1,310,000	360,000	86,000	810,000	310,000	
Steel:								
Bars and rods		940,000	400,000	90,000	430,000	10,810,000	10,050,000	
Billets & forgins	151,134	190,000	280,000	120,000	186,000			
Seamless pipe		1,140,000	680,000	110,000	1,170,000	1,170,000	1,390,000	
Shapes & sections		370,000	180,000	60,000	186,000	3,940,000	1,660,000	
Sheets & plates		6,010,000	2,380,000	2,040,000	2,650,000	13,950,000	9,080,000	

Source: Mining Annual Review, 1989, 1990, 1991, 1992, 1993, 1994, 1995 and 1996.
Figures for export and import in 1992 were estimated from those of first three quarters divided by 0.7.

subcommittee under the China Chamber of Commerce of Metals, Minerals, and Chemicals Imports and Exports is being set up to co-ordinate import and export prices, marketing, business contracts and product quality. The subcommittee ordered the steel companies to reduce their production of rolled steel products by 10 per cent in the second half of 1994. It also announced that the China National Metals and Minerals Import and Output Corp. (Minmetal) was the sole company that could negotiate on steel imports with Japanese steel export companies.

Joint venture activities between China's iron and steel enterprises and foreign companies have been encouraged by the Chinese government since the 1980s. These foreign companies are mainly from Australia, Germany, India, Italy, Japan, New Zealand, Peru, and the United States.

Non-ferrous metals

According to the country's economic development plan, China planned to produce 4.5 to 5 Mt of its 10 non-ferrous metals – aluminium, antimony, copper, lead, magnesium, mercury, nickel, tin, titanium, and zinc, in the year 2000. China produced 2.35 Mt, 2.52 Mt and 3.27 Mt of these metals in 1990, 1991 and 1993 respectively.

Aluminium

Like all the other countries in the world, China uses bauxite as the most important raw material for the production of metallic aluminium. China has bauxite retained reserves of a few billion tonnes, 90 per cent of which has been proved to be concentrated in Shanxi (39 per cent), Guizhou (20 per cent), Henan (17 per cent) provinces, and Guangxi Zhuang Autonomous Region (14 per cent).

Unlike bauxite deposits abroad (which were largely formed in the Cenozoic metallogenetic epoch and are dominated by lateritic weathering deposits of good-quality ores with gibbsite as the principal mineral in composition and high Al/Si ratios), China's bauxite deposits were mostly generated during the older Carboniferous metallogenetic epoch and are essentially paleo-weathering crust deposits composed of diaspore ores with low Al/Si ratios. China's bauxite ores, therefore, are unfit for opencast mining and have to use more energy for processing. China's bauxite deposits are commonly accompanied, paragenetically, by various other mineral resources. These include, coal beds and limestones in the overlying formations, clay, iron and pyrite in the host rock series, and a multitude of elements of Ga, V, Li, rare earths, Nb, Ta, Ti and Sc in the bauxite ores associated with Al.

China is the world's seventh largest alumina producer and sixth largest aluminium metal producer. The major producers of alumina in China are the Great Wall Aluminium Corp. (the Zhengzhou Aluminium Plant and the Zhongzhou Aluminium Plant), the Guizhou Aluminium Plant, the Shandong Aluminium Plant, and the Sanxi Aluminium Plant. The production of Alumina in China has increased steadily since the late 1980s reaching 1.9 Mt in 1994 from 1.5 Mt in 1988.

The demand for the alumina in China over the same period has been very strong, resulting in a much sharper increase in the import of aluminium, reaching 1.91 Mt in 1994 from 0.16 Mt in 1988. About 50 per cent of the alumina consumed at China's 69 smelters was imported in 1994. To change this situation, China planned to increase the import tax on alumina from 7 per cent to 20 per cent in 1995. The CNNC's Aluminium Plants planned to lower their second grade alumina price from 1,750 yuan/t to 1,650 yuan/ton in 1995. At the same time, China reintroduced restrictions on the use of the metal and its alloy in the construction industry to moderate aluminium demand. In recent years, China has remained a net importer of aluminium metal in spite of its increased exports. (table 43).

The main tasks facing China's aluminium industry are to reduce energy consumption in smelting, to recover the other minerals in an integrated way, and to explore for high-grade bauxite ores.

CNNC, Aluminium Co. of America (Alcoa) of the U.S., and Kobe Steel Ltd. of Japan have agreed to co-operate in developing China's aluminium sector. The agreement included upgrading China's technology in the aluminium sector and construction of greenfield refineries.

Table 43. Production and trade of bauxite ore and alumina/aluminium of China, 1987-1995

Year	*1987*	*1988*	*1989*	*1990*	*1991*	*1992*	*1993*	*1994*	*1995*
Production (in t)									
Bauxite		3,500,000	4,000,000	2,400,000	2,600,000	2,700,000	3,500,000	3,700,000	4,000,000
Alumina		1,500,000	1,440,000	1,460,000	1,600,000	1,600,000	1,820,000	1,900,000	2,100,000
Metal	615,000	713,000	750,000	854,000	860,000	1,100,000	1,250,000	1,450,000	1,680,000
Export (in t)						-Sep/0.7			
Bauxite	515,849	364,235		680,315	541,554	970,000	520,000	380,000	
Alumina								40,000	
Metal and alloys:									
Unwrought			13,411	65,125	68,556	63,377	67,784	130,901	
Semi-manufactures	20,157	28,891	6,299	17,764	27,300	34,911	30,281	39,815	
Import (in t)						-Sep/0.7			
Alumina	322,149	159,534	298,002	582,390	685,525	643,000	970,000	1,910,000	
Metal and alloys:									
Unwrought			175,510	71,771	43,711	119,540	165,694	168,853	
Semi-manufactures	148,130	75,381	70,812	43,858	65,028	105,698	125,030	177,194	

Source: Mining Annual Rview, 1987, 1988, 1989, 1990, 1991, 1992, 1993, 1994, 1995 and 1996.
Figures for export and import were estimated from those of first three quarters divided by 0.7.

Antimony

Antimony resources are abundant in China. China possesses 9 out of 17 large antimony deposits in the world, with reserves of more than 100,000 tonnes. The world's largest antimony mine, Hunan's Xikuangshan mine, is in China.

The antimony deposits in China are mostly large and medium size, and its antimony ores are simple in composition and relatively high in grade. China has four industrial commodity types of antimony ores, i.e. single antimony type (68 per cent of China's total reserves), tungsten-gold-antimony type, lead-antimony type and mercury-antimony type. China's antimony resources are mainly concentrated in Hunan, Guangxi, Guizhou, Yunan, Ganshou and Sanshi provinces (regions), which accounted for 90 per cent of the total antimony reserves in China.

China's output of antimony, about 60,000 tonnes, accounted for about 75 per cent of the world's total in 1994. In the past several years, the world supply of antimony exceeded demand. In 1993, the State Council and MFTEC approved the formation of the China National Tungsten and Antimony Import and Export Corp. to oversee the export of antimony and to clamp down on illegal exports. At the same time, the government removed subsidies and increased the tax burden on antimony enterprises. Chinese antimony producers began to cut back on production and reduced the export of antimony at low prices. Heavy rainstorms caused great difficulty in shipping products from the Guanxi and Hunan producing areas in the summer of 1994 and most western antimony consumers faced shortages of antimony in the third quarter of 1994. In April 1994, the price of antimony was about 15,000 yuan/t in China, but in August, went up to 36,000 yuan/t.

China's antimony production and export in the last ten years is shown in table 44.

Table 44. Production and trade of antimony ore and metal in China, 1986-1995

Year	*1986*	*1987*	*1988*	*1989*	*1990*	*1991*	*1992*	*1993*	*1994*	*1995*
Production										
Antimony	18,000	20,000	25,000	25,000	33,000	35,000	59,000	81,000	101,000	91,000
Export										
Antimony metal unwrought				32,993	33,368	29,373	18,827	12,997	12,927	

Source: Mining Annual Review, 1987, 1988, 1989, 1990, 1991, 1992, 1993, 1994, 1995 and 1996.
Figures for export in 1992 was estimated from that of the first three quarters divided by 0.7.

Copper

Copper is the second major non-ferrous metal in China. However, copper has been in short supply in China for many years. China produced 843,000 t of refined copper in 1995, and imported 480,000 t of copper ore from other countries in the same year (table 45). The net import of unwrought copper metal reached nearly 320,000 t in 1993. The government issued a ban on copper exports in 1988 and restricted domestic use.

China has retained copper reserves of a few tens of million tonnes, which are distributed in all the provinces, municipalities and autonomous regions of China except Tianjin Municipality. Over 70 per cent of reserves are concentrated in five regions, i.e. the middle and lower Changjiange river valley, Changdu of Xizang (Tibet), Sichun-Baiyinchang, Ganshou Jingchuan-baiyan and the Zhongtiao Mountains. More than 84 per cent of the reserves belong to more than 100 copper deposits with reserves over 100,000 tonnes. China has a nearly complete range of copper deposit types: porophry type (41 per cent), skarn type (27 per cent), stratiform type (11 per cent), volcano-sedimentary type (5.5 per cent) and copper-nickel sulfide type (6.4 per cent). Low grade ore makes up 80 per cent of the total copper deposits in China.

According to the country's economic development plan, China will need about 1.2 Mt of copper in the year 2000. To meet the demand for copper, Chinese geologists have actively prospected for copper for years and have discovered new resources recently, including six large, four medium and 26 small copper deposits in Jiangxi; a

Table 45. Production and trade of copper minerals and metals of China, 1986-1995

Year	*1986*	*1987*	*1988*	*1989*	*1990*	*1991*	*1992*	*1993*	*1994*	*1995*
Production (in t)										
copper, refined	350,000	400,000	510,000	520,000	560,000	560,000	659,000	730,000	684,000	843,000
Export (in t)										
Metal and alloys:							-Sep/0.7			
Unwrought				11,770	18,003	8,711	13,631	6,443	16,306	
Semi-manufactures		22,298	44,252	16,479	25,229	33,655	41,940	43,296	60,624	
Import (in t)							-Sep/0.7			
Ore		199,572	172,999	178,166	242,197	309,786	285,700	240,000	250,000	480,000
Metal and alloys:										537,897
Unwrought	171,118	75,490	84,370			113,966	313,577	363,544	121,396	
Semi-manufactures						52,167	148,190	208,317	283,057	

Source: Mining Annual Review, 1987, 1988, 1989, 1990, 1991, 1992, 1993, 1994, 1995 and 1996.
Figures for export and import were estimated from those of first three quarters divided by 0.7.

large copper zone near Toba in the Changdu region, Xizang; 47 areas containing copper occurrences in Nei Monggol; and six copper occurrences in the Zhongtiao Mountain area in the northern and north-eastern part of China. More than $5 million annually has been spent to explore for domestic copper prospects.

China has built four copper-producing bases: the lower and middle Yangtze River valley, Sichuan-Yunnan, Jinchun-Baiyinchang and Zhongtiao Mountain regions. Founded in 1981, the Jiangxi Copper Co. is the largest and most efficient copper producer in China, the production capacity of which was expanded from 70,000 t/y to 140,000 t/y in 1993. The company's copper reserves account for one third of the total reserves of the country and its operations include six copper mines scattered around Poyang Lake. The operation also recovers gold, silver, molybdenum and tungsten as by-products.

China has three copper processing bases, i.e. Luoyang in Henan province, Shenyang in Liaoning province and Lanzhou in Ganshu province. Now the fourth base is developing in Wusong of Shanghai, the largest consuming centre in the country.

CNNC has been seeking cooperation with foreign countries on the joint development of mining operations, including projects on the Chengmenshan copper mine in Jiangxi province and the copper mines in Heilongjiang and Sichuan provinces. Joint ventures have also been formed with foreign countries for expansion of refined copper production from Chinese smelters.

China has also sought copper resources abroad, such as purchase of the Altamira copper deposits in Chile. China North Industries Corp. signed an agreement with Philcopper Gold Mining Co., of the Philippines, in which 30 per cent of the copper concentrates of the deposits with contained gold, silver, pyrites and rare metals will supply China for a period of 30 years. China reached an agreement with Udokan Mining Co. (UMC) of Russia on the construction of a large copper mine in Chita region in Russia. UMC will sell no less than 260,000 t and up to 400,000 t of contained copper annually for 25 years, to CNNC when it comes on stream in 1997. The Gold Mines of Australia Ltd. (GMA) signed a letter of intent with CNNC and Asia Resource Capital Ltd. on GMA's Mount Lyell copper-gold mine in Tasmania, Australia. CNNC will purchase 100 per cent of the Mount Lyell mine's output for four years from January 1, 1996.

Gold

According to the Gold Administration Bureau of MMI, China produced 90.2 t and 105 t of gold in 1994 and 1995, which were well below the figure of 130 tonnes estimated by western gold observers. China is the sixth of the top ten gold producing countries in the world. According to the World Gold Council, China was the fourth largest gold consumer in the world after India, the United States and Japan.

The overwhelming majority of gold mine properties in China belong to the state, i.e. to the different levels of government – central, provincial, or local governments (district, municipality, or country). Only a few small gold mines belong to rural collectives.

Lead and Zinc

The retained reserves of lead and zinc metal in China accounted for 33 and 84 million tonnes of metal respectively by the end of 1990. They are extensively distributed in China, but more concentrated in Yunan, Inner Mongolia, Guangdong, Gansu, Hunan, Jiangxi provinces and the Guangxi autonomous region, which are responsible for 74 per cent of the country's total reserves. Major types of lead-zinc deposits in China include sedimentary telescoped type (52 per cent), hydrothermal type (25 per cent), volcanic type, and skarn type (8 per cent). The sulfide type ore is dominant in China, which contributes 80 per cent of the total reserves. The grade of China's lead-zinc ore is intermediate; the average grade of lead is 1.42 per cent and that of zinc 3.15 per cent; the Pb/Zn ratio is 1:2.4.

China is one of the world's principal sources of lead and zinc. Over 100 lead and zinc mines have been constructed throughout the country, producing up to 10 million tonnes of ore each year. China's lead and zinc production reached 268,000 tonnes and 530,000 tonnes respectively in 1990, and has ranked fourth in the world.

Nickel

The nickel reserves of China total several million tonnes. The nickel industry in China was established in 1957. In 1958, a huge Cu-Ni deposit, Baijiazuizi deposit, was discovered in Jinchuan, Gansu province, and a mine was soon constructed there. The deposit has a total retained nickel reserve of 5.09 million tonnes and an average nickel grade of 1.06 per cent.

The nickel production in China accounts for about 30,000 tonnes annually, 80 per cent of which is from the Jinchuan, Gansu province with the rest being derived mainly from the Hongqiling nickel deposit in Jilin province.

A joint venture was announced in December 1994 between the Jinchuan Nonferrous Metals Corp., a subsidiary of CNNC of China and INCO Ltd. of Canada. The joint venture plans to construct a nickel chemical plant in the Shanghai area at a cost of about $10 million to produce nickel salts for the Asian market. After completion, Jinchuan will be able to increase its output to 50,000 t/y of refined nickel. The Jinchuan Nonferrous Metals Corp. is the largest nickel and cobalt producer in China now. Its annual output accounts for more than 80 per cent of the country's total output.

Rare Earths

China has the world's largest resources of rare-earths with reserves of 3.6 Mt, equivalent to four times that in the rest of the world. The largest occurrence of rare-earths is associated with the magnetite deposits at Bayan Obo, Nei Monggol. In addition to the light rare earths in Nei Monggol, China's heavy rare earths i.e. weathered-crust deposits in the Nanling region (Hunan, Jiangxi, Guangdong, Guangxi and Fujian provinces) are also well-known.

China is also one of the largest producers of rare-earths in the world. China's production was believed to equal or to exceed U.S. production of rare-earths in 1988. The 'global rare-earths rush' in the late 1980s brought vitality to the rare earths industry in China. Since then, China's productive capacity has leaped to being first in the world along with its processing capacity. Baotou is China's largest rare-earths producer, accounting for 80 per cent of its total output. The Bayan Obo ore is reduced to pig iron at Baotou, and the slag is treated for by-product rare-earths recovery.

China is ranked second in the world, behind the U.S., in rare earths consumption.

China is a major rare earths exporting country and holds a decisive position in the world market, exporting high-purity rare-earths and yttrium compounds, most notably to Japan. There is keen competition in international marketing of rare earths.

The State Planning Commission continues its ban on foreign investment in domestic rare earth mining projects. However, foreign investments in rare earth metallurgical processing using, advanced technology for extraction and separation, are exceptions to the ban.

Tin

China has rich tin resources and its reserves may account for as much as 20 per cent of the total world reserves. China's tin provinces belong to the Circum-Pacific belt, and the minerlizations are Mesozoic and Cenozoic in age. China's tin reserves are highly concentrated in Guangxi and Yunnan (60 per cent), and Hunan and Guangdong (20 per cent) provinces. Another characteristic is that the primary tin ores are dominant, making up 84 per cent of the country's total reserves. Placer tin holds a subordinate position.

60 per cent of China's tin production comes from the Yunnan province with its Gejiu tin province. Another important tin province is the Dachang tin deposit in Guangxi, which accounts for about 20 per cent of the country's tin production output. China's tin smelter capacity in 1988 was estimated to be 25,000 t.

Although China is a major global tin producer, most of its output is for domestic consumption. China's tin production and export is shown in table 46. Exports of tin decreased from 15,894 t in 1987 to 10,717 t, valued at $71.04 million, in 1988. Nearly 60 per cent of China's export of tin metal in 1988 was to the U.S.

Table 46. Tin production and export of China, 1986-1995

Year	*1986*	*1987*	*1988*	*1989*	*1990*	*1991*	*1992*	*1993*	*1994*	*1995*
Production										
Tin, smelter	20,000	25,000	29,000	40,000	30,000	38,000	39,000	52,000	54,000	60,000
Export										
Tin metals/alloys unwrought		15,894	10,717	9,874	10,135	15,692	28,524	40,703	44,379	

Source: Mining Annual Review 1987, 1988, 1989, 1990, 1991, 1992, 1993, 1994, 1995 and 1996.
Figures for export in 1992 was estimated from that of the first three quarters divided by 0.7.

In June 1994, China officially joined the Association of Tin Producing Countries (ATPC) and pledged to reduce its tin exports to 20,000 t in 1994. In an effort to meet its ATPC quota, the government published new regulations concerning tin exports in mid-1994. Under the new rules, foreign traders wishing to export tin must coordinate with the China Chamber of Commerce of Imports and Exports of Metals.

Titanium

By the end of 1990, the retained reserves of rock ilmenite (TiO_2) amounted to nearly 600 Mt, placer ilmenite (mineral) over 40 Mt, rock rutile (TiO_2) over 7 Mt and placer rutile (mineral) 0.8 Mt. Rock ilmenite reserves in China are concentrated in Sichuan (95 per cent) and Hebei provinces; and the placer ilmenite reserves in China are concentrated in Hainan (50 per cent), Yunan, Guanxi and Guangdong provinces.

Annual production capacity of titanium in China is only 3,000 t generated by three plants. In comparison, the smallest plants worldwide have at least a 5,000 t/y capacity for sponge, while the more efficient plants have an economy of scale of 10,000 t. China has been seeking technology to treat the titanium-ferrous slag produced at Ma'anshan, Anhui, and Panzhihua, Sichuan for the production of pigment and metal.

Tungsten

China has large resources of tungsten. About 70 per cent of China's tungsten reserves are concentrated in the Nanling region, i.e. Hunan (30 per cent), Jiangxi (22 per cent), Guangdong, Guangxi and Fujian. 64 per cent of the reserves are scheelite ore and 34 per cent, wolframite ore.

China produced over 60,000 tonnes of tungsten ore concentrates in 1990, ranking first in the world. More than 95 per cent of the country's tungsten production comes from the five provinces mentioned above with 45 per cent of the production from Jiangxi.

China continues to dominate the world market for tungsten. Its tungsten exports are mainly concentrates, not value-added products. In 1988, China's exports of ore totalled 27,290 t with a value of $96.8 million, and 408 t of metal with a value of $3.2 million. China's tungsten production and export is shown in table 47.

Table 47. Tungsten production and export of China, 1988-1995

Year	*1988*	*1989*	*1990*	*1991*	*1992*	*1993*	*1994*	*1995*
Production (in t)								
Tungsten, content of mine output	25,000	30,000	35,000	30,000	25,000	22,000	27,000	21,000
Export (in t)								
tungstate		791	335	420	6,000	9,226	12,946	
ore			16,450	6,071	1,581	699	1,133	

Source: Mining Annual Review, 1989, 1990, 1991, 1992, 1993, 1994, 1995 and 1996.
Figures for export in 1992 was estimated from those of the first three quarters divided by 0.7.

1.1.1.2 Democratic People's Republic of Korea

Ferrous metals

The Democratic People's of Korea's iron ore reserves are estimated at 3,300 Mt. High-grade ore deposits have been found in the Munrakpyong and Pungsan areas. Iron ore has developed at Musan, Unryul, Songhung, Komdok, Tokonsong and Schaeri. Promising deposits of magnetite ore occur in South Hamgyong province. Figure 54 shows the administrative map of the Democratic People's Republic of Korea.

Production of iron ore in the late 1980s is estimated at 8.5 Mt/y. The Musan iron ore complex, which is the largest producer, produced some 10 Mt of iron in 1993. Production of steel was 4.5 Mt in 1987 and 4.7 Mt in 1988. The main steel producer, Kim Chaek steel works, which has a capacity of 4 Mt/y, was negotiating with the Republic of Korea's Pohang Iron & Steel Co. to modernize its steel-making facilities.

Non-ferrous metals

High-grade non-ferrous ore deposits have been located in ten districts in the northern highlands, notably in Munrakpyong, Ryongam, Sinpa, Huchang and Pungsan. Non-ferrous metals produced in 1988 were about 800,000 t (against 700,000 t in 1987). During the third Seven Plan period (1987-1993), a target of 1.7 Mt/y was envisaged for non-ferrous metals.

The Ministry of Mining & Industry has formulated plans to upgrade technology for extracting non-ferrous ores.

MAP NO. 3399 Rev. 1 UNITED NATIONS
SEPTEMBER 1993

Figure 54. Administrative map of the Democratic People's Republic of Korea

Copper

The country has some copper deposits, estimates being around 2.2 Mt. The country also imports copper ore from Peru.

Gold

Some Japan-based Korean businessmen have established a company in Tokyo to assist the Democratic People's Republic of Korea with a gold mining project at Unsan, 100 km north of Pyongyang. Production may commence in the 1990s with an initial output of 2 t/y.

Lead and zinc

The deposits of lead and zinc are estimated at 6 Mt and 12 Mt respectively. At Komdok, where a lead-zinc complex with a capacity of 100,000 t/y began production in 1992, further modernization is under way.

Rare earths

Rare earth ores have also been discovered in the North Hamgyang province. The South Hamgyang province has promising deposits of magnetite, which are proposed to be exploited shortly.

Wollastonite and molybdenum

Exports of wollastonite and molybdenum concentrates have started, by Inphung Trading Corp. The company has already been exporting copper concentrates and kaolin.

1.1.1.3 Japan

Ferrous metals

The production of pig iron has for years been in second position in the world. Japan's production of crude steel reached 99.55 million metric tonnes in 1992, becoming the second in the world before the United States, after the former USSR. Japan produced only 0.040 million tonnes in actual weight of iron ore but imported 113.743 million tonnes in 1992, accounting for 30 per cent of the world total. Japan has ranked as top iron ore importer for many years.

Non-ferrous metals

Since non-ferrous metal prices maintained high levels in 1988, 1989 and 1990, the Japanese mining industry on the whole has enjoyed stable and sound conditions. There were no shut-downs or reductions of production at major mines as seen in 1986 and 1987.

However, the continuing low price of lead and zinc in the early 1990s resulted in the shutdown or in production cutbacks in some Japanese metal mines. Nippon Mining Co.'s Toyota decreased its zinc mine production by 12 per cent in 1993; Dowa Mining Corp. closed its Hanoka mine in March in 1994. The total production of the Kamioka mine of Mitsui Metal Mining, which accounts for about 40 per cent of the total Japanese primary zinc output, decreased by 15 per cent in 1993. Japan's domestic base metals mine production was decreased in 1993, and was cut sharply in 1994: 41 per cent for copper; 40 per cent for lead and 7 per cent for zinc. By 1994, there were only three major mines still operating in Japan.

During the late 1980s and in 1990 and 1991, the production levels in the non-ferrous smelting industry were high, supported by strong demand. Although the domestic consumption of metals decreased significantly in 1992, refined metal production increased for copper, and stayed at the same level for lead and zinc.

MAP NO. 3487 Rev. 1 UNITED NATIONS
NOVEMBER 1992

Figure 55. Administrative map of Japan

Reflecting the slow-down of manufacturing in Japan, consumption of base metals decreased again in 1993 (table 48). Copper consumption decreased by 2.4 per cent, lead by 7.9 per cent, zinc by 7.8 per cent and aluminium by 6.2 per cent. The consumption of each non-ferrous metal in 1994 moved in different ways reflecting their own individual demands: Copper increased by 1.5 per cent, lead decreased by 3.4 per cent and zinc decreased by 2.0 per cent. The production of lead and zinc in 1993 and 1994 reduced by over 5 per cent, but the production of copper increased by 2.4 per cent in 1993 then decreased by about 6 per cent in 1994 (tables 49 and 50).

Imports of non-ferrous metals greatly increased in 1988, 1989, 1990 and stayed in 1991. The economic recession in Japan resulted in a drastic decrease in the import of refined metals. Importation of refined copper, refined zinc and primary aluminium decreased by 42 per cent, 28 per cent and 11 per cent in 1992. This situation continued for copper and zinc in 1993 and 1994, but the import of refined lead increased in 1994.

Gold

Despite the poor market situation, the gold production in Japan has increased since 1989 (table 51). Sumitomo Metal Mining, one of the six major non-ferrous smelters in Japan, is a large global gold producer. Its large Hishikari mine produced more than 80 per cent of total gold mine output in Japan in 1992. Hinshikari

Table 48. Copper production and import of Japan, 1988-1995

Year	*1988*	*1989*	*1990*	*1991*	*1992*	*1993*	*1994*	*1995*
Production (in ton)								
copper, mine	16,666	14,650	12,927	12,413	12,047	10,277	6,043	2,376
copper, refined	955,108	989,566	1,007,976	1,076,283	1,160,859	1,188,776	1,119,168	1,187,959
Import (in ton)								
copper, ores/concs.	n/a	913,309	955,550	1,092,455	1,012,825	1,098,626	1,046,232	1,093,067
copper, refined	n/a	487,191	618,785	625,114	366,569	363,635	354,869	n/a

Source: Mining Annual Review, 1989, 1990, 1991, 1992, 1993, 1994, 1995 and 1996.

Table 49. Lead production and import of Japan, 1988-1995

Year	*1988*	*1989*	*1990*	*1991*	*1992*	*1993*	*1994*	*1995*
Production (in ton)								
lead, mine	22,889	18,595	18,727	18,329	18,839	16,470	9,946	9,659
lead, refined	267,395	260,230	261,076	272,592	270,296	258,128	234,497	226,564
Import (in ton)								
lead, ores/concs.	n/a	194,910	195,142	205,101	203,475	209,505	185,025	129,706
lead, refined	n/a	59,689	70,423	69,834	50,200	42,318	52,369	n/a

Source: Mining Annual Review, 1989, 1990, 1991, 1992, 1993, 1994, 1995 and 1996.

Table 50. Zinc production and import of Japan, 1988-1995

Year	*1988*	*1989*	*1990*	*1991*	*1992*	*1993*	*1994*	*1995*
Production (in ton)								
zinc, mine	147,217	131,794	127,273	133,004	134,510	118,510	100,653	95,274
zinc, refined	678,175	663,807	687,461	730,829	729,454	695,687	665,502	663,562
Import (in ton)								
zinc, ores/concs.	n/a	523,105	586,323	572,291	586,819	555,977	507,720	544,836
zinc, refined	n/a	132,746	140,470	142,194	102,978	8,439	64,156	n/a

Source: Mining Annual Review, 1989, 1990, 1991, 1992, 1993, 1994, 1995 and 1996.

Table 51. Gold production of Japan, 1988-1995

Year	*1988*	*1989*	*1990*	*1991*	*1992*	*1993*	*1994*	*1995*
Production (in ton)								
Gold mine	7.31	6.10	7.30	8.30	8.89	9.35	9.55	9.19
Gold refined	92.09	110.33	108.15	103.02	107.96	108.77	102.78	103.15

Source: Mining Annual Review, 1989, 1990, 1991, 1992, 1993, 1994, 1995 and 1996.

increased its production by 4 per cent in 1993, and 2 per cent in 1994. It is also engaged in exploration activities in the Otoge area in Yamagata Prefecture, where a gold bonanza was believed to be found.

Geological surveying by MMAJ in 1992 revealed the existence of gold in Mitsui's mining area in Oita Prefecture in the southern part of Japan. Analysis suggests that the bonanza may be similar to the one in Hishikali.

1.1.1.4 Mongolia

Mongolia is a major producer and exporter of copper and molybdenum, and is a leading world producer of fluorspar. Some 20 per cent of Mongolia's GNP or 40 per cent of the country's total exports flow from the mining industry. Copper, molybdenum and fluorspar account for over 35 per cent of Mongolia's exports.

Figure 56. Administrative map of Mongolia

Mongolia has large deposits of copper, molybdenum, fluorspar, gold, zinc, nickel, tin, wolfram, coal and phosphorites. About 4,000 mineral deposits and ore occurrences are known in Mongolia. Mongolia has focussed upon mineral and mining development since 1975. Exploration prospects include known and potential porphyry copper-molybdenum systems, lead-zinc skarns, gold skarns, tin-tungsten in granites and placers. Large scale mining has focussed on the development of copper, molybdenum, fluorspar and coal. In addition, hundreds of deposits of non-ferrous, rare and precious metals and limestones are also mined in small scale operations. The distribution of major mineral deposits and mines in Mongolia is shown in figure 57.

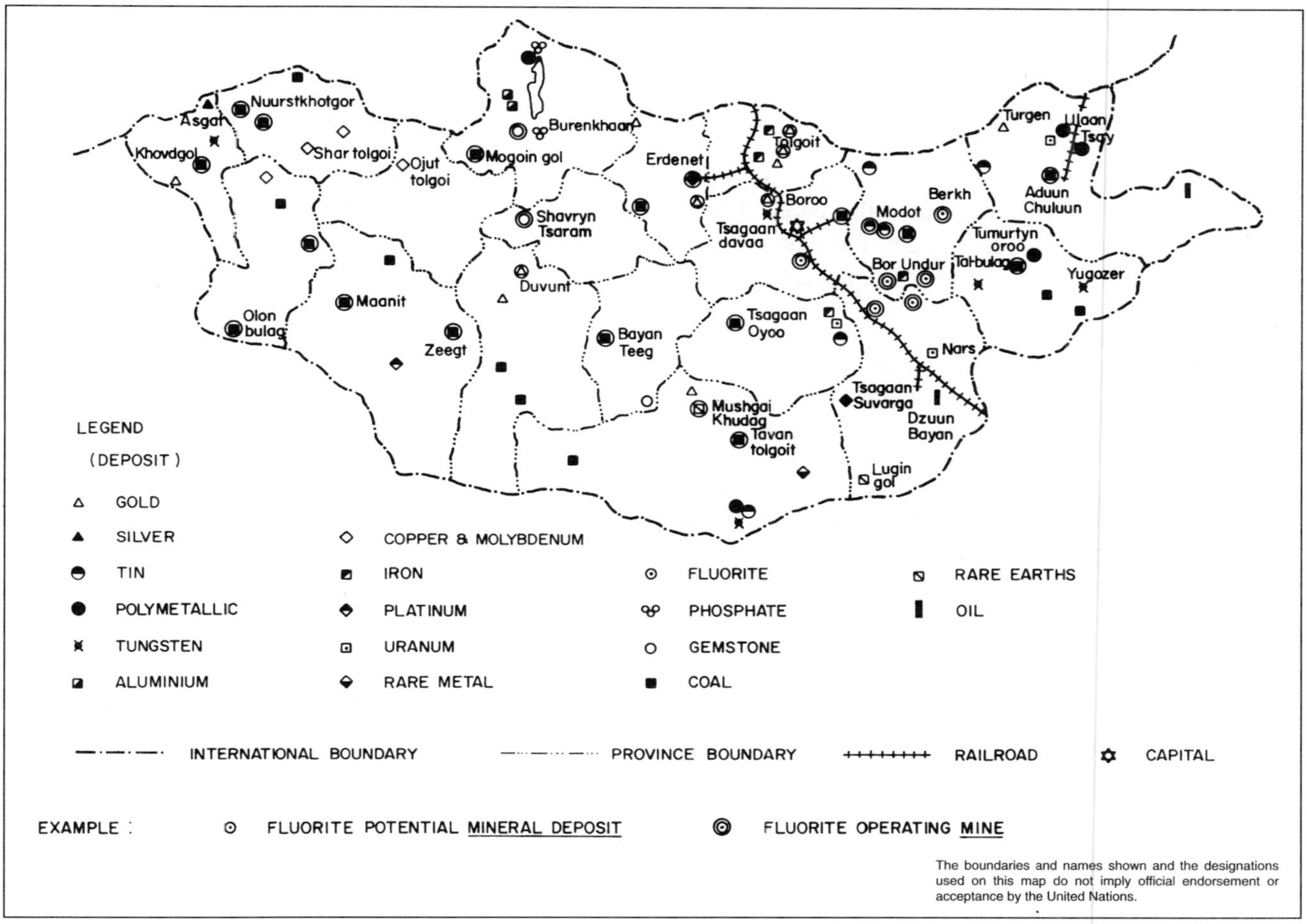

Figure 57. Distribution of major mineral deposits and mines in Mongolia

Copper and molybdenum

The Erdenet mining complex is the biggest mining operation in Mongolia, accounting for 16 per cent of its industrial production and one-third of its total earnings. The mining complex at Erdenet about 400 km northwest of Ulaanbaatar, is a former Soviet-Mongolian joint venture which has been mining a large porphyry copper deposit for over 18 years.

The Erdenet dressing plant produced 300,200 t of copper concentrate and 3,238 t of molybdenum concentrate in 1992, 117,000 t and 4,000 t in 1993, and 340,000 t and 3,700 t in 1994. It processed its copper concentrate in plants in Kazakhstan, producing 40,000 t of cathode copper, 6-9 t of refined silver and 500 t of molybdenum oxide in 1994. In 1995, shortages of energy and materials prevented the Erdenet plant from operating at its full capacity. Power cuts alone during 1992 accounted for the joint venture's $US40 million financial losses.

Copper and molybdenum exported to Russia, China, Switzerland, Germany, and Finland accounted for nearly half of the country's exports.

The copper-molybdenum deposit at Tsagaan-Suvarga in Dornogovi province of the eastern Gobi desert needs to be developed, where the estimated reserves are 240 Mt with average content of 0.53 per cent copper, 0.018 per cent molybdenum, plus 0.084 g/t gold, 0.39 g/t rhenium, 8.3 g/t selenium, 2.63 g/t silver and 12.8 g/t tellurium.

Gold

Mongolia is rich in gold resources with major concentrations in the eastern and central parts of the country. The Boroo, Zaamar and other gold deposits in Mongolia have substantial proven gold reserves in both primary and placer deposits. These deposits are a continuation of gold-bearing belts and zones in the former Soviet Union and display certain similarities. More than a dozen gold deposits have been identified, with total estimated reserves of 3,000 t of gold. Some 60 per cent of all gold reserves prospected so far are from placers.

Gold ventures include Tolgoit, which is the largest gold producing district, where Mongol-Sovtsvement operates two Russian built dredges with a capacity of 350-400 kg/y. The other important gold producing operations are those in the Zaarmar district 250 km west of Ulaanbaatar with production of 250-300 kg/y, and Duvunt gold operations in Bayan-Khongor province in central Mongolia.

The Mongolian Government approved the 'GOLD' project in 1992 to increase the country's production tenfold during the following decades. The government is interested in developing gold refining facilities, employing state-of-the-art electroplating and other advanced technology to refine gold to international standards (99.99 per cent purity) and to market value-added products with assistance from foreign countries.

Following the recent government policy to encourage gold mining, the country's gold production has increased by 700 kg to almost 2,000 kg in 1994. To support the development of the private sector, the government has granted mining licenses to many small-scale local mining enterprises, and currently about 60 mostly small-scale mining enterprises are producing alluvial gold. The Oloon Ovoot hard rock gold deposit located in the South Gobi province has been acquired by the Mongolian private company Omnogovi, which is planning to start its mining operations at the beginning of 1995.

In December 1994, the Mongolian corporation Mongol Alt and the Canadian company Mongolia Gold Resources agreed to develop Bumbat gold fields, which are situated 200 km west of Ulaanbaatar. The property contains 200 quartz veins, many of which contain significant gold mineralization. The results of the exploration work in 1993-1994 showed that the area licensed by the joint venture has seven quartz veins which host at least 29,300 kg of gold.

Among the attractive mineral prospects for potential investors are the big undeveloped gold deposits in the Zaamar area which include about 150 gold bearing quartz veins (estimated reserves – 100 t Au); the hard rock deposit, Boroo (30 t Au), located 150 km north of Ulaanbaatar in the area of developed infrastructure; the well prospected Tuul river alluvial gold placers; the Erdenet (or Tavt) gold prospects located 200 km north-west of Erdenet with about 100 gold bearing veins; the Olon Ovoot gold veins in central Gobi discovered in 1990 (currently being prospected by a Mongolian-Japanese joint venture) and the Bayan-Khongor mining district hosting several interesting hard rock and placer deposits which need further detailed investigation.

Lead and zinc

Polymetallic (Pb-Zn-Ag) ore deposits at Ulaan, Tsav in eastern Mongolia have estimated reserves of about 75 Mt. The Mineral Development Corp. of Mitsui Co. of Japan has, since 1992, agreed to undertake the extensive Mongolian-Japanese joint project on exploration and development of these base metal prospects, as a part of the Japanese technical development aid to Mongolia. Following the agreement between the Mongolian Ministry of Trade and Industry, Mongolian State Geological Centre and the Japanese International Cooperation Agency (JICA), the pre-feasibility study for the development of the Tsav Pb-Zn deposit was expected to be completed during the first half of 1995. The two sides has agreed to extend the agreement and to carry out a more detailed study by March 1996.

Silver

Mongolia's silver deposit at Asgat near the western border of Mongolia with large reserves of proven mineable 3.5 Mt, and possible 21.4 Mt.

Tin and Tungsten

Mongolia's tin and tungsten are mined at the Modot, Tsagaandavaa and Ulaan-Ul deposits. An electro-metallurgical plant was being constructed in 1991 near Darhan with Japanese assistance. Iron ore from the Tumurtin Ovo deposit would be used for this plant.

Mongolia's tungsten concentrate (60 per cent WO_3) has been around 6,300 t/y, the entire quantity being shipped and processed at Nal'chgik hydrometallurgical plant. The Hungarian-assisted Caganova tungsten mining and dressing project was commissioned in 1988.

Uranium

A joint Mongolian-American-Russian venture Gurvan Saikhan ('Three Beauties'), established in 1994 to prospect for uranium in the Mongolian Gobi region, has reported several new occurrences with uranium deposit in the Central Gobi province and is testing the possibility of leaching uranium ore underground.

Figure 56 shows the administrative map of Mongolia, whilst figure 57 shows the minerals and locations of Mongolian mines.

1.1.1.5 Republic of Korea

Ferrous metals

Figure 58 shows the administrative map of the Republic of Korea. In early December 1989, the Republic of Korea became the tenth country to reach 20 Mt/y steel production level. As a result, the country's per capita steel consumption reached 440 kg. Its self-sufficiency rate in iron and steel products rose from 76 per cent to 83 per cent in 1989. Pohang Iron & Steel Co. (POSCO) registered an estimated turnover of $7.6 billion in 1991 and has maintained its position as the largest steel maker of the country.

The steel industry of the Republic of Korea is composed of approximately 200 companies, including one fully-integrated steel company, 12 electric furnace companies and about 70 rolling-reduction mills. As of January 1991, the major steel capacities nationwide were: iron making 17.8 Mt (eight blast furnaces), steel making 25.4 Mt (11 BOFs and 33 electric furnaces) and rolling 30 Mt (flat-rolling product of 21.6 Mt and long products of 8.4 Mt). By the year of 1993, POSCO of Republic of Korea had a total of 21 Mt/y capacity, including 9.4 Mt at Pohang and 11.4 Mt at Kwangyang, ranking second only to that of NSC of Japan.

By 1992, the Republic of Korea had iron ore reserves of 177 Mt of 35 per cent Fe grade, the Yangyang iron mine in Kangwondo and the Shinyaemi iron mine, being the two major iron ore producers. Iron ore production has, however, continuously decreased: from 390,000 t in 1988; 344,294 t in 1989; 298,292 t in 1990; 221,525 t in 1991; 221,502 t in 1992; 218,663 t in 1993; to 191,313 t in 1994.

Conversely, imports of iron ore have risen rapidly with Pohang Iron and Steel Co's expansion. 26.7 Mt of hematite iron ore and 1.01 Mt of magnetite iron ore were imported in 1991, mostly from Australia. In 1992 iron ore imports were 33.8 Mt. Australia supplied 15 Mt; Brazil shipped 7.7 Mt and India about 4 Mt.

The Republic of Korea's crude steel demand increased more than 50 per cent in two years from 24.5 Mt in 1991 to 37.2 Mt in 1993. The Republic of Korea's crude steel demand rose 10.1 per cent to 42 Mt in 1994, of which domestic demand was 31 Mt.

The Republic of Korea's total crude steel output increased about 54 per cent in the five years from 1989 to 1994. Crude steel production amounted to 21.9 Mt in 1989 (of which 70 per cent from POCSO); 16 Mt in 1991; 27.2 Mt in 1992 (of which nearly 37 per cent was from POSCO); 33 Mt in 1993 (of which nearly 67 per cent was from POSCO, and the rest from electric furnace); 33.7 Mt in 1994 (of which 64 per cent was from POSCO and 36 per cent from electric furnace).

Figure 58. Administrative map of the Republic of Korea

The Republic of Korea exported 6.8 Mt of steel products in 1991, 8.9 Mt in 1992, 11.8 Mt in 1993 and 10.1 Mt in 1994. Sharp increases in 1992 and 1993 were due to strong export demand in the South-East Asian market. Exports declining in 1994 from the previous year was due to increased domestic demand coupled with stagnation in overseas sales. Steel product imports amounted to 2.8 Mt in 1992, 4.2 Mt in 1993 and 8 Mt in 1994.

Non-ferrous metals

Since local ore production of non-ferrous metals had not grown as much as the demand of the refineries, the Republic of Korea imported 463,175 t of copper concentrates, 449,999 t of zinc concentrates and 31,320 t of aluminium in 1989.

Total demand for non-ferrous metals has increased recently. The demand in 1989 was 500,000 t for refined aluminium, 241,616 t for refined zinc, 114,784 t for refined lead, and 193,644 t for refined copper. At the end of 1993, demand for aluminium stood at 611,649 t, for zinc 295,340 t, for lead 172,058 t, and for electrolytic copper 381,474 t. (1995)

Total non-ferrous production in 1993 consisted of 221,570 t of copper; 85,160 t of lead and 11,330 t of utility nickel based on imported nickel oxide. In 1993, the Republic of Korea imported 418,456 t of aluminium ingot, 131,307 t of billet; 29,580 t of DC ingot and 32,306 t of scraps. Refined copper imports in 1993 reached 156,000 t as compared with 129,749 t in 1992.

The Republic of Korea's total non-ferrous metal production in 1994 consisted of 224,000 t of copper, 76,000 t of lead and 272,000 t of zinc. In 1994, the Republic of Korea imported 503,437 t of aluminium ingot, 240,445 t of copper cathode, 42,801 t of refined zinc and 42,444 t of refined lead.

Aluminium

Aluminium of Korea Ltd. produced 16,172 t of refined aluminium metal in 1989. The Republic of Korea has depended totally on imports, amounting to 95 per cent of its 400,000 t of aluminium metal demand. It has imported aluminium for domestic demand since March 1990 when Aluminium of Korea Ltd. closed down the operation of its Ulsan aluminium smelter with local output of refined aluminium amounting to only 178 t.

The Republic of Korea imported 300,104 t of aluminium ingots and 71,389 t of billets and 57,805 t of aluminium scrap in 1991. In 1992, Aluminium of Korea Ltd. was constructing a new rolling mill with capacity of 150,000 t/y hot and 100,000 t/y cold.

Copper

Copper reserves were estimated by 1991 to be only 20 Mt, averaging 1.13 per cent Cu. The Republic of Korea's total copper refining capacity by 1991 was 200,000 t/y at two smelters: one at Onsan with capacity of 150,000 t/y and the other at Changhang with capacity of 50,000 t/y, both operated by Lucky Metals Corp.

The Republic of Korea imported 429,317 t of copper concentrates in 1991. The Republic of Korea imported 122,046 t of cathodes and 51,348 t of copper scrap in 1991 for remelting purpose. Lucky Metal Corp. consumed 244,173 t of imported copper concentrates in 1992.

The Republic of Korea's total refined copper production amounted to 179,890 t in 1989; 185,563 t in 1990; 201,911 t in 1991; and 209,757 t in 1992. Refined copper production from Lucky Metal Corp. reached 167,000 t of cathode in 1993 and 168,700 t of cathode in 1994 from its Onsan refinery, and 52,000 t in 1993 and 53,330 t in 1994 from it's Changhang smelter.

Lead and Zinc

The reserves of lead and zinc in the Republic of Korea are estimated at 37 Mt averaging 3.9 per cent lead and 7.3 per cent zinc. The bulk of the deposits occur at the Yeonwha Mine, Kungsang North Province (1991).

The Republic of Korea's total lead and zinc concentrate production amounted to 44,077 t of lead and 25,265 t of zinc in 1991; 43,766 t of lead and 27,255 t of zinc in 1992; 14,818 t of lead and 27,616 of zinc in 1993; down to only 4,345 t of lead and 14,243 t of zinc in 1994.

The biggest producer is Youngpoong Mining Co. Ltd. In 1992, it produced 30,393 t of zinc concentrates, accounting for nearly 70 per cent of the country's total, and 13,305 t of lead concentrates, accounting for nearly 50 per cent of the country's total, from its Yeonwha mine, Kyungpuk province. The mill capacity is at present 30,000 t/month of run-of-mine ore. The other significant producer is Dongbang-Janggun Mine, located in Sochun, Bongwha-gun.

Domestic demand for zinc ingots had increased by 9.4 per cent annually to 220,000 t in 1991 due to increased demand for galvanizing and die-casting. Total zinc smelting capacity in 1990 was 250,000 t/y at two smelters, one at Onsan Smelter of 165,000 t/y operated by the Korea Zinc Co. Ltd. (KCZL) and the other one at Yeonwha Mine site of 80,000 t/y capacity operated by Youngpoong Mining Co.

Onsan Refinery produced 248,612 t of refined zinc and 30,600 t of refined lead in 1990; 171,523 t of zinc and 32,534 t of lead in 1991; 169,844 t of zinc and 61,524 t of lead in 1992; 182,594 t of zinc and 85,161 t of lead in 1993; 186,770 t of zinc and 86,480 t of lead in 1994. The Republic of Korea's production of refined zinc was 258,830 t in 1991 and 255,911 t in 1992.

KCZL exported 48,911 t of refined zinc in 1989 and 74,683 t in 1990. In 1991, the Republic of Korea imported 468,242 of zinc concentrates and 9,456 t of lead concentrates.

Gold and silver

The Republic of Korea's total demand for gold in 1992 was estimated at 55 t, of which 32 t were imported. The Republic of Korea's gold production in 1992 was reported to be 23,263 kg from imported gold concentrates and 1,399 kg from domestic ore. The Republic of Korea produced 10,443 kg gold in 1993, including 1,348 kg of gold from domestic ores, and produced 12,328 kg of gold in 1994, including 1,370 kg gold from domestic ores.

The demand for silver in 1992 was 598 t, of which 276 t were imported valued at $36 million. Silver production in 1992 amounted to 332,791 kg from imported concentrates and 4,723 kg from domestic ores. The Republic of Korea produced 215,100 kg silver in 1993, including 4,725 kg of silver from domestic ores, and produced 257,500 kg silver in 1994, including 8,600 kg silver from domestic ores.

Nickel

The Korea Zinc Co. Ltd. (KCZL) produced nickel in a joint venture with Inco of Canada with an annual capacity of 15,000 t/y at its Onsan Industrial complex. KCZL exported 1,144 t of nickel in 1989.

Korea Nickel Refining Co., a joint venture of Korea Zinc, Inco and Posco, produced 15,950 t of utility nickel in 1994.

Tungsten

As estimated in 1992, the Republic of Korea had tungsten ore reserves of 63 Mt of 0.5 per cent WO_3 grade. Korea Tungsten Mining Co. (KTMC) is the largest tungsten producer in the Republic of Korea and its Sangdong mine in Kangwon Province is one of the largest producers in the world.

Total tungsten ore output amounted to 2,451 t (21 per cent down from 1988) in 1989, 2,401 t in 1990, 1,405 t in 1991, and only 445 t in 1992.

KTMC produced 2,731 t of APT, 1,851 t of tungsten oxide, and 1,444 t of tungsten metal powder in 1990. Other by-products were 79 t of bismuth metal, 136 t of moly oxide and 138 t of ferro-moly. Under a technical assistance programme with Lamp Metals of the U.K., KTMC commenced production of tungsten wire in 1990 with a production capacity of 40 t/y.

The Republic of Korea imported 3,412 t of tungsten concentrates for further processing at the KTMC's plant in Daegue in 1991. The Republic of Korea imported 633 t in 1992 of tungsten concentrates from China.

Table 52 and table 53 present the consumption, production and trade of nonferrous-metallic ores of the Republic of Korea from 1992 to 1994.

Table 52. Consumption and production of metallic ores of the Repblic of Korea, 1992-1994

Mineral	*Grade (%)*	*Unit*	*1992*		*1993*		*1994*	
			Consumption	*Production*	*Consumption*	*Production*	*Consumption*	*Production*
Gold	Au 99.9	Kg	15,275	1,399	18,942	1,438	19,872	11,374
Silver	Ag 99.9	Kg	250,990	4,723	343,510	4,712	420,118	8,600
Copper	Cu 25-29	t	245,067	–	591,278	–	394,235	–
Lead	Pb 50	t	78,439	27,255	137,407	14,818	101,626	4,345
Zinc	Zn 50	t	528,940	43,766	490,482	27,616	573,649	11,243
Iron	Fe 56-65	mt	32,394	222	36,026	219	33,980	191
Tungsten	WO_3 70	t	3,028	445	633	–	–	–
Molybdenum	MoS_2 90	t	1,294	9	2,055	–	2,165	4
Manganese	Mn 307	t	405,093	–	422,123	–	436,636	–
Titanium		t	408,118	348,040	374,262	425,200	347,174	360,264
Tin		t	374	–	32	–	–	–
Bismuth	Bi 99.9	t	8	7	–	–	–	–
Antimony		t	–	–	–	–	–	–
Chromium		t	319	–	449	–	537	–
Zirconium		t	11,823	3	9,842	1	13,109	–
Aluminum		t	42,980	–	58,058	–	63,312	–
Arsenic	As 30	t	–	–	268	–	635	–
Iron Sulphide	FeS 45	t	–	1	59	1	166	–

Source: KIGAM, 1995. Geological Investigation and Mineral Resources Exploration & Development in the Republic of Korea, ESCAP, May 1995.

Table 53. Export and import of metallic ores of the Republic of Korea, 1992-1994

Mineral	*Grade (%)*	*Unit*	*1992*		*1993*		*1994*	
			Consumption	*Production*	*Consumption*	*Production*	*Consumption*	*Production*
Gold	Au 99.9	kg	22,687	32,284	26,346	34,435	49,591	58336
Silver	Ag 99.9	kg	220,161	276,343	108,033	236,023	151,937	302,527
Copper	Cu 25-29	t	–	244,173	–	571,187	4,466	426,421
Lead	Pb 50	t	–	60,842	–	67,338	–	121,078
Zinc	Zn 50	t	–	471,588	–	462,011	–	576,231
Iron	Fe 56-65	mt	–	31,784	–	35,549	–	34,250
Tungsten	WO_3 70	t	–	2,637	–	41	77	–
Molybdenum	MoS_2 90	t	–	1,249	–	2,038	336	2,498
Manganese	Mn 30	t	–	397,923	–	406,125	–	457,636
Titanium		t	87,485	108,845	147,341	97,453	108,040	88,248
Tin		t	–	250	–	20	–	–
Bismuth	Bi 99.9	t	–	–	–	–	–	–
Antimony		t	–	–	–	–	–	0
Chromium		t	–	319	–	449	–	537
Zirconium		t	–	11,820	–	9,841	–	13,109
Aluminum		t	–	42,908	–	58,058	–	63,312
Arsenic	As 30	t	–	–	–	268	–	635
Iron Sulphide	FeS 45	t	–	–	–	58	–	165

Source: KIGAM, 1995. Geological Investigation and Mineral Resources Exploration & Development in the Republic of Korea, ESCAP, May 1995.

1.1.1.6 Russian Far East

The Russian Far East (RFE) region includes the Sakha Republic (Yakutia), Chukotskiy Okrug, Magadanskaya Oblast, Kamchatskaya Oblast, Amurskaya Oblast, Yevareiskiy Aut, Oblast, Sakhalinskaya Oblast, Khabarovskiy Krai, and Primorskiy Krai. The region, covering 6,215,900 sq km or more than 35 per cent of Russia's territory, is richly endowed with natural resources and is positioned geographically along the north-west Pacific rim. Figure 59 shows the administrative map of the Russian Federation.

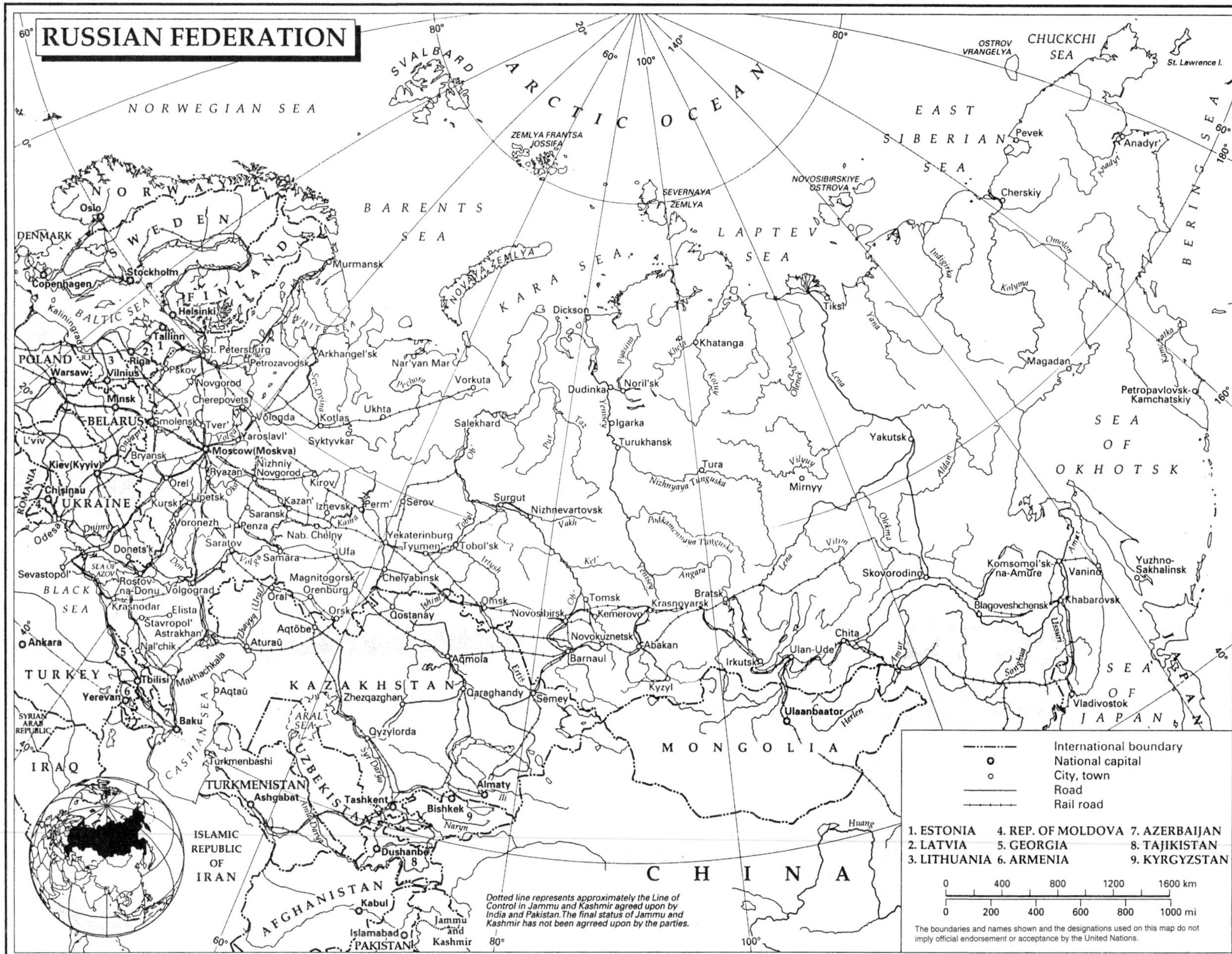

Figure 59. Administrative map of the Russian Federation

The leading industry sectors are non-ferrous metals extraction, timber exploitation and processing, and fishing. The extraction of mineral resources makes up more than 25 per cent of the total volume of industrial output in the region. Raw materials, including timber, furs, fish, coal, oil products and non-ferrous metallic ores, comprise 93 per cent of the Far East's export. 40 per cent of the former Soviet's lumber export were from its Far East to such markets as Japan, the Republic of Korea and Australia. Most of the raw materials produced in the Far East region are exported unprocessed.

The Russian Far East contains more than 70 types of minerals, including gold, diamonds, tin, iron ore, oil and gas, and coal. More than 90 per cent of Russian diamonds are extracted in the Far East or Yakutia, making Russia the world's fifth largest diamond producer and a major world exporter. The region processes 95 per cent of Russia's tin reserves and produces nearly the same percentage of Russia's tin production. The Far East is also a significant supplier of zinc (with 4 per cent of Russia's total reserves and 14 per cent production), of fluorspar (with 41 per cent reserves and 91 per cent production) and of coal (with 9 per cent reserves and 14 per cent production). Though with some notable exceptions, only a limited number of deposits have been sufficiently identified, evaluated and developed in the Far East. In fact, the Russian Far East region represents one of the world's last great frontiers for minerals assessment and development.

The Russian Far East can be divided into four mining zones:

a) the South Zone incorporating parts of the Khabarovskiy Krai and the Amurskaya Oblast for tin, gold and coal;

b) the Pacific Zone including the Primorskiy Krai, the Sakhalinskaya Oblast and Kamchatskaya Krai for polymetalic ores and tungsten;

c) the Northern Central Zone including the Sakha Republic (Yakutia), Magadanaskaya Oblast, and part of the Khabarovskiy Krai for precious and non-ferrous metals, iron ore, diamonds, coal, and natural gas; and

d) the Far North Zone including the Chukotskiy Okrug for non-ferrous metals, gold and diamonds.

Ferrous metals

Reportedly, Russia produced 97.3 Mt of iron ore in 1992. A 'National Programme to Modernize Metallurgy in 1993-1995 and up to the year 2000' called for production to be increased in some of the iron mining regions including Siberia. About 85 per cent of Russia's iron and steel works had been privatized by 1994.

Russia has a large steel industry, much of which is in need of modernization. Production of crude steel reportedly to be 48.8 Mt in 1994 compared with 93 Mt in 1989, rolled steel 35.8 Mt in 1994 compared with 65.4 Mt in 1989, and pig iron 36.1 Mt in 1994 compared with 61.5 Mt in 1989. The special steel producer Sibelektrik Stal in Krasnoyarsk, East Siberia, with the capacity to produce 110,000 t/y of crude steel, closed in 1994 owing to a major cutback in orders from defence industries.

In the Russian Far East, prospecting activities have been concentrated in the Amur region and Khabarosk Territory, where ferruginous quartzites and magnetites have been discovered, and in southern Yakutia, where the old southern Aldanski district of magnetite ores are found. The most probable sites for development are the Svobodny town in the Armur region and Chulman township in southern Yakutia. The prospected reserves at southern Aldanski are estimated at 1.5 bn tonnes. West of this region and near the BAM railway, a major iron-ore district Charo-Tokkinski are estimated at 3.7 bn tonnes. However, the Charo-Tokkinski reserves are almost inaccessible and the mining conditions extremely difficult.

Copper

According to preliminary reporting of the Russian Committee for Metallurgy, Russia produced 505 000 t of refined copper in 1994. Norilsk enterprise, as the biggest producer of copper in Russia produced under 300,000 t in 1994, down from its peak output of 400,000 t at the end of the 1980s. Norilsk reportedly has two-thirds of Russia's copper reserves. The Krasnouralsk copper smelter in the Urals produced 57,000 t of copper in 1994. Russia exported 469,700 t of copper in 1994 compared with 163,800 t in 1993.

Russia used to receive a large percentage of its copper from Kazakhstan and other republics, but with the break-up of the Soviet Union it began experiencing copper shortages. To alleviate these shortages, Russia has proceeded with plans to develop the Udokan copper deposit in the east of Lake Baikal in Chita Oblast. On the other hand, Russia received 97,000 t of copper in concentrate form in 1994 from the Erdenet in Mongolia. According to a trade agreement signed by the two countries, Russia continued to receive this level of copper export from Mongolia in 1995.

Large copper reserves are possible in Yakutia, where stock work deposits of copper-molybdenum and copper-tungsten ores can be found in the northern Stankovaya metallogenic zone and the Tompo-Bryungadinski ore district, as well as copper sandstone in western Aldanskaya metallogenic zone, and other deposits. Similarly large deposits are predicted in the Magadan and Kamchatka regions, although of copper-porphyry type (Dorian, 1993).

Gold

According to the chairman of the Russian State Committee for Precious Metals and Stones (Roskomdragmet), Russia's gold reserves as of January 1, 1995, were 321.8 t, including 189 t deposited with the Central Bank of Russia and 132.8 t held by Roskomdragmet. In May 1993, control of Russia's gold mining enterprises was placed officially under the jurisdiction of Roskomdragmet.

According to Roskomdragmet, Russia's gold production was reported to be 146.6 t in 1994 compared with 149.5 t in 1993. Of this production of 146.6 t in 1994, 135.5 t was mined directly as primary gold while the remaining production presumably was gold produced as a by-product of polymetallic ores or from scrap. Russia's largest gold producing region, the Magadan Oblast including the Chukotka autonomous district, produced 27.4 t of gold in 1994 compared with 43.5 t in 1992. Production in Yakutia, Russia's second largest gold producing region, reportedly decreased to 29.7 t in 1994 compared with 33.4 t in 1993. Russia's prospecting groups, being about 250 such groups in total in 1992, reportedly produced over 50 per cent of all placer gold produced in Russia.

In January 1993, a new law 'On Concession and Profit Sharing Agreement with Foreign Investors' passed by the Russian Parliament, provided a legal basis for foreign investment in the Russian industry. In 1993, western companies for the first time appeared to gain actual rights to develop Russian gold deposits. One of the first companies to acquire such rights was Australia's Star Technology which negotiated with Lenzoloto to mine reserves. Other firms active in Russian gold development include the Cypriot company Transpacific Resources, Canada's Echo Bay Minerals Corp., and the U.S. company Cyprus-Amax Minerals.

At year-end of 1993, the Russian President signed two decrees to reorganize the country's marketing of precious metals and stones. According to the terms of first decrees, the Central Bank of Russia acquires the right, in accordance with the Ministry of Finance, to issue licenses to other banks to conduct sales of precious metals and stones. The decree also calls for the establishment of a Russian precious metals exchange and for drawing up proposals whereby it will be possible to sell gold to foreign participants in gold mining enterprises in accordance with their share of the profits. The second decree provides for less VAT expenditures when selling precious metals and stones from the State Reserve.

In March 1995, it was announced that the state-owned company Severovostokzoloto was being changed into a public stock company, offering preferred and common shares of stock. Management, the employees and the state were each being allotted a certain percentage of the shares.

Lead and zinc

The Far East produces a small amount of these metals and will not be able effectively to increase its production in the next decade or so. Dalnegorski is expected to remain the only lead-zinc producing enterprise in the region until at least 2000. Predicted reserves of lead and zinc rank high in Yakutia. New deposits can be found in the Mamsko-Kyllakhskaya zone and the southern Verkhoyanski, western Verkhoyanski and Tastayakhatasski districts, all of which are nearly inaccessible and underdeveloped. Polymetallic deposits have been discovered in the Magadan region, where the Taskano-Omulskaya zone (stratibound deposit) shows the greatest promise.

Some of Russia's richest zinc deposits have been discovered in the Far East, particularly within the Sikhote-Alin mountains of the Maritime Territory, the southern reaches of the Khabarovsk territory, in Yakutia and in the Chukchi Peninsula.

Mercury

Nearly half of Russia's estimated mercury reserves occur in the Far East. Mineralisation of mercury is abundant in Yakutia and the Magadan and Kamchatka regions. Several deposits have been discovered in these regions, though none are being developed for economic reasons. In recent years, surveying and evaluation of the largest discovered deposits – Zvezdochka (Yakutia) and Tamvatpeiski (Magadan region) – were carried out in the upper and middle Kolyma regions and the Koryakskoye highlands.

Tin

The Far East is the principal supplier of tin in Russia and contains the bulk of the country's reserves. Many deposits are being developed in the Magadan region, Yakutia, the Khabarovsk and the Maritime Territories. Placer deposits account for more than 40 per cent of the national output of tin; however, remaining placer reserves are insignificant.

Production of tin in the Far East continues to decline as older deposits are depleted and new regions are developed slowly. The national demand for tin is only half met, and imports are becoming more costly. The future of the tin-producing industry in the Far East depends on the opening up of new regions. Yakutia (Republic of Sakha) has ample reserves of tin, as do the central and southern Yanski districts. However, physical environmental conditions in the Yanski area difficult. The Magadan region also holds promise for tin development, second only to Yakutia in estimated reserves, while leading in tin mining and production from concentrates.

Tungsten

Forty deposits have been identified in the Magadan region, the Khabarovsk and the Maritime Territories, and the Republic of Sakha, accounting for less than 10 per cent of the nation's reserve base of tungsten trioxide. A majority of the deposits are small and insignificant placers. The major Iultinski and Primorski processing facilities are currently operational; tungsten is extracted during tin-ore enrichment at the Solnechny and Deputaski facilities. Prospects for continued production in the Far East are largely linked with the huge Primorski ore-enrichment plant.

1.1.2 Non-metallic minerals

1.1.2.1 China

By the end of 1990, China had explored 86 non-metallic minerals with known reserves. According to China National Nonmetallic Minerals Industry Corp (CNNMIC), out of the 86 explored minerals, China's reserves of flake graphite, fluorspar and gypsum rank first in the world; asbestos, bentonite, glauber salt, talc and wollastonite rank second, and perlite and zeolites rank third. China also has large mineral resources for attapulgite, bauxite, celestite, diatomite, kaolin, kyanite, pyrophyllite and sepiolite. China is also rich in high quality granite and marble resources. China is one of the few countries in the world that have a more complete set of known non-metallic minerals with great potential resources.

Non-metallic minerals are geographically distributed throughout China. Most of them are concentrated in the more economically developed eastern and central regions of China. Phosphates occur in Yunan, Guizhou, and Sichuan; pyrite in Guangdong; fluorspar in Fujian, Nei Mongggol and Zhejiang; barite in Hunan; graphite in Heilongjiang and Shandong; magnesite in Liaoning; asbestos in Qinnghai and Sichuan; and salt minerals in Qaidam Basin, Qinghai (figure 60).

China is a major world producer of a variety of non-metallic minerals. Production of such mineral products as cement, graphite, talc, fluorite, barite and magnesite rank first in the world. The output value of non-metallic mineral products grew at an average annual rate of 10-15 per cent during the decade 1979-1989, which far exceeds the growth of output value of metallic mineral products. The output value of the non-metallic mineral industry reached 103.8 billion Chinese yuan in 1988, accounting for about 30 per cent of the total of the country's mining industry. The non-metallic mineral industry is playing an important role in the national economy of China.

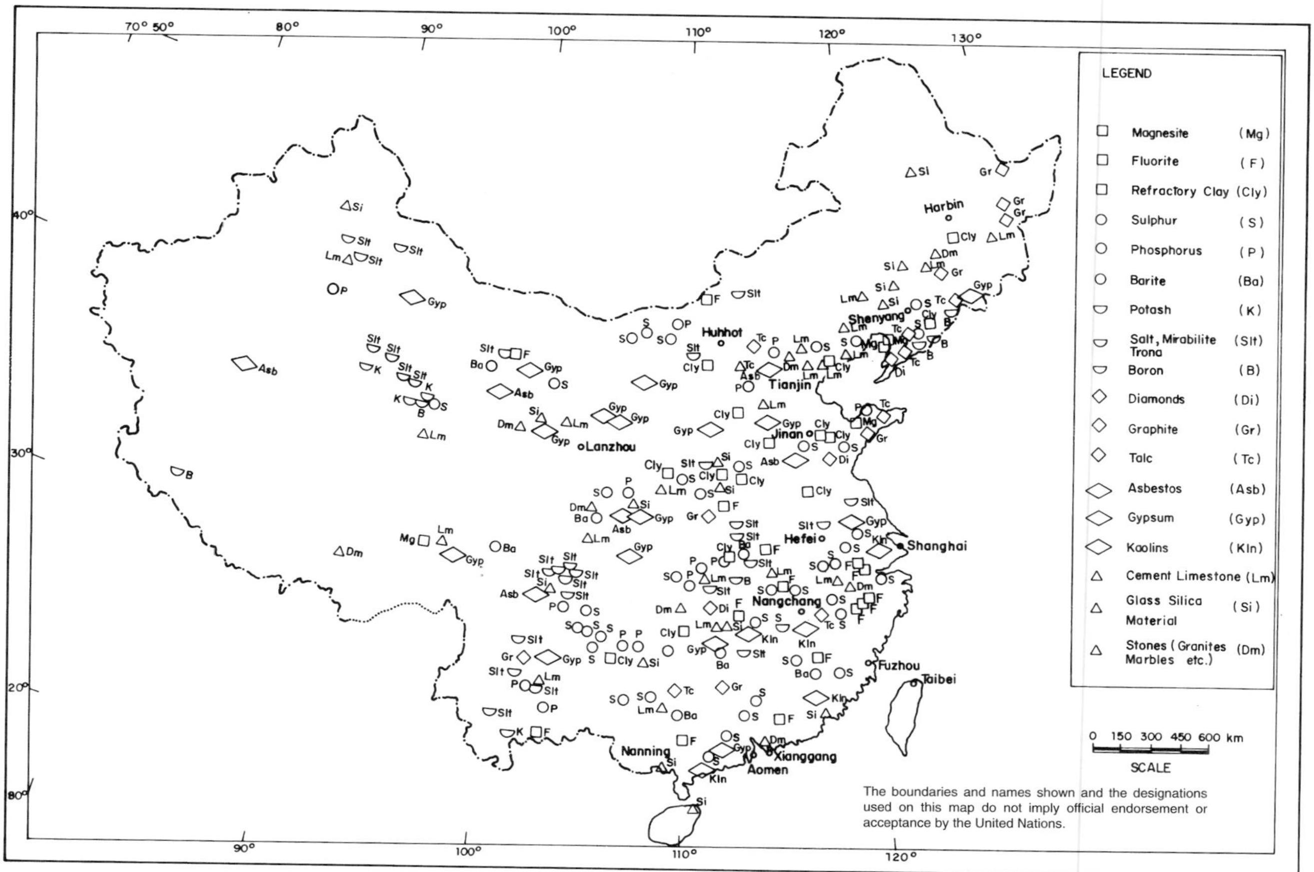

Source: Minerel Resources of China, 1992.

Figure 60. Distribution map of non-metallic mineral resources in China

During the Seventh Five-year Plan (1986-1990), CNNMIC began to integrate its operations by the establishment of industrial mineral production centres. The initial ten resource bases are set out in table 54. The important producers are CNNMIC, Ministry of Chemical Industry, Ministry of Metallurgical Industry and the State Administration for Building Materials. Other producers include the China National Nonferrous Metals Industry Corp., Ministry of Geology and Mineral Resources, and provincial entities.

Barite

China's barite reserves account for about 300 Mt, ranking the first in the world, followed by the former USSR, the US, India and Morocco. China's barite resources are relatively concentrated in Guizhou (30.4 per cent), Hunan and Guangxi. China's barite deposits are mainly sedimentary type (accounting for 60 per cent), volcano-sedimentary, hydrothermal and alluvial types.

China is the world's largest producer of barite with an annual output at least twice that of Russia, the world's second largest producer. Mine output in 1988 may well be around 1.5 Mt, with Guanxi accounting for about two thirds of the total. China's barite production has grown more rapidly since 1975 due to the sharp increase in the consumption of oil and gas drilling mud and in export.

China exported 1.1 Mt of barite, valued at $28.1 million, in 1988. During 1985-1988, Chinese shipments accounted for close to 60 per cent of U.S. receipts of crude barite.

Table 54. Production centres of industrial minerals in China

	Adiminstrative Division: Principal location (other location)		*Pricipal commodity (other commodity)*
Talc Centre	Liaoning:	Haichnang (Gansu, Yingkoy and Yiuyan)	Talc (magnesite and stone)
Talc Centre	Shangdong:	Qixia (Haiyan, Pingduo and Yiexian)	Talc (granite and marble)
Talc Centre	Guangxi:	Longshen	Talc
Graphite Centre	Shangdong:	Nanshu (Beishu, Laixi and pingduo)	Graphite (granite and marble)
Graphite Centre	Heilongjiang:	Liumao (jixi and Luobei)	Graphite
Graphite Centre	Nei Monggol:	Xinghe	Graphite
East China Clay Base	Fujian: Anhui: Zhejiang:	Longyan Jiashan Linan	Kaolin Attapulgite Bentonite
Wollastonite Centre	Jilin:	Lishu and Panshi Jiutai Chanbai Pershi and Pingtong	Wollastonite Bentonite Diatomite Granite
Sillimanite Centre	Heilongjiang:	Jixi	Sillimanite
North China, East Mountain and South-Central Stone Material Base	Shangxi: Hebei: Shangdong: Henan: Hubei: Guangdong:	Linqui Quyan Jinan, Rushan and Taishan Yanshi Huangang Liangian	Granite Marble Granite Granite Granite Marble

Source: China National Non-metallic Minerals Industry Corp. (CNNMIC).

Cement

Total limestone resources in China are estimated at 12-13 trillion tonnes, of which 3.3 trillion are cement limestone. They are extensively distributed in all the provinces, autonomous regions and municipalities except Shanghai Municipality. The most abundant reserves are in Shangxi (11.7 per cent), Sichuan, Guangxi, Haibei and Shandong.

China has been the world's largest producer of cement since 1985 with an output of 146 Mt in that year. China produced 400 Mt in 1994. In the late 1980s, China was a net importer of cement, but since 1990 China has been a net exporter of cement (table 55).

Table 55. Cement production and trade of China, 1987-1995

(in '000 t)	*1987*	*1988*	*1989*	*1990*	*1991*	*1992*	*1993*	*1994*	*1995*
Production	181,000	203,000	207,000	203,000	252,000	304,000	356,000	400,000	450,000
Export	168	152	436	6,829	10,736	*8,085	2,450	4,520	n/a
Import	2,106	1,518	1,230	400	160	–	–	–	–

Source: *Mining Annual Review 1988, 1989, 1990, 1991, 1992, 1993, 1994, 1995 and 1996.*
** Figure was estimated from that of first three quarters divided by 0.7.*

Fertilizers

By the end of 1990, China's explored phosphate rock reserves of 15.7 billion tonnes, (which are concentrated in Yunnan, Guizhou, Sichuan, Hubei and Hunan) accounted for nearly 80 per cent of China's total reserves. In contrast to phosphate resources which are relatively abundant, China's potash resources are poor. As of the end of 1990, its explored reserve of KCl was 396 Mt, most of which occurs in the saline lakes of the Quaidam Basin in Qinghai province.

China is the third largest producer and user of fertilizers after the U.S. and the former USSR in 1988. Production of chemical fertilizers totalled 17.7 Mt including about 14.9 Mt for phosphorus (P), and 40,000 t for potash (K). There are about 1,900 plants producing fertilizers – 1,200 for nitrogen (N), 700 for P, and 2 for K. Based solely on domestic production, the nutrient ingredient of chemical fertilizer output is remarkably skewed to nitrogen, resulting in a N to P to K ratio of 375:70:1.

To supplement domestic production, China imports fertilizers for N and P, and imports virtually all its requirements for K. Total imports of fertilizers in 1988 were 14.7 Mt composed of 8.6 Mt for N, about 3.4 Mt for P and about 2.7 Mt for K. Total imports of fertilizers in 1993 and 1994 are 20.16 and 22.76 Mt respectively.

Fluorspar

Fluorite reserves in China are estimated at 130 Mt, ranking first in the world. Fluorspar occurs in numerous provinces of China, but the largest reserves are believed to be in the Nei Monggol and Zhejiang provinces.

China is probably the world's largest producer of fluorspar, outpacing production by Mexico and Mongolia. China's fluorspar products are mainly for export (table 56). In 1993, China produced 2.4 Mt and exported 1.38 Mt of fluorspar.

Table 56. Fluorspar production and export of China, 1988-1995

(in ton)	*1988*	*1989*	*1990*	*1991*	*1992*	*1993*	*1994*	*1995*
Production	1,400,000	1,700,000	1,700,000	1,700,000	1,700,000	2,400,000	2,400,000	2,400,000
Export	911,976	1,183,553	1,093,847	933,694	*1,071,000	1,380,000	1,220,000	n/a

Source: Mining Annual Review 1989, 1990, 1991, 1992, 1993, 1994, 1995 and 1996.
* Figure was estimated from that of first three quarters divided by 0.7.

Salt

China is rich in salt resources. As of 1990, China had 118 known salt occurrences totalling 364.5 billion tonnes in reserves. The explored reserves are located mainly in the Qaidam Basin in Qinghai, accounting for four-fifths of the country's total. China's other salt resources are distributed throughout the country, particularly in Sichuan and Hubei.

China is the world's second largest producer of salt after the United States. Salt production from evaporates and rock deposits increased 24 per cent, reaching 21.9 Mt in 1988. Table 57 shows the production of salt in China in recent years. China's major salt operations are in Anhui, Hainan, Jiangsu, Jiangxi, Liaoning, Nei Monggol, Qinghai, Sichuan and Xinjiang.

Table 57. Salt production of China, 1987-1995

(in '000 t)	*1987*	*1988*	*1989*	*1990*	*1991*	*1992*	*1993*	*1994*	*1995*
Production	17,697	22,000	28,000	19,842	24,100	28,100	29,530	32,000	24,990

Source: Mining Annual Review 1988, 1989, 1990, 1991, 1992, 1993, 1994, 1995 and 1996.

China exported about 1.1 Mt of salt each in 1986 and 1987. However, because of the growing demand by the domestic chemical sector, exports in 1988 were only about a third of that in prior years.

Talc

The retained reserves of talc in China were 250 Mt at the end of 1990, ranking second in the world after the former USSR. The reserves are mainly in the Jiangxi, Liaoning, Shandong, Qinghai and Guangxi provinces.

China is a major world producer of both industrial- and cosmetic-grade talc. The talc producing bases are in Guangxi, Liaoning and Shandong. China's annual talc production is over 2 Mt. Output is sufficient to meet domestic demand as well as provide substantial quantities for export. In 1988, China exported close to 750,000 t of talc, at which time the three talc producing bases were undergoing expansion. China was expected to be a major supplier in world talc commerce by the early 1990s. Table 58 shows China's talc production and export.

Table 58. Talc production and export of China, 1988-1995

(in ton)	*1988*	*1989*	*1990*	*1991*	*1992*	*1993*	*1994*	*1995*
Production	1,000,000	1,500,000	1,500,000	2,600,000	2,650,000	2,650,000	2,700,000	2,500,000
Export	746,197	943,381	869,755	948,196	1,014,000*	1,060,000	1,240,000	n/a

Source: Mining Annual Review 1988, 1989, 1990, 1991, 1992, 1993, 1994, 1995 and 1996.
* Figure was estimated from that of the first three quarters divided by 0.7.

1.1.2.2 Democratic People's Republic of Korea

Magnesite

The Democratic People's Republic of Korea is the world biggest supplier of high quality dead-burnt magnesite, with a MgO content of 95 per cent and silicon and calcium content of less than 2 per cent. The annual output was estimated at around 1.5 Mt in 1988. The Democratic People's Republic of Korea also produces barites and large tonnages are shipped to the Soviet Union.

Monazite

Reserves of monazite estimated at 500,000 t exist at the Chelsan mine in the northwest of the country, near the border with China. The monazite is processed at a 1,500 t/y plant operated by the International Chemical Joint Venture Corp., owned by a Japanese company and the state-owned Korea Ryongaksan General Trading Co. The product is sent to Japan.

Phosphate

A crushing and processing plant was established at the Taedaeri phosphate mine in the central western region in 1993. The plant enables the complex to increase its ore handling capacity by 20 per cent.

1.1.2.3 Japan

Limestone

The construction boom in the late 1980s spurred the growth of the non-metallic minerals sector though the activity in the metal and coal mining industries was curtailed. The mining of non-metallic minerals including limestone, has, since 1990 come to represent the major part of the mining industry in Japan, in terms of both quantity and value. Outputs of industry decreased slightly after 1991 but high levels of limestone production were maintained.

Limestone production reached a historic high of 207 Mt in 1991 and kept above 200 Mt in the following years (table 59). Contributing factors were the high levels of demand for cement and industrial minerals in the private construction sector. The economic measures taken by the government to salvage the Japanese economy helped to sustain the private construction industry.

Table 59. Limestone production of Japan, 1988-1995

(in ton)	*1988*	*1989*	*1990*	*1991*	*1992*	*1993*	*1994*	*1995*
Production	182,468	190,859	198,224	206,780	203,789	200,451	202,488	201,089

Source: Mining Annual Review 1989, 1990, 1991, 1992, 1993, 1994, 1995 and 1996.

Silica stone

Table 60 shows the production of silica stone in Japan.

Table 60. Silica stone production of Japan, 1988-1995

(in ton)	*1988*	*1989*	*1990*	*1991*	*1992*	*1993*	*1994*	*1995*
Silica stone	16,214	17,223	17,925	18,472	19,319	18,854	18,479	18,334
Silica sand						3,884	3,944	3,737

Source: Mining Annual Review 1989, 1990, 1991, 1992, 1993, 1994, 1995 and 1996.

1.1.2.4 Mongolia

Cement

Mongolia's cement production, which was around 165,500 Mt in 1983, had risen to about 425,000 t in 1988.

Fluorspar

Mongolia produces nearly 15 per cent of the world's fluorspar. The largest fluorspar deposits include Zuun-Tsagaan-Del Chuluut-Tsagaan-Del, Orgon, Delgerkhaan, Berkh and Adag. (1995) The biggest mining and processing complex is in Bor-Ondor, situated 290 km south-east of Ulaanbaatar. A joint Mongolian-Russian enterprise Mongolrostsvetmet Corp., which also owns the Tolgoit gold alluvial venture, operates most of the fluorspar mines in the country.

Mongolia produced 110,000 t of 95-97 per cent fluorspar concentrate together with 96,000 t of metallurgical grade fluorspar in 1993. Bor-Ondor mining and processing plant, Berkh, Khajuu Ulaan and Khar Ariag opencast produced 17,000 t of metallurgical-grade fluorspar in 1994. Almost all fluorspar production was exported to the former Soviet Union for consumption by its iron and steel industry.

There are several potential deposits to be prospected and evaluated fully. The deposit of phosphorite at Buren Khan in W. Mongolia has estimated reserves of 400 Mt of ore. The Buren Khan deposit is situated along the western shores of Mongolia's biggest lake, Khovsgol, and the ore is part of the Khovsgol phosphorite-bearing basin – the largest so far in Central Asia. The environmental problems, which may rise from the exploitation of the phosphorite deposit, have to be assessed.

1.1.2.5 Republic of Korea

The Republic of Korea's non-metallic sector became the most important mining industry for the first time in 1990, valued at $745 million, compared to $663 million for anthracite coal, with coal mining production declining. In 1994, Republic of Korea's non-metallic mineral sector was valued at $US1,021 million, equal to 64 per cent of the value of all mineral production.

Cement

The Republic of Korea's cement production amounted to 30.2 Mt in 1989, 33.6 Mt in 1990, 39.2 Mt in 1991, 42.6 Mt in 1992, 46.9 Mt in 1993, and 51.6 Mt in 1994. Of the major producers in 1994, Sangyong Cement produced 14.2 Mt, Tongyang Cement 8.7 Mt, Hanil Cement 5.6 Mt and Hyundai Cement 6.4 Mt.

The Republic of Korea exported 1.9 Mt of cement in 1990 and imported 9 Mt to satisfy the strong domestic demand. The situation then reversed. The Republic of Korea exported 5.1 Mt of cement and imported 1.1 Mt in 1993, exported 4.3 Mt and imported 2.3 Mt in 1994.

Graphite

Production of amorphous graphite continuously dropped from 107,782 t in 1988 to 100,282 t in 1989, 98,897 t in 1990 and down to 75,239 t in 1991.

Limestone

Limestone output continuously increased from 46 Mt in 1988 to 48 Mt in 1989, 48.8 Mt in 1990, 59.3 Mt in 1991, 65.4 Mt in 1992, 75.8 Mt in 1993 and up to 82.8 Mt in 1994.

Kaolin

Kaolin production continuously increased from 0.8 Mt in 1988 to 1.2 Mt in 1989, 1.4 Mt in 1990, 1.7 Mt in 1991, 1.8 Mt in 1992, 2.3 Mt in 1993 and up to 2.6 Mt in 1994.

Silica stones

Silica stone production continuously increased from 1.4 Mt in 1990 to 1.6 Mt in 1991, 1.9 Mt in 1992, 2.5 Mt in 1993 and up to 2.4 Mt in 1994. Silica sand production amounted to 1.1 Mt in 1993 and 1.4 Mt in 1994.

Talc

Talc production increased from 141,800 t in 1990 to 170,563 t in 1991.

Table 61 and 62 presents consumption, production and trade of non-metallic minerals of the Republic of Korea in 1992-1994.

1.1.2.6 Russian Far East (RFE)

Apatites and phosphorites

Russia's major phosphate producer is the apatite complex at the Kola Peninsula, which during the 1980s was mining 60 Mt/y of ore and producing between 19-20 Mt/y of apatite concentrate with a P_2O_5 content of over 39 per cent. Over 3 Mt/y of the concentrate exported.

The Russian Far East (RFE) mineral fertilizers, particularly phosphorous, are insignificant at present. Reserves at Udsko-Selemdjinski in the Khabarovsky Territory are estimated at 1.3 billion tonnes of ore, or 100 million tonnes of P_2O_5. Prospecting is generally concentrated in the western portion closer to the BAM zone. Exploration work for apatites has also been done in the Aldanski district in Yakutia, where the Seligadarskoye deposit holds the greatest promise. Other apatite reserves have been identified in the Khaninski district near BAM in Yakutia and in the Khabarovsk Territory. In recent years, several districts have been selected for phosphate raw materials production, including Yudomo-Maiski, Ayanski, Meginski, Malo-Khinganski, Zeya-Selemdjinski, Oldoe-Urushinski, some of them lying in the BAM zone.

Table 61. Consumption and production of non-metallic ores of the Republic of Korea, 1992-1994

Mineral	*Unit*	*1992*		*1993*		*1994*	
		Consumption	*Production*	*Consumption*	*Production*	*Consumption*	*Production*
Talc	t	347,060	149,862	319,330	53,923	359,857	35,340
Pyrophyllite	t	373,124	602,580	412,348	644,890	479,697	707,951
Feldspar	t	278,067	281,093	279,115	321,964	327,829	319,658
Kaolinite	t	1,883,450	1,856,157	2,263,374	2,328,921	2,655,610	2,675,485
Limestone	'000 t	65,208	65,446	76,521	76,851	82,944	82,809
Quartzite	t	1,787,950	1,870,216	2,488,162	2,509,828	2,369,640	2,360,265
Silica sand	t	1,957,845	1,226,280	1,835,141	1,116,664	2,385,810	1,451,793
Guano	t	89,449	76,775	84,227	67,324	92,121	82,738
Phosphate	'000 t	1,671	–	1,646	–	1,681	–
Sulphide	t	527,223	–	375,416	–	441,022	–
Serpentine	t	623,297	606,206	690,216	714,591	631,395	658,500
Wollastonite	t	2,943	–	5,272	280	6,250	–
Zeolite	t	108,909	104,725	116,104	111,657	128,155	122,411
Scaly Graphite	t	24,066	–	9,027	–	9,067	–
Earthy Graphite	t	37,229	8,412	27,371	5,910	27,325	4,300
Marble	t	83,587	62,470	9,604	34,372	52,587	45,988
Asbestos	t	97,761	2,308	82,836	–	83,190	–
Fluorite	t	93,399	70	90,277	299	89,198	100
Mica	t	11,086	7,732	21,123	28,462	53,199	37,470
Andalusite	t	3,973	–	6,821	–	8,503	–
Kyanite	t	2,383	38	2,299	18	2,183	–
Barite	t	40,793	40	39,848	–	53,916	85
Magnesite	t	16,781	–	21,324	–	24,998	–
Gypsum	t	377,863	–	489,553	–	640,840	–

Source: KIGAM, 1995. Geological Investigation and Mineral Resources Exploration & Development in the Republic of Korea, ESCAP, May 1995.

Diamonds

The Sakha Republic (Yakutia) of the RFE has most of the diamond mines in Russia. There was a reported 10 per cent decrease in diamond extraction in 1992 against 1991, and a 25 per cent decrease in comparison with 1990, when production peaked. The Sakha Government forecast in 1992 that diamond output would further decrease an additional 14 per cent in 1993. Decreased production is attributed to deteriorating conditions at existing diamond mines as well as the renovation work being conducted at these mines. Another possible factor is the exodus of the skilled workers from Yakut-Sakha because of deteriorating economic conditions.

According to President of Almazy Rossii-Sakha, Russia's main diamond producing firm, diamond output reportedly increased by 7 per cent in 1994 compared with 1993. Almazy Rossii-Sakha's goals for 1995 included completing development of the Yubeleyny diamond mine near the Arctic Circle and commissioning the new Botubobinskaya diamond pipe that had been discovered in March 1994.

Table 62. Export and import of non-metallic ores of the Republic of Korea, 1992-1994

Mineral	*Unit*	*1992*		*1993*		*1994*	
		Export	*Import*	*Export*	*Import*	*Export*	*Import*
Talc	t	9,011	202,962	806	261,595	1,622	306,374
Pyrophyllite	t	216,759	652	223,593	2,663	247,285	4,578
Feldspar	t	24,049	9,435	23,631	8,560	15,817	4,288
Kaolinite	t	131,389	224,253	172,001	249,251	262,493	267,674
Limestone	t	354,287	241,165	362,350	269,384	367,879	227,982
Quartzite	t	1,393	8,521	1,686	15,515	8,172	11,999
Silica sand	t	2,164	716,201	38	774,750	19	875,550
Guano	t	440	5,382	359	8,927	250	9,349
Phosphate	'000 t	–	1,671	10	1,646	–	1,681
Sulphide	t	4,966	532,189	16,225	391,641	27,942	438,964
Serpentine	t	–	–	–	–	–	–
Wollastonite	t	5	2,948	–	4,992	–	6,250
Zeolite	t	737	109	787	398	719	452
Scaly Graphite	t	142	24,189	10	9,037	58	9,125
Earthy Graphite	t	11,421	33,275	7,738	28,049	5,565	28,551
Marble	t	136	21,673	36	913	67	1,413
Asbestos	t	23	95,476	18	82,854	86	83,276
Fluorite	t	18	74,017	–	80,987	–	93,198
Mica	t	149	5,714	932	6,072	51	8,666
Andalusite	t	–	3,973	–	6,821	–	8,503
Kyanite	t	–	2,383	–	2,299	–	2,079
Barite	t	477	40,557	10	39,758	13	53,859
Magnesite	t	69	16,850	92	21,417	63	25,061
Gypsum	t	1,849	379,712	3,280	492,833	5,923	646,764

Source: KIGAM, 1995. Geological Investigation and Mineral Resources Exploration & Development in the Republic of Korea, ESCAP, May 1995.

In December 1992, the Russian Supreme Council approved a decree, which transfers the exclusive right to market uncut diamonds to the new firm Diamonds of Russia-Sakha. In December 1994, Almazy Rossii-Sakha officially began distributing gem diamonds through the new National Diamond Foundation (NDF) that had been created to try to form a western-style diamond market in Russia. Besides Almazy Rossii-Sakha, the new firm's stockholders reportedly include the Russian commercial bank Yakimanka, the Russian-Angolan joint venture Angoros, the Central Bank of Russia, the ITAR-TASS news agency, and Belgium's Stavanger International SA. The firm reportedly will offer for sale exclusive jewellery and select specially-cut gem diamonds at a price which the NDF believes will be 30-40 per cent cheaper than other stones available on the Russian market.

In 1992, the Japanese diamond importer, A. Alder of Tokyo and Toimada Diamond, the diamond producer of the Sakha Republic (Yakutia), established a joint venture diamond polishing company, based in Sakha on a 50:50 equity share basis. Sakha's diamond output has been estimated at 18 m carats/y, being one of the world's largest producers.

The Sakha-Japanese venture expected to ship 200,000 carats of polished diamonds to Japan in 1993 and aimed for an eventual 300,000 carats, or 10 per cent of the Japanese market, the largest single market in the world. Alder had invested $8.06 million to build three factories in Sakha and planned to construct 12 more, each employing 200 workers. The Sakha-Japanese operations expected to export 50,000 carats of polished stones per year. Traders view the new operation as a challenge to De Beers group, which controls nearly 80 per cent of the international rough-diamond trade.

Under a five-year deal signed in 1990, Russia continued to sell 95 per cent of its output through De Beers' Central Selling Organization (CSO), which purchased over $1 billion of Russian rough diamonds in 1994. Russia retained 5 per cent of its output to sell freely.

Fluorspar

In the Russian Far East, fluorspar reserves have been discovered in the Khinganski and Jewish Autonomous Region in the Khobasrovsk Territory, and in the Voznesenski and Pogranichny deposits in the Maritime Territory. Minor amounts of fluorspar concentrate are produced in combination with tin concentrate at the Khinganolovo plant and transported to the Ural area. The Voznesenski deposit is the principal supplier for the major Yarolski ore-dressing facility.

In 1994, Russia continued to import fluorspar from its joint venture in Mongolia which was formed during the Soviet period. In 1994, Russia reportedly imported 88,000 t of flotation fluorspar concentrate and 17,000 t of lumpy fluorite ore from Mongolia.

1.2 Fuel minerals

1.2.1 Oil and gas

1.2.1.1 China

Geological setting

As a result of the collision of the Indian plate, the Pacific plate and the Siberian plate, the Chinese continent has three different types of sedimentary basins (China, 1994).

In the west, due to the repeated collisions and subductions of the Indian and Siberian plates, the earth's crust contracted, thickened, and swelled, forming simultaneously a series of foredeeps (the Tarim Basin) and the intermountain basins (the Chaidamu, Turpan-Hami) under the consistent pressures. In these basins, the structures in the sedimentary cover are not entirely controlled by the basement, the surface folding is intensive, the lineament of the structures is apparent and always accompanied with thrust faults along their boundaries. The simple flexures or faulted blocks are formed in places where the pressure slackened.

In the central part of China, the basement of the sedimentary basins, such as Ordos and Sichuan basins, are a stable part of the continental landmass, which were not strongly affected by the forces of the interaction of the three crustal plates. In general, the structures are asymmetrical. Thrust fault belts often developed along the western margin of the basins. The centres of subsidence in the western parts of the basins also formed during Mesozoic and Cenozoic times.

In eastern China, under tensile stress, the faulted rifting type basins and the basins with wedged shapes were formed with a NNE direction. The prominent characteristics of these basins are block faulting, tensional faulting, faulted subsidence, juxtaposition of multi depressions and multiswells, and the controlling of surface structures by basement. On the other hand, due to the upswelling of the mantle and the deep cutting of faults, the calci-alkalic magma was active, with high geothermal gradient throughout the basins.

The Yellow Sea and East China Sea shelf sedimentary basins are separated by the submerged Fukian-Reinan massif or swell, and Taiwan-Sinzi folded zone respectively, which were believed to act as a dam behind which sediments of the Yellow Sea and East China Sea basins successively accumulated (Emery et al., 1969). The East China Sea exemplifies the structural characteristics of a trench-arc-basin system with the East China Sea Shelf

Basin and Okinawa Trough Basin, folded zone and island arcs, and Okinawa Trench. The recent tectonic framework of sedimentation in the northern part of the South China Sea is similar to that of the East China Sea and Yellow Sea. A series of north-east trending ridges, in the basin and on basin slopes, have been recognized as providing sediment dams (UNESCAP/CCOP, 1977).

Petroleum basins

A total area covering as much as 6,690,000 km^2 is underlain by sedimentary rocks, in which 5,220,000 km^2 are onshore (78 per cent) and 1,470,000 km^2 offshore (22 per cent). There are about 363 sedimentary basins in China, covering a total of 3,530,000 km^2. Of these basins, 10 basins are classified as large with sizes of more than 100,00 km^2, 45 are in the medium category from 10,000 to 100,000 km^2, and the remaining 308 are of a small size with less than 10,000 km^2 (figure 61). Most of the basins in China were formed in Meso-Cenozoic time.

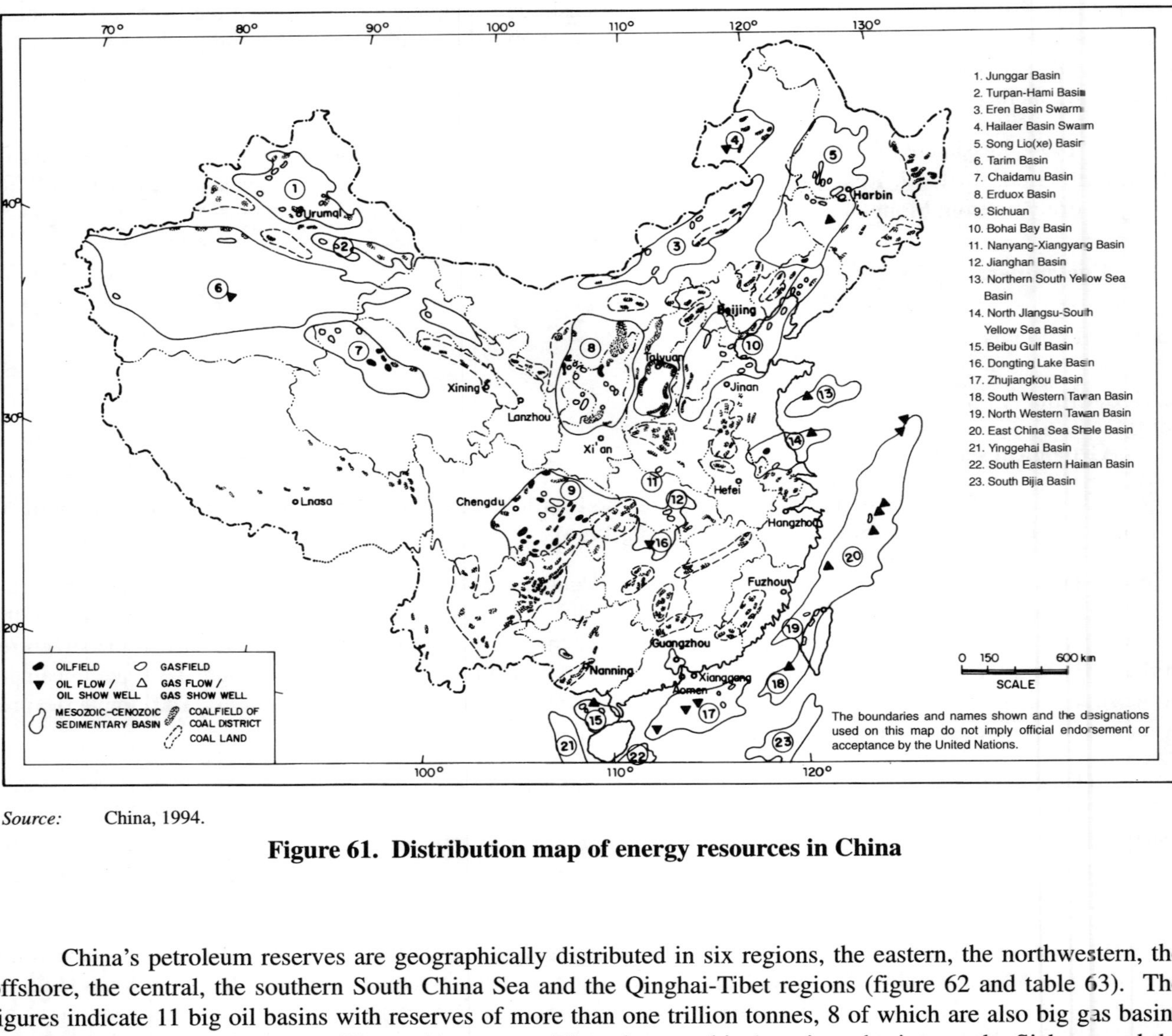

Source: China, 1994.

Figure 61. Distribution map of energy resources in China

China's petroleum reserves are geographically distributed in six regions, the eastern, the northwestern, the offshore, the central, the southern South China Sea and the Qinghai-Tibet regions (figure 62 and table 63). The figures indicate 11 big oil basins with reserves of more than one trillion tonnes, 8 of which are also big gas basins with reserves of more than one trillion cubic metres. The other two big 'pure' gas basins are the Sichuan and the south-eastern Hainan basins.

The eastern part of China contains five productive basins, namely Songliao, Huabei, Subei, Jianghan and Nanxiang, totalling an area of more than 600,000 km^2. The crude oil and gas resources in this province are estimated at 22 billion tonnes and 5,880 billion cubic metres respectively. Oil and gas production from this province accounts for 84 per cent and 47 per cent of China's total respectively.

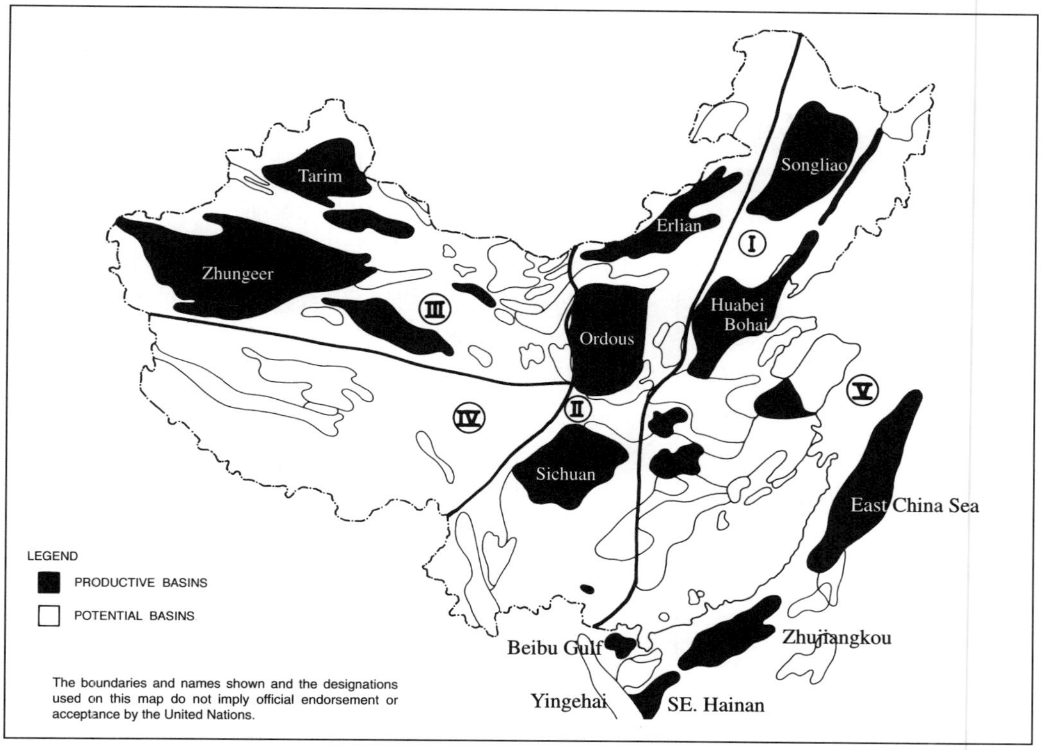

Source: Mineral Resources of China, 1992.

Figure 62. Distribution of petroliferous basins in China

The northeastern part of China contains four main productive basins, namely the Zhungeer, Tarim, Turpan-Hami, and Chaidamu, with a total area of 1,900,000 sq km. The crude oil and gas resources in this province are estimated at more than 20 billion tonnes and 11,400 billion cubic metres. In 1993, the crude oil production in the Zhungeer, Tarim and Turpan-Hami basins reached 10.35 million tonnes. These basins are the main target areas for future exploration and development plans in China.

The central petroleum province includes the Hailaer, Erlian, Ordos, Sichuan and Baise productive basins. The gas resources in this province are estimated at nearly 12,000 billion cubic metres. Gas production is mainly from Sichuan and Ordos basins, accounting for about 50 per cent of China's total.

China's offshore continental shelf contains six petroleum provinces of Cenozoic age. They are, from north to south, the Bohai Bay, East China Sea, Zhujiangkou, southeast Hainan, Beibu Wan and Yinggehai basins. By 1992, 68 oil and gas bearing structures had been found with more than 1 billion tonnes of crude oil and 133 billion cubic metres of natural gas. It is expected that China's annual offshore oil and gas production may reach as high as 8-10 million tonnes in 1997.

As indicated in table 64, the total proved reserves of natural oil in China at the end of 1994 was 24 thousand million barrels, the oil R/P ratio 22.6 years, with oil production 2,905 thousand barrels daily. The total proved reserves of gas in China at the end of 1994 was 59 trillion cubic feet, the gas R/P ratio 102.7 years with annual gas production 14.9 Million tonnes oil equivalent (M toe), or about 0.58 trillion cubic feet.

However, there are different estimations shown in table 64. China's estimated proven oil reserves are 96.8 billion barrels, which is four times that estimated by British Petroleum. The Chinese derived this figure from the estimated total conventional resources of crude oil in China being 81.47 billion tonnes, of which the proven oil reserves accounted for 16.2 per cent. The proven gas resource was estimated by the Chinese as 35.2 trillion cubic feet, which is between the figures estimated by Wood Mackenzie and the Petroleum Economist. In this estimation, the Chinese counted the gas resources as only 2.0 per cent of their total gas reserves i.e. 51.7 trillion cubic metres (China, 1994).

Table 63. Distribution of petroleum reserves of China

	Oil		*Gas*		*Formation*
	Thousand million tonnes	*%*	*Billion cubic metres*	*%*	
Total China	81.47	100.00	51,708.05	100.00	
Total Eastern	22.02	27.03	5,880.21	11.37	
Huabei	12.10	14.85	2,496.10	4.83	Meso-Cenozoic, Paleozoic
Songliao	8.33	10.22	1,872.60	3.62	Mesozoic
South Huabei	0.28	0.34	477.12	0.92	Meso-Cenozoic
Subei	0.19	0.23	61.60	0.12	Meso-Cenozoic
Nanxiang	0.19	0.23	16.90	0.03	Cenozoic
Jianghai	0.11	0.14	47.60	0.09	Meso-Cenozoic
Others	0.82	1.00	828.29	1.60	
Total Central	3.96	4.86	11,934.57	23.08	
Ordous	2.04	2.50	3,536.20	6.84	Mesozoic, Upper Paleozoic
Erlian	1.02	1.25	49.80	0.10	Mesozoic
Hailaer	0.41	0.50	159.70	0.31	
Sichuan	0.13	0.16	5,771.40	11.16	Mesozoic, Upper Paleozoic
Others	0.36	0.44	2,417.47	4.68	
Total Northwestern	20.70	25.41	11,400.10	22.05	
Tarim	10.10	12.40	8,128.30	15.72	Meso-Cenozoic, Paleozoic
Zhungeer	8.22	10.09	2,755.00	5.33	Meso-Cenozoic, Upper Paleozoic
Chaidamu	0.64	0.79	346.10	0.67	Meso-Cenozoic
Turpan-Hami	0.28	0.33	38.09	0.07	Meso-Cenozoic, Paleozoic
Others	1.46	1.79	132.61	0.26	
Total Qinghai-Tibet	0.28	0.34			
Total Offshore	34.51	42.36	22,493.17	43.50	
Zhujiangkou	5.60	6.87	1,015.90	1.96	Meso-Cenozoic
Bohai	5.30	6.51			Meso-Cenozoic
East China Sea	3.28	4.02	5,106.80	9.88	Meso-Cenozoic
Yinggehai	1.40	1.72	2,524.00	4.88	Meso-Cenozoic
Beibu Gulf	1.10	1.35	304.00	0.59	Meso-Cenozoic
S. Yellow Sea	0.44	0.54	19.00	0.04	Meso-Cenozoic
SE. Hainan			3,970.00	7.68	Meso-Cenozoic
Southern S. China Sea	15.22	18.68	8,269.97	15.99	
Others	2.17	2.66	1,283.50	2.48	

Source: China, 1994, The Petroleum Resources and Cooperation Prospectivity in China, presented by Chinese delegation in 1st Exploration Promotion Forum, 2-4 June, Stavanger, Norway.

Table 64. Estimated petroleum reserves in China

Estimated by	*China[1]*	*BP[2]*	*Wood Mackenzie[3]*	*Petroleum Economist[3]*
Oil, billion barrels	96.8	24	36	17
Gas, trillion cubic feet	35.2	59	41	30

Source: 1. Chinese delegation, 1994: The Petroleum Resources and Cooperation Prospectivity in China, presented in 1st Exploration Promotion Forum, 2-4 June, Stavanger, Norway.
2. BP, June 1995, Statistical Review of World Energy.
3. Petroleum Economist, June 1995, Vol. 64, No. 6, pages 14-16.

Unconventional petroleum resources also occur in China. The estimated oil shale reserves were 452 billion tonnes with estimated shale oil resources of 14.20 billion tonnes. China's coal-bed methane resources at a depth of less than 1,000 metres is about 6.4 trillion cubic metres. They are mainly distributed in the Huabei and northwestern regions of China.

Crude oil production

Oil and gas exploration has been conducted in more than 80 basins. Commercial oil and gas fields have been found in 22 onshore and 5 offshore basins. China's major producing areas, as of year end 1992 are shown in table 65.

Table 65. Major oil and gas producing areas in China, as of year end of 1992

Producing area	*Production (1,000 metric tonnes)*	*Development cost (Yuan/ton)*	*Total development outlays (million yuan)*	*Finding cost (Yuan/ton)*	*Cost of increasing productive capacity (yuan/ton)*	*Decline rate (%)*
Daqing	54,312	217.90	5,239.99	11.2	690	3.82
Shengli	32,330	251.91	5,851.10	8.8	862	10.19
Liaohe	13,184	261.32	3,023.36	6.7	795	11.25
Xinjiang	7,106	515.75	2,926.16	8.7	1,185	5.77
Zhongyuan	5,654	381.08	1,712.50	25.0	1,743	13.75
Huabei	4,530	459.68	1,823.08	15.9	1,106	7.55
Dagang	3,764	357.50	1,631.36	7.3	790	4.80
Jilin	3,154	284.85	1,362.87	8.0	1,439	6.16
Henan	2,175	444.24	625.74	36.8	1,055	9.64
Changqing	1,583	360.50	1,410.96	4.0	1,589	3.84
Yumen	1,107	554.63	2,187.17	5.8	2,472	1.43
Qinghai	990	364.00	774.40	22.3	1,574	6.08
Tarim	833	404.88	2,468.05	23.0	2,443	20.00
Jiangsu	825	278.74	385.08	17.4	1,251	9.43
Jianghan	733	465.15	519.82	20.0	1,181	7.09
Jidong	361	396.31	429.90	20.0	1,464	5.67
Sichuan	139	220.97	1,466.79	23.7	980	8.69
Yunnan, Guizhou & Guanxi	107	374.49	241.93	26.5	1,101	11.59
Anhui	37	407.00	63.19	12.9	1,189	5.85

Source: Oil & Gas Journal Special, May 9, 1994.

In 1995, China produced 2.99 million barrels of oil daily and is the world's seventh largest oil producer after Saudi Arabia, the United States, the Russian Federation, the Islamic Republic of Iran, Mexico and Norway. In 1994, about 70 per cent of China's oil production was produced from three complexes, Daqing (nearly 38 per cent), Shengli (20.5 per cent) and Liahoe (9.8 per cent), all in northeast China. Offshore production reached only 4.44 per cent in 1994.

China produced 17.6 billion cubic metres of gas in 1995, 98 per cent of which was produced on shore. Sichuan province produced about 7 billion cublic metres of gas in 1991, accounting for 50 per cent of China's annual gas production.

Offshore oil and gas fields

China's offshore fields flowed nearly 1 million tonnes of crude in 1989, 1.24 million tonnes in 1990, 2.4 million tonnes in 1991, and over 3 million tonnes in 1992. 4 million tonnes of crude is expected from offshore in 1995 and 8-10 million tonnes by 1997. According to CNOOC officials, however, it will be difficult to maintain that output after 1997 (Petromin, September, 1992).

In the South China Sea, the Yacheng 13-1 gas field, about 95 km south of Sanya of Hainan province, is expected to yield at least 3.4 billion cublic metres of gas a year starting in 1996. A pipeline was built to link the field with Hong Kong, China offering 2.9 billion cublic metres annually. The confirmed reserves of the Yacheng gas field are around 100 billion cublic metres, about half of the proven reserves of the Sichuan gas field.

The Bohai field, discovered in 1975, is the second largest offshore field in China, comprising three oil fields i.e. Chengbei, Bozhong 28-1 and Bozhong 34-2/4. Their 1991 production amounted to 950,000 tonnes, making up 40 per cent of China's offshore total. The three fields were jointly explored and developed in the central area of the Bohai Sea by the CNOOC and the Japan-China Oil Development Corporation in 1989 and 1990.

Two thirds of the Bohai sea area (57,500 km^2) has been opened to foreign co-operative enterprise. The other one-third of the Bohai sea, in northwest Liaodong Bay, has been the subject of self-financed E & D by BOC. BOC's Jinzhou 20-2 gas/condensate field in the Liaodong Bay area is expected to produce gas of about 500 million cublic metres, and crude oil of one million tonnes annually. A 48.6 km, 12 inch gas pipeline to transport gas from Jinzhou 20-2 to the Jinxi petrochemical plant in Liaoning province was completed in Summer 1991. So far it is China's longest offshore pipeline. Also under development is the Suizhong 36-1, an oilfield covering 6 km^2 with estimated crude reserves of 200 million tonnes. It is about 20 km south of the Jinzhou 20-2 field.

In the East China Sea, two oil fields i.e. Pinghu and Canxue, were jointly explored by the CNOOC and the Chinese Ministry of Geology and Mineral Resources (MGMR). A total of 20 oil and gas production wells were completed by September 1992. Pinghu field was expected to start production at a capacity of 800,000 to 1.2 million cubic metres per day in 1995. Explorationists have estimated the field's reserves at 14.56 billion cublic metres of gas and 8.26 million tonnes of condensate. A 400 km underwater gas pipeline linking Shanghai and the two fields was to start construction in 1993 (Petromin, September 1992).

Refined products

China's refined oil products output in recent years is shown in table 66. The output of gasoline in 1993, with the fastest growing rate in China's refined products, is 2.4 times that in 1983.

China's refining capacity was 3.4 million b/d in 1994, giving China the world's fourth largest crude processing capacity after the US, Russia, and Japan. China's refining capacity is expected to reach 4.6 million b/d by the year of 2000, with its refiners operating at a utilization rate of about 85 per cent because of domestic crude supply constraints (table 67). China's fluid catalytic cracking (FCC) capability is about 900,000 b/d, which is higher than Japan's 700,000 of FCC capacity, and fairly high relative to crude distillation capacity – 35 per cent in China compared with 18 per cent in Asia (PE, March 1995).

The distribution of refineries in China is uneven, with the southern coastal provinces having few large-scale refineries, while the northeastern regions have more capacity than they need. The refineries for crude oil distillation of Sinopec are shown in table 68, noting that Sinopec operated about half the country's refineries but at 2.8 million b/d more than 85 per cent of the nation's refining capacity.

Table 66. Output of refined oil products of China

Unit in 1,000 barrels daily (b/d)

	1983	*1990*	*1991*	*1992*	*1993*
Gasoline	247.0	401.2	469.0	526.6	598.8
Kerosine	81.2	77.0	81.2	78.6	74.0
Diesel	376.4	507.4	553.0	612.2	661.4
Lubricants	26.8	33.8	34.6	37.4	37.2
Solvents	6.8	10.2	10.0	11.0	9.6
Fuel oil	579.0	643.4	635.4	635.6	610.0
Waxes	9.4	13.8	13.4	13.8	14.4
Petroleum coke	17.4	27.2	31.6	33.0	37.6
Asphalt	35.4	54.6	62.4	68.6	65.4
Detergent feedstock	N/A	0.6	0.8	0.8	0.6
Light oil feedstock for chemical industry	92.6	157.0	160.2	166.6	163.0

Source: Oil & Gas Journal Special, May 9, 1994.

Table 67. Outlook of oil refining capacity in China in 1992, 1995 and 2000

Unit: 1,000 b/d

	1992	*1995*	*2000*
Capacity	3,253.6	3,600.0	4,600.0
Throughput	2,131.2	2,900.0	3,900.0
Utilization (%)	65.0	80.0	85.0

Source: Oil & Gas Journal Special, 1994, May 9.

Consumption

China became the world's third largest oil consumer at 3,310 thousand barrels daily in 1995, overtaking the Russia Federation. Commercial oil and gas consumption in China in 1992 was 133 Mtoe and 14 Mtoe respectively, accounting for 18.7 per cent and 1.9 per cent in the energy mix, with growing rates of 5.9 per cent/year and 7.3 per cent/year in the period of 1971-1992.

The consumption of petroleum downstream products by various sectors in 1988 is shown in table 69. Industry is the major sector for consuming of fuel (50 per cent), gasoline (31 per cent), diesel (30 per cent) and Liquefied Petroleum Gas (LPG, 40 per cent). Transportation is another major sector for consumption of gasoline (32 per cent) and diesel (27.8 per cent). However, the figures for the transport sector could be underestimated as more than 90 per cent of the vehicles were not controlled by the Ministry of Transportation. Power generation accounted for 41 per cent of the fuel oil, household sector 57 per cent of LPG and agriculture 30 per cent of diesel.

Consumption of LPG in China has increased at a rate of 5,000 metric tonnes/year during the past several years. Of total production of about 2.6 million tonnes/year, three fourths is consumed by local resident utilities. Growth in demand for LPG is shifting from north China toward southeastern coastal areas, where town gas (coal gas) is expensive to produce and demand is growing rapidly. Typical of residential consumption patterns for heating and cooking is this breakout for Beijing households: 560,000 town gas, 220,000 natural gas, and 1 million LPG metric tonnes annually (Oil & Gas Journal Special, May 1994).

Table 68. Sinopec's 1992 refining configuration by region, refineries for crude distillation

Unit: 1,000 b/d

Northeast	*806*	
Linyuan		30
Harbin		30
Daqing		110
Qingguo		30
Fushan		174
Dalian		142
Jinxi		100
Jinzhou		110
Liaoyang		30
Anshan		50
North	*588*	
Yanshan		140
Tianjin		74
Shijiazhuang		70
Cangzhou		24
Luoyang		60
Jinan		60
Qilu		160
Lower Yangzi	*616*	
Anqing		60
Jinling		134
Yangzi		60
Gaoqiao		146
Jinshan		106
Zhenhai		110
Middle Yangzi	*300*	
Baling		100
Wuhan		50
Jingmen		100
Jiujiang		50
South	*324*	
Guangzhou		104
Maoming		170
Fujian		50
Northwest	*117*	
Lanzhou		100
Lanzhou Pet.		17
West	*116*	
Urumuqi		50
Dushanzi		66
Total	*2,867*	

Source: Sinopec.

Table 69. Petroleum consumption in various sectors of China in 1988

	Fuel Million tonnes	*Fuel* %	*Gasoline Million* tonnes	*Diesel* %	*Diesel Million* tonnes	*LPG* %	*LPG Million* tonnes	%
Power generation	13.5	41.2	–	–	1.9	7.5	–	–
Industry	16.3	49.8	5.6	31.3	7.6	29.5	0.9	40.0
Transportation	1.8	5.6	5.7	31.6	6.9	26.8	–	–
Agriculture	0.04	0.1	1.5	8.6	7.7	29.8	–	–
Household	–	–	0.2	0.9	0.1	0.2	1.3	57.0
Others	1.1	3.3	4.4	27.4	1.6	6.2	0.1	3.0
Total Consumption	32.8	100.0	17.9	100.0	25.8	100.0	2.3	100.0

Source: China Statistic Publishing House, Energy Statistical Yearbook, 1989.

Unlike the urban residents who consume mostly commercial energy, the rural residents in China consume commercial energy accounting for only one fourth of their total energy consumption. Large quantity of biomass energy, for example about 208 million tonnes of oil equivalent (Mtoe) in 1988, has to be consumed there (Qiu, 1991).

Trade

China had been a net energy export country for many years prior to 1994. Energy exports were 21.4 and imports 1.8 Mtoe in 1980, compared to exports of 40.4 and imports of 6.4 Mtoe in 1988. Nevertheless, domestic energy supply has fallen short of demand for a long time. In the late 1980s, the deficit of energy supply was about 15-20 Mtoe per year, in which the gap of oil was 5 million tonnes and electricity 50 TWh (Qiu, 1991).

China became a net oil importer in 1993, the first time since the 1960s. Its net export of crude oil in 1993 was 3.60 million tonnes, but its net import of refined oil products in the same year was 17.32 million tonnes (table 70). China's imports of crude oil and refined products have skyrocketed recently. Crude imports were 17.10 million tonnes in 1995, nearly 22 times that of 0.78 million tonnes in 1988. China's exports of crude oil have continually declined; 18.80 million tonnes in 1995, 27.5 per cent less than the 25.94 million tonnes in 1988. In connection to this, the average growth rate of China's oil production in the period 1985 to 1995 was about 1.9 per cent, but the average growth rate of its oil consumption in the same period was more than 7.4 per cent.

China imports crude oil from the Middle East, West Africa, and South-East Asia. About 65 per cent of China's crude imports are 30-45° gravity, low sulfur crudes from Indonesia, Oman, and Malaysia. The balance comes form Viet Nam, Russia, Islamic Republic of Iran, and Dubai. As traditional suppliers such as Indonesia face the prospect of becoming net oil importers, China is looking to Middle East countries for crude in the years to come. Emblematic of that change is the scramble to revamp, upgrade and expand its domestic refineries to process the Middle East's medium to low gravity, sour crudes (Oil & Gas Journal, May 1994).

China exported crude oil and its refined products of 22.1 million tonnes total in 1995, of which over 52 per cent went to Japan, nearly 35 per cent to other Asian countries and nearly 12 per cent to the United States (BP, 1995). The major buyers of China's crude oil are Brazil, Japan, the Democratic People's Republic of Korea, the Republic of Korea, the Philippines, Singapore, and the USA (OGJ Special, 1994, May 9, p. 34, 54). The major trading countries for China's refined products in 1993 included Singapore, the Republic of Korea and Japan (table 71).

China's major coastal crude tanker terminals are listed in table 72. Most of the new tankers built or to be built in the 1990s are large with more than 150,000 dwt. As estimated by China's Ministry of Transport, transportation costs per ton of crude is about US$17, 14, 12, 11 and 10.6 for tankers in 60, 100, 150, 200 and 250 thousand dwt respectively. The newest development is with Guandong and Zhejiang provinces, both of whom are in southeast China.

Table 70. Oil balance in China, 1985-1995-2000

Unit in Million tonnes

	Crude Oil				Refined Products	
	Production	*Consumption*	*Imports*	*Exports*	*Imports*	*Exports*
1985	124.9	90.3	–	29.92	0.09	6.58
1986	130.7	100.0	0.45	28.38	2.16	5.96
1987	134.1	105.3	–	27.12	2.01	5.92
1988	137.1	110.2	0.78	25.94	3.18	4.77
1989	137.6	112.3	3.25	24.29	5.43	4.79
1990	138.3	110.3	2.91	23.89	3.18	5.42
1991	141.0	117.9	5.95	22.51	4.64	5.92
1992	142.0	129.0	11.31	21.42	7.64	5.37
1993	144.0	140.5	15.58	19.36	17.32	3.60
1994	146.1	149.6	12.30	19.70	14.40	2.70
1995	149.0	157.5	17.10	18.80	14.40	3.30
2000	159.4-164.3	219.1	49.98	9.96		

Source: The Production and Consumption figures for 1985-1993 and all the figures for 1994 and 1995, from 1995 and 1996 BP Stastical Review of World Energy; All the other figures, derived from Oil & Gas Journal Special, May 9, 1994.

1.2.1.2 Democratic People's Republic of Korea

The Democratic People's Republic of Korea has offshore oil and gas in the Korea Bay but is in need of technology and equipment to explore and exploit them. The Democratic People's Republic of Korea has two refineries with a capacity of 3 Mt/y.

The Democratic People's Republic of Korea's oil imports fell from 3.3 Mt in 1989 to only 2.5 Mt in 1990, due to cut backs in Russian exports. The supply further dropped sharply in 1991 as Russia cut its supply by more than 90 per cent and demanded hard currency for deliveries. With this sharp cut-back in supply, the Democratic People's Republic of Korea is desperate for oil. The oil supply is heavily dependent on imports from China now.

1.2.1.3 Japan

Japan consumed 268.7 million tonnes oil equivalent in 1994, ranked the second in the world. However, domestic oil production can only provide 0.4 per cent of Japanese total consumption with the balance from imports.

1.2.1.4 Russian Far East

Production

The former Soviet Union produced 20 per cent of the world's oil in 1990. The Russian oil industry comprised 840 oil fields with 148,000 oil wells, 48,300 km of main oil pipeline, and 28 oil refineries that can refine more than 300 Mt/y of oil. The oil industry and its auxiliary services employed about 900,000 people. Oil production in Russia was 552 Mt in 1989, 395 Mt in 1992, 357 Mt in 1993, and 315.7 Mt in 1994 Russia reportedly exported 79.7 Mt of oil outside the CIS in 1993. Russia's exports of oil to other CIS states was 40.4 Mt in 1993, only 53 per cent of its 1992 level (Mining Annual Review).

Natural gas production in Russia was 616 billion cubic metres in 1989, 640 billion cubic metres in 1992, 617.4 billion cubic metres in 1993, and 581.02 billion cubic metres in 1994. Russia's main producing area of

Table 71. Oil refined products imports and outputs of China, the third quarter of 1993

	Imports from		*Exports to*	
	Volume (b/d)	*Value (1,000$)*	*Volume (b/d)*	*Value (1,000$)*
Singapore	217,548	466,861	53,194	107,957
Republic of Korea	59,258	113,298	12,981	30,402
U.S.	38,741	105,199	9	45
Russia	30,753	63,969	21	191
Japan	13,062	34,452	8,873	24,200
Mexico	6,462	14,457		
Hong Kong, China	6,339	17,273	4,801	9,868
Romania	5,742	13,944		
Croatia	4,804	11,948		
Indonesia	3,951	8,059	37	255
Saudi Arabia	2,301	2,642		
India	1,906	1,958		
Thailand	1,411	3,484	995	4,186
Philippines	989	2,748	17	40
Taiwan Province of China	832	4,141	11	81
Netherlands	807	1,772		
U.K.	551	2,095		
Italy	510	2,792		
Bangladesh	436	525	15	102
People's Democratic Republic of Korea	403	1,044	841	2,839
Kuwait	109	154		
Belarus	79	220		
Macau			283	627
Malaysia			117	518
Myanmar			112	919
Pakistan			9	56
Yemen			8	71
Others	1,076	3,769	24	182
TOTAL	392,279	876,810	82,348	182,540

Source: East-West Centre, in Oil & Gas Journal Special, May 9, 1994.

Table 72. Major coastal crude tanker terminals in China

Location	*Maximum tanker (1,000 dwt)*	*Year of completion*	*Crude capacity (1,000 b/d)*	*Owener*
Huangdao, Shandong	200	1988	340	Ministery of Transport
Aoshan, Zhejiang	200	1992	320-400	Sinochem, Zhejiang province
Zhenhai, Zhejiang	150	1990	320	Zhenhai Refinery
Dalian, Liaoning	100	1976	200	Ministry of Transport
Xiaocuo, Fujian	100	1992	200	Fujian Refinery
Dalian, Liaoning	50	1976	100	Ministry of Transport
Qinhuangdao, Hebei	50	1984	100	Ministry of Transport
Huangdao, Shandong	50	1976	200	Ministry of Transport
Zhenhai, Zhejiang	50	1977	100	Zhenhai Refinery
Zhanjiang, Guangdong	50	1974	90	Ministry of Transport
Bohe Bay, Guangdong	250	1994	400 plus	Maoming Refinery
Qidong, Jiansu	250	To be approved	400 plus	Jinling Refinery
Cezi Island, Zhejiang	200	feasibility study	300-400	Sinopec, EIF, Shanghai
Daya Bay, Guangdong	100	1997	200	Huizhou Refinery

Sources: Oil & Gas Journal, May 9, 1994.

natural gas is West Siberia. The Gazprom controls practically all gas production in Russia. Russia exported 99 billion cubic metres in 1992, and 96 billion cubic metres of natural gas in 1993. About one third of the 1993 exports went to the countries of the former Soviet Union, and the rest to market economy countries.

The Russian Eastern Siberian and Pacific basins, remain immature in terms of exploration maturity. About 10 per cent of Russian oil production comes from these basins. New or additional oil and gas reserves will be found in the region, but are not expected to make an impact until the turn of the century.

As for current production of hydrocarbons, the nationwide comparisons are even less in favour of the Russian Far Eastern region (RFE). The region, producing less than 40,000 barrels daily of crude and condensate in Sakhalin, accounts for only 0.4 per cent of Russian oil output. The RFE with several developed gas fields located on Sakhalin island and in central Yakutia, produced gas around 300 mmcfd, or 0.5 per cent of Russian gas production.

Consumption and imports

Crude oil produced in Sakhalin is mainly transported to the mainland via a double pipeline that connects the oilfield of the island's northern area to a refinery at Komsomolsk-on-Amur with capacity of 10,000 b/d. The needs of the RFE's only other refinery at nearby Khabarovsk (85,000 b/d) are provided for by feedstock shipped in and by railroad and river tankers from Siberia.

The insufficient indigenous production is wholly consumed within the region's bounds – mainly at local power plants (53 per cent), in industry (31 per cent), and for household/municipal needs (10 per cent) in the cities of Yakutsk, Okha, Komsomolsk-on-Amur, and in neighbouring industrial towns.

In 1990, the region's dependence on imported liquid fuels amounted to 80 per cent in the case of crude oil and to 55 per cent, on an average, for all oil products, including 65 per cent for diesel fuel, 58 per cent for jet kerosene, 55 per cent for residual fuel oil, and 37 per cent for motor spirit.

In recent years, more than 200,000 b/d of petroleum products and 130,000 b/d of crude oil have been transported there by sea and railway, at an annual cost of more than R 300 m. Russia's plan for comprehensive development of the Far Eastern region to year 2000 highlighted the need to sharply increase oil and gas prospecting. The volume of surveying activity is to increase 2.5-3 times in the next two five-year development periods.

1.2.2 Coal

1.2.2.1 China

By the end of 1990, China had 972.4 billion tonnes of explored coal reserves and 954.4 billion tonnes of retained coal reserves, of which anthracite accounted for 12 per cent, coking coal 28 per cent, non-coking bituminous 47 per cent and lignite 13 per cent. It is estimated that there are a total of 4.5 trillion tonnes of coal resources. China's coal reserves are distributed mainly in northern and western China (figure 63). Nearly 70 per cent of them are in Shanxi, Shaanxi and Inner Mongolia.

The coal industry in 1994

China, the world's largest coal producer, produced 1,110 Mt in 1992. China consumed 527.1 Mt of oil equivalent in 1992, which accounted for a share of 24.4 per cent of the world total and ranked number one. China produced 1,227 Mt of raw coal in 1994 accounting for 34 per cent of the world total.

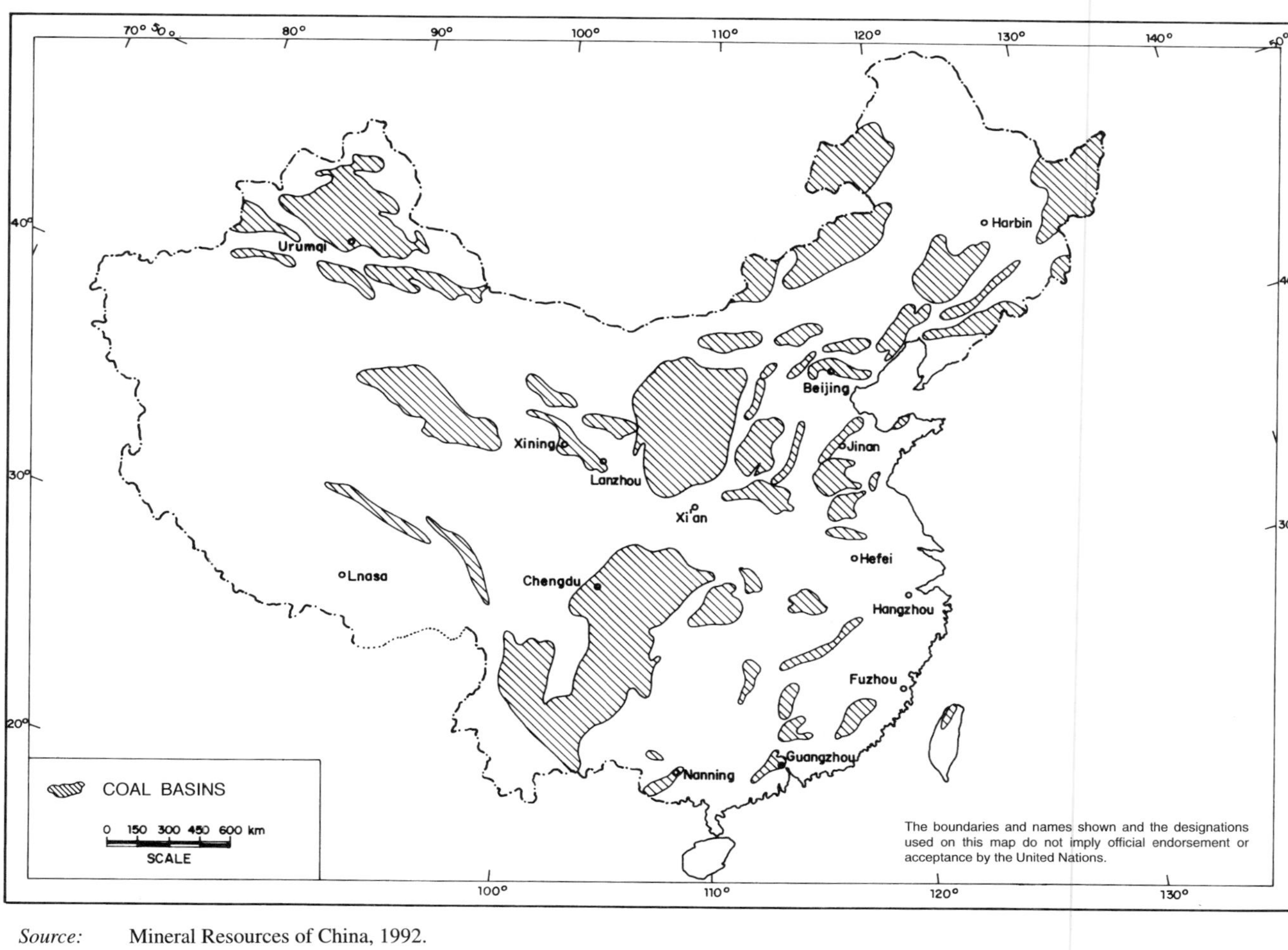

Source: Mineral Resources of China, 1992.

Figure 63. Distribution map of coal resources in China

According to the PRC YEAR BOOK'95, 1994 was the best year for China's coal export, which came to 25 Million tonnes with foreign exchange earnings amounting to US$785 million. This success highlighted the effort to expand sales of Chinese coal in southeast and south Asia and Europe while continuing to consolidate the Japanese and other traditional markets.

China's coal industry had been losing money for several years due to reasons such as increased production costs resulting from inflation. But there were signs of improvement in 1994. Of the 93 state-owned key mines, 29 were profitable, 11 more than in 1993. State owned mines throughout the country incurred deficts of 1.974 billion yuan, 26 million yuan less than permitted under state planning and compared with the 1993 deficit of 3.25 billion yuan.

In 1994, supply of coal continued to exceed demand. Altogether 1.186 billion tonnes of coal were burned, 33 million tonnes more than in 1993. Meanwhile, 78.16 million tonnes more coal were produced. This, plus inadequate railway transport capacity, caused unsold coal to continue to be stockpiled. In view of this, the coal industry was urged to supply the market with coal of better quality and increased variety. In 1994, 76.5 million tonnes of washed coal were produced, 7 per cent more than in 1993.

China's 1994 coal exports accounted for 2 per cent of its production. It was estimated that China's coal exports in 2000 would account for 0.8 per cent of its production; and China would became a coal importer in 2010, imported coal accounting for 6.2 per cent of its production (Fujime, 1996).

Coal exports in 1995

China exported 27.5 Mt of hard coal in 1995, ranked sixth after Australia 136.1 Mt, the United States 80.3 Mt, South Africa 59.7 Mt, Canada 34.0 Mt, Poland 31.9 Mt and Indonesia 31.6 Mt.

China exported 23.2 Mt of steam coal in 1995. Australia and South Africa remained the largest exporters, with about 61.2 Mt and 53.5 Mt respectively and Indonesia exports reached 29.4 Mt in 1995.

China exported 4.3 Mt of coking coal in 1995. Australia and the United States remained the largest coking coal exporters, with about 41 per cent and 26 per cent of the total market, respectively.

1.2.2.2 Democratic People's Republic of Korea

The Democratic People's Republic of Korea's annual energy consumption is about 42 Mt of oil equivalent, with coal meeting 75 per cent and oil 10 per cent. (1995) Coal deposits in the country are estimated at 12,000 Mt, in which metallurgical coal is limited. The main coal region is the south Pyongyang province covering the Tokchon, Kangdon, Kangso, Kowon and other areas. (1994) The Anju area, about 70 km north of Pyongyang, completed with Soviet assistance, has reserves to support extraction of 100 Mt/y for 100 years.

1.2.2.3 Japan

The Japanese coal industry has been in a long downward trend with production of steam coal at 6.9 Mt in 1994 (compared with about 10.4 Mt in 1988), with an average decreased rate of 5.6 per cent. Japan produced 110,000 Mt of coking coal in 1990, falling 87 per cent from the previous year, and has not been a coking coal producer since then (table 73).

Due to growing demand for energy, import of steam coal has been in a long trend upward reaching 57.1 Mt in 1995, compared with 50 Mt in 1994 and 31 Mt in 1989. Overall, import of coking coal is in a downward trend, reflecting the steel mills shifting to other energy sources because of increased prices. Japan imported 63 Mt of coking coal in 1994, compared with about 71 Mt in 1988. Within the Asia-Pacific region, Japan accounts for about two thirds of the coking coal market, importing 65.3 Mt in 1995.

According to the Democratic People's Republic of Korea's press, coal production in 1988 was placed between 48 Mt and 70 Mt in 1988, sharply reduced to 43.1 Mt in 1991, and placed at 90 Mt/y in 1994. Output of coking coal has remained stagnant at around 1 Mt/y, coming from the Kukdong and Yangjong collieries.

Table 73. Coal production and imports of Japan, 1988-1995

Year	*1988*	*1989*	*1990*	*1991*	*1992*	*1993*	*1994*	*1995*
Production (in '000 t)		-8.4%	-13%	-2.5%	-6%	-5%	-4%	
coal, steam		9,330	8,150	7,790	7,600	7,200	6,900	
coal, coking	1,036,000	860	110	0				
Import (in '000 t)		+8.7%	+10%	+9%	+12%	+3%	+13%	
coal, steam		31,000	34,200	37,200	42,200	43,000	50,000	
coal, coking		68,600	68,500	69,600	64,400	65,000	63,000	
		-3.5%	-1.7%	+3%	+7%	+1%	-3%	

Source: Mining Annual Review, 1989, 1990, 1991, 1992, 1993, 1994, 1995 and 1996.

In recent years, the Democratic People's Republic of Korea imported about 1-2 Mt/y of steam coal from China and the former Soviet Union, while Australia has supplied prime coking coal.

1.2.2.4 Mongolia

Mongolia is rich in coal with identified reserves of 10,000-15,000 Mt, and 15 large coal mines, both open and underground. Mongolia's coal production today is predominantly brown coal from Nalaith, an underground mine 35 km from Ulanabaatar, and from Raganur, an open cast mine (120 km southeast of the capital) with 627 Mt reserves. Mongolia produced 6.9 Mt coal in 1993 and was expected to produce 7.2 Mt in 1994.

A top priority project for the Mongolian energy sector is the development of Tavantolgoi in southern Mongolia, the country's largest known coal deposit. The coal mine, covering 90-150 km^2, is located 90 km east of Dalandzadgad, the capital of south Gobi province. The deposit has been explored for more than 20 years and has coal reserves of 5,000-7,000 Mt, of which 2,000 Mt are metallurgical quality coking coal. The deposit is large and readily mined by open cast methods. A capacity of 20 Mt/y of coking coal for an open-pit mine is a realistic possibility.

1.2.2.5 Republic of Korea

Energy

Energy production in 1991 decreased by 2.2 per cent to 22.7 Mt of oil equivalent (TOE). Total energy imports were 9.6 Mtoe – an increase of 39.4 per cent against 1990. The Republic of Korea's energy consumption amounted to 103 Mtoe in 1991, an increase of 1.1 per cent over 1990. Petroleum dominated energy consumption with 424 Mbbl, up 19.1 per cent over 1990. Imports of crude oil were approximately 399 Mbbl and 2.8 Mt of LNG from Indonesia.

The Republic of Korea's energy bill amounted to $10,900 million, up from $7,500 million in 1989. The relatively sharp increase in energy demand was attributed to a growing demand for bituminous coal and naphtha triggered by facility expansions in the steel and petrochemical industries. The year's import of crude oil for domestic consumption totalled $6,386 million, compared with $4,933 million in 1989.

Coal production

Coal is a major indigenous energy resource in the Republic of Korea. It is found in the form of anthracite, with a low calorific value of 4,500 kal/kg. The Republic of Korea's anthracite reserves are estimated at 1,520 Mt, of which 699 Mt are mineable. The government owns 23 per cent of the reserve through Dai Han Coal Corp (DHCC) with three big mines: Changsong, Whasoon and Dogae. Private mine operators hold 77 per cent of the reserves with three coal producers having capacity of over 1 Mt/y: the Dongwon coal mine: the Samchuck coal mine; and the Kyungdong coal mine.

Some 85 per cent of local anthracite coal is used as household fuel in coal briquette form. The balance is used for power generation. Coal consumption in the residential sector will continue to decrease both in real terms and as a proportion of total consumption as domestic consumers switch to gas and electricity. The government provided a $410 million subsidy loan to coal production in 1994.

Due to falling coal demand and rising labour costs, the Ministry of Energy and Resources implemented a Coal Mining Rationalization Plan. Tens of small coal mines were closed annually, with about 210 operating mines in 1990, 170 in 1991, 130 in 1992, and down to 26 in 1994. In connection to this, anthracite coal production has decreased continuously in recent years, from a total of 23.8 Mt in 1988 to 20.8 Mt in 1989, 17.2 Mt in 1990, 15.1 Mt in 1991, 11.9 Mt in 1992, 9.4 Mt in 1993 and down to 7.44 Mt in 1994.

Coal imports

Total imports of coal in 1992 amounted to 30.1 Mt, an increase of 3 per cent over the previous year. The Republic of Korea's imports of anthracite coal declined drastically, with 2.8 Mt in 1988, 1.5 Mt in 1989, 808,629 t in 1992 (477,014 Mt from China and 301,613 Mt from Viet Nam), 780,000 t in 1993 and 700,000 t in 1994.

On the other hand, the use of bituminous coal, which is imported, has increased significantly and it now provides around 60 per cent of primary coal consumption. Bituminous coal imports totalled 27.8 Mt (15.5 Mt for metallurgical coal and 12.3 Mt for thermal coal) in 1991, 29.3 Mt (16.4 Mt/metallurgical and 12.9 Mt/thermal) in 1992, 34.2 Mt in 1993 and 37.8 Mt in 1994.

Coking coal imports for steel use amounted to 11.6 Mt in 1989, 11.3 Mt in 1990, 17.3 Mt in 1993, 16.9 Mt in 1994 and 17.2 Mt in 1995. Steaming coal imports for power generation, cement and boiler industry amounted to 11.8 Mt in 1989, 11.6 Mt in 1990 and 16.9 Mt in 1993. In 1994, the Republic of Korea imported 20.9 Mt of steaming coal for power (24 per cent) and 26.7 Mt in 1995.

Overseas coal investment

By 1990, six local companies had invested in overseas coal mines on a joint-venture basis in Australia, Canada, USA, and Indonesia with a total investment of $US189 million since 1982.

1.2.2.6 Russian Far East (RFE)

Russian coal production was reported to be 337 Mt in 1992; 300 Mt in 1993; and 271 Mt in 1994. The Kuznetsk basin (Kuzbas) is Russia's largest coal producing region, producing one third of the Russia's total coal production. The basin has a total of 60,000 Mt of coal reserves, of which 33,000 Mt is coking coal. Almost all Russian coal mines are run by the firm Rosugal, which produced 261.2 Mt of coal in 1994 or about 96.4 per cent of the country's total.

The condition in the coal mines was reportedly deteriorating with both the danger and labour intensity of mining increasing. Out of 270 open pit mines in operation, 60 have been mined for over 20 years without any renovation. In 1994 it was stated that the industry faces the problem of shutting down 42 non-viable open pit mines with a total capacity of over 11 Mt/y and a work force of 48,000. The serious environmental problems facing the Russian coal industry must be addressed.

Russia exported 23.6 Mt of coal in 1991; 17.9 Mt in 1992; 25 Mt in 1993; and 19-20 Mt in 1994. In 1993, coal exports to countries of the former Soviet Union were about 5 Mt, and those to countries outside the former Soviet Union 17.6 Mt. The Kuznetsk basin in Siberia accounted for most of Russia's exports to the west. Ukraine, the second largest coal producer in the former Soviet Union, was a major importer of Russian coal for its steel industry.

Proved coal and lignite reserves top 13 bn tonnes in the Russian Far East, while speculative reserves stand at 15 times more. The majority of coal reserves (75 per cent) are concentrated in the Republic of Sakha, Amur region, and the Maritime Territory. Four major coal-bearing areas identified include the Okhotsk, Anadyr, Omsukchan and Arkagala districts. The coals range in rank from brown to anthracite. Most of the coal deposits have a complex geological structure, limiting surveying and development activities. During the 1975-1985 period, coal output expanded from 35.2 Mt to 51.3 Mt, or 1.5 times. Russia's policies resolve to bring its Far Eastern coal production to 82-85 Mt by 2000 through the retooling and expansion of existing facilities and the building of new mines.

In 1991, the output of the region's 25 coal mines and 14 coal quarries dropped to less than 47 Mt against its 1988 peak of 57 Mt. This caused a severe energy shortage and necessitated increasing imports of solid fuels from neighbouring Siberia, from Irkusk and even Kansk – Achinsk coal basins. The intraregional supplies of solid fuels are mainly produced in the Maritime Territory (or Primorsky krai), and in southern Yakutia which exports its hard coal also to Japan and the Republic of Korea. The Kamchatka and the Khabarovsk Territory have to rely on imported coal.

However, RFE's gas and coal potential can be regarded as an adequate source for both the now insufficient domestic supplies and future exports of excess fuel output. As far as coal prospects are concerned, they are now mainly bound up with the projected development of the Bureya and south Yakutia basins and, in particular, of the Elga coal field, which contains about 3 billion tonnes of high-quality coal. The Elga project is considered as a promising source of sizeable coal exports to the Republic of Korea and, perhaps to Japan, but the pace of its realization will largely depend on foreign participation (Petromin, Oct. 1992).

An Australian firm has set up a joint venture to develop and operate a brown coal open pit at the Khurmulinskoye field near Komsomolsk-na-Amur in Khabarovsk of the Far East. Output from the Khurmulinskoye deposit is projected to be 1 Mt/y.

2.0 NATIONAL DEVELOPMENT POLICIES

2.1 China

2.1.1 Economy, energy and environment

Economy

According to the Ninth National Development Plan for 1996-2000, China's strategic goal of national economic development is to raise the GNP per capita to about $1,300. The average annual growth rate (AAGR) of the GNP will be controlled at 8-9 per cent. The transition from central planned economy to socialist market economy is expected to be complete by the year 2000 (World Daily, 3 July 1995). If China's development goal by 2050 was set to quadruple GNP per capita based on the value in 2000, it would then reach about US$5,000 and rank China the middle level of development in the world countries.

The Institute of Social Science of China stated that the coming 16 years would be a golden period of economic development for China, with an AAGR of the GNP not less than 7.5 per cent. By the year 2010, the overall scale of the economy of China would rank as the third in the world (World Daily, 6 July 1995). There would be six economic regional groups in China, the centres of which would be the cities of Guangzhou, Shanghai, Beijing, Harbian, Urumuqi and Kunming (World Daily, 13 July 1995).

Energy

Energy demand

In 1994, China overtook Russia to become, after the United States, the second largest energy market in the world (BP, 1995). Energy consumption over the past ten years is at an average rate of 5.2 per cent/year. According to the forecast by Chinese INET/ITEESA, Tshnghua University, China's total primary energy demand by 2000 is about 1,000 Mtoe and the AAGR of the energy demand between 1985 and 2000 is about 5.6 per cent. The World Energy Outlook predicted in 1995 that China would demand primary energy of 1460 Mtoe by the year 2010 with an average growth rate of 4.1 per cent/year between 1992-2010.

Energy mix and distribution

Coal shares about 70 per cent in China's primary energy mix, and this situation seems likely to be unchanged for many years to come. There are plenty of coal resources in China, of which the R/P was estimated at the end of 1994 to be more than 100 years. The R/P ratio for oil, less than 23 years, is much less than that for coal. The substitution of coal for oil will be an important long-term policy in China.

The structural scenario for primary energy for 1971, 1992, 2000 and 2010 is shown in table 74. China is one of a few countries in the world which uses coal as the principal energy source. China has been the world's largest consumer since 1989, followed by the USA. Together they produced and consumed about half the world's coal in 1994. The outlook estimated that China's solid energy would still share nearly 70 per cent of the total primary energy by 2010 although it will have declined from 77.7 per cent in 1992.

The geographical distribution of energy resources in China is uneven. About 80 per cent of the coal and more than 64 per cent of the oil reserves are concentrated in northern China. 70 per cent of hydro-reserves are concentrated in south-west China. The rest of China's petroleum resources is mainly located offshore. Lack of energy resources in southeast China has exerted greater pressure on the economic development there. Coal transportation accounts for 40 per cent of the total volume of railway freight and more than 50 per cent of the handling capacity of harbors in China (Qiu, 1991).

Table 74. Primary energy structure of China, 1971-1992-2000-2010

	Levels (Shares,%)				*Growth Rates (% p.a.)*			
	1971	*1992*	*2000*	*2010*	*1971 -1992*	*1992 -2000*	*2000 -2010*	*1992 -2010*
PRIMARY ENERGY (MTOE)	236 (100.0)	710 (100.0)	991 (100.0)	1,460 (100.0)	5.4	4.3	3.9	4.1
Solids	190 (80.7)	552 (77.7)	730 (73.7)	1,018 (69.7)	5.2	3.6	3.4	3.5
Oil	40 (16.9)	133 (18.7)	204 (20.6)	325 (22.3)	5.9	5.5	4.8	5.1
Gas	3 (1.3)	14 (1.9)	29 (2.9)	57 (3.9)	7.3	9.8	7.1	8.3
Nuclear	0 (0.0)	0 (0.0)	5 (0.5)	18 (1.2)	–	–	13.1	–
Hydro	3 (1.1)	11 (1.6)	23 (2.3)	42 (2.9)	7.3	9.2	6.3	7.6
Geo/Others	0 (0.0)	0 (0.0)	0 (0.0)	0 (0.0)	–	–	–	–
Electricity Output (TWh)	138 (100.0)	754 (100.0)	1,272 (100.0)	2,169 (100.0)	8.4	6.8	5.5	6.0
Solids	99 (71.6)	559 (74.1)	919 (72.2)	1,540 (71.0)	8.6	6.4	5.3	5.8
Oil	9 (6.7)	61 (8.1)	61 (4.8)	61 (2.8)	9.4	0.0	0.0	0.0
Gas	0 (0.0)	3 (0.3)	7 (0.5)	11 (0.5)	–	13.1	4.8	8.4
Nuclear	0 (0.0)	0 (0.0)	20 (1.6)	69 (3.2)	–	–	13.1	–
Hydro	30 (21.7)	131 (17.4)	265 (20.8)	488 (22.5)	–	9.2	6.3	7.6
Geo/Others	0 (0.0)	0 (0.0)	0 (0.0)	0 (0.0)	–	–	–	–
Carbon Dioxide	864	2,562	3,540	5,101	5.3	4.1	3.7	3.9
Emissions (Mt) % change since 1990		7.9	49.1	114.9				
GDP (Billion 1987 US$)	–	491	988	1,943	–	9.1	7.0	7.9
GDP per capita (Billion 1987 US$)	–	423	771	1,409	–	7.8	6.2	6.9
Energy per capita (toe)	0.28	0.61	0.77	1.06	3.8	3.0	3.2	3.1
Energy Intensity (toe/1,000 $)	–	1.44	1.00	0.75	–	-4.5	-2.9	-3.6

Source: World Energy Outlook, the IEA's 1995 edition.

Energy intensity and efficiency

China's energy consumption level has been rather low. In 1992, its energy consumption per capita was 0.61 toe, about 42 per cent of the world average, and only 13.4 per cent of the OECD's average. The level is expected to rise to 1.06 toe by 2010, about 71 per cent of the world average, or 21 per cent of the OECD's average.

China's energy intensity (energy consumption/GNP), however, was among the highest in the world, which was 1.44 toe/1,000 USD in 1992, 3.3 times that of the world average or nearly 8.5 times of the OECD's. Therefore there is plenty of scope for China to save energy. This should be an important measure for China to overcome its shortage of energy supply. The energy intensity of China is expected to be reduced to 0.75 toe/1,000 USD by the year of 2010, which would be about twice the world average, or about triple the OECD's average.

Domestic energy supply in China has fallen short of demand for a long time and this situation is likely to be unchanged up to early next century. China now carries out both the principle of energy conservation and of exploration.

Achievements have been made in energy conservation in China. The energy intensity was down from 1.40 kgoe/USD in 1980 to 1.01 Kgoe/USD in 1988 (1980 values), the average annual decline was about 5 per cent. The per hundred km gasoline truck consumption declined on average from 8.7 litres to 7.5 litres and diesel consumption elasticity was only 0.44 in 1980-1988 (Qiu, 1991). To increase the efficiency of energy production, transportation and consumption are important objects to save energy resources and to maintain environmental quality as well.

Environmental impacts

China's CO_2 emission in 1988 was nearly 3,000 million tonnes, in which fossil energy combustion accounted for 87 per cent. Following the USA (23 per cent) and former USSR (18 per cent), China's CO_2 emission comprised 11.7 per cent, ranking third in the world. On the other hand, the per capita CO_2 emission of China from fossil fuel combustion was about 2.4 tonnes in 1988, which is not only much lower than the 18.5 tonnes in USA, but also lower than the world average of 4 tonnes (Qiu, 1991, p. 317). The CO_2 emission of China is expected to rise to 5,101 Mt by the year 2010, about double the 1992 emission, accounting for 16.6 per cent of the contribution of the world total, or nearly 23 per cent of the OECD.

The discharged volume of waste water from energy consumption was 37.0 billion cubic metres in 1988 and 20 billion tonnes polluted water from oil drilling and refinery. The solid waste from industry was doubled from 1981 to 1988, but the processing rate was only 11.5 per cent. The accumulated waste was 6.6 billion tonnes, and the annual growth rate 6.6 million tonnes (Qiu, 1991).

2.1.2 Petroleum sector

Petroleum demand

With each percentage increase in China's gross domestic product, China's consumption of oil products is raised by a corresponding 0.63 per cent, which could be derived from the figures of China's GDP average annual growth rate of 9.1 per cent in the period of 1980-1992 (reported in 1995 Far Eastern Economic Review), and of China's average annual growth rate 6.66 per cent consumption of oil products in the period of 1984-1994 (recorded in 1995 BP Statistical Review of World Energy). Therefore, with an average GDP growth rate of 8-9 per cent/year in the years to come, oil consumption is expected to grow at a rate of 5.3-6.0 per cent/year. The 1995 Outlook by IEA estimated that China's GDP growth rate would be 7.9 per cent/year in the period 1992-2010 and the oil growth rate in the period 5.1 per cent/year as primary energy, or 5.8 per cent/year as final energy. China's demand for primary oil is estimated to be 204 Mtoe (nearly 4.1 million b/d) by the year 2000, and 325 Mtoe (about 6.5 million b/d) by the year 2010; the demand for gas, 29 Mtoe (3.1 billion cf/d) by 2000 and 57 Mtoe (6.1 billion cf/d) by 2010.

The consumption of oil and gas in China is currently running at around 3.3 million b/d and 1.7 billion cf/d. The Petroleum Economist estimated in 1995 that China would have an oil supply deficit of 1 million b/d by the turn of the century, reaching 2 million b/d by 2005. The demand is expected to be 3.7 billion cf/d of gas by 2005, driven mainly by the power and industrial sectors. With domestic supply forecast at only 3.1 billion cf/d then, there will be a supply deficit of around 600 million cf/d by 2005, which could grow to 1.5 billion cf/d by 2010, creating a possible liquefied natural gas market in China in the medium to longer term (Petroleum Economist, June 1995).

Petroleum is one of the key energy resources which plays an important role in the development of China, especially in transportation, the petro-chemical industry, modernization of agriculture, residential living quality, as well as power generation. As indicated in table 74, oil and gas demand accounted for 18.2 per cent in 1971 and 20.6 per cent in 1992 in the energy mix of China and will account for 23.5 per cent in 2000 and 26.2 in 2010.

There will be only 5.3 per cent of power generation in China consuming oil and gas by the year 2000 which will shrink to 3.3 per cent by 2010. In the transportation sector, however, it was estimated that nearly 90 per cent of the total energy demand by 2000 would depend on oil consumption, accounting for 84.3 Mtoe (Qiu, 1991).

Petroleum supply

With its recent economic reforms and vast population, the sharply stimulated demand has opened a huge domestic market for petroleum products in China. Oil and gas production, however, is relatively minimal and hardly meets the needs for its products. This inevitably results in rising imports and dwindling exports of crude oil and its products. This situation will most likely remain for many years to come. The petroleum industry has entered a new development phase which stresses efficiency and conservation. This was a consensus opinion of the directors of China's on-shore oil fields when they attended a meeting held in Beijing in January 1995.

China's major oil producing fields namely Daqing, Dagang and Shengli oil fields, concentrated in the northeastern part of China, are facing the challenge of production. Average water cut has risen to 70-80 per cent, and the field's decline rate has risen to 7-8 per cent annually. The production costs have skyrocketed. However, enhancing exploration for new discoveries in those old oil fields and improving the recovery ratio of its old parts is the way to maintain or increase petroleum production in China.

China has abundant potential of natural oil and gas resources but proven reserves (estimated in 1987), are only 16 per cent for conventional oil resources and 2.6 per cent for gas resources. This estimation that exploration of petroleum resources was still in an initial stage and needs to be promoted. Exploration progress has resulted in new discoveries of petroleum resources: the Talimu basin of Xinjiang province in northeastern China, the Zhujiangkou basin and Baibu gulf basin offshore Guangdong and Hainan provinces in southern China, and East China Sea basin in the continental shelf of the East China Sea. These new oil fields, after development, will play important roles in the economic development of China early next century. This also will ease the pressure from the uneven distribution of energy/petroleum in China.

The development of natural gas will be paid more attention. The potential for its development is quite high in China since its proven resource is only 2.6 per cent of its total reserves and its share of total energy consumption is only 2 per cent. Over 10 per cent of gas resources in China is concentrated in Shichuan province in southwest China where there is a lack of petroleum resources.

As China becomes a net crude importer, China's refining sector is being pressed to accommodate the rapid changes. The refiners have to deal with the higher gravities and higher sulfur contents of the Middle East crudes. Construction of big crude tanker terminals to accommodate 100,000-250,000 dwt oceangoing tankers becomes a priority. Terminals linked to storage facilities and interior rail/highway infrastructure are needing to be improved.

The long-time pursuit of output quotas under a planned economy has rendered the organizational structure of oil enterprises irrational; their operational mechanism lacks flexibility; their management is lax; they are loaded with heavy burdens; and they have not quite adapted to the development of a market economy (Petroleum Economist, February 1995).

Institutional and legal aspects

Institutional arrangement

The State Council of China holds periodic administrative meetings for energy efficiency and conservation, and other problems concerning policies and principles. Relevant divisions have been set up in the State Planning Commission and the Ministry of Energy. Departments for energy administration have also been set up in various ministries and commissions of the central provincial and local Governments, and in energy management organizations in large and medium size enterprises. A comprehensive energy efficiency and conservation system has been formed in China.

There are three state owned corporations engaged in the petroleum industry in China, namely China National Petroleum Corporation (CNPC), China National Offshore Oil Corporation (CNOOC) and China Petrochemical Corporation (Sinopec). The establishment of these institutions is intended to separate on-shore and offshore exploitation, and to separate up-stream and down-stream development of the petroleum industry.

Sinopec was established in 1983 and given responsibility for managing most petroleum and petrochemical processing facilities in the country. At the same time, CNPC was given control of a smaller number of oil field refiners. In addition to Sinopec, there are other state trading companies including China National Chemicals Import & Export Corp. (Sinochem), China International United Petroleum & Chemical Corp. Ltd. (Unipec, a 50-50 joint venture of Sinochem and Sinopec, established in 1993), and China National United Oil Corp. (Sinoil, an independent government owned oil trading company with a 50-50 equity split by Sinochem and CNPC, established in 1993) (table 75).

Table 75. China's 1993 oil trade by company

Unit: barrel/day

Company	*Crude oil*		*Refined products*	
	Imports	*Exports*	*Imports*	*Exports*
Sinochem	140,000	200,000	n.a.	54,440
Unipec	100,000	n.a.	200,000	19,800
Sinoil	n.a.	180,000	n.a.	n.a.
Others	60,000	n.a.	n.a.	n.a.
China's Total	**312,860**	**388,686**	**347,820**	**72,240**

Source: Oil & Gas Journal Special, May 9, 1994.

The management strategy of the petroleum sector has been considered to set a balance between central planning and reasonable decentralization, to foster globalization through import of foreign cooperation, capital, technology transfer, and crude supplies while pursuing investments overseas and promoting "share-holding" at home and abroad.

Legal framework

The State Council issued the Provisional Regulation for Energy Conservation Management in 1986. A set of technical criteria for energy conservation project design were stipulated. Special funds for saving electricity, coal, oil and water were recommended. Energy prices were rationally readjusted step by step, and its effect as an economic lever was enhanced in energy conservation. Energy standardization and energy monitoring were enhanced. Law of energy conservation was stated to be studied and formulated.

Laws and regulations have been set up in China to facilitate and promote foreign financial investment in the development of the Chinese petroleum industry. The two basic governing regulations are "Regulations of the People's Republic of China on the Exploration for Offshore Petroleum Resources in cooperation with Foreign Enterprises (January 30, 1982)" and "Regulations of the People's Republic of China on the Exploration for Onshore

Petroleum Resources in Cooperation with Foreign Enterprises (October 7, 1993)". Others include a law on economic contracts involving foreign interests (July 1, 1985), regulations on the payment of royalty for exploitation of offshore (December 5, 1988), and on-shore (January 13, 1990) petroleum resources, income tax law for enterprises with foreign investment and foreign enterprises (April 9, 1991) and detailed rules for the implementation of the income tax law (June 30, 1991).

Production sharing contract (PSC)

A Production Sharing Contract (PSC), Joint Study Agreement (JSA) and Geophysical Survey Agreement (GSA) are the forms for the Chinese upstream oil and gas development to attract foreign investment and technical expertise.

The Chinese Model Contract is in the form of a Production Sharing Contract (PSC) with tax and royalty plus 51 per cent State participation. The foreign companies will run the sole risks during the exploration period. The Chinese side has the right to participate in financing during the development period. Both parties will share the remainder of the total oil output after subtracting industrial and commercial consolidated tax, lease mining royalty, operation costs, exploration costs and development costs recoveries, and shall pay the income tax to the Chinese Government according to their respective shares during the production period. The major competitive items for foreign companies to bid are the exploration commitments and profit sharing ratios (China, 1994).

Bidding for offshore development

The first round of bidding for offshore oil exploration in cooperation with foreign companies was accomplished during 1982-1984, in which the CNOOC signed 18 contracts with 27 companies from 9 countries covering a total block area of 39,200 km^2. In the second round of bidding during 1984-1989, five oil contracts were signed with foreign companies plus a geophysical prospecting agreement. The third round of bidding in 1989-1992 offered seven blocks covering more than 32,000 km^2 in the Pearl River Mouth. The fourth round of bidding in 1992-1993 covered a total acreage of 72,800 km^2 in the East China Sea, consisting of 20 blocks.

Since the fourth round, all the seas surrounding China's mainland have been opened to foreign enterprises, covering more than half a million square kilometres. The last full offshore bidding round was in 1994, offered 13 blocks in the offshore Hainan and Pearl River Mouth areas totalling 39,550 sq km. Five blocks had been licensed by the middle of 1995, with negotiations still proceeding for one more.

The Bohai Oil Corporation (BOC), a subsidiary of the CNOOC, has now opened two-thirds of the 57,500 km^2 Bohai Sea area to foreign co-operative enterprise, the remaining one-third of the sea, in northwestern Liaodong Bay, the BOC has been conducting self-financed exploration and development (E & D). By August 1992, six contracts in the Bohai Sea were under implementation by the CNOOC and companies from Japan, the US, Britain, Australia and Norway.

China's offshore basins have seen higher levels of foreign activity since EIF became the first foreign company to operate in 1980. Today there are 47 PSCs in operation, 38 of which are offshore, mainly in the South and East China Seas (Petroleum Economist, June 1995).

In June 1995, the ACT Operators Group – Agip (overseas), Chevron Overseas Petroleum, Texaco China – started production from the Huizhou 32-3 and Huizhou 32-3 oil-fields of Pearl River Mouth Basin in South China Sea. According to Texaco, the first oil was pumped five months ahead of schedule and the development cost was 10 per cent below budget. Initial output of the 30-39° API oil is 50,000 b/d, with a peak of 60 000 b/d expected in late 1995. The ACT/CNOOC consortium will then be producing over 100,000 b/d from the Pearl River Mouth Basin, making it the largest offshore producer in China (Petroleum Economist, July 1995).

Bidding for onshore development

Foreign operators have been generally far less active in the onshore sector than offshore. The first round of bidding for onshore oil exploration was initiated in March 1993. The areas for bidding included five blocks located in the southeast part of Tarim basin, where two contracts have been signed with the Esso and AGIP groups, respectively. A total of eight contracts and six agreements have been signed.

The second round of bidding was announced by CNPC on January 18, 1994, covering 26 blocks located primarily in east and northeast China, inside or closely to existing large oil and gas fields and industrial centres. Also included in the second round were 11 enhanced oil recovery (EOR) projects for the mature fields. Up to the middle of 1995, only one of the original blocks was signed and no bids received for the EOR projects.

The third round was announced for international bidding on 9 June 1994, offering 12 blocks with a total area of 112,739 km^2 in the Tarimu and Zhungeer basins in Xinjiang. It seems that most of the western oilmen in Beijing for the launch of the round expressed disappointment with the blocks on offer, particularly as the two previous on-shore rounds, and especially the second, were not successful (Petroleum Economist, July 1995).

Overseas operation

China's foreign exploration and production (E & P) activities began in 1991. Because of lack of experience in international E & P, China's approach is one of "many channels, many blocks, and small shares". That means a broad portfolio of investment and an emphasis on sharing risk with international partners. CNPC's overseas focus, then is to: (a) purchase reserves – the safest approach, which limits risk but also limits profit; (b) purchase interest from concessionaires or operators of integrated license blocks; and (c) bid or apply for development rights in targeted blocks (Oil & Gas Journal Special, May 1994).

CNPC's units involved in foreign E&P include China National Oil Development Corp., CNPC International Ltd., CNPC Canada Ltd., CNPC Latin America Ltd., MC & CNPC Oil (Hong Kong) Ltd./CNPC Central Asia CO., and CNPC Asia-Pacific Ltd. Some of the subsidiaries have close relationships with the local agencies that operate China's domestic oil field. For example, the president of CNPC Central is the former head of Liaohe oil producing complex in Liaoning province, and the president of CNPC Latin America is the former head of Shengli oil producing complex in Shandong province.

China's foreign E & P activities have not gone as smoothly as Chinese officials had hoped. Simplifying approval procedures is required so that opportunities aren't missed while CNPC awaits government approval. Favorable tax treatment was requested to be given to foreign projects so that possible financial losses from exploration could be offset by reduced taxes. In response to the call from the industry officials for a centralized administration of foreign operations, CNPC is planning to form an international cooperation bureau to oversee its foreign projects.

On the other hand, some domestic oil producing agencies are taking the initiative for ventures in foreign E & P on their own. For example, Daqing oil production complex was negotiating in 1994 for the rights to develop two oil fields in Russia's Irkusk region.

2.1.3 Non-fuel minerals

China's mining industry contributes significantly, to its social and economic development. Lack of investment for pursuing its ambitious plan and lack of up-to-date technologies are major problems facing China attempting to develop and reform its mining sector. China has introduced market mechanism into minerals development since the late 1970s when the government decided to adopt an open policy. Tens of legal rules and regulations have been overridden to facilitate international cooperation in various forms including direct foreign investment (FDI), joint venture and technology transfer.

China's per capita mineral reserve endowment, mining product output and consumption are at rather lower levels compared to other countries. The geographical distribution of mineral resources between mining production and mineral consumption places are sometimes poorly placed. In addition to China's consuming of huge amounts of resources and energy with rather low efficiency in its mining, processing, refining and manufacturing sectors, China releases huge quantities of solid wastes, waste water, and waste gases into the environment. While raising its mining outputs, conservation of mineral resources has been set by the government as a priority policy.

China's mineral reserves are among the richest in the world. China possesses one of the largest mining industries in the world. China is a leading producer in some minerals, and also a big consumer. China's import and export of minerals has significant influence on the world market. Along with its high GDP growth rate, marketing and technological progress, China's mining activities and policy direction will play an even more significant role in the world's mining sector in the years to come.

A report in the "China Daily" said the country needs to attract new resources of investment if its mineral industry is to keep up with demand. If funding levels remain unchanged, China will run short of a third of its main mineral products by 2000, deteriorating to two-thirds by 2010, the paper said. China plans to allow state-owened mining companies to trade mining rights. New amendments to the Mineral Resources Law will allow such rights to be sold in the future. The amendments have been approved by the State Council, China's Cabinet, and have been submitted to the standing of parliament for approval. At present, only the oil and natural gas industries allows foreign companies to explore and exploit resources they find. China's gold mining, for example, allow foreign companies to explore for the metal, but they must hand over production control to local companies (Asian Journal of Mining, July/August, 1996).

Institutional arrangements

The relevant government agency in charge of the non-ferrous metal industries in China is the China National Nonferrous Metals Industry Corp. (CNNC). China's National Tungsten and Antimony Import and Export Corp., a joint venture between China National Non-ferrous Metals Import and Export Corp. (CNIEC) and China National Metals and Minerals Import and Export Corp. (Minmetals), was approved by the State Council and the Ministry of Foreign Economic Relations and Trade in 1993.

CNNC was said to be waiting for governmental approval in order to become a holding company in 1995. Then, a national association would be set up to manage the non-ferrous industry in China and CNNC would focus on diversifying its business and tapping overseas resources and capital to secure the supply of raw materials needed for expansion. In order to attract foreign investors into the non-ferrous sector, CNNC was to restructure several existing enterprises into share holding firms. The foreign investment regulations were expected to be published in 1994.

China's 1986 mining regulations are strongly nationalistic and not an incentive to foreign investment. In February 1991, the State Council listed tungsten, tin, antimony and rare earths as special minerals under State protection. The State Planning Commission (SPC), with the help of the CNNC and the committee in charge of rare earths utilization and exploration under the State Council, has the exclusive authority for exploration of these four minerals. Foreign firms and individual foreigners were not allowed to exploit or obtain data, mineral samples or technology on rare earths. In 1994, the State Planning Commission announced a halt on investments for aluminium and copper semi-manufactured products.

However, Chinese authorities have, since the late 1980s, realized that the development of minerals requires a large mount of capital and advanced technology, both of which China lacks. The priority minerals for exploration and development include copper, lead, zinc, iron, coal, diamonds, rare earths, uranium and possibly gold.

International cooperation

Trade

The Ministry of Foreign Trade and Economic Cooperation (MFTEC) published regulations for quota controls on export commodities in July 1994. Minerals and metals under quota control were antimony, caustic soda, cement, coal, coke, crude oil, iron alloy, limestone, magnesium, refined oil, rolled steel, soda ash, talc, tin tungsten, and zinc. The tariff Policy Commission of the State Council announced provisional rates for 255 items which were subject to change beginning March 1, 1995. China reduced export tariffs on lead ore and concentrates, zinc ore and concentrates, and unforged lead and scrap lead to 0 per cent; tin ore and concentrates from 50 per cent to 20 per cent; phosphate from 20 per cent to 10 per cent; and forged zinc and scrap zinc from 20 per cent to 5 per cent.

Major commodities which China imported in 1993 were oil products, fertilizer, copper ore, metal processing machinery, communication equipment, and aeroplanes. By 1994, mineral and metal commodities under import quota control were crude oil, oil refinery products and chemical fertilizers such as potash and phosphate.

According to the General Administration of Customs, China reduced the import tariff rate beginning December 31, 1993 on chromium, copper, germanium, petroleum, vanadium and other raw materials, machinery and equipment which the government considered to be in short supply.

China cut tariffs on imported crude and refined products in 1994 which was the second time after 1992. Along with the abolition of the crude import licenses, tariffs cuts will spur further efforts by state trading companies to import crude and refined products. New regulations to clarify responsibilities among China's importing and exporting companies were also issued (Oil & Gas Journal, May 1994).

Metals Exchange Market

China opened its metals exchange market at Shenzhen, Guangdong in January 1992. The exchange is co-sponsored by CNNC and several other stated-owned enterprises. The exchange was designed to inter-relate China's planned commodity economy to a more advanced world market mechanism in the production and consumption of non-ferrous metals. The Shenzhen metal exchange lists eight metals for trade – aluminium, antimony, copper, lead, magnesium, nickel, tin and zinc. In May 1992, China opened a second Metals Exchange in Shanghai. It is under the supervision of both the Ministry of Materials and Equipment and the Shanghai People's Government. Seven metals are listed – aluminium, copper, lead, nickel, pig iron, tin and zinc. The two exchanges are used by domestic producers and consumers, at least in the initial stage.

China National Nonferrous Metals Industry Corp. has two subsidiaries traded on the Stock Exchange of Hong Kong: Onfem Holdings Ltd. and Oriental Metals (Holdings) Co. PRC companies can also list in New York and Singapore, and may soon be able to list in Tokyo and London too. CNNC allowed metal suppliers in the PRC to buy stakes in Onfem and Oriental Metals for the first time in 1996 by selling shares in the unlisted Hong Kong holding company that controls them (Asian Journal of Mining, July/August, 1996).

Joint Venture

CNNC and Macquarie Bank Ltd. of Australia have formed a 50-50 joint venture, Asia Resource Capital Ltd., incorporated in Australia. The initial assets will reportedly be $3.8 million. The new company will also provide a service in economic and financial viability studies on mines and production facilities for international investors.

However, international mining companies doing business in China face some obstacles. State agencies battle over jurisdiction on mining ventures. Rules on taxation, operating policies, and project examination procedures need to be clarified.

2.2 Democratic People's Republic of Korea

The country's long term annual production goals set for 1989 are shown in table 76.

The country's economy is caught in a downward spiral with the GNP declining at an annual rate of 5 per cent in the early 1990s. The Soviet bloc collapse ended the barter trade that supplied the country with cheap and vital oil imports. The former USSR accounted for about 50 per cent of total trade with the Democratic People's Republic of Korea. The Democratic People's Republic of Korea's planning system and a large heavy industrial sector working with outdated equipment have also greatly debilitated the economy.

In 1991, the GDP was over 50 per cent in exports from the mining sector. In 1993, the GNP by the mining sector of the Democratic People's Republic of Korea accounted for 8.2 per cent.

Coal deposits in the Democratic People's Republic of Korea were estimated at 12,000 Mt. Annual coal consumption of the country was estimated at 31.5 Mtoe in 1994. The R/P ratio was over 100 years. However, the metallurgical coal is limited in the country. Other mineral deposits in the Democratic People's Republic of Korea include iron ore reserves at 3,300 Mt, lead 6 Mt, zinc 12 Mt, copper 2.2 Mt, magnesite (the number one position in the world as a supplier of dead-burnt magnesite), and monazite 0.5 Mt.

Table 76. Democratic People's Republic of Korea's long term annual production goals set for 1989

Sector	*Goals set in 1980 (to be completed by 1989)*	*Achievements made by 1986*	*Goals revised in 1989 (to be completed by 1993)*
Electricity	100 bn KWh	60 bn KMh	100 bn KWh
Coal	120 mn tonnes	70 mn tonnes	120 mn tonnes
Grain products	15 mn tonnes	10 mn tonnes	15 mn tonnes
Steel	15 mn tonnes	1.9 times increase (no figures given)	10 mn tonnes
Chemical fert.	7 mn tonnes	5 mn tonnes	7.2 mn tonnes
Cement	20 mn tonnes	12 mn tonnes	22 mn tonnes
Marine products	5 mn tonnes	3.1 mn tonnes	11 mn tonnes
Textiles	1.5 bn metres	800 mn metres	1.5 bn metres
Non-ferrous metals	1.5 mn tonnes	1.5 mn tonnes	1.7 mn tonnes
Tideland reclamation	300,000 ha	(no figures given)	300,000 ha (150,000 ha by 1990)

Source: UNDP 1994 Annual Development Cooperation Report for the Democratic People's Republic of Korea.

The Democratic People's Republic of Korea has long been dependant on imports of oil from the former Soviet Union and from China. The country imported 3.3 Mt of oil in 1989. It was estimated that the Democratic People's Republic of Korea had offshore oil and gas in the Korea Bay, but the country is in need of investment and technology to develop them.

In the export categories of the Democratic People's Republic of Korea in 1993, metals and minerals were the largest, accounting for 40 per cent. In the import categories in the same year, mining products, inclusive of oil, were also the largest, accounting for 22 per cent.

The Democratic People's Republic of Korea was successful in establishing trading relationships with the Republic of Korea, coal being the principal item. It also contracted to sell 200 t of copper valued at $660,000 to an the Republic of Korea firm. The Islamic Republic of Iran had agreed to export 80,000 t of lead concentrate and 250,000 t of zinc concentrates in late 1980s, which the Democratic People's Republic of Korea would convert into bars on a toll basis. The Democratic People's Republic of Korea would also export to the Islamic Republic of Iran 200,000 t of steel products, and offer assistance in exploration and extraction of bauxite, phosphate and other minerals in the Islamic Republic of Iran.

The government is now trying to attract foreign investment by revising its laws on the model of current practices in China. The Democratic People's Republic of Korea proposed an economic zone on the north-east coast, based on the ports of Najin and Sonbong, allowing 100 per cent ownership by foreign firms. Free economic zones have been opened, but they are located in sparsely populated regions.

2.3 Japan

Before 1991, the economic boom in Japan had lasted over four years at an average GNP growth rate of around 5 per cent annually. The production index for the mining and manufacturing industries also showed a high growth rate of 6 per cent for 1989 and 4.6 per cent for 1990. Housing and equipment investment by private companies were the most outstanding elements of the economic boom. The number of new housing had exceeded 1.6 million annually and reached 1.7 million units in 1990, the highest number over the past 14 years.

Consumption of construction-related copper alloy semi-products including wires, had been high. Personal consumption had also been favourable, pushing up production of automobiles (13.55 million units in 1990) and electric appliances, resulted in an increase in the consumption of copper, zinc and nickel.

Owing to the very active economic trends, Japanese imports increased by 12.5 per cent on a dollar basis (20.7 per cent on a yen basis) in 1989 over the previous year. Imports of mining and construction machinery increased by 45 per cent in yen value compared with 1988. The imports of crude oil, fuel coal, refined copper and aluminium in 1989 increased by 8.2 per cent, 8.7 per cent, 15.7 per cent and 3.6 per cent respectively. Japanese imports of refined copper and primary aluminium in 1989 were the largest in the world. Imports of refined copper increased by 26.7 per cent and of aluminium by 10.8 per cent in 1990.

1991 was a turning point for the Japanese economy. The GNP growth rate for the last quarter of 1991 was-0.2 per cent, this being the first negative figure in 30 months. In October 1991, the Index of Industrial Production (IIP) recorded a negative outcome to the previous year for the first time in 33 months. Japan's GNP gross rate was 1.5 per cent for 1992, 0.1 per cent for 1993 and 0.6 per cent for 1994.

The sharp slow-down in the economy was mainly due to a drop in domestic consumption. For instance, private housing starts fell from 1.71 million units in 1990 to 1.37 million units in 1991. One of the major factors behind the stunted growth of the Japanese economy is private company investment, which decreased by 4 per cent in 1992, by 8.4 per cent in 1993 and a further 8.8 per cent in 1994. However, housing investment by the private sector increased by 3.7 per cent in 1993 following a decrease of 5.4 per cent in 1992, and increased by 9.6 per cent in 1994.

The appreciation of Japanese currency in the exchange market is another negative factor affecting the Japanese mining and smelter industry as the domestic price, thus their main revenue, is based on the LME price (i.e. $US). Compared with the previous year, the average $US/Yen exchange rate had appreciated by nearly 6 per cent in 1992, and by 14 per cent in 1993. It broke through the barrier of 100 yen/$US and reached the 98.99 yen/$US in 1994.

The manufacturing industry has been severely damaged by the economic recession. Indices of industrial production marked a minus 4.5 per cent growth rate in 1993 after a 6.1 per cent negative rate in 1992. The automobile industry, which was one of the main contributors to the economic boom, lost momentum drastically. The electric goods industry, another player in the economy, also showed a 13.1 per cent drop in 1992. Naturally, the slowdown in the manufacturing sector depressed the consumption of raw materials.

The Japanese mining industry has changed drastically since the early 1950s. The metal and coal mines of the 1950s were very active and were providing ores to Japanese industry in reconstruction of the economy after World War II. During the period from 1950 to 1990, the ores produced from comparatively small-scale underground mining in Japan could not compete with the low-cost ores produced by large-scale mining of open pits in other countries. Figure 64 shows how sharply was the decrease in the number of Japanese working mines in this period, despite the sharp increase in the consumption of the metals, for example copper. Today, almost all of the ores for Japanese industry are imported.

According to the Metal Mining Agency of Japan (MMAJ), Japanese non-ferrous metal policy consists of the following:

- Promotion of exploration within Japan and overseas,
- Financial support to operating companies,
- Bilateral technical cooperation with developing countries,
- Stockpiling,
- Technical research and development for mineral exploration etc.,
- Prevention of mine-related pollution.

A recent increase in the involvement of Japanese mining companies in world-class mining projects is becoming obvious – Nippon Mining in Australia, Sumitomo Metal Mining in Chile, Dowa Mining in Mexico, and many others. MMAJ is also active in worldwide geological surveys and exploration through its governmental grants to developing countries.

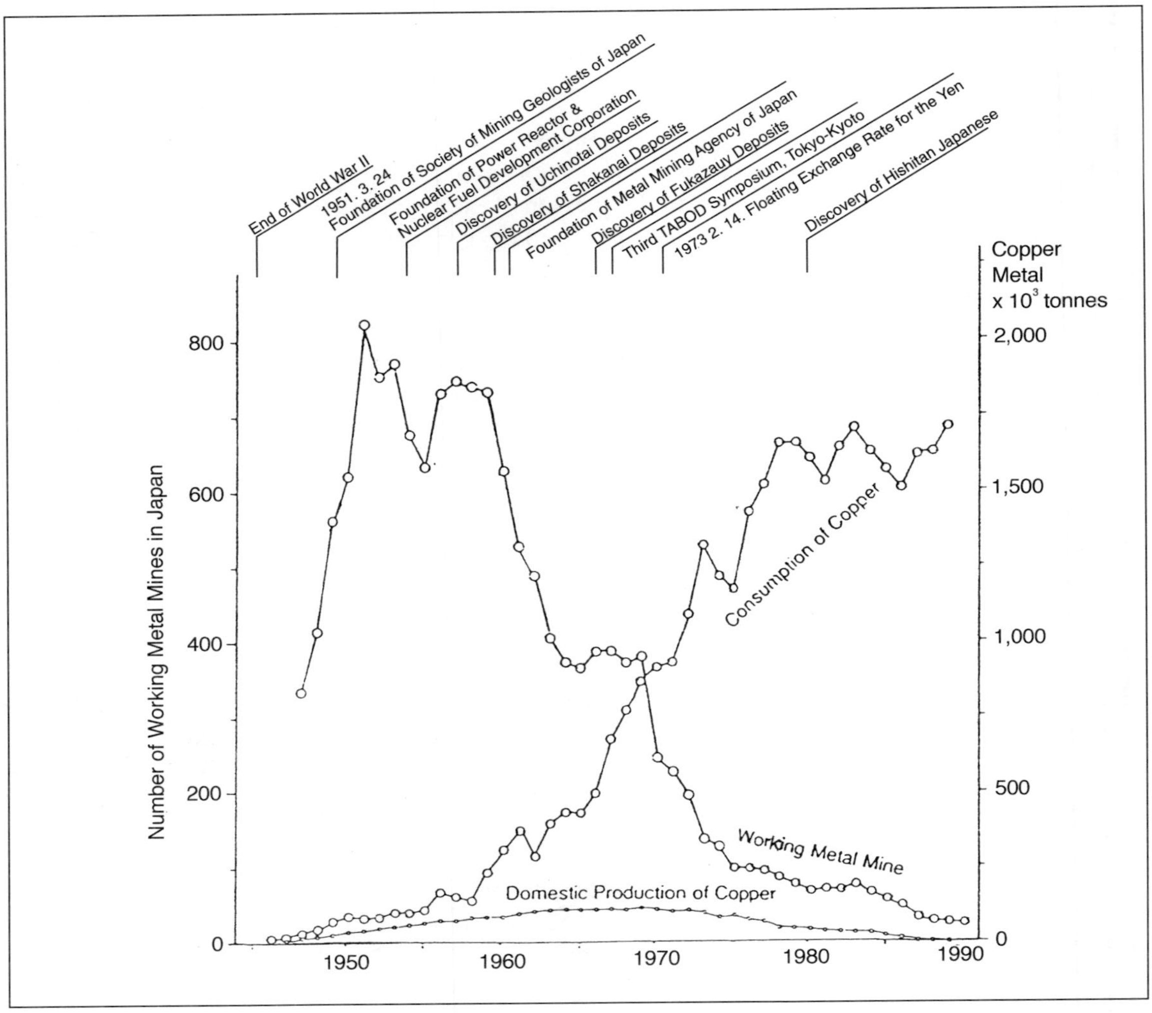

Source: Shunso Ishihara, 1993.

Figure 64. Trends in Japanese mining over the last 40 years, coincident with the history of SMGJ, using Cu as an examples

2.4 Mongolia

Mongolia is a traditional agricultural country dominated by animal husbandry. The livestock population of the country is ten times the human one. Primary industries in Mongolia include the processing of animal products, the manufacture of building materials, and mining. The former Soviet Union accounted for more than 86 per cent of Mongolian foreign trade, which reached $1 billion in 1988.

Mining is the top priority sector in Mongolian industry today. The mining sector includes mining and processing of non-ferrous, precious and rare metals, coal and fluorspar. The government regards the country's rich mineral resources as the way to economic recovery; it planned in 1992 to increase production in the mining industry by 5 per cent in 1993.

Mongolia is a major producer and exporter of copper and molybdenum in Asia and a leading world producer of fluorspar. There are nearly 200 mines operating in Mongolia, for the exploitation of copper/molybdenum, fluorspar, lignite, coal, gold, tin and tungsten ore. Table 77 shows the main minerals production of Mongolia.

Under the new government, the two former governmental agencies in charge of mining and geology (the State Geological Centre and the Mining Bureau) have been merged, and are under the newly established Ministry of Agriculture and Industry.

Table 77. Main minerals production of Mongolia

Commodities	*1993*	*1994*	*1995**
Coal ('000 t)	5,557.8	5,790.0	6,130.0
Copper concentrate ('000 t)	334.317	343.300	337.100
Copper in concentrate ('000 t)	117.0	120.2	118.0
Molybdenum concentrate (t)	4,367.0	4,396.0	3,808.0
Molybdenum concentrate (t)	2,052.0	2,066.0	1,790.0
Gold t/oz	35,916.4	54,662.4	100,964.6
Fluorite: ('000 t)			
– Concentrate (96% CaF_2)	77.0	88.0	118.0
– Ore	87.0	17.0	32.2

Source: The Ministry of Energy, Geology and Mining, Asian Journal of Mining, May/June 1995.
* expected to produce.

Widespread mineral occurrences in geologically promising areas have not been fully evaluated. There is a comprehensive mineral data base in the National Geo-information Centre although this is predominantly in the Russian language. The major difficulties for development of the mineral resources in the country are logistics, market conditions and transportation costs. The nearest sea port is through China to Tianjiang with a distance of 1,340 km, and the other one is through Russia to Vladivostok with a distance of 2,050 km.

The country is now in the process of transferring to a political democracy and market-oriented economy. A new revised constitution was adopted by Parliament in mid-1991, becoming a legislative guarantee of a free economy system in Mongolia. The government was planning to establish free-trade zones by the end of 1993, mainly in the eastern and southern parts of the country.

To encourage the activities of foreign investors and to make the establishment of joint ventures easier, a new foreign investment law was adopted in 1990 and updated in 1992. Other legislation such as the new mining law and the revised customs law were also designed. The government has since 1990 started the process of privatization and was planning to privatize 70 per cent of holdings which hitherto had been owned entirely by the state.

From 1992, the World Bank was assisting in improving management within the mining industry by undertaking the 'Mining and Minerals Policy Development' technical assistance project. The activities of the project include reviews of the 'Regulatory and Fiscal Regime', 'Present and Potential Mining Operations' and 'Coal Mining', and on the basis of these preparing, and recommending the 'Policy Development Review'. World Bank experts also assisted with the drafting of the 1993 'Mining Code' draft, which was expected to be approved by the spring session of the Mongolian Parliament. The 'Mining Code' covers mining, exploration, processing and trading all mineral resources excluding oil and gas by Mongolian and potential foreign investors.

In 1993 the UNDP started a 'Management Development Assistance for Specific Key Economic Sectors' programme, which includes training of mining industry personnel in economic management, foreign trade policy and market competition by Australian experts and study tours abroad.

2.5 Republic of Korea

The Republic of Korea's economy has recorded a high growth rate of 7 to 12 per cent since 1986 due to continued export expansion and rising domestic demand. It's GDP ranked 11th in the world in 1994, recorded as US$379.5 billion, 2.2 times higher than the US$169 billion in 1988. It's per capita GNP was $4,040 in 1988, $4,968 in 1989, $5,477 in 1990, $6,498 in 1991, $6,749 in 1992, $7,513 in 1993 and reached $8,482 in 1994, ranking 32nd in the world.

According to the 1996 WTO Secretariat Report on Trade Policy Review, the Republic of Korea has continued to be one of the world's most dynamic economies. Following a mild downturn in 1992-93, the Korean economy has expanded strongly with real GDP growth exceeding 8.5 per cent in both 1994 and 1995. Per capita income now exceeds US$10,000. The stimulus came from both buoyant exports, in the wake of the appreciating Japanese yen, and strong domestic investment for equipment.

While industrial production increased at two-digit rates, inflationary pressures have remained under control; consumer prices increased by 6.2 per cent in 1994 and 4.5 per cent in 1995. Construction-related investment was sluggish in 1994, but recovered in 1995, buoyed by public infrastructural expenditures. Core manufacturing industries (motor vehicles, electronics, steel and shipbuilding) have expanded rapidly and modern services sectors have started to emerge, contrasting with stagnation or even decline in light industries and agriculture (table 78).

Table 78. Production structure of the Republic of Korea, 1990-1995[a]

	1990	*1992*	*1994*	*1995*
Agriculture, forestry and fishing	8.7	7.4	7.0	6.6
Mining, quarrying and manufacturing	29.7	28.1	27.2	27.2
Manufacturing	29.2	27.8	26.8	26.9
Heavy and chemical industry	19.2	19.3	19.5	20.5
Light industry	10.0	8.5	7.3	6.4
Electricity, gas and water	2.2	2.2	2.3	2.2
Construction	11.5	13.7	13.5	14.1
Commercial services	38.2	38.2	39.4	39.4
Government services & private non-profit services	9.7	10.4	10.5	10.5

Source: Government of the Republic of Korea, WTO Secretariat Report on Trade Policy Review of Korea.
[a] at current prices.

Strong economic growth, deregulation, trade liberalization and an easing of foreign investment restrictions are helping in the Republic of Korea to become not only more competitive but also to take a more forceful approach to domestic reform.

Referred to in the 1996 WTO Secretariat report as "one of the world's most dynamic economies," the Republic of Korea is now in the end phase of a five-year programme which seeks to reduce state control, abolish unnecessary regulations and restrictions, increase transparency of trade-related policies and align domestic laws with international regulations.

The Republic of Korea is very poorly endowed with mineral resources. Anthracite coal is its main mineral resource, with reserves estimated in 1989 at 1,615 Mt. According to the Korea Institute of Geology, Mining and Materials (KIGAM), the Republic of Korea's mineral production in 1991 only met with about 26.6 per cent of Korea's mineral demand and the balance was achieved from imports. The Republic of Korea has to import oil, coking coal, iron ore and non-ferrous metal ores such as copper, zinc, etc.

With the exception of the large tungsten mine and a number of anthracite coal producers, the Republic of Korea is essentially a country of small and medium-size mines with 794 mines in operation. Few mines reported to have production over $12 million. At end-December 1994, there were 410 mines in operation: 26 anthracite coal mines, 15 metal mines and 369 non-metal mines.

According to the Ministry of Trade, Industry and Energy, the 1994 value of minerals production in the Republic of Korea was estimated to be $1,588 million, about 0.4 per cent of its total GDP. The value of non-metal mineral product of $1,021 million, the highest value product, accounted for 64 per cent of all mineral values; the value of anthracite coal at $379 million accounted for 24 per cent; and the value of metal minerals at $188 million accounted for 12 per cent.

In comparison, the value of mineral production in 1988 was more than 16 per cent higher amounting to $1,870 million. In 1988, anthracite coal was the highest value product, at $1,235 million, accounting for 66 per cent of all mineral values; the non-metal group at 391 million accounted for 21 per cent and the metal group at $235 million accounted for 13 per cent (table 79).

Table 79. The value of mining sector of the Republic of Korea, 1988-1994

unit – US$, million

	Mineral Production	*Anthracite Coal*	*Non-metal Group*	*Metal Group*	*GNP*	*GDP per capita*
1988	1,870	1,235	391	235	4,040	169
1989	1,804	1,158	554	92	4,968	
1990	1,791	663	745	117	5,659	242
1991	1,797	786	648	330	6,518	280
1992	1,767	625	692	370	7,007	295
1993	1,610	494	926	190	7,513	
1994	1,588	379	1.021	188	8,482	380

Sources: Mining Annual Review 1989, 1990, 1991, 1992, 1993, 1994 and 1995.

According to the Office of Customs, the Republic of Korea imported $9,299 million worth of mineral products in 1989, including $4,932 million of crude oil, $1,289 million of coal, $493 million of iron ore, $437 million of copper concentrate and $216 million of zinc concentrates.

The Republic of Korea imported $16,958 million worth of mineral products during 1993, including $9,150 million of crude oil, $1,710 million of coal, $887 million of iron ore, $341 million of copper ore and $103 million of zinc concentrates.

The Republic of Korea imported $17,325 million worth of mineral products during 1994, including $8,878 million of crude oil, $1,751 million of coal, $796 million of iron ore, $300 million of copper ore and $132 million of zinc ore concentrates.

The Republic of Korea exported $961 million worth of minerals in 1989, including $620 million of oil products; $620 million of coal; $65 million of granite stone; $150 million of cement and $3.4 million of kaolin.

The Republic of Korea exported $2,027 million worth of minerals in 1994, including $1,698 million of oil products, $35 million of granite stone and $160 million of cement.

2.6 Russian Federation

To improve economic conditions in its far eastern region and increase access to international markets, a development programme known as 'The Long-Term State Programme for the Economic and Social Development of the Far East Region to the Year 2000' was initiated by the Soviet government in August 1987.

Business cooperation is now getting under way between Yuzhno-Sakhalinsk, the island's regional centre, and its sister-city Yanji in China. Vladivostok has been building up links with Pusan, the Republic of Korea's second largest political and economic centre with a population of around four million. Both cities boast large ports, complete with fishing and fishing-manufacturing capacities, and shipbuilding and repair yards. Pusan government officials consider it necessary to promote long-term economic cooperation, trading, and cultural links between the two cities. Sakhalin's cooperation with the Japanese island of Hokkaido is developing along the same lines.

In December 1988, the USSR Council of Ministers issued the resolution 'On Further Measures to Develop the External Economic Activities of State-Run Establishments, Co-operatives and Other Non-Governmental Entities', giving the Far East special status in developing external economic relations.

In August 1991, a law on foreign investment was issued by the Russian Federation, which governs all types of property and investment in the form of both joint ventures and fully foreign-owned enterprises. In 1991, Russia established a Committee for Precious Metals and Stones, which monitors all related operations and oversees a state depository. The Committee established principles for setting precious metals and precious stone prices.

In 1992, Russia issued two laws which govern mineral exploration and development activities – the Law on Subsurface Resources (sometimes referred to as the Law on Mineral Resources) adopted 21 February 1992, and the Regulations on Procedures for Licensing the Use of Mineral Resources adopted 15 July 1992. The Law on Mineral Resources outlines three important subjects, i.e. the State Mineral Fund, the licensing of mining activity, and a system of payments and fees for the use of mineral resources. The State Mineral Fund comprises all of the mineral resources of the Russian Federation.

The investment is being sought for the exploitation of export-oriented minerals, construction and renovation of processing facilities, and environment restoration and pollution control.

Investors have been generally reluctant to initiate projects in the Far East region owing to its inadequate infrastructure base, the non-convertibility of the rouble currency and inefficient bureaucracy. In addition, the dissolution of the former Soviet Union has promoted many Far Eastners to call for economic independence from Moscow, raising concerns among foreign companies about the ownership rights of mineral resources in the region.

As to the mineral resources development in the Russian Far East region, problems include aging production and processing plants, inefficient centralized planing, outmoded managerial techniques, a shortage of capital, poor transport and infrastructure, and labour disputes. Nonetheless, the abundance of resources in the region continues to attract many of the world's largest mining companies including US-based Cyprus Minerals Company, Canada's Placer Dome, Inc and Teck Corporation, and Australia's Broken Hill Proprietary Co. Ltd. and CRA Ltd.

Australia-based CRA Ltd. established a joint venture with the Far Eastern Institute of Raw Materials (FEIMS) and the regional branch of the Russian Ministry of Geology and Metallurgy (Dalgeologia) to pursue development of precious metals and other commodities in Khabarovskiy Krai. CRA signed its preliminary joint venture agreement in March 1992 received two separate licenses for undertaking exploration activities for a variety of mineral commodities (including gold) over an extended period of time. These were the first exploration licenses given to a major international company in the RFE.

In 1992, the Japanese diamond importer, A. Alder of Tokyo and the diamond producer of the Sakha Republic (Yakutia) named Toimada Diamond, created a joint venture diamond polishing company, based in Sakha on a 50:50 equity share basis. The Sakha-Japan venture aimed for an eventual 300,000 carats/y for the Japanese market and 50,000 carats/y for other exports.

A joint venture agreement with a Norwegian company, Norsk Hydro, was established to develop a bauxite deposit in Khabarovskiy Krai. Joint venture negotiations were also under way to explore and develop coal and other resources by firms from Republic of Korea and China, and to mine building materials by a Swiss company.

The partners from the Russian side include the regional branch of the Russian Ministry of Geology and Metallurgy (Dalgeologia) at Khabarovsk, the Far Eastern Institute of Raw Materials (FEIMS) at Kahabarosk; the Institute of Mining at Khabarovsk; Primorski Geological Production Amalgamation (Primorgeologia) at Vladivostok; Sakhalingeologia at Yuzhno-Sakhalinsk; Taezhgeologia at Blagovecsengk; Yakutskgeologia, Yakutsk; Yakutalmaz (yakutia Diamond), Yakutsk; and Primorzoto (Primorski Gold) at Vladivostok.

Foreign investors have repeatedly shown interest in working oil and gas deposits off Sakhalin shelf, as well as in organizing large-scale gas and chemical production. In fact, three of four projects negotiated in the oil sector in Russia in 1994 were related to Sakhalin Island. The Sakhalin-1 project comprised the U.S. firm Exxon and the Japanese firm Sodeco; the Sakhalin-2 project comprised the U.S. firms Marathon and McDermott, the Japanese companies Mitsui and Mitsubishi, and the Anglo-Dutch firm Shell; and the Sakhalin-3 project comprised the U.S.

firms Exxon, Mobil and Texaco. However, only the Sakhalin-2 agreement had actually been concluded and provided for investment of $8 billion.

With the approval of Moscow's Rosgeolcom, a Canada-UK consulting group prepared and organized the first licensing round for Sakha (Yakutia), presented in Tokyo in March 1994. Further licensing rounds are planned, probably for 1996, for onshore northern Sakhalin and the Irkustsk Region (Petromin Economist, May 1995).

The greatest hopes for a better energy future are pinned on tapping the vast gas resources of central and western Yakutia and Sakhalin Island, within the ambitious development scheme called the Wostok ("East") project. The project envisages the construction of gas pipelines, a 4,200 km trunkline system from Sakhalin and Yakutia gas fields through the Khabarovskiy Krai, the Amurskaya Oblast, and the Primorskiy Krai of the RFE, and down through the Korean Peninsula and cross the Tsushima Straits to southern Japan (figure 68).

Overall, the proposed project intended to provide foreign countries up to 17 bcmy (billion cubic metres per year) of exportable gas by the year of 2000, and up to 20 bcmy by 2005 as well as to satisfy all the gas needs from the RFE region, which were expected to grow to 2 bcmy by 2000 and 5 bcmy by 2010 (table 80). The forecast 2005 exportable surplus of some 20 bcmy was actually earmarked to the Democratic People's Republic of Korea for 2 bcmy, and to Republic of Korea and Japan each for 9 bcmy.

Table 80. Estimated natural gas exports from the Far Eastern Region of the Russian Federation in 2005

		in: bcmy (billion cubic metres per year)	*in: mmfd (million cubic feet per day)*	*Estimated by others*
Export from:	Russian Far East	20	1,900	
Import to:	Democratic People's Republic of Korea	2	190	
	Republic of Korea	9	855	550
	Japan	9	855	700

In addition to the technical and huge capital requirements for the implementation of this ambitious export-oriented plan, its expectation on the potential gas demand might be too optimistic. The Republic of Korea's requirements, which are currently not provided with firm LNG supplies, are unlikely to exceed 550 mmcfd, while Japan's incremental import needs may amount to only 700 mmcfd. In this regard, the project will probably have to be rescheduled, scaled down, or replanned, taking into account obviously-needed Chinese participation (Petromin, Oct. 1992).

A major study of Yakutia's gas reserves has recently commenced, underpinned by Russian, Yakutian and the Republic of Korea's governments. If independent assessment of the Russian figures confirms enough proven and probable gas reserves, say 70 trillion cf in eastern Russia, then a pipeline to the Pacific coast would be an economic reality and suddenly the whole region would be no more a frontier (Petromin Economist, May 1995).

3.0 INTERNATIONAL COOPERATION ACTIVITIES

An integrated approach and participatory strategies for sustainable development should include economic, social and environmental aspects of growth.

Meeting the basic human needs of all and eradicating absolute poverty is an objective of the highest priority that has been regarded as such in all the recent United Nations conferences since the World Summit for Children in 1990, including the World Summit for Social Development, Copenhagen, 6-12 March 1995, the Fourth World Conference on Women, Beijing, 4-15 September 1995, and the United Nations Conference on Human Settlements (Habitat II), Istanbul, 3 June 1996. Governments have been urged to formulate or strengthen, as a matter of urgency and preferably by the end of 1996 the International Year for the Eradication of Poverty, national strategies to eradicate absolute poverty.

It has been recognized that poverty is a complex multidimensional problem with origins in both the national and international domains. To resolve the problem all the components of sustainable development – economic development, social development and environmental protection – should be addressed since they are interdependent and mutually reinforcing. Therefore, it would be inappropriate to relax environmental laws, regulations and standards or their enforcement in order to encourage foreign direct investment or to promote exports.

3.1 Minerals assessment, development and trade

Basic differences exist in the economies of the countries in North-East Asia. Some of these are reflected in table 81.

3.1.1 Complementary factors of production

The potential complementary factors of production are summarized in table 82 (Jung and Hishida, 1995). The implications, drawn from the table may be summarized as follows:

(a) Japan and the Republic of Korea have more capital, higher technology supply and managerial skills for development and investment. The Russian Federation may also provide technology for the subregion. However, Japan and the Republic of Korea are dependent on imports of raw materials including minerals and oil and gas.

(b) China and the Democratic People's Republic of Korea have sufficient human resources, especially low-cost unskilled labour. Skilled labour may also be available from the Russia's Far East and China's north-east provinces.

(c) Mongolia, the Russian Far East, China and the Democratic People's Republic of Korea are rich in natural endowments including mineral resources. However, capital, technology and managerial skills are needed for development of these resources.

(d) A high income country like Japan has a big market. However, the Republic of Korea as a newly industrialized country with higher income, and China as a fast economically growing country with a huge population, have also considerably large markets. Markets, domestic and/or international, are the key to all the countries in the subregion for the development of their raw materials and/or processed products.

Other specific factors include the following:

(e) The Far Eastern part of the Russian Federation and Mongolia have severe weather and geographical conditions and, together with the Democratic People's Republic of Korea, have less developed infrastructure. On the other hand, the northeast of China, Japan, and the Republic of Korea have rather mild weather and more favorable geographical conditions. The transportation and communication facilities are more developed in China and the Republic of Korea, and highly developed in Japan.

(f) The countries/areas in the sub-region are in different stages of economic development: Japan is a highly developed country, China and the Democratic People's Republic of Korea are developing countries, the Republic of Korea is a newly industrialized country, Mongolia and the Russian Far East belong to countries with economies in transition.

(g) Mongolia and the Russian Federation have their own specific culture and languages. Rather similar culture exists among China, the Democratic People's Republic of Korea, the Republic of Korea and Japan with, however, three different languages.

The complementary factors, if they are properly dealt with, could facilitate the development of economic and technological cooperation between and among the countries/areas in North-East Asia. However, all the factors in common or in difference among the countries/areas also imply challenges and competition.

Table 81. North-East Asia countries' statistics

Items listed	*China*	*Democratic People's Republic of Korea*	*Japan*	*Mongolia*	*Republic of Korea*	*Russian Federation*
Population (million)	1,210.0		125.8		45.2	
Population growth rate (average annual %, 1985-1994)	1.4		0.4		1.0	
Foreign exchange reserves (Sep. 1995, billion)	95.0		215.0		36.6	
Gross foreign debt (as of the end of 1995, billion)	106.6		1,974.0		86.2	
GNP per capita at purchasing power parity	2,510.0		21,350.0		10,540.0	
Nominal GDP (billion, 1995)	691.4		5,120.0		454.0	
Nominal GDP per capita	569.7		40,879.0		10,128.0	
Growth in real GDP (%)	10.2		2.3		9.0	
Trade						
Total exports (billion)	148.8		442.9		125.0	
Total imports (billion)	132.1		336.1		135.0	
Merchandise trade balance (billion)	16.7		106.8		-4.8	
Current account balance (billion)	17.1		104.6		-9.0	
Direction of trade (% of total, Jan.-June 1995), as: exports/imports						
United States	16.9/11.7		27.0/23.0		17.4/22.5	
European Union	13.6/14.4		16.0/14.0		12.5/13.8	
Japan	20.9/20.9				12.4/21.0	
Asia excluding Japan	40.0/38.6		44.0/36.0		37.7/17.6	

Source: Asian Economic Survey, 1996-1997, the Asian Wall Street Journal, Oct. 22, 1996.

Major exports:

China: electrical-machinery products, garments, yarn and textile products, rubber and leather shoes, tourism products.
Japan: automobiles, semiconductors, computers, auto parts, steel.
The Republic of Korea: semiconductors, petrochemicals, steel, motor vehicles, machinery.

Major imports:

China: electrical-machinery products, steel, raw plastic, chemical fertilizer, grain and grain powder.
Japan: crude oil, office equipment, clothing, seafood, semiconductors.
The Republic of Korea: petroleum, machinery, steel products, electric and electronic goods, chemical products.

Table 82. Complementarity of production in North-East Asia

	Labour		*Natural Resources*				
	Skilled	*Unskilled*	*Energy*	*Others*	*Capital*	*Technology*	*Managerial Admin*
China – 3 NE provinces	I	S	M	M	I	I	I
Japan	S	I	I	I	S	S	S
Democratic People's Republic of Korea	I	S	I	M	I	I	I
Republic of Korea	M	I	I	I	M	M	M
Mongolia	I	I	I	S	I	I	I
Russia Far East	M	I	S	S	I	M	I

S = sufficient; M = marginal; I = insufficient

Source: ESCAP, 1996. Studies in Trade and Investment (ST/ESCAP/1640).

3.1.2 Mineral resource assessment

Mineral resources classification

Resource assessment pertains to the estimation and evaluation of minerals in the ground, both discovered and undiscovered. Attention centres on the form, concentration, and location of the minerals in order to determine whether they might be extractable under foreseeable economic and technological conditions. In practical terms, there is no such thing as an all-purpose resource assessment. Diverse groups of interested people – mineral exploration planners, economic analysts, land use planners, or policy makers look for the aspects that are most pertinent to their particular fields and time-frames.

For more than a century, efforts have been made to establish definitions that would be widely accepted and applied. Unfortunately, the result has been a multitude of definitions, none of which has won general acceptance, largely because of differences in purpose and emphasis. Even the commonly used expressions "crustal abundance", "mineral endowment", "resources", and "reserves" convey different meaning to different people.

In resource assessments, the type of classification used will depend on the type of assessment. Resource assessments are of two main types, i.e. primarily geoscientific, and geoscientific/economic. The geoscientific/economic assessments aim to provide policy analysts with estimates of the magnitude (currently perceived) of the sources of short-range and longer-range mineral supply, so that appropriate efforts may be made to explore and develop them and technology developed to aid in their recovery. Sources of supply that can be quantified reliably are mostly those from which supplies may come within the next 15-20 years.

The assessments require a multi-disciplinary approach, which covers the degree of assurance about an estimated quantity actually being present in the earth's crust and the degree of economic attractiveness it offers. Given these two concerns, the resources can be subdivided into four main categories: discovered and economic; discovered but sub-economic; undiscovered and economic; and undiscovered and sub-economic. These classifications are dynamic, changing with discoveries and with fluctuations in cost and price. The Australian mineral resource classification system is presented in figure 65.

The level of assurance about the physical existence of the resources is generally expressed in discrete gradations such as proved, measured, probable, indicated, possible, inferred, speculative, and so on. The dividing lines between such descriptive terms are difficult to define satisfactorily, so that what is "proved" to one person may be "probable" to another.

The distinction between economic and sub-economic is also largely marginal. It can be made only on the basis of mining feasibility studies that consider all the details of mining methods, costs, and revenues. The explicit criterion used for economic subdivisions is generally one of the following: price; cost; cost/price ratio; or

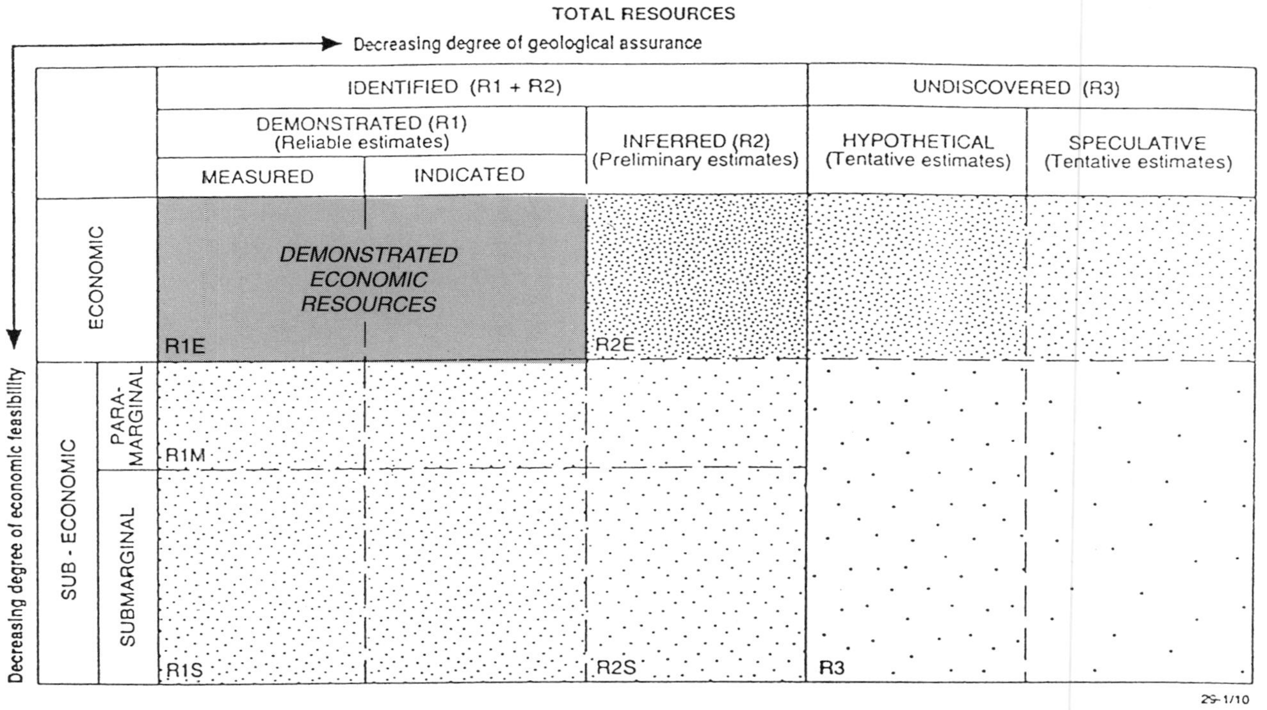

Key
R – *in situ* resources

R1 –	known deposits (reliable estimates)	R2 –	extensions of known and newly discovered deposits (preliminary estimates)	R3 –	undiscovered deposits (tentative estimates)
R1E –	economically exploitable	R2E –	economically exploitable		
R1M –	marginally economic	R2S –	sub-economic		
R1S –	sub-economic				

Sources: Australian Bureau of Mineral Resources (1984); see "BMR refines its mineral resources classification system", *Australian Mineral Industries quarterly,* vol. 36, No. 3 (1984).

Note: Sub-economic categories also represent unusual concentrations of the relevant elements, and can be reasonably well defined for particular deposit types.

[a] Modified McKelvey system, also showing United Nations resources categories.

Source: E/C.7/1996/11.

Figure 65. Australian resource classification system

probability of becoming economically mineable within a given time period. For some metals, economic subdivision is complicated by association with other metals in the same deposits, which may enhance the economics of exploitation.

Detailed knowledge about discovered and economic reserves – and about past production – is invaluable as a basis for extrapolation in resource assessment. However, the cost of upgrading knowledge about the exact tonnage of new discoveries to the "reasonable assurance" level of reserves is high, so is the cost of economic analysis. Therefore, the drilling and analysis necessary to establish reserves is normally undertaken only to the extent needed for production planning.

Especially imprecise is the outer limit of resources determined by foreseeable technology and economics, concepts that respond closely to one another. Some of the technological improvements that would extend the economic limit of resources are techniques for locating deposits at greater depths and methods for better extraction of ores and recovery of the mineral commodities from them. These represent changes in the technology of raw material production. The economic limit of the resources of a commodity can also be affected by the technology of material use, which, for example, might lower the demand for that commodity (and thus its price) through the use of another material in its place.

Discovered resources

DER and R/P ratio

Time series of production, demonstrated economic resources (DER) and resource/production (R/P) ratios illustrate the availability of mineral resources through time, and indicate the effects of past changes in demand. The stock of DER is continually being renewed either by discovery of new economic resources or by transfer from the large pool of known but previously sub-economic resources, as a result of technological advances or price rises induced by scarcity.

In general, the global annual production of mineral commodities has increased greatly and steadily this century, but DER have also increased, so that R/P ratios have been maintained. In the case of bauxite, however, the R/P ratio has declined because of greatly increased annual production but maintained DER.

Large tonnage commodities, such as coal, iron ore, bauxite and phosphate (roughly 30 kg per capita of phosphate consumed globally), have large R/P ratios, in the hundreds of years. However, such minerals occur in near-surface deposits: the capacity to continue to renew the stock of their DER is in some doubt, and their mining also has the largest immediate environmental impact.

For most of the metallic minerals, R/P ratios are much smaller, in the tens of years, but it has been possible to maintain their ratios. It has been noted that the time-lag between investment and the establishment of new DER for metallic resources is typically 10 years or more.

Undiscovered resources

The supply of mineral and petroleum resources over time-scales of a few decades is well assured, based essentially on the assessments of identified resources. For petroleum, there are also reliable global estimates for undiscovered resources; but this is not the situation for metallic minerals. There are many different types of deposits for metallic minerals and the processes that generate them are very complex and less well-known than those that generate petroleum deposits, and methodologies for estimating such undiscovered resources are much less reliable.

Mineral resources are essentially non-renewable, and the economic resource potential has been severely depleted during the twentieth century at an ever increasing rate. The trend towards utilization of lower-grade ores is already well-established. However, the issues involved in assessing the sustainability of mineral supplies in the long term have long been neglected and current knowledge of the global potential for discovering new deposits is very limited. There are also increasing pressures on land use, which may make it increasingly difficult to develop available mineral resources. There is a question of ensuring the optimum and efficient use of available resources with minimum environmental impact.

Resource assessment activities may begin with an evaluation of identified deposits and their reserves in a region or country and may or may not proceed to an estimation of undiscovered resources by analogy. Reserve estimation methodologies make quantified estimates available, but they require many data that are frequently unavailable. The methodologies used in computing reserves may be conventional or geostatistical.

For assessments of undiscovered mineral resources, various estimation methodologies have been developed and are utilized internationally. In general, there are six of them, plus a multitude of techniques based on their fundamental concepts. The six methodologies are based on areal value, crustal abundance, volumetric analysis, deposit, Delphi surveys, and integrated synthesis. They are described below along with the analytical procedures required to implement them. Cost and data requirements, the product's strengths and weaknesses, and country applications of the methodologies are summarized in table 83.

The selection of a specific resource assessment methodology is an important and sometimes difficult task. The first step is to decide on the type of information needed from a particular resource assessment activity. If a broad-based resource assessment is desired, the areal value, crustal abundance, or volumetric estimation procedures may be adequate for the purpose. If a more detailed characterization is needed, the deposit modelling procedure may be more appropriate. Cost, time, and personnel requirements, the availability of geological, geochemical, and geophysical data, and the acceptability of biases and inaccuracies in the assessment must all be considered in the selection.

Perceived mineral resource potential is the most important single criterion for the mining industry in assessing the investment environment for exploration. The most important step for the assessment of resource potential is the adequate geological definition of permissive tracts, that is, geological regions which permit the occurrence of various mineral deposit types. This information is an essential ingredient in considering wider issues of integrated land management and land use. A major contribution to the long-term management and sustainable development of mineral resources is to develop a global knowledge base, at appropriate scales, of the potential for mineral resource exploration and development.

Among several approaches, most notably a three-part method of quantitative assessment of undiscovered mineral resources has been applied by the United States Geological Survey since 1975. It provides quantitative resource information in a form consistent with other competing uses of land. More recently, there has been a proposal to make a national three-part assessment for the entire United States.

As a first step in this programme, it was also proposed to carry out two-year preliminary quantitative national assessments based on existing national data to produce maps showing the outlines of tracts that are promising for the types of deposits concerned. Such a preliminary assessment would be of even greater value for planners in developing countries with substantial resource potential, which would involve only three steps, namely:

- compile existing data,
- apply limited mineral-deposit models, and
- construct maps of delineated promising tracts.

The maps delineating promising tracts would provide a more realistic assessments of the mineral resources base which is beyond the horizon currently provided by only identified resources. The endowment of mineral resources can then be effectively managed and utilized. Such maps would also assist the consideration of minerals issues within an integrated approach to land-use planning. These maps could play a crucial role in strengthening the capacity to integrate mineral resources information into the decision-making processes.

Mineral resources assessment programmes

China

With its complex geology and huge land area (China ranks second after the Russian Federation in area), China is well-endowed in most major metallic, non-metallic and mineral fuel resources. However, with a population of more than 1,100 million, China's mineral and energy allotment per capita is less than that of most other resource-rich countries. In addition, many of the deposits of non-fuel minerals are of poor quality or in remote locations, making them very costly and sometimes difficult to exploit.

All the mineral exploration activities in China are organized by the State. Out of the 1.1 million people engaged in geological exploration in China at present, nearly 400,000 belong to the Ministry of Geology and Mineral Resources (38.2 per cent of the total), and the rest belong to other related ministries. There are more than 900 field geological exploration teams in the country, reporting to the Ministry of Geology and Mineral Resources (51.9 per cent), the Ministry of Energy Resources (21.8 per cent), the Ministry of Metallurgical Industry (8.6 per cent), China Nonferrous Metals Industry Corporation (12.0 per cent), the Ministry of Chemical Industry

Table 83. Mineral resources assessment methodologies

Methodology	*Estimation procedures*
1. Area value estimation	Extrapolation of estimated mean unit value of well-explored developed area to similar, less explored, underdeveloped region of interest
Data requirements	Products
Data on rock types and estimate of production data by year and commodity for both study area and analog area	National, regional, or provincial resource potential for known and unknown deposits on a commodity-by-commodity basis
Methodology	Estimation procedures
2. Crustal abundance	Estimation of the amount of recoverable resource from a representative mean abundance, normally through an empirical function
Data requirements	Products
Data on rock types and geochemical values	Regional, local, or specific estimates on an elemental basis. Empirical estimation of production capability, and recoverable resources formulas
Methodology	Estimation procedures
3. Volumetric estimation	Extrapolation of representative estimated or determined mean concentration of a known unit volume to the volume of interest
Data requirements	Products
Data on quantity and size of prospects, stratigraphy porosity/permeability, yield, areas, and thickness	Estimate of resources on a regional, local, or specific geologic unit, based on analogy
4. Methodology	Estimation procedures
Deposit modelling	Estimation of resources in a specific geologic environment based on an analysis of known deposits in geologically similar environments
Data requirements	Products
Geologic, geochemical, geophysical, mineral commodity occurrence, and remote-sensing data. Also, genetic information and results of previous exploration	Area-specific or local estimate of tonnage and grade distribution by deposit type and at .05, .5, and .9 confidence levels
Methodology	Estimation procedures
5. Delphi estimation on analysis of all available	Resource estimation based on a collection of opinions from expert geologists. Such opinions are based data and individual experience
Data requirements	Products
No specific data requirements except knowledge of the geology of the area of interest	Subjective estimation by area of deposit type. May estimate tonnage, grade, and deposit size distribution
Methodology	Estimation procedures
6. Integrated synthesis	Estimation of resource potential based on a combination or integration, or both, of some or all of the above methods

Data requirements	Products
Data from all of the above methodologies and, in some cases, data files of estimates produced as basic inputs to other methodologies	Specific resource estimation by deposit, commodity, tonnage, grade, and area (highly disaggregated). Estimates can be used by national econometric models
Methodology	Strengths
7. Areal value estimation	Relatively simple to use for mineral resource planning; universally applicable to developed and developing countries; low cost; short evaluation time
Weaknesses	Selected country applications
Basic assumption that geologically similar areas contain similar mineral resources; requires a reliable geological map of area of interest; dependent on availability and accuracy of commodity data	Australia, Canada, China (in preparation), Mexico, New Zealand, Papua New Guinea, United States, Zimbabwe
Methodology	Strengths
8. Crustal abundance	Quick and moderately reliable for resource planning; can be easily updated with new data and relies on fairly accurate analytical inputs
Weaknesses	Selected country applications
Estimates are subject to wide range of error, data-dependent; totally dependent on availability of good geological maps; assumes close genetic relationship between rock types and associated mineral deposits	Canada, China, Soviet Union (former), Turkey, United States
Methodology	Strengths
9. Volumetric estimation	Relatively simple to use and requires minimum amount of data; a standard method for petroleum and natural gas basin estimation; excellent for deposits with simple and uniform geometry
Weaknesses	Selected country applications
Paucity of information used is not apparent to the uninitiated; misleading to use volumetric estimates to guide major economic or national policy decisions; assumes geologic similarity between regions	Canada, Indonesia, United States (Alaska)
Methodology	Strengths
10. Deposit modelling	Incorporates all available data and permits the incorporation of geological concepts; resource estimates reflect the quality of the data used; delineates exploration targets
Weaknesses	Selected country applications
Data for deposit model construction is limited and can lead to misapplication of specific	Bolivia, Canada, Colombia, Costa Rica, Cyprus, Finland, Norway, Papua New Guinea, United States
Methodology	Strengths

11. Delphi estimation	Rapid and low cost; applicable anywhere in the world, provided geological experts are available; provides disaggregated estimate
Weaknesses	Selected country applications
Very easy to introduce bias either intentionally or accidentally; resource estimates tend towards group mean; totally dependent on knowledge and experience of geological experts	Canada, Costa Rica, Mexico, Papua New Guinea, United States
Methodology	Strengths
12. Integrated synthesis	Incorporates all available data, concepts and geological experts, provides disaggregated, commodity specific estimates; useful on delineation of exploration targets and mineral resource planning
Weaknesses	Selected country applications
Quite expensive and time Consuming; requires substantial amount of data, hence limited to local application; requires complex mixture of skilled personnel	Bolivia, Canada, China, Colombia, Costa Rica, Papua New Guinea, Puerto Rico, Scandinavia, United States, Venezuela

Source: A.L. Clark, "The importance of data banks for resource assessment", in New Paths to Mineral Exploration. Proceedings of the Third International Symposium (Hannover, Federal Republic of Germany, 27-79 October 1982), F. Bender, ed., pp. 97-98.

(2.6 per cent), the State Bureau of Building Materials Industry (2.8 per cent) and the Ministry of Light Industry (0.2 per cent).

There are different degrees of intensity of geological investigation in China. It was not until the end of 1985 that the Chinese Government undertook the first detailed and comprehensive nationwide survey of mineral and energy resources and mines. The 1985 survey was supervised by the State and involved such organs as the Ministry of Geology and Mineral Resources. It examined the number of mines and mineral and energy resources discovered in China, and the national production, consumption, and employment levels. According to it and other surveys since, geologists in China have discovered more than 160 types of mineral and energy resources and have verified reserves of most types. These include five types of ferrous minerals, 20 types of non-ferrous and precious metallic minerals, 76 types of non-metallic and geothermal and groundwater minerals, and six types of mineral fuels. Mineral ores in China are found in over 200,000 locations, although distribution is fairly uneven due to the complicated and varied geological conditions.

The reserves of many types of mineral and energy resources in China are the largest in the world, and several minerals, including rare earths and tungsten, occur in an abundance that not only satisfies domestic demand but also allows for exports. China's mineral reserves can be considered relatively small in chromite, platinum-group metals, titanium (rutile), and zirconium.

By the end of 1990, regional geological surveying at a scale of 1:1,000,000 was completed on land; that at a scale of 1:200,000 was completed on two thirds of the territory (table 84). The first round of nationwide mineral prospecting was completed. More than 200,000 deposits (occurrences) of various varieties of minerals were found, with 15,000 of them having proved reserves.

China's eighth five-year plan (1991-1995) promotes accelerating the pace of geological prospecting so as to ensure sufficient mineral reserves to facilitate continued economic growth. The recently announced 12-year exploration programme by China's Ministry of Geology and Mineral Resources put emphasis on a search for ores of aluminium, chromium, iron, lead and zinc, and manganese, and locating new deposits in the interior and western provinces of China. Exploration activities for industrial minerals required by China's chemical and construction industries will include phosphate, potash and soda ash.

Table 84. Major geologic maps and atlases of China published during 1989-1992

Title	*Scale*	*Date*	*Remarks*
Mineral Resources Map of China	1:5,000,000	1992	Set of 3 maps
Map Series of Geology and Geophysics of China Seas and Adjacent Areas	1:5,000,000	1992	Set of 9 maps
Landsat Images of China	1:6,000,000	1992	1 sheet
	1:4,000,000	1992	2 sheets
	1:2,500,000	1992	6 sheets
	1: 500,000	1992	15 sheets
Progress in Geology of China	–	1992	93 papers
Mineral Deposits of China	–	1992	Volume 2
	–	1990	Volume 1
Stratigraphy and Paleontology of China	–	1991	Volume 1
Land Use Map of China	1:1,000,000	1991	Set of 64 maps
Atlas of Landsat Imagery of Main Active Fault Zones in China	–	1990	135 pages
Geological Map of China	1:5,000,000	1991	With explanatory text
Geological Hazard Type Map of China	1:5,000,000	1991	With explanatory China text
Quaternary Geologic Map of China and Adjacent Sea Areas	1:2,500,000	1990	9 sheets with explanatory text
Quaternary Geology and Environment in China	–	1991	Volume 1
Loess Plateau	–	1991	374 colour plates and 19 maps
Geomorphologic Map of Huang-Huai-Hai Plain (North China Plain) in China	1:1,000,000	1990	1 map
Quaternary Map of Huang-Huai-Hai Plain in China	1:1,000,000	1990	2 maps
Quaternary Lithofacies and Paleogeographic Map of Huang-Huai-Hai Plain	1:2,000,000		
Geological Map of Shenzhen	1: 500,000	1989	1 map
Geological Map of Qinghai-Xizang (Tibet) Plateau and Adjacent Areas	1:1,500,000	1989	1 map
Plate Tectonic-Lithofacies Map of Xizang (Tibet)	1:1,500,000	1989	1 map

As part of China's expanding efforts to search for and discover new mineral deposits, the Metal Mining Agency of Japan has recently signed an agreement with the China National Nonferrous Metal Industry Corporation to begin a six-year base and rare metals exploration and development project in south central China. The focus will be on the Western Yangtze Para Platform area, and the survey will cover some 150,000 km^2 in the Mian-Lue-Ning area of Shanxi Province and the Lu-Wa area of Yunnan. The potential for copper, lead, zinc, nickel and cobalt minerals is thought to be particularly promising. The survey began by examining existing data. Geological mapping, geophysical and geochemical surveying, drilling and underground exploration will be implemented in the subsequent years of the programme.

Russian Federation

As a means of attracting foreign investment into its mining industry, in 1989 the Government of the USSR released a list of 120 mineral deposits found throughout the country suitable for overseas investment. The Robertson Group of the United Kingdom was hired to undertake a joint cooperation project to produce a paleogeographic geological atlas of the shelf regions of Eurasia. The terms of the agreement were that the Geological Institute of the USSR Academy of Sciences (GINAS) would provide the company with geological maps and supporting technical data of the offshore and adjacent inshore area of the Soviet Union. Those maps, compiled by a team of more than 100 Soviet geologists over the past seven years, are now being made available for the first time.

Between 1990 and 1992, the Group used digital production techniques to prepare high quality colour maps and also assisted the Soviets in preparing technical reports to accompany the atlas. To complete the coastline coverage, GINAS gathered information from counterpart organizations in countries outside the USSR which border the Eurasia landmass, stretching from Japan, China and Viet Nam through India and the Mediterranean region. The project was valued at nearly $US750,000, financed by oil and mining companies. The maps contain petroleum- and mineral-oriented geological information which can be used to identify which offshore and adjacent coastal areas are the most promising for development.

In 1991, following an agreement with the Soviet Ministry of Geology, the Robertson Group acquired for sale and distribution limited quantities of current geological and mineral maps of the Soviet Union, previously unavailable in the west. The maps have an English legend. Commodities shown include oil and gas, metallic minerals, industrial minerals, diamonds and gold.

With the dissolution of the Soviet Union in 1991, the business relationship between the British contractor and the Soviet Government disintegrated. Today, the Russian Committee on Ecology and Rational Use of Natural Resources has assumed much of the role of the former Soviet Ministry of Geology and, as such, has access to current geological maps of the former Soviet Union.

A 1991/92 Memorandum of Understanding on priorities in Sino-Soviet cooperation in the fields of geology and mineral resources was signed in Beijing in April 1991 between the geology ministries of the two countries. According to the document, China and the Soviet Union would begin geological cooperation in the border areas, exchanging scientific and technological information and new technology and materials.

Chinese oceanologists started working with Soviet geologists in 1991 in studies ranging from marine physics and geology to information exchange and instrument development. A five-year contract, the result of a bilateral agreement signed in September 1990 in Moscow, was drawn up. This Sino-Soviet cooperation was based on mutual interest in the study of the western part of the north Pacific, which both countries border in the east, and of the Kuroshio, a northward-moving warm current originating near the tropics. It had been separately reported that the former Soviet Union had planned to exploit manganese nodules on the Pacific Ocean floor to a depth of about 5 km.

Today, the Russian Federation is the largest country in the new Commonwealth of Independent States (CIS). Geologically complex, much of the former Soviet Union's mineral abundance is to be found in Russia, and the newly independent states contain the largest mining industry in the world. In 1992, Russia was a major world producer of a variety of mineral and energy resources, including aluminium, cement, copper, coal, diamonds, gold, iron ore, lead, manganese ore, natural gas, petroleum, tin, and zinc.

Explored, developed and assessed reserves of major minerals in Russia are estimated (using world prices) to be worth about $28.5 trillion. Altogether the figure is speculative. Russian leaders realize the importance of the country's mineral industry to long-term economic growth. A draft federal programme, released in 1993, outlines Russia's intention to ensure a reliable mineral resource base to the year 2000 and beyond. The basic principles for forming a reliable base are:

(a) Promoting accelerated development by applying advanced technologies to production and processing;

(b) Accelerating investment in the discovery and development of deposits;

(c) Quantifying and acknowledging the potential economic contribution of Russia's mineral industry in the year 2000 and 2010;

(d) Disassembling the State monopoly in geology and promoting the development of subsurface resources through a licensing procedure; and

(e) Privatizing enterprises and reorganizing them into stock companies specializing in geological and geophysical services.

Clearly, the Geological Survey of Russia will face tremendous difficulties in promoting the discovery and development of mineral resources in the country during the next few years. The State budget is in disarray, and legal issues such as ownership rights are still uncertain. Regardless of these problems, however, the Russian Government will continue trying to attract foreign investment to assist in geological surveying activities and project development.

3.1.3 Mineral reserves

Non-fuel minerals reserves

Proven reserves of a number of mineral and energy resources in China are among the richest in the world, including antimony, barite, coal, fluorspar, rare earth metals, tin, tungsten, and vanadium.

The minerals reserves in North-East Asian countries are as follows:

Antimony

Antimony resources are abundant in China. China possesses 9 out of 17 large antimony deposits in the world, with reserves of more than 100,000 tonnes. The world's largest antimony mine, Xikuangshan mine of Hunan, is in China.

Barite

China's barite reserves account for about 300 Mt, ranking the first in the world, followed by the former USSR, the US, India and Morocco.

Copper

China has a retained copper reserves of a few tens of million tonnes, the Democratic People's Republic of Korea has some copper deposits, estimated in 1989 as being around 2.2 Mt. The Republic of Korea's copper reserves were estimated by 1991 to be only 20 Mt at an average of 1.13 per cent Cu. Large copper reserves are possible in Yakutia of the Far Eastern part of the Russian Federation, where stock work deposits of copper-molybdenum and copper-tungsten ores can be found.

Diamond

The Sakha Republic (Yakutia) of the RFE mines practically all diamonds in Russia.

Fluorite

Fluorite reserves in China were estimated at 130 Mt, ranking first in the world. Fluorspar occurs in numerous provinces of China, but the largest reserves are believed to be in Nei Mongol followed by Zhejiang.

Gold

According to the chairman of the Russian State Committee for Precious Metals and Stones (Roskomdragmet), Russia's gold reserves as of January 1, 1995, were 321.8 t, including 189 t deposited with the Central Bank of Russia and 132.8 t held by Roskomdragmet. In May 1993, control of Russia's gold mining enterprises was placed officially under the jurisdiction of Roskomdragmet. China has a total proven reserve of 4,500 t of gold.

Iron ore

China had a total iron ore reserve of 50.1 billion tonnes ranking the second in the world in 1990. The Democratic People's Republic of Korea's iron ore reserves are estimated at 3,300 Mt. By the estimation in 1992, the Republic of Korea had iron ore reserves of 177 Mt of 35 per cent Fe grade.

Lead and zinc

The retained reserves of lead and zinc metal in China accounted for 33 and 84 million tonnes of metal respectively by the end of 1990. The deposits of lead and zinc in the Democratic People's Republic of Korea are estimated at 6 Mt and 12 Mt respectively. Reserves of lead and zinc in the Republic of Korea are estimated at 37 Mt averaging 3.9 per cent lead and 7.3 per cent zinc and the bulk of the deposits occur at the Yeonwha mine, Kungsang North Province. Russia's major lead-zinc mining complexes are Dalpolimetal in its far east region, Nerchinsk in the trans-Baikal region, and Salair and Altay in Western Siberia in the Kemerovo and Altay regions respectively.

Limestone

The total limestone resource in China is estimated to be 12-13 trillion tonnes, of which 3.3 trillions are cement limestone. They are extensively distributed in the country.

Mercury

Nearly half of Russia's estimated mercury reserves occur in its far eastern region. Mineralization of mercury is abundant in Yakutia and the Magadan and Kamchatka regions.

Monazite

The Democratic People's Republic of Korea's reserves of monazite, estimated at 500,000 t, exist at the Chelsan mine in the northwest of the country, near the border with China.

Phosphate rock

By the end of 1990, China's explored phosphate rock reserves were 15.7 billion tonnes, concentrated in Yunnan, Guizhou, Sichuan, Hubei and Hunan, which totally accounted for nearly 80 per cent of China's total. In contrast, China's potash resources are poor. As of the end of 1990, its explored reserve of KCl was 396 Mt, most of which occur in the saline lakes of the Quaidam Basin in Qinghai province.

Rare earths

China has the world's largest resources of rare-earths with reserves of 3.6 Mt, equivalent to four times that in the rest of the world. China is one of the largest producers of rare-earths in the world.

Tin

China has rich tin resources and its reserves may account for as much as 20 per cent of the total world reserves.

The Far East part of the Russian Federation contains the bulk of the country's reserves. Many deposits are being developed in the Magadan region, Yakutia and the Khabarovsk and the Maritime Territories. Placer deposits account for more than 40 per cent of the national output of tin; however, remaining placer reserves are insignificant.

Talc

The retained reserves of talc in China were 250 Mt at the end of 1990, ranking second in the world after the former USSR.

Fuel minerals reserves

Oil

According to the estimation by British Petroleum in 1996, the total world proved reserves of natural oil at end 1995 was 1.009 trillion barrels, the world oil production in 1995 was 67,535 thousand barrels daily, and the oil reserves/production ratio (R/P) was 42.8 years (table 85). The total world proved reserves of natural gas at end 1995 was 4,933 trillion cubic feet, the world production in 1995 was 1,908.0 million tones of oil equivalent, and the gas R/P was 64.7 years.

Table 85. Oil reserves, production and R/P ratio in Asia and the Pacific at the end of 1995

Oil	*Reserves (Thousand Million barrels)*	*Reserves (share of total, %)*	*Production (Thousand barrels daily)*	*R/P ratio (years)*
Australia	1.6	3.6	575	7.9
Brunei Darussalam	1.4	3.2	175	21.8
China	24.0	54.0	2,990	22.0
India	5.8	13.1	785	20.9
Indonesia	5.2	11.7	1,575	9.3
Japan, 1994	0.1	0.3	<0.05	9.0
Malaysia	4.3	9.7	735	16.1
New Zealand, 1994	0.1	0.3	<0.05	8.9
Papua New Guinea	0.4	1.0	100	11.0
Viet Nam	0.5	1.1	150	9.2
Others	0.9	2.0	185	13.4
Total Asia & Australasia	44.3	100.0	7,270	17.0

Source: British Petroleum, 1996. Statistical Review of World Energy.

The total proved reserves of natural oil in Asia and the Pacific at end 1995 was about 44.1 thousand million barrels, the oil production in the region in 1994 was 7,270 thousand barrels daily, and the R/P ratio of the region was estimated as 17.0 years. The total proved reserves of natural gas in the region at end 1995 was 328.6 trillion cubic feet, the gas production in the region in 1995 was 182.7 million tonnes oil equivalent (about 7 trillion cubic feet), and the gas R/P ratio was nearly 45.8 years.

As indicated in table 85, the total proved reserves of oil in China at the end of 1994 was 24 thousand million barrels, then the oil R/P ratio would be 22.6 years with oil production 2,905 thousand barrels daily. The total proved reserves of gas in China at end 1994 was 59 trillion cubic feet, then the gas R/P ratio would be 102.7 years with gas production annually of 14.9 million tonnes oil of equivalent (Mtoe), or about 0.58 trillion cubic feet.

The Russian Eastern Siberian and Pacific basins, remain immature to virgin in terms of exploration maturity. New or additional oil and gas reserves will be found in the region, but are not expected to make an impact until the turn of the century.

The Democratic People's Republic of Korea has offshore oil and gas in the Korea Bay but is in need of technology and equipment to explore and exploit them. The Democratic People's Republic of Korea has two refineries with a capacity of 3 Mt/y.

The reserves, production and R/P ratio for oil in North-East Asia are shown in table 86.

Natural gas

The total proved reserves of natural gas in Asia and the Pacific region at the end of 1995 was 328.6 trillion cubic feet, the gas production in the region in 1995 was 182.7 million tonnes of oil equivalent (about 7 trillion cubic feet), and the gas R/P ratio was nearly 45.8 years (table 87). Table 88 shows the related information in the North-East Asian countries.

Table 86. Oil reserves, production and R/P ratio in North-East Asia, Asia and the world at the end of 1995

Oil	*Reserves (Thousand Million barrels)*	*Reserves (share of total, %)*	*Production (Thousand barrels daily)*	*R/P ratio (years)*
China	24.0	2.4	2,990	22.0
Japan	0.1	<0.05	<0.05	9.0
Democratic People's Republic of Korea				
Republic of Korea				
Mongolia				
Russia	49.0	4.8	6,200	21.9
Total NE Asia				
Total Asia	42.5	4.18	6,695	17.4
Total WORLD	1,016.9	100.0	66,695	42.8

Source: British Petroleum, 1996. Statistical Review of World Energy.

Table 87. Natural gas reserves, production and R/P ratio in Asia and the Pacific at the end of 1995

Gas	*Reserves (Trillion cubic feet)*	*Reserves (share of total, %)*	*Production (Million tonnes oil equivalent)*	*R/P ratio (years)*
Australia	20.1	6.1	26.7	19.2
Bangladesh	10.1	3.1	6.6	38.9
Brunei Darussalam	14.0	4.2	9.3	38.4
China	59.0	18.0	15.8	94.9
India	25.0	7.6	16.8	37.8
Indonesia	68.9	21.0	52.6	33.4
Malaysia	68.0	20.7	26.1	66.5
Pakistan	27.0	8.2	12.1	56.9
Papua New Guinea	3.0	0.9	–	–
Thailand	5.9	1.8	8.3	18.1
Viet Nam	4.0	1.2	–	–
Others	23.6	7.2	8.4	73.4
Total Asia & Australasia	328.6	100.0	182.7	45.8

Source: British Petroleum, 1996. Statistical Review of World Energy.

Table 88. Natural gas reserves, production and R/P ratio in North-East Asia, Asia and the world at the end of 1995

Gas	*Reserves (Trillion cubic feet)*	*Reserves (share of total, %)*	*Production (Million tonnes oil equivalent)*	*R/P ratio (years)*
China	59.0	1.2	15.8	94.9
Japan	1.0	<0.05	2.0	12.0
Democratic People's Republic of Korea				
Republic of Korea				
Mongolia				
Russia	1,700.0	34.5	499.9	82.1
Total NE Asia				
Total Asia	308.5	6.3	156.0	54.7
Total WORLD	4,933.6	100.0	1,908.0	64.7

Source: British Petroleum,1996. Statistical Review of World Energy.

Coal

Coal deposits in the Democratic People's Republic of Korea are estimated at 12,000 Mt, in which metallurgical coal is limited. The main coal region is the South Pyongyang province covering the Tokchon, Kangdon, Kangso, Kowon and other areas.

Mongolia is rich in coal with identified reserves of 10,000-15,000 Mt.

The Far Eastern region of the Russian Federation accounts for 9 per cent of coal reserves and 14 per cent of coal production of the Russian Federation . With some notable exceptions, only a limited number of deposits have been sufficiently identified, evaluated and developed in the Far East. The coal prospects are now mainly bounded up with the projected development of the Bureya and South Yakutia basins and, in particular, of the Elga coal field, which contains about 3 billion tonnes of high-quality coal. The Elga project is considered as a promising source of sizeable coal exports to the Republic of Korea and, perhaps to Japan, but the pace of its realization will largely depend on foreign participation (Petromin, Oct. 1992).

The reserves, production and R/P ratio for coal in North-East Asia are showed in table 89. Coal production and consumption in North-East Asia is shown in table 90.

3.1.4 Characteristics of mineral development

Mineral demand

The metals demand per unit of GDP varies based on the stage of development of the nation and for individual commodities. Figure 66 shows that nations experience a high rate of increase in metals demand during the economic transition from a developing nation to a newly industrialized economy and a developed nation. Having reached mature development status, however, the intensity of use of metals is assumed to decrease, although overall demand may continue to increase. This dichotomy occurs because although the rate of growth has slowed, continued population growth requires that more metals be provided for more people.

Table 89. Coal reserves, production and R/P ratio in North-East Asian, Asia and the world at the end of 1995

Coal	*Reserves (Million tonnes)*	*Reserves (share of total, %)*	*Production (Million tonnes oil equivalent)*	*R/P ratio (years)*
China	114,500	11.1	655.5	88
Japan	821	0.1	4.2	130
Democratic People's Republic of Korea	12,000			>100
Republic of Korea	183		3.0	32
Mongolia				
Russian Far East				
Total NE Asia				
Total Asia	220,550	21.4	864.9	>100
Total WORLD	1,031,610	100.0	2,225.1	228

Source: British Petroleum, 1996. Statistical Review of World Energy.

Table 90. Coal production and consumption in North-East Asia, Asia and the world at the end of 1995

Coal	*Production (Million tonnes oil equivalent)*	*Consumption (Million tonnes oil equivalent)*	*Import (Million tonnes)*	*Export (Million tonnes)*
China	655.5	640.3		
Japan	4.2	85.9		
Democratic People's Republic of Korea		31.5*		
Republic of Korea	3.0	27.3		
Mongolia				
Russian Federation	116.9	119.4		
Total Asia	864.9	959.1		
Total WORLD	2,225.1	2,210.7		

Source: British Petroleum, 1996. Statistical Review of World Energy.
* Mining Annual Review, 1990s.

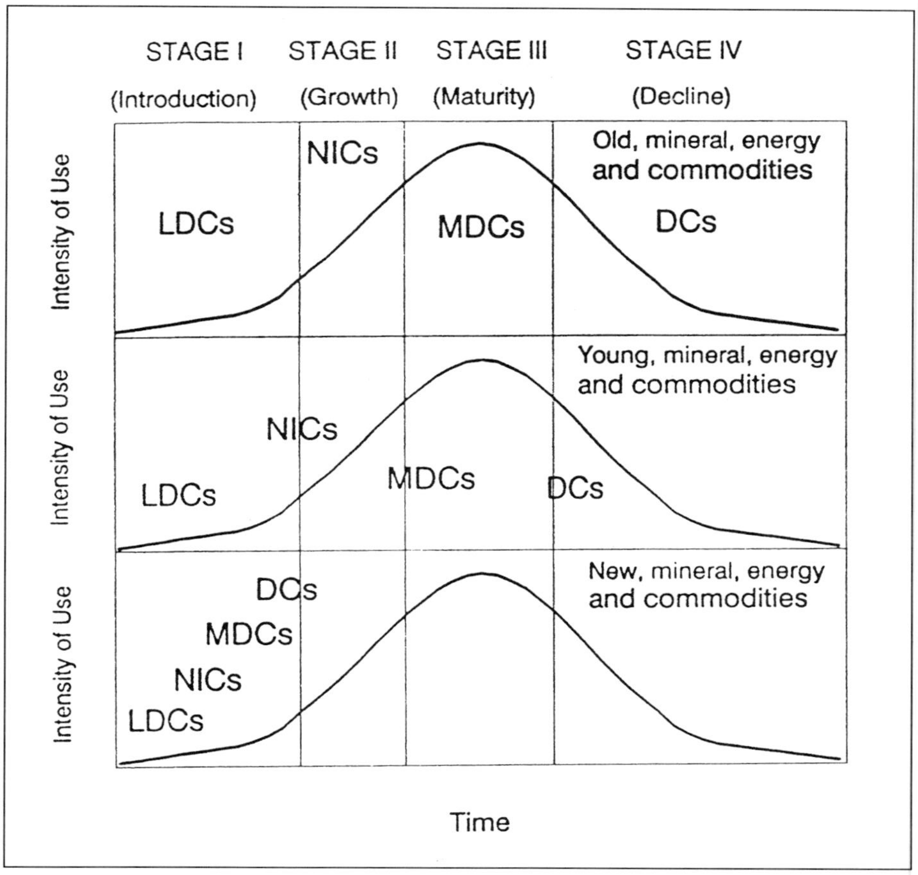

Source: Clark and Jeon, 1990.

Figure 66. Theoretical lifecycle curve of minerals consumption for groups of countries

Additionally, the figure shows the relationship between demand and the relative age of the metal being used. Old metals such as base metals, copper, lead, zinc, tin and iron are those most commonly used for industry expansion, development and infrastructure during initial development. Young metals such as nickel, chromium, manganese, cobalt, molybdenum, and vanadium are those associated with a combination of industrial development, specialized metals uses and economic diversification. New metals such as aluminium, rare earths, platinum, titanium and germanium are those associated with highly specialized uses, broad economic diversification and strong technological development.

Developing and newly industrialized countries in North-East Asia are experiencing and can be expected to continue experiencing high rates of growth in metals demand, particularly for the old and young mineral commodities. Conversely, Japan being more developed, is expected to have a declining demand for the old metals, a level or a declining demand for young metals and an increasing to level demand for new metals.

Demand for metals, industry minerals and energy minerals in the Asia-Pacific region will continue to expand as population growth and economic development take place.

Metallic and non-metallic minerals

The metallic minerals consumption in developed countries has declined in recent decades. The intensity of use of aluminium, for example, (measured in kilogram (kgs) per millions of dollars of gross national product (GNP)), has declined since 1975. Substitution by non-metallic materials, combined with efficiency in the use of the metals and the great increase in value-adding in finished products also contribute the lower intensity of use of metals in industrialized countries. However, there has not been significant decline in the absolute quantities of metals used although the relative importance of metals has declined in industrial economies.

In recent years, exploration for metallic mineral resources has been increasingly dominated by international mining companies with the resources and expertise to find and develop world-class deposits thus following the earlier trend in the petroleum industry. As a result, the development of such mineral resources is very largely a matter for private industry under the overall control of national governments and local governments.

In contrast, non-metallic minerals may contribute quickly and inexpensively to a country's development. These minerals require less capital investment for their extraction and processing than the metallic minerals. Many non-metallic minerals also lend themselves to processing and marketing inside the country, which encourages entrepreneurship, creates local value adding and leads to the improvement of regional transportation systems. In contrast to metals projects where price is generally the critical variable, industrial minerals projects tend to be more sensitive to volume considerations.

Mineral development stages

Mineral development can be divided into four stages, i.e. exploration stage for collection of baseline data, development stage for mine application and environmental impact assessment, production stage for mine production and environmental impact mitigation, and the post-mining stage for mine closing and land reclamation. The exploration activity usually takes 5 to 10 years, following a much shorter development phase of a year or so, and then a much longer production phase with a period of over 20 or 40 years. The post-mining stage may also take years to complete. A typical time frame for world-class mineral developments is shown in figure 67.

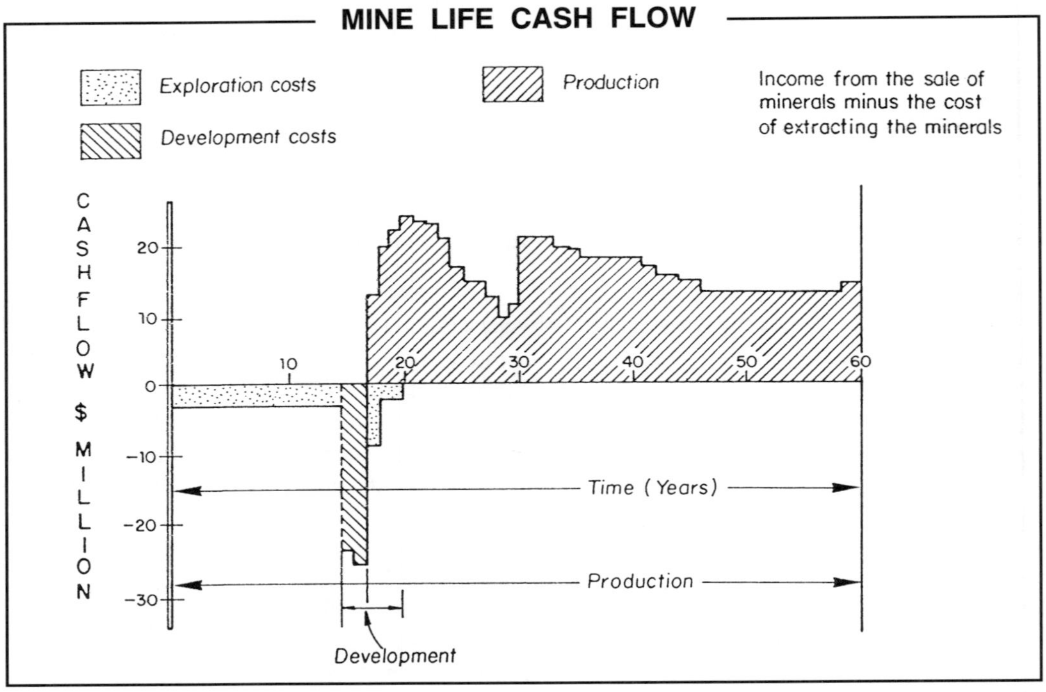

Source: Ritahie, 1992.

Figure 67. Mine life cash flow

Exploration is an extremely high-risk activity. In a study of Canadian exploration, the chances of an economic discovery resulting from any one programme was 1:100 to 1:1,500. Exploration is also expensive since it is usually carried out with sophisticated equipment in remote areas with little existing infrastructure. It is estimated that the industry spends US$35 million to $40 million on exploration for every economic mineral discovery.

Even more capital is needed for the development phase. However, the following production period is the time of high profit. The post-mining stage can require much less cost if the environment impact assessment and the relevant mitigation measures are conducted very well during the mining phase.

In a word, the characteristics unique to the modern mining industry, are, inter alia, that the industry is high risk, capital and technologically intensive, a price taker and usually remotely located; that mines have finite lives and therefore restoration requirements; and that there is generally State ownership of a resource. Therefore, developing countries and economies in transition are highly dependent on attracting and maintaining foreign investment and expertise to help in the development of their mineral resource potential. And the transnational companies are very careful to investigate the investment risks before making an investment decision in the mining sector.

The mining sector towards the 21st century

Clark indicated that there could be little doubt that the mining industry of today will undergo substantial change before and after the beginning of the 21st century (Clark, 1996). Among these ongoing and impending transitions are the following:

(a) The development of a global mining industry. The industry is slowly being transformed from one of few international companies and many large national companies to one of many international companies and smaller national companies.

(b) The privatization of national mining activities. The success of privatization efforts in China and Russia, with two of the largest mining industries, and Brazil, Zaire and Zambia remain among the biggest question marks on the horizon.

(c) Geographic shifts in exploration and development expenditures. In the first decade of the 21st century, emerging international trends in exploration and development indicate that exploration and development expenditures will shift, probably not dramatically, to North Asia (the far east part of the Russian Federation, Kazakhstan, Mongolia), Africa (Ghana, Namibia), Central Europe (Poland), and Indochina (Myanmar).

(d) Structural changes in mining policy and legislation. In the Asia and Pacific region alone, during the last three years (1994-1996) new mining laws have been legislated in Mongolia, Myanmar, Philippines, Russia and Viet Nam. Minerals policy and legislation will continue to evolve in the future, particularly with respect to ancillary issues such as environment and social-cultural issues and fiscal regimes.

(e) Reallocation of decision-making power and authority. The role of indigenous communities and the provincial or local government levels will be largely strengthened in the decision making process.

(f) Emerging dominance of social cultural issues. The ever increasing role of these issues in mineral development is both a complex and a constantly evolving challenge. The socio-cultural impacts have local, regional, national and in some cases international impacts that are largely not anticipated nor understood by government and industry.

(g) Risk mitigation and new risks in mining. To reduce traditional political and economic risks the major mining companies have adopted a policy of risk reduction through contracts and consortia that are so constructed that an adverse action against the mining enterprises will result in unacceptable political and economic repercussions to the nation.

Contingent liability, which asserts that the purchaser of an existing or pre-existing mining activity also purchase all of the associated environmental problems, is being reduced through (i) government indemnification and (ii) insurance. Neither approach is expected to suffice in the future.

(h) Global constraints and NGO's. More treaties and agreements, lead by the NGOs' "action agendas" and based on international standards, will be promulgated, enforced and will impact on the international minerals industry.

(I) Increasing environment-economic efficiency. The international mining industry is making great strides in the development of technologies and procedures which make mining a much more environmentally acceptable activity. To a great extent these developments represent a win-win situation for the environment and for the mining industry.

The nine major trends outlined above represent some of the most important opportunities and challenges within the international minerals industry. The extent to which the industry is able to anticipate and respond to these changes will determine, to a large extent, it's success in the 21st century.

3.1.5 Non-fuel minerals development and trade in North-East Asia

The mineral sector plays an important role in the economic growth and development of developing countries and economies in transition. The extraction and export of minerals provides countries with a valuable stream of government revenues, in both the short and the long term, through royalties and taxes, much needed foreign currency, added employment, in particular for rural workers, infrastructure investment, and the establishment of ancillary industries. The country also benefits from technology transfers through foreign investment activities.

The mining sector of the Democratic People's Republic of Korea in 1993 accounted for 8.20 per cent of GNP. The relevant GDP figures for China's three northeast provinces, was 9.39 per cent, for Japan 0.26 per cent, for the Republic of Korea 0.34 per cent, for Russia's Far East 3.8 per cent, and for Mongolia 26.44 per cent. Accordingly the mining sector has played an important role in the development of economies in the North-East Asian countries (table 91).

Iron and steel

Iron ore

China's production of iron ore has continuously increased from 105 Mt in 1988 to 245 Mt in 1995. In contrast, the Republic of Korea's production of iron ore has continuously decreased from 0.370 Mt in 1988 to 0.184 Mt in 1995. Russia produced 78.3 Mt of iron ore in 1995. The Democratic People's Republic of Korea's production of iron ore in late 1980s is estimated at 8.5 Mt/y. Production of iron ore, pig iron and crude steel in North-East Asia in 1993 is shown in table 92.

China, the Democratic People's Republic of Korea, Japan and the Republic of Korea have been importers of iron ore in North-East Asia. The biggest importer in North-East Asia is Japan, which imported 68.25 million tonnes of iron ore in 1992. The second and third importers in the subregion in 1992 were the Republic of Korea and China, who imported 19.07 and 15.94 million tonnes of iron ore respectively. These two countries only imported 9.96 and 7.62 million tonnes in 1987. They have increased imports of iron ore sharply in recent years. China imported 41 million tonnes of iron ore in 1995. The only country that exports iron ore in the sub-region is the Russian Federation. It exported 14.25 Mt of iron ore in 1995, 18 per cent of its production of 78.3 Mt in the same year.

Table 91. The contribution of mining sector to GDP of the North-East Asian countries in 1993

(in millions of US$)	*GDP*	*Mining*	*per cent*
China, 3 northeast provinces	61,579	5,782	9.39
Japan	4,207,468	10,939	0.26
Democratic People's Republic of Korea	20,500	1,681	8.2
Republic of Korea	330,585	1,124	0.34
Mongolia	1,100	291	26.44
Russian Federation, the far eastern part	14,288	543	3.8

Source: Trade and investment complementarities in North-East Asia (ST/ESCAP/1640).

Table 92. Iron ore production and trade of the North-East Asian countries, 1987-1995

(in '000 t)	*1987*	*1988*	*1989*	*1990*	*1991*	*1992*	*1993*	*1994*	*1995*
Production									
China	50,407	54,173	56,735	59,378	61,354	68,578	82,155	84,000	85,750
Democratic People's Republic of Korea	4,000	4,000	4,500	4,750	4,750	4,750	5,000		
Japan	168	158	136	112	108	22			
Mongolia									
Republic of Korea	286	370	193	199	174	330	156	191	184
USSR (former)	137,995	148,800	144,600						
Exports									
China									
Democratic People's Republic of Korea									
Japan									
Mongolia									
Republic of Korea									
USSR (former)	24,079	25,838	23,928	21,793	16,433	16,207			
Imports									
China	7,622	8,127	7,934	9,036	11,688	15,938	20,908	37,340	
Democratic People's Republic of Korea	310	364	364	364	364	364	364		
Japan	67,220	74,026	76,625	75,174	76,312	68,246	68,691		
Mongolia									
Republic of Korea	9,964	9,887	13,674	13,500	17,064	19,067	21,319		
USSR (former)									

Pig iron

China produced 87.3 Mt of pig iron in 1993 and 101.71 Mt in 1995. Russia's production of pig iron was 36.16 Mt in 1994, and 39.76 Mt in 1995 compared with 61.5 Mt in 1989 (table 93).

According to the data from 1989 to 1993, Japan and the Republic of Korea were net importing countries of pig iron and China a net exporting country. Japan and the Republic of Korea imported pig iron of 1,102 and 590 million US dollars respectively and China exported pig iron of 480 million US dollars in 1993 (table 94).

Crude steel

In 1993, China produced 88.7 Mt of crude steel and replaced the United States to become the second largest crude steel producer in the world after Japan. China produced 92.97 Mt of crude steel in 1995. The demand for steel in China was expected to reach 120 Mt by the year 2000 and the Ministry of Metallurgical Industry (MMI) planned to expand it to 105 Mt by the year 2000.

In early December 1989, the Republic of Korea became the tenth country to reach the 20 Mt/y steel production level. The Republic of Korea's total crude steel output increased about 67.6 per cent in six years from 1989, amounting to 36.7 Mt in 1995.

Table 93. Pig iron and crude steel production and trade of the North-East Asian countries, 1987-1995

(in '000 t)	*1987*	*1988*	*1989*	*1990*	*1991*	*1992*	*1993*	*1994*	*1995*
Pig iron production									
China	55,030	57,040	58,200	62,606	67,614	73,438	86,332		101,710
Democratic People's Republic of Korea	5,900	5,900	5,900	5,900	6,000	6,000	6,000		
Japan	73,419	79,221	80,121	80,144	79,909	73,086	73,738		
Mongolia									
Republic of Korea	11,057	12,578	14,840	15,390	18,510	19,323	21,776		
USSR (former)	113,920	114,558	113,921	110,184	90,953				* 39,760
Crude steel production									
China	56,280	59,431	61,430	66,349	70,436	80,037	89,453		92,970
Democratic People's Republic of Korea	6,730	6,830	6,830	7,000	7,000	7,000	7,000		
Japan	98,513	105,681	107,909	110,331	109,649	98,131	99,623		
Mongolia									
Republic of Korea	16,782	19,118	21,873	23,126	26,001	28,054	33,027	33,732	36,700
USSR (former)	161,935	163,037	160,092						* 51,320

Source: UNCTAD Commodity 1995 Yearbook for the data of 1988-1993; Mining Annual Review, 1995 and 1996 for the data of 1994 and 1995.
* data for Russian Federation.

Russia produced 48.812 Mt and 51.323 Mt crude steel in 1994 and 1995 respectively compared with 93 Mt in 1989. The special steels producer Sibelektrik Stal in Krasnoyarsk, East Siberia, with the capacity to produce 110,000 t/y of crude steel, closed in 1994 owing to a major cutback in orders from the defence industries.

Table 94 shows the crude steel imports and exports in North-East Asia.

China net-exported crude steel of nearly 136 million US dollars in 1991 and Japan and the Republic of Korea net-imported crude steel of about 678 and 754 million US dollars in the same year. In 1993, however, China net-imported crude steel of 3,475 million US dollars and Japan and the Republic of Korea net-exported crude steel of about 443 and 112 million US dollars. Exports of steel products from China were expected to increase from the present level of 2 Mt to 4 Mt in the next few years.

The Republic of Korea exported 6.8 Mt of steel products in 1991, 8.9 Mt in 1992, 11.8 Mt in 1993 and 10.1 Mt in 1994. Sharp increases in 1992 and 1993 were due to strong export demand in the Southeast Asian market. Exports declined in 1994 from the previous year, due to increased domestic demand coupled with stagnation in overseas sales. The Republic of Korea's steel products imports amounted to 2.8 Mt in 1992, 4.2 Mt in 1993 and 8 Mt in 1994.

Non-ferrous metals

China produced 2.35 Mt, 2.52 Mt, 3.27 Mt, 4.25 Mt of ten non-ferrous metals (aluminium, antimony, copper, lead, magnesium, mercury, nickel, tin, titanium and zinc) in 1990, 1991, 1993 and 1995 respectively. The production of non-ferrous metals in China is mainly for the domestic market and this will continue into the next decade. According to the country's economic development plan, China planned to produce 4.5 to 5 Mt of its 10 non-ferrous metals in the year 2000.

The Democratic People's Republic of Korea's production of non-ferrous metals was about 800,000 t in 1988 (against 700,000 t in 1987). During the third Seven Year Plan period (1987-1993), a target of 1.7 Mt/y was envisaged for non-ferrous metals.

Table 94. Pig iron and crude steel trade in the North-East Asian countries, 1989-1993

('000 US Dollars)

Group 671 PIG IRON ETC of SITC REVIEW 2					
	1989	*1990*	*1991*	*1992*	*1993*
Pig iron imports					
China	111,655	202,197	57,743	33,853	16,509
Democratic People's Republic of Korea					
Japan	1,650,817	1,457,423	1,551,642	1,054,560	1,198,614
Mongolia					
Republic of Korea	424,383	366,843	400,891	447,021	592,464
USSR (former)					
Pig iron exports					
China	246,442	342,297	440,027	423,983	496,865
Democratic People's Republic of Korea					
Japan	57,825	49,508	52,053	82,321	96,336
Mongolia					
Republic of Korea	7,506	3,921	1,024	1,722	1,982
USSR (former)					
Group 672, IRON STEEL PRIMARY FORMS, of SITC REVIEW 2					
Iron steel imports					
China	51,044	104,181	69,497	779,436	3,481,362
Democratic People's Republic of Korea					
Japan	1,552,359	1,358,230	1,781,765	1,078,762	1,034,460
Mongolia					
Republic of Korea	1,258,518	1,331,318	1,850,655	1,170,962	1,134,215
USSR (former)					
Iron steel exports					
China	21,287	128,849	205,226	204,472	6,343
Democratic People's Republic of Korea					
Japan	1,474,963	1,087,595	1,103,737	1,013,556	1,477,003
Mongolia					
Republic of Korea	1,103,047	923,612	1,097,126	1,177,902	1,246,374
USSR (former)					

Source: Foreign trade statistics of Asia and the Pacific, 1989-1993. (ST/ESCAP/1589).

Japan's non-ferrous metal industry has encountered particular management difficulty in recent years caused by the high appreciation of the Japanese currency yen. Japan's non-ferrous metal industry includes mine production of gold, silver, copper, lead and zinc, and refined metal production of these metals plus aluminium, tin, cadmium and selenium.

The Republic of Korea's total non-ferrous metal production in 1994 and 1995 consisted of 224,000 t and 233,000 t of copper, 76,000 t and 175,000 t of lead and 272,000 t and 285,000 t of zinc. In 1994, the Republic of Korea imported 503,437 t of aluminium ingot, 240,445 t of copper cathode, 42,801 t of refined zinc and 42,444 t of refined lead.

In Russia's Far East region, non-ferrous metals extraction is one of the three leading industry sectors, with the other two being timber exploitation and processing, and fishing. More than 90 per cent of Russian diamonds are extracted in the Far East or Yakutia, making Russia the world's fifth largest diamond producer and a major world exporter. The region processes 95 per cent of Russia's tin reserves and produces nearly the same percentage of Russia's tin production. The Far East is also a significant supplier of zinc with 4 per cent of Russia's total reserves and 14 per cent of production. The region also has 41 per cent of the country's fluorspar reserves and 91 per cent of Russia's total fluorspar production.

Antimony

China's output of antimony, about 60,000 tonnes, accounted for about 75 per cent of the world's total in 1994.

Russia is in the process of developing its own antimony metal producing facilities. All antimony metal in the former USSR was produced at the Kadamzhay plant in Kyrgyzstan with much of the raw material coming from Russia's Sakha/Yakutia Republic. Production of antimony metal was scheduled to begin in 1996 at the Yuzhural nickel plant in the southern Urals, with a production level of 3,000 t/y.

Bauxite/alumina/aluminium

China produced 1.105 million tonnes of bauxite in 1993, ranked sixth in the world after Australia (9.586 Mt), Ghana (4.242 Mt), Guinea (3.919 Mt), Brazil (3.436 Mt) and the former USSR (1.452 Mt) and closely followed by India (0.999 Mt). China is the only country in the subregion that exported bauxite with a small amount (0.106 Mt) in 1993. Japan plus the former USSR imported 1.277 Mt of bauxite in 1993 (table 95).

China has become a net importer from a net exporter of alumina in 1991 with a net import of 0.436 Mt in 1993 although being the world's seventh largest alumina producer with production of 2.1 Mt in 1995, compared to 1.5 Mt in 1988. All the other countries in North-East Asia except Japan import alumina. Japan is the only net exporter of alumina in the subregion, with net export of alumina of 0.037 Mt in 1993.

China, Japan and the Republic of Korea are net importers, and the former USSR is exporter of aluminium metal in North-East Asia, although China is the sixth largest aluminium metal producer in the world. Japan and the Republic of Korea have depended totally on imports of over 99 per cent and 95 per cent for their aluminium metal demand. Japan imported aluminium metal of 2.604 Mt and the former USSR exported of 1.8 Mt in 1993.

Bauxite, alumina and aluminium in the North-East Asian countries in the year of 2000 is presented in table 96.

Copper

Except the Republic of Korea, all the countries including Japan in North-East Asia produced copper ore in 1993. Former USSR, China, Mongolia, the Democratic People's Republic of Korea and Japan produced 0.865 Mt, 0.325 Mt, 0.096 Mt, 0.012 Mt and 0.010 Mt of copper ore respectively in 1993.

Mongolia is the only country in North-East Asia that exports copper ore. Mongolia exports 95 per cent of its copper ore production. It produced 0.0957 Mt and exported 0.1 Mt of copper ore in 1993. All the other countries in North-East Asia are importers of copper ore. They imported 1.295 Mt of copper ore Cu content in total in 1993, in which Japan was the largest importer with 1.058 Mt followed by the Republic of Korea 0.145 Mt (table 97).

In fact, Mongolia is a major producer and exporter of copper and molybdenum, and is a leading world producer of fluorspar. Some 20 per cent of Mongolia's GNP or 40 per cent of the country's total exports came from the mining industry in 1991. Copper, molybdenum and fluorspar account for over 35 per cent of Mongolia's exports. The Erdenet dressing plant produced 300,200 t of copper concentrate and 3,238 t of molybdenum concentrate in 1992, 117,000 t and 4,000 t in 1993, and 340,000 t and 3,700 t in 1994. It processed its copper concentrate in plants in Kazakhstan, producing 40,000 t of cathode copper, 6-9 t of refined silver and 500 t of molybdenum oxide in 1994.

Table 95. Bauxite/alumina/aluminium production and trade in the North-East Asian countries, 1987-1993

Bauxite – Al content (000 Mt)	*1987*	*1988*	*1989*	*1990*	*1991*	*1992*	*1993*
Bauxite production							
China	550.0	595.2	629.0	621.3	1,007.0	1,200.0	1,105.0
Democratic People's Republic of Korea							
Japan							
Mongolia							
Republic of Korea							
USSR (former)	1,140.0	1,300.0	1,300.0	1,500.0	1,260.0	1,220.0	1,452.0
Bauxite exports							
China	128.5	91.0	87.5	103.6	82.5	216.1	130.1
Democratic People's Republic of Korea							
Japan		0.5	0.3	0.2	0.3	0.5	0.8
Mongolia							
Republic of Korea							
USSR (former)							
Bauxite imports							
China						12.4	24.0
Democratic People's Republic of Korea							
Japan	402.5	461.9	487.9	495.0	440.5	400.6	463.0
Mongolia							
Republic of Korea	7.0	5.7	7.2	7.9	8.6	9.9	14.5
USSR (former)	1,150.0	1,196.0	1,430.0	962.0	728.0	696.0	775.0
Alumina production							
China	650.0	650.0	700.0	600.0	600.0	800.0	950.0
Democratic People's Republic of Korea							
Japan	355.5	389.0	431.5	445.0	432.0	357.0	352.0
Mongolia							
Republic of Korea							
USSR (former)	2,900.0	3,059.0	2,990.0	2,819.5	2,638.0	2,620.0	2,578.0
Alumina exports							
China	138.7	434.6	461.3	278.1	323.2	74.4	50.1
Democratic People's Republic of Korea							
Japan	98.9	99.1	124.8	139.3	126.0	98.3	101.0
Mongolia							
Republic of Korea							
USSR (former)							
Alumina imports							
China	161.0	79.8	149.1	291.5	343.1	400.4	486.2
Democratic People's Republic of Korea	10.0	10.0	10.0	10.0	10.0	10.0	10.0
Japan	21.0	36.0	38.5	36.7	41.8	41.7	64.0
Mongolia							
Republic of Korea	53.5	43.3	69.1	57.5	66.8	78.5	79.0

Table 95. *(continued)*

Bauxite – Al content (000 Mt)	*1987*	*1988*	*1989*	*1990*	*1991*	*1992*	*1993*
USSR (former)	500.0	550.0	600.0	650.0	252.5	250.0	400.0
Aluminium Production							
China	640.0	713.0	758.0	854.0	963.0	1,080.0	1,210.0
Democratic People's Republic of Korea	10.0	10.0	5.0	11.0	9.5	10.0	10.0
Japan	40.6	35.3	35.0	34.2	32.4	18.9	18.3
Mongolia							
Republic of Korea	16.8	16.1	17.4	2.0			
USSR (former)	3,400.0	3,500.0	3,450.0	3,523.0	3,251.0	3,220.0	3,065.0
Aluminium exports							
China	60.7	130.4	13.4	65.1	68.6	55.8	68.0
Democratic People's Republic of Korea	7.0	5.8	5.6	3.8	2.9	3.0	3.0
Japan	1.9	3.4	2.9	4.8	4.6	5.2	8.6
Mongolia							
Republic of Korea	2.9	6.2	3.8	6.3	1.8	15.9	20.3
USSR (former)	612.0	550.0	640.0	650.0	935.0	950.0	1,800.0
Aluminium imports							
China	148.1	75.4	175.5	71.8	43.7	229.6	165.6
Democratic People's Republic of Korea	10.0	5.0	7.0	6.0	1.0	1.0	1.0
Japan	1,835.3	2,292.0	2,363.2	2,653.1	2,831.0	2,541.0	2,604.4
Mongolia							
Republic of Korea	200.2	217.7	281.9	361.9	391.8	425.5	579.5
USSR (former)	2.0	1.0	1.0	1.0	1.0	1.0	1.0
Aluminium consumption							
China	727.0	658.0	920.0	861.0	938.0	1,253.8	1,318.0
Democratic People's Republic of Korea	55.0	60.0	60.0	50.0	40.0	30.0	19.0
Japan	1,696.8	2,123.2	2,211.6	2,415.2	2,431.6	2,297.7	2,174.8
Mongolia							
Republic of Korea	207.9	268.0	287.6	345.4	383.3	397.0	557.1
USSR (former)	2,830.0	2,900.0	2,700.0	2,790.0	2,409.0	1,770.0	1,185.0

Source: UNCTAD Commodity Yearbook, 1994 and 1995.

Table 96. Bauxite mine, alumina refinery and aluminium smelter capacity of the North-East Asian countries, 1995-2000

(in '000 mt)	*1995*	*1996*	*1997*	*2000*
Type–metallurgical grade (MG), Bauxite mine capacity 1995-2000				
China	5,000	5,500	5,500	6,500
Democratic People's Republic of Korea				
Japan				
Mongolia				
Republic of Korea				
Russian Federation	4,700	4,700	4,700	4,700
Alumina refinery capacity 1995-2000				
China	2,640	2,860	3,420	3,420
Democratic People's Republic of Korea				
Japan	215	215	215	215
Mongolia				
Republic of Korea				
Russian Federation	1 900	1 900	1 900	1 900
Primary aluminium smelter capacity 1995-2000				
China	1,866	1,866	1,880	2,120
Democratic People's Republic of Korea	40	40	40	40
Japan	64	64	64	64
Mongolia				
Republic of Korea				
Russian Federation	2,979	2,979	2,979	3,287

Source: United Nations Conferrence on Trade and Development – Recent and planned changes in production capacity for bauxite, alumina and aluminium by Olle Ostensson, Resources Development Section, Commodities Division, 1996.

Table 97. Copper production/consumption and trade of the North-East Asian countries, 1987-1993

(in '00 Metric tonnes)	*1987*	*1988*	*1989*	*1990*	*1991*	*1992*	*1993*
Copper ore* production							
China	278.2	281.9	299.1	295.9	304.0	334.3	324.7
Democratic People's Republic of Korea	12.0	12.0	12.0	12.0	14.0	12.0	12.0
Japan	23.8	16.7	14.7	13.0	12.4	12.1	10.3
Mongolia	120.9	121.6	123.4	123.9	90.1	105.1	95.7
Republic of Korea	1.7	0.1		0.1			
USSR (former)	1,010.0	990.0	1,135.0	1,135.0	1,030.0	935.0	865.0
Copper ore* exports							
China							
Democratic People's Republic of Korea							
Japan							
Mongolia	117.2	128.0	131.5	136.5	120.0	140.0	100.0
Republic of Korea							
USSR (former)							
Copper ore* imports							
China	59.9	43.3	44.5	68.6	77.5	85.3	60.6
Democratic People's Republic of Korea	2.5	3.0	3.5	3.9	4.0	4.2	1.2
Japan	804.9	927.7	892.6	933.5	1,028.1	983.4	1,058.3
Mongolia							
Republic of Korea	147.6	111.8	128.9	100.6	109.0	62.0	144.7
USSR (former)	117.2	128.0	131.5	136.6	120.0	140.0	30.0
Unrefined-copper production							
China	348.7	335.2	346.3	358.5	385.0	418.0	443.8
Democratic People's Republic of Korea	30.0	29.0	27.0	22.0	22.0	17.0	16.9
Japan	871.0	858.6	938.2	973.0	1,018.0	1,103.0	1,105.5
Mongolia							
Republic of Korea	143.8	123.5	159.6	160.5	148.6	174.0	142.0
USSR (former)	1,140.0	1,110.0	1,135.0	1,135.0	1027.0	920.0	900.0
Unrefined-copper exports							
China							
Democratic People's Republic of Korea		0.5	0.4	1.5	1.9	0.8	0.4
Japan	12.7	48.6	32.5	29.0	22.1	7.3	9.2
Mongolia							
Republic of Korea							
USSR (former)							
Unrefined-copper imports							
China	18.2	5.9	1.2	1.9	6.1	113.2	102.6
Democratic People's Republic of Korea							
Japan	27.9	22.9	21.9	27.4	33.6	30.1	39.8
Mongolia							
Republic of Korea	12.5	21.8	20.4	37.9	51.5	45.4	50.8
USSR (former)							

Table 97. *(continued)*

(in '00 Metric tonnes)	*1987*	*1988*	*1989*	*1990*	*1991*	*1992*	*1993*
Refined-copper production							
China	496.4	517.0	555.1	561.6	560.0	659.2	690.7
Democratic People's Republic of Korea	40.0	40.0	40.0	30.0	30.0	28.0	27.9
Japan	980.3	955.1	989.6	1,008.0	1,076.3	1,160.9	1,188.8
Mongolia							
Republic of Korea	157.9	170.5	178.7	192.2	200.8	209.8	219.5
USSR (former)	1,310.0	1,370.0	1,330.0	1,160.0	1,040.0	875.0	800.0
Refined-copper exports							
China	22.3	44.3	11.3	16.5	6.8	11.4	5.9
Democratic People's Republic of Korea	1.0	1.0	1.2	1.5	1.0	1.0	1.0
Japan	60.4	29.9	36.1	52.7	65.0	104.6	162.2
Mongolia							
Republic of Korea	2.7	20.8	35.1	1.1	0.1	13.7	6.5
USSR (former)	120.0	140.0	180.0	200.0	200.0	220.0	363.0
Refined-copper imports							
China	76.0	78.5	68.9	38.4	107.8	265.7	259.7
Democratic People's Republic of Korea							
Japan	352.0	425.2	492.9	625.2	635.6	374.6	376.4
Mongolia							
Republic of Korea	107.0	125.9	94.8	137.9	148.0	161.2	183.2
USSR (former)		1.0					
Refined-copper consumption							
China	540.0	550.0	528.0	512.0	449.0	369.3	243.9
Democratic People's Republic of Korea	22.0	25.0	25.0	30.0	590.0	882.0	942.3
Japan	1,284.2	1,330.7	1,446.4	1,577.5	1,613.2	1,411.1	1,384.1
Mongolia							
Republic of Korea	259.0	266.3	244.8	324.2	343.3	353.5	399.8
USSR (former)	1,200.0	1,240.0	1,140.0	1082.0	859.0	770.0	572.0

Source: UNCTAD Commodity Yearbook, 1994 and 1995.
* Copper ore - Cu content.

As to refined copper, Japan is the largest producer and consumer. Japan produced 1.189 Mt refined copper and consumed 1.384 Mt in 1993. The balance was met by net import: Japan imported 0.376 Mt of refined copper and exported 0.162 Mt. The Republic of Korea produced and imported refined copper of 0.220 Mt and 0.183 Mt and consumed 0.400 Mt in 1993. The figure for the consumption of refined copper of 0.942 Mt for the Democratic People's Republic of Korea in 1993, recorded in an UNCTAD statistics table, is most likely an error. China produced refined copper of 0.691 Mt, consumed 0.244 Mt, exported 0.06 Mt and imported 0.260 Mt in 1993. If the figures were correct, it means that China stockpiled refined copper of about 0.65 Mt in 1993. According to the country's economic development plan, China will need about 1.2 Mt of copper in the year 2000. The Russian Federation exported 0.164 Mt, 0.470 Mt and 0.464 Mt of refined copper in 1993, 1994 and 1995 respectively. There were no figures on production, exports, imports and consumption of refined copper for Mongolia reported in 1993 although Mongolia was the seventh largest exporter of copper ore in the world after Chile, Canada, the United States, Indonesia, Portugal and Papua New Guinea.

According to preliminary reports of the Russian Committee for Metallurgy, Russia produced 505,000 t of refined copper in 1994.

Gold

Total world gold production from 1986 to 1995 is shown in table 98. Asia accounted for 8.2 per cent of the world total production of gold in 1995, 2 per cent higher than that in 1986. Among the top 20 gold producing countries in 1995, Russia ranked the 5th, China 6th, and the Democratic People's Republic of Korea 20th (table 99).

According to the Russian Committee for Precious Metals and Stones (Roskomdragmet), Russia's gold production was reported to be 132.17 t in 1995, 146.6 t in 1994 and 149.5 t in 1993. Of these, 112.37 t in 1995 and 135.5 t in 1994 were derived as primary gold while the remaining production presumably was gold produced as a by-product of polymetallic ores or from scrap metal. Russia's largest gold producing region, Magadan Oblast including the Chukotka autonomous district, produced 24 t of gold in 1995, 27.4 t in 1994, 43.5 t in 1992, compared to 120 to 130 t in the late 1980s. Production in Yakutia, Russia's second largest gold producing region, reportedly decreased to 29.7 t in 1994 compared with 33.4 t in 1993. About 250 Russia's prospecting groups reportedly produced over 50 per cent of all placer gold in 1992.

According to the Gold Administration Bureau of MMI, China produced 90.2 t and 105 t of gold in 1994 and 1995, which were well below estimates by western gold observers. According to the World Gold Council, China was the fourth largest gold consumer in the world after India, the United States and Japan.

Despite the poor market situation, the gold mine production in Japan has increased since 1989, reaching to 9.55 and 9.19 t in 1994 and 1995.

The Republic of Korea's total demand for gold in 1992 was estimated at 55 t, of which 32 t were imported. The Republic of Korea's gold metal production in 1992 was reported to be 23,263 kg from imported gold concentrates and 1,399 kg from domestic ore.

Mongolia is rich in gold resources with major concentrations in the eastern and central parts of the country including the Boroo and Zaamar deposits.

Lead and zinc

The major countries of lead ore production in North-East Asia in 1993 were China with 0.357 Mt, the former USSR with 0.203 Mt and the Democratic People's Republic of Korea with 0.070 Mt of Pb content. The largest importer of lead ore in 1993 was Japan with 0.200 Mt, followed by the Republic of Korea with 0.037 Mt. Two exporters of the lead ore in the subregion in 1993 were China (who exported only 0.012 Mt) and the Democratic People's Republic of Korea (0.001 Mt) (table 100).

The major producers of primary lead metal in North-East Asia are China, Japan and Russia. In 1993, China's and Japan's lead production were 0.344 Mt and 0.212 Mt respectively, ranking first followed by the US, and fourth after the former USSR. They are also the leading consumers of lead metal in the world. The net exporters of lead metal in North-East Asia in 1993 were China with 0.089 Mt and former USSR with 0.026 Mt; the net importers were the Republic of Korea with 0.086 Mt and Japan with 0.057 Mt.

All the countries in North-East Asia except Mongolia are zinc ore producers: China ranked first in 1993 with production of zinc ore of 0.775 Mt; followed by the former USSR with 0.452 Mt, Japan at 0.119 Mt, the Democratic People's Republic of Korea at 0.110 Mt, and the Republic of Korea at 0.014 Mt. Only China is an exporter of zinc ore, exporting 0.059 Mt in 1993. The largest importers are Japan and the Republic of Korea, who imported 0.591 Mt and 0.231 Mt respectively in 1993 (table 101).

In 1993, China took the lead among zinc metal producers in North-East Asia with 0.857 Mt output. The Democratic People's Republic of Korea ranked fifth in the subregion in 1993 with production of zinc metal of 0.115 Mt. The Democratic People's Republic of Korea has been an exporter of zinc metal, exporting 0.095 Mt in 1993. China was a net exporter of the zinc metal with net exports of 0.167 Mt in 1993. The former USSR became an exporter in 1992 and 1993. Japan and the Republic of Korea had been importers before 1993.

Table 98. World gold mine production

Unit: Metric tonnes

	1986	*1987*	*1988*	*1989*	*1990*	*1991*	*1992*	*1993*	*1994*	*1995*
Asia										
Indonesia	8.4	12.2	12.3	10.8	17.6	24.4	45.9	52.2	55.3	74.1
Papua New Guinea	36.1	33.9	36.6	22.8	33.6	60.8	71.2	61.5	60.5	54.8
Philippines	38.7	39.5	39.2	38.0	37.2	30.5	27.2	29.8	31.0	28.4
Japan	10.3	8.6	7.3	6.1	7.3	8.3	8.9	9.4	9.6	9.2
Saudi Arabia	–	–	1.7	2.9	3.5	4.8	6.0	6.8	7.6	8.0
Malaysia	2.6	2.6	3.2	3.0	2.9	2.8	3.5	4.5	4.1	3.2
India	2.1	1.6	1.8	1.7	2.4	2.1	1.8	2.0	2.2	2.5
Viet Nam	–	–	–	1.0	1.0	1.0	1.0	1.5	1.5	1.7
Republic of Korea	3.5	3.8	4.4	4.9	2.2	2.1	2.0	0.2	1.0	0.8
Thailand	–	–	–	–	–	–	0.1	2.1	0.3	0.4
Others	0.6	0.2	0.2	0.2	0.7	1.8	1.8	171.5	2.5	2.9
Total Asia	102.3	102.5	106.7	102.4	108.4	138.6	169.4	730.8	175.8	186.9
Total Oceania	79.2	114.7	163.6	213.0	254.3	246.5	257.8	262.3	269.0	268.4
Total Africa	693.1	670.9	688.6	675.2	675.2	690.9	715.4	730.8	701.8	652.8
Total Europe	16.7	18.5	21.7	30.2	35.1	32.0	25.2	25.2	26.7	27.7
Total North America	224.0	271.4	335.8	425.2	425.2	471.3	489.9	485.2	472.4	479.6
Total Latin America	181.0	205.6	234.6	236.9	236.9	210.1	214.9	228.7	251.6	274.4
Western World Total	1,296.3	1,383.66	1,551.0	1,682.9	1,755.3	1,789.5	1,872.2	1,903.7	1,897.3	1,889.8
Other Countries										
Russia	–	–	–	–	–	–	151.7	164.5	158.1	142.1
China	65.0	72.0	78.0	86.0	95.7	104.2	113.1	121.0	124.1	136.4
Uzbekistan	–	–	–	–	–	–	64.5	66.6	64.4	63.6
North Korea	–	–	–	9.5	13.0	13.0	17.0	15.0	14.0	14.0
Mongolia	0.8	1.0	0.9	1.1	1.0	0.8	1.0	1.4	2.1	4.9
Other CIS	–	–	–	–	–	–	13.5	17.6	20.0	21.2
Soviet Union (former)	275.0	277.0	280.0	285.0	270.0	252.0	–	–	–	–
Total Other Countries	340.8	350.0	358.9	381.6	379.7	370.0	360.8	386.1	382.7	382.2
World Total	1,637.1	1,733.6	1,909.0	2,064.5	2,135.0	2,159.5	2,233.0	2,289.8	2,280.0	2,272.0

Source: Asian Journal of Mining, May/June 1996.

Table 99. World's top 20 gold producing countries in 1995

Ranking		*tonnes (1994 figures in brackets)*		
1	(1)	South Africa	522.6	(583.9)
2	(2)	United States	329.3	(326.0)
3	(3)	Australia	253.5	(254.9)
4	(5)	Canada	150.3	(146.4)
5	(4)	Russian Federation	142.1	(158.1)
6	(6)	China	136.4	(124.1)
7	(10)	Indonesia	74.1	(55.3)
8	(7)	Brazil	67.4	(73.4)
9	(8)	Uzbekistan	63.6	(64.4)
10	(9)	Papua New Guinea	54.8	(60.5)
11	(11)	Ghana	52.2	(44.5)
12	(13)	Peru	51.5	(39.3)
13	(12)	Chile	48.5	(43.3)
14	(14)	Philippines	28.4	(31.0)
15	(16)	Zimbabwe	26.1	(22.5)
16	(15)	Colombia	24.1	(25.5)
17	(19)	Mexico	20.3	(13.9)
18	(20)	Venezuela	17.1	(13.7)
19	(17)	Bolivia	16.0	(14.7)
20	(18)	Democratic People's Republic of Korea	14.0	(14.0)

Source: Asian Journal of Mining, May/June, 1996.

Table 100. Lead ore/metal production/consumption and trade in the North-East Asian countries, 1987-1993

Unit: Thousand metric tonnes

	1987	*1988*	*1989*	*1990*	*1991*	*1992*	*1993*
Lead ore production (Pb content)							
China	267.2	312.6	341.4	315.3	319.7	286.9	356.9
Democratic People's Republic of Korea	90.0	90.0	80.0	70.0	80.0	70.0	70.0
Japan	27.9	22.9	18.6	18.7	18.3	18.8	17.0
Mongolia							
Republic of Korea	13.9	14.5	16.5	14.9	12.6	16.0	7.0
USSR (former)	260.0	270.0	260.0	240.0	230.0	225.0	203.0
Lead ore exports (Pb content)							
China	80.0	58.8	19.2	0.1	2.5	7.4	12.0
Democratic People's Republic of Korea		1.0	1.0	1.0	1.0	1.0	1.0
Japan							
Mongolia							
Republic of Korea	5.0		4.6	12.3	0.8		
USSR (former)							
Lead ore imports (Pb content)							
China					0.1	0.4	
Democratic People's Republic of Korea	4.0			3.0			
Japan	180.0	162.0	178.0	178.0	194.0	186.0	200.0
Mongolia							
Republic of Korea	3.0	1.0			5.2	33.4	36.9
USSR (former)	40.0	6.8	17.9	7.4	8.0	8.0	8.0
Lead metal production (primary)							
China	210.3	206.0	269.0	265.7	276.2	308.4	343.9
Democratic People's Republic of Korea	90.0	85.0	80.0	70.0	80.0	70.0	55.0
Japan	218.8	217.8	207.7	204.9	220.1	218.8	212.2
Mongolia							
Republic of Korea	52.5	46.0	48.7	34.0	40.0	60.0	36.0
USSR (former)	310.0	310.0	310.0	270.0	250.0	230.0	343.9
Lead metal exports (Pb content)							
China	23.5	11.9	1.8	38.0	15.9	89.9	90.8
Democratic People's Republic of Korea	14.1	11.5	6.0	4.8	14.0	14.0	5.0
Japan	17.1	1.4	1.4	1.4	0.9	4.3	8.8
Mongolia							
Republic of Korea	8.7	3.5	5.5	4.6	5.9	8.6	20.6
USSR (former)	45.0	33.0	52.5	69.6	60.0	90.0	71.0
Lead metal imports (Pb content)							
China	4.6	4.6	36.4	1.7	0.2	1.7	2.3
Democratic People's Republic of Korea	8.9	4.4	2.1	1.6	1.6	1.6	1.5
Japan	49.4	75.8	83.1	99.0	112.5	79.7	65.6
Mongolia							
Republic of Korea	76.5	87.2	97.0	107.4	138.1	143.3	106.3

Table 100. *(continued)*

	1987	*1988*	*1989*	*1990*	*1991*	*1992*	*1993*
USSR (former)	15.0	25.0	18.0	7.0	8.0	70.0	45.0
Lead metal consumption							
China	256.0	250.0	250.0	250.0	250.0	240.0	290.0
Democratic People's Republic of Korea	28.0	30.0	35.0	38.0	40.0	40.0	40.0
Japan	378.0	406.5	405.7	416.4	422.2	401.9	371.3
Mongolia							
Republic of Korea	112.4	146.0	176.0	147.4	164.3	163.9	177.0
USSR (former)	435.0	410.0	400.0	380.0	310.0	260.0	200.0

Source: UNCTAD Commodity Yearbook, 1994 and 1995.

Table 101. Zinc ore/metal production/consumption and trade of the North-East Asian countries, 1987-1993

Unit: Thousand metric tonnes

	1987	*1988*	*1989*	*1990*	*1991*	*1992*	*1993*
Zinc ore production (Zn content)							
China	458.2	527.3	620.4	618.9	710.0	705.9	775.0
Democratic People's Republic of Korea	130.0	150.0	110.0	100.0	120.0	120.0	110.0
Japan	165.7	147.2	131.8	127.3	133.0	134.5	119.0
Mongolia							
Republic of Korea	24.2	22.7	23.2	22.8	22.6	21.9	14.0
USSR (former)	640.0	670.0	650.0	610.0	580.0	506.0	452.0
Zinc ore exports (Zn content)							
China	18.6	54.1	57.5	38.1	38.5	32.7	58.8
Democratic People's Republic of Korea	2.5	1.0	1.0				
Japan							
Mongolia							
Republic of Korea				0.7			
USSR (former)							
Zinc ore imports (Zn content)							
China					0.1	3.6	4.0
Democratic People's Republic of Korea	9.8	8.3	18.2	3.8	15.0		
Japan	487.0	532.0	523.0	586.0	578.0	617.0	591.0
Mongolia							
Republic of Korea	205.0	228.0	225.0	243.0	234.0	236.0	231.0
USSR (former)	70.0	38.0	52.0	52.0	10.0	10.0	10.0
Metal production (primary)							
China	383.1	425.4	450.9	526.3	576.7	648.3	857.0
Democratic People's Republic of Korea	130.0	160.0	120.0	110.0	120.0	120.0	115.0
Japan	665.6	678.2	664.5	687.5	730.8	729.5	696.0
Mongolia							
Republic of Korea	186.0	226.0	241.7	257.0	232.0	255.9	258.0
USSR (former)	705.0	710.0	684.0	644.0	575.0	540.0	502.0

Table 101. *(continued)*

	1987	*1988*	*1989*	*1990*	*1991*	*1992*	*1993*
Zinc metal exports							
China	95.3	13.8	20.1	16.7	6.3	84.9	205.6
Democratic People's Republic of Korea	68.9	99.9	74.5	54.4	90.0	80.0	95.0
Japan	53.2	26.7	21.9	25.7	22.4	34.4	36.1
Mongolia							
Republic of Korea	38.0	70.7	73.6	56.3	27.6	58.4	47.7
USSR (former)	28.0	28.0	18.0	19.0	5.0	80.0	149.0
Zinc metal imports							
China	68.2	62.0	19.2	4.1	11.7	42.2	40.1
Democratic People's Republic of Korea							
Japan	116.9	130.3	150.7	154.8	154.5	115.4	93.1
Mongolia							
Republic of Korea	36.4	20.5	20.3	25.3	22.4	25.1	46.5
USSR (former)	35.0	80.6	48.9	30.1	7.0	5.0	3.0
Zinc metal consumption							
China	409.0	385.0	390.5	500.0	530.0	551.0	530.0
Democratic People's Republic of Korea	39.0	40.0	40.0	40.0	40.0	36.0	35.0
Japan	728.7	774.2	768.7	814.3	845.5	783.7	718.7
Mongolia							
Republic of Korea	178.7	173.0	188.0	230.0	269.0	257.0	292.0
USSR (former)	750.0	750.0	710.0	640.0	520.0	400.0	330.0

Source: UNCTAD Commodity Yearbook, 1994 and 1995.

Rare earths

China's production was equal or exceeded U.S. production of rare-earths in 1988. China has ranked second in the world, after the U.S. in rare earths consumption.

China is a major rare earths exporting country and dominates in the world market. More than one-half of China's rare-earth ore production is exported. China exports high-purity rare-earths andytrium compounds to Japan.

Tin

Asia today is still the leading region in the world for its tin mine production (over 50 per cent of the total) although some major shifts have taken place in the relative importance of the countries that dominate 90 per cent of world tin mine production. In the Asia-Pacific region, China produced 51,000 tonnes of tin ore in 1993, ranked number one in the world, followed by Indonesia at 28,600 tonnes ranked number two, Malaysia at 10,400 tonnes ranked sixth, the former USSR at 10,000 tonnes ranked seventh, Australia with 8,100 tonnes production ranked eighth, and Thailand at 7,500 tonnes ranked tenth (table 102). Trade in tin ore was insignificant in recent years.

By December 1991, Asia's primary tin smelter capacity was over 60 per cent of the world total, with which Malaysia, China, Thailand and Indonesia ranked first, third, fourth and fifth in the world (table 103). China is the largest primary tin metal producer in North-East Asia and in the world with a production of 52,000 tonnes in 1993, followed by the former USSR at 11,200 tonnes.

Table 102. Tin mine production in Asia-Pacific countries, 1970-1993

Unit: Thousand tonnes

	1970	*1980*	*1990*	*1991*	*1992*	*1993*
Australia	8.8	11.6	7.4	5.4	6.1	8.1
China	22.0	16.0	44.0	39.0	45.0	51.0
Indonesia	19.1	32.5	31.7	30.1	28.3	28.6
Malaysia	73.8	61.4	28.5	20.7	14.3	10.4
Thailand	21.8	33.7	14.4	10.1	11.5	7.5
USSR (former)	10.0	16.0	15.0	12.0	10.0	10.0

Source: UNCTAD, 1995: Market situation and outlook for tin, 1994 (UNCTAD/COM/54).

Table 103. Major tin mine and primary smelter capacity in Asia and the Pacific, December 1991

Unit: Metric tonnes/year

	Mine capacity	*Smelter capacity*
Australia	10,000	2,000
China	45,000	45,000
Japan		2,000
Indonesia	35,000	32,000
Republic of Korea	100	2,000
Lao People's Democratic Republic	1,000	
Malaysia	40,000	120,000
Myanmar	2,000	1,000
Thailand	30,000	44,000
USSR (former)	20,000	20,000
Viet Nam	1,000	1,000

Source: UNCTAD, 1995: Market situation and outlook for tin, 1994 (UNCTAD/COM/54).

Asia is also the leading trade region in the world for its exports of tin metal, which accounted for 70 per cent of the world total exports in 1993 (table 104). Asia's imports of tin metal accounted for about 21 per cent of the world total in 1993, in which Japan's import for 15.5 per cent ranked number two, and Republic of Korea's 5.1 per cent ranked number six. Tin production, consumption and trade in the North-East Asian countries in 1993 are shown in table 105.

Tungsten

China produced over 60,000 tonnes of tungsten ore concentrates in 1990, ranking first in the world. China continues to dominate the world market for tungsten. Its tungsten exports are mainly concentrates. China's tungsten production and exports in 1993 were 23,000 tonnes and 20,000 tonnes respectively. The Russian Federation was another country which produced large amounts of tungsten ore and concentrates with production of 6,500 tonnes in 1991, but it has to import tungsten ore to meet its domestic demand. Its production is recovering.

All the other countries in the North-East Asian region including Japan and the Republic of Korea are producers of tungsten ore. The Democratic People's Republic of Korea has produced between 500 and 1,000 tonnes of tungsten ore in recent years. The Republic of Korea, Japan and Mongolia are also producers of

Table 104. Major tin trade partners, 1981 and 1993 – share of world tin metal trade (%)

	1981	*1993*		*1981*	*1993*
Tin exporters	%	%	*Tin importers*	%	%
Malaysia	35.5	21.1	USA	26.9	19.0
China	3.0	20.9	Japan	17.5	15.5
Singapore	1.4	15.8	Germany	9.9	10.2
Brazil	2.6	12.7	Hong Kong, China	1.0	9.3
Bolivia	9.6	10.7	UK	4.7	7.2
Indonesia	17.1	10.3	Republic of Korea	1.2	5.1
Belgium	9.4	3.6	France	5.1	4.3
Thailand	17.1	1.9	Tawan province of China	0.0	3.8
USA	1.3	1.0	Italy	2.8	3.0
Mexico	0.0	0.3	USSR (former)	8.2	2.8

Source: UNCTAD, 1995: Market situation and outlook for tin, 1994 (UNCTAD/COM/54).

Table 105. Tin ore/metal production and trade of the North-East Asian countries, 1987-1993

Unit: Thousand metric tonnes

	1987	*1988*	*1989*	*1990*	*1991*	*1992*	*1993*
Tin ore production (Sn content)							
China	28.0	30.0	44.0	44.0	39.0	43.8	49.1
Democratic People's Republic of Korea							
Japan							
Mongolia	0.1						
Republic of Korea							
USSR (former)	15.0	15.0	15.0	15.0	13.5	10.5	8.5
Tin ore exports (Sn content)							
China	10.4	14.8	21.8	15.8	10.5	5.2	2.2
Democratic People's Republic of Korea							
Japan							
Mongolia							
Republic of Korea							
USSR (former)							
Tin ore imports (Sn content)							
China		0.1	0.1	0.1	0.1	0.7	0.9
Democratic People's Republic of Korea							
Japan							
Mongolia							
Republic of Korea	2.6	2.6	2.3	2.6	1.8	0.2	0.0
USSR (former)	1.3	0.4					

Table 105. *(continued)*

	1987	*1988*	*1989*	*1990*	*1991*	*1992*	*1993*
Tin metal production (primary)*							
China	25.0	24.0	28.3	28.0	28.0	39.6	52.0
Democratic People's Republic of Korea							
Japan	0.9	0.8	0.8	0.8	0.7	0.8	0.8
Mongolia							
Republic of Korea	1.8	2.5	2.4	2.4	2.3	0.4	0.4
USSR (former)	18.0	17.0	16.5	16.0	14.0	10.5	11.2
Tin metal exports							
China	17.6	10.7	9.9	10.1	15.7	30.3	40.7
Democratic People's Republic of Korea							
Japan		0.1	0.5	0.3	0.3	0.3	0.3
Mongolia							
Republic of Korea		0.6	0.3	0.2	0.4	0.2	0.0
USSR (former)							
Tin metal imports							
China		0.1	0.1	0.1	0.1	0.4	0.9
Democratic People's Republic of Korea							
Japan	33.6	35.5	35.7	36.0	38.7	31.1	29.4
Mongolia							
Republic of Korea	4.3	4.9	10.4	6.1	6.6	7.4	9.4
USSR (former)	11.8	8.2	10.5	10.0	9.0	6.2	5.0
Tin metal consumption (primary)							
China	12.5	13.3	15.0	17.0	18.0	17.0	21.1
Democratic People's Republic of Korea							
Japan	32.6	31.3	33.8	34.8	34.9	31.0	26.4
Mongolia							
Republic of Korea	4.0	6.7	11.6	7.4	6.0	6.0	9.0
USSR (former)	31.0	26.0	27.0	26.0	20.0	15.0	15.0
Tin metal consumption (secondary)							
China							
Democratic People's Republic of Korea							
Japan		0.8	4.9	5.1	4.9	4.1	6.6
Mongolia							
Republic of Korea			0.9	0.6		0.6	0.1
USSR (former)		5.0	3.0	3.0	3.0	3.0	3.0

Source: UNCTAD Commodity Yearbook, 1994 and 1995.
* Japan produced about 700 metric tonnes of secondary tin metal yearly in 1988-1993.

tungsten concentrates. As estimated in 1992, the Republic of Korea had tungsten ore reserves of 63 Mt of 0.5 per cent WO_3 grade. Korea Tungsten Mining Co. (KTMC) is the largest tungsten producer in the Republic of Korea and its Sangdong mine in Kangwon Province is one of the largest producers in the world. However, the Republic of Korea's total tungsten ore output declined continuously in recent years from 2,451 t (21 per cent down from 1988) in 1989 to 2,401 t in 1990, 1,405 t in 1991, and 445 t in 1992. Japan produced 347 t and Mongolia 300 t in 1992.

The former USSR, the Republic of Korea and Japan imported tungsten concentrates in the amount of 2,000 t, 1,000 t and 700 tonnes respectively in 1991. Table 106 shows the production, trade and consumption of tungsten ore in the Democratic People's Republic of Korea, Japan, Mongolia and the Republic of Korea in 1993. Exports of the ore from China in 1993 is also shown in the table for comparison.

Table 106. Tungsten ore production and trade in the North-East Asian countries, 1987-1993

Unit: Thousand metric tonnes

	1987	*1988*	*1989*	*1990*	*1991*	*1992*	*1993*
Tungsten ore production (W content)							
China	28.0	36.0	41.0	32.0	27.0	23.0	23.0
Democratic People's Republic of Korea	0.5	0.5	0.5	1.0	1.0	0.5	0.5
Japan	0.3	0.3	0.3	0.3	0.3	0.3	0.1
Mongolia	1.0	1.0	1.0	0.5	0.3	0.3	0.3
Republic of Korea	2.2	1.8	1.6	1.7	1.4	0.8	0.2
USSR (former)	9.0	8.0	7.0	7.0	6.5		
Tungsten ore exports							
China	12.9	14.1	15.9	8.5	3.1	1.2	0.4
Democratic People's Republic of Korea							
Japan							
Mongolia							
Republic of Korea							
USSR (former)							
Tungsten ore imports							
China							
Democratic People's Republic of Korea							
Japan	1.1	1.7	1.4	1.6	0.7	0.4	0.3
Mongolia							
Republic of Korea			0.5	0.7	1.1	1.2	0.0
USSR (former)	6.0	6.0	5.0	5.0	2.0		
Tungsten ore consumption (W content)							
China	16.0	22.0	25.0	22.5	21.0	21.0	20.0
Democratic People's Republic of Korea	0.5	0.5	0.5	1.0	1.0	0.5	0.5
Japan	2.1	2.0	1.5	1.4	1.3	0.9	0.8
Mongolia							
Republic of Korea	2.0	1.6	2.0	2.0	1.8	1.4	0.2
USSR (former)	15.0	12.0	12.0	11.0	7.0		

Source: UNCTAD Commodity Yearbook, 1994 and 1995.

Non-metallic minerals

By the end of 1990, China had exploited 86 non-metallic minerals with known reserves. According to China National Nonmetallic Minerals Industry Corp. (CNNMIC), out of the 86 explored minerals, China's reserves of flake graphite, fluorspar and gypsum rank first in the world; asbestos, bentonite, glauber salt, talc and wollastonite ranked the second, and perlite and zeolites did the third. China also has large mineral resources for attapulgite, bauxite, celestite, diatomite, kaolin, kyanite, pyrophyllite and sepiolite. China is also rich in high quality granite and marble resources. China is one of the few countries in the world that have a more complete set of known non-metallic minerals with great potential resources.

China is a major world producer of a variety of non-metallic minerals. Production of such minerals products as cement, graphite, talc, fluorite, barite and magnesite rank first in the world. The output value of non-metallic mineral products grew at an average annual rate of 10-15 per cent during the decade 1979-1989, which far exceeds growth of the output value of metallic mineral products. The output value of non-metallic mineral industry reached 103.8 billion Chinese yuan in 1988, accounting for about 30 per cent of the total of the country's mining industry. The non-metallic mineral industry is playing an important role in the national economy of China.

The Republic of Korea's non-metallic sector became the most important mining industry for the first time in 1990, valued at $745 million compared to over $663 million for anthracite coal, as coal mining production declined. The Republic of Korea's non-metallic mineral sector accounted for a value of US$1,021 million in 1994, equal to 64 per cent of the value of all mineral production. The Republic of Korea's non-metallic mineral sector had a value of US$1,102 million in 1995, an increase of 6.3 per cent over the previous year, equal to 44.2 per cent of all mineral value.

Barite

China is the world's largest producer of barite with an annual output at least twice bigger then that of Russia, the world's second largest producer. China exported 1.1 Mt of barite valued at $28.1 million in 1988. During 1985-1988, Chinese shipments accounted for close to 60 per cent of U.S. receipts of crude barite.

Diamonds

The Sakha Republic (Yakutia) of the RFE mines practically all diamonds in Russia. There was a reported 10 per cent decrease in diamond extraction in 1992 against that of 1991 or a 25 per cent decrease in comparison with 1990 when production peaked. In 1996, Russia and De Beers reached an agreement acknowledging De Beer's rights to again buy Russian uncut diamonds for export.

Fluorite

China is probably the world's largest producer of fluorspar, outpacing production by Mexico and Mongolia. China's fluorspar products are mainly for export. In 1993, China produced 2.4 Mt and exported 1.38 Mt of fluorspar.

Mongolia produces nearly 15 per cent of the world's fluorspar.

In 1994, Russia continued to import fluorspar from its joint venture in Mongolia which was formed during the Soviet period. In 1994, Russia reportedly imported 88,000 t of flotation fluorspar concentrate and 17,000 t of lumpy fluorite ore from Mongolia.

Graphite

The Republic of Korea's production of amorphous graphite continuously dropped from 107,782 t in 1988 to 100,282 t in 1989, 98,897 t in 1990, and down to 75,239 t in 1991.

Kaolin

The Republic of Korea's kaolin production continuously increased from 0.8 Mt in 1988 to 1.2 Mt in 1989, 1.4 Mt in 1990, 1.7 Mt in 1991, 1.8 Mt in 1992, 2.3 Mt in 1993, 2.6 Mt in 1994 and 2.8 Mt in 1995.

Limestone and cement

China has been the world's largest producer of cement since 1985 with an output of 146 Mt in that year.

China produced 356, 400 and 450 million tonnes of cement in 1993, 1994 and 1995 respectively. In late 1980s, China was a net importer of cement, but since 1990, China has been a net exporter.

Japan's limestone production reached a historic high of 207 Mt in 1991 and remained above 200 Mt in the following years. Japan produced 201.089 Mt of limestone in 1995.

Limestone output continuously increased from 46 Mt in 1988 to 48 Mt in 1989, 48.8 Mt in 1990, 59.3 Mt in 1991, 65.4 Mt in 1992, 75.8 Mt in 1993, 82.8 Mt in 1994 and 87.3 Mt in 1995.

The Republic of Korea's cement production amounted to 30.2 Mt in 1989, 33.6 Mt in 1990, 39.2 Mt in 1991, 42.6 Mt in 1992, 46.9 Mt in 1993, 51.6 Mt in 1994 and 55.1 Mt in 1995. The Republic of Korea exported 1.9 Mt of cement in 1990 and imported 9 Mt to meet the strong domestic demand. The situation then reversed. The Republic of Korea exported 5.1 Mt and 4.3 Mt of cement and imported 1.1 Mt and 2.3 Mt in 1993 and 1994 respectively.

Mongolia's cement production, which was around 165,500 Mt in 1983, has risen to about 425,000 t in 1988.

Magnesite

The Democratic People's Republic of Korea is the world biggest supplier of high quality dead-burnt magnesite with a MgO content of 95 per cent and silicon and calcium content of less than 2 per cent. The annual output was estimated at around 1.5 Mt in 1988.

Liaoning Magnesite and Refractories Corp., the largest magnesite company in China was undergoing extensive restructuring in 1995. The corporation has four operating units including the Liaoning magnesite plant which has a capacity of 50,000 t/y of high grade dead burnt magnesia.

Monazite

The Democratic People's Republic of Korea's monazite is processed at a 1,500 t/y plant operated by the International Chemical Joint Venture Corp., owned by a Japanese company and the state-owned Korea Ryongaksan General Trading Co. The product is sent to Japan.

Phosphate

China was the third largest producer and user of fertilizers after the U.S. and the former USSR in 1988. China produced 22 million tonnes of phosphate rock in 1993 and imported 0.112 million tonnes. China has become a net importer of the ore since 1993.

In the Democratic People's Republic of Korea, a crushing and processing plant was established at the Taedaeri phosphate mine in the central western region in 1993. The plant enables the complex to increase its ore handling capacity by 20 per cent.

Russia's apatite concentrate output was estimated to be 8.8 Mt in 1995, compared to 8 Mt in 1994. Almost all of the apatite comes form the Khibinsky complex on the Kola Peninsula. About 20 per cent of apatite concentrate production was exported with 75 per cent going to countries in Western Europe and the remainder to the former Soviet bloc countries in Eastern Europe.

Salt

The Asian-Pacific region accounted for nearly 30 per cent of salt production in the world as shown in table 107. China is the world's second largest producer of salt after the United States (table 108).

Table 107. World production of salt, 1992

Unit: Thousand metric tonnes

W. Europe	38,852
E. Europe	21,060
N. America	54,894
S. America	8,432
Asia-Pacific	54,828
Africa	3,409
Others	2,025
World Total	183,500

Source: Industrial Minerals, Feb. 1992.

Table 108. Top 20 salt producing countries, 1988 and 1992

Unit: million metric tonnes

	1988	*1992*
USA	35.326	36.400
China	22.640	28.370
Soviet Union	14.800	14.000
Canada	10.686	11.099
Germany	12.576	10.366
Sub-total	**96.028**	**100.235**
% of World Total	**52.05**	**54.62**
India	9.204	9.500
Australia	6.979	7.693
Mexico	6.788	7.395
France	6.308	6.516
UK	6.130	6.250
Brazil	4.386	4.800
Italy	4.579	4.111
Poland	6.179	3.800
Netherlands	3.693	3.628
Spain	3.878	3.870
Romania	5.353	2.490
Turkey	1.358	1.750
Chile	1.043	1.672
Japan	1.363	1.390
Israel	0.386	1.100
Total	**163.655**	**166.200**
% of World total	**88.70**	**90.57**

Source: Industrial Minerals, Feb. 1992.

Salt production from evaporates and rock deposits in China increased at 24 per cent reaching 21.9 Mt in 1988. The production of salt in China reached levels of 22.64, 29.53, 32.00 Mt and 24.99 Mt in 1992, 1993, 1994 and 1995 respectively. China exported about 1.1 Mt of salt annualy in 1986 and 1987. However, because of the growing demand by the domestic chemical sector, exports in 1988 were only about a third of that in prior years.

Silica stones

The Republic of Korea's silica stone production increased from 1.4 Mt in 1990 to 1.6 Mt in 1991, 1.9 Mt in 1992, 2.5 Mt in 1993, 2.4 Mt in 1994 and 2.7 Mt in 1995. Silica sand production amounted to 1.1 Mt in 1993, 1.4 Mt in 1994 and 1.7 Mt in 1995.

Talc

China is a major world producer of both industrial and cosmetic-grade talc. The talc producing bases are in Guangxi, Liaoning and Shandong provinces. China also exported nearly 0.75 Mt of talc in 1988. China has become a major influence in world market for talc.

The Republic of Korea's talc production increased from 141,800 t in 1990 to 170,563 t in 1991.

3.1.6 Fuel minerals development and trade

World energy in 1991-1994

The Energy Statistics Yearbook issued by the Statistics Division of the Department for Economic and Social Information and Policy Analysis (DESIPA) of the United Nations, indicated that global commercial energy production increased by 3.9 per cent between 1991 and 1994, reaching 8,325 million metric tonnes of oil equivalent, of which liquid fuels accounted for 39 per cent, solid fuels 28 per cent, natural gas 23 per cent and electricity 10 per cent. The United States (21 per cent), Russian Federation (12 per cent) and China (10 per cent) continued to lead in global energy production.

World energy consumption grew by 3 per cent between 1991 and 1994 reaching 7,881 million metric tonnes of oil equivalent in 1994. Figures for 1991 through 1994 reflect increased demand for primary electricity (8 per cent), natural gas (2 per cent), solid fuels (5.4 per cent) and liquid fuels (1.6 per cent).

The production of 3,032 million metric tonnes of world crude oil in 1994 reflected a 2 per cent increase since 1991. Among the regions, North America showed a significant decrease of 4.7 per cent between 1991 and 1994. Since 1986, the United States has consistently reduced production of crude oil, reaching 336 million metric tonnes in 1994 – its lowest output level in three decades. Global demand of 3,031 million metric tonnes in 1994 showed a modest increase of 1.3 per cent over figures for 1991. However, Asia displayed a marked increase in consumption in 1994, taking 32 per cent of world-wide demand – up from 27 per cent in 1991.

Asia remained the world's largest regional exporter of crude petroleum with 51 per cent of total exports. Europe was the largest importer with 36 per cent of total imports. Total energy products from refineries increased slightly by 0.8 per cent to 2,757 million metric tonnes between 1991 and 1994, with motor gasoline and gas-diesel oil accounting for 28 per cent and 32 per cent of total production respectively.

World coal production increased to 3,580 million metric tonnes in 1994, a 3.3 per cent increase over 1991. About 80 per cent of the world output is accounted for by seven countries: Australia, China, Germany, Poland, the Russian Federation, South Africa and the United States. In 1994, China and the United States remained the leading coal producers, with 35 and 24 per cent of global production respectively. Australia, the sixth largest producer in world-wide production of coal in 1994, exported the largest amount of hard coal in that year – 129 million metric tonnes or 30 per cent of world exports. Japan was the leading importer of coal with 116 million metric tonnes or 26 per cent of world imports.

Natural gas remained the world's consistently growing energy source. While production of natural gas, which provides 23 per cent of global energy, rose by 4 per cent from 1991 to 1994, consumption per capita declined 2.5 per cent during the same period. Although the United States and the Russian Federation – leaders in production of this commodity – each produced 20 million terajoules of natural gas or 26 per cent of the global total production in 1994, there was a marked difference in the consumption between the two countries, with the United States registering 29 per cent and the Russian Federation 17 per cent of world-wide demand.

From 1991 to 1994, the generation of electricity worldwide increased by 5.5 per cent to 12,681 billion kilowatt/hour, of which 63 per cent was generated by fossil fuels, 19 per cent by hydropower and 17 per cent by nuclear power. In 1994, regional electricity production ranged from a low of 215 billion kilowatt/hour in Oceania to a high of 4,015 billion kilowatt/hour in Europe. Consumption per capita was lowest in Africa – 494 kilowatt/hour, and highest in North America – 8,990 kilowatt/hour. Among individual countries, Norway, including Svalbard and Jan Mayen Islands, showed the highest consumption figures per capita of 26,205 kilowatt/hour. Significantly, almost 100 per cent of Norway's electricity – 113,389 million kilowatt/hour – was generated from hydropower.

Total production of nuclear energy remained at 17 per cent of global electricity production. Five countries – the United States, France, Japan, United Kingdom and the Russian Federation – accounted for 66 per cent of all nuclear energy produced in the world in 1994. In France, nuclear energy production of 359,981 million kilowatt/hour represented 76 per cent of total electricity production in that country (down 2.2 per cent from the previous year); in the United States, 640,491 million kilowatt/hour marked 20 per cent of the country's total electricity production. Japan's 269,126 million kilowatt/hour of nuclear energy generated in 1994 not only reflected 28 per cent of Japan's total electricity production (up 1 per cent from 1993) but also represented 70 per cent of all nuclear energy generated in Asia in the same year.

World oil and gas demand in 1995 and 2010

In 1994 and 1995, oil remained the world's most important commercial fuel, accounting for about 40 per cent of the energy market. Global oil demand grew by 1.2 per cent in 1995. The demand growth for oil in 1995 was still strong in Asia and the Pacific, at 4.6 per cent, but declining from 6 per cent in 1994. The Republic of Korea's oil consumption passed the UK in 1994 with a demand growth rate of 9.9 per cent, and overtook France in 1995 with a demand growth rate of 9.0 per cent to catch up with Italy.

The demand for gas in the world grew at a healthy rate of 2.5 per cent. The demand growth for gas in the Asia-Pacific was even stronger at a rate of 4.8 per cent. World trade in gas by pipeline grew by nearly 11 per cent, and for shipped liquefied natural gas (LNG) more than 5 per cent in 1995 (BP Statistical Review of World Energy, 1996).

The IEA's 1995 edition of the World Energy Outlook analysed how world energy markets might develop to the year 2010. Two alternative scenarios are examined, i.e. Capacity Constraints (CC) and Energy Savings (ES). The noteworthy implications derived from the IEA's analysis are:

(1) In both cases, world energy demand will grow over the next 15 years, but more slowly than the rate of economic growth. By 2010, the world is projected to consume between 34 and 45 per cent more primary energy than it did in 1992.

(2) By 2010, the global energy system is likely to remain overwhelmingly based on the burning of fossil fuels. The fossil fuels will still account for nearly 90 per cent of global energy consumption by 2010. Within fossil fuels, while the share of oil in the world energy mix will continue to decline, the volume of oil consumed will continue to increase. Oil is still the single most important primary energy form.

(3) By 2010, the OECD will consume half or less of the world's energy and oil. In terms of the expected annual increase in energy demand between 1992 and 2010 the figures even more stark, with the OECD accounting for less than a third of the increase in total energy and only around a quarter of the increase in oil demand.

(4) Technical and environmental factors, as well as the restructuring which is now under way in the energy industry, are likely to lead to a significant increase in the use of gas, especially as a power generation fuel. This will lead to a sharp expansion in gas trade, both by pipeline and by ship as LNG.

Coal, which has also traditionally not been traded internationally on a large scale, is increasingly becoming part of world trade, with international trade coal having doubled since 1973 and projected to more than double again by 2010, leading to a large increase in necessary export facilities and infrastructure. Gas and coal are expected to have a greater influence on world energy prices.

(5) A potential shortage of financing for the capital intensive energy sector could arise. As a result of the institutional changes and financial pressures evident in many regions, private capital is playing an increasingly important role in the development of the energy sector. Both oil and gas supplies are likely to increase as a result of the introduction of private capital, the more competitive environment and, in many countries, the likely opening up to foreign or privately owned companies (IEA, 1995, p. 4).

(6) Global energy-related CO_2 emissions in 2010 are projected to increase by between 30 and 40 per cent over their 1990 level under the two sets of scenarios. Global annual emissions in 2010 would still exceed those of 1990 by 25 per cent. The OECD will account for nearly 45 per cent of the world total. The increase in the rest of the world (Africa, China, East Asia, South and Central America, the Middle East and South Asia) share will be more than three times the growth rate in the OECD, though the per capita CO_2 emissions in these countries will still be only a fraction of those in the OECD. These results emphasise both the global nature of the greenhouse gas debate and its long term nature.

In this period, China and East Asia are estimated to have a GDP increment or 2,883 billion US dollars, which would account for 21.7 per cent in world total increment and would be more than one third of that of the OECD. The dynamic economies of East Asia together with China will account for a greater increase in annual oil demand than the whole of the OECD (tables 109-110). These shifts, which are not restricted to the energy world, have important strategic implications. Global cooperation in the energy sector will become more necessary than ever.

Table 109. The increase of world energy demand between 1992 and 2010

(Unit: Mtoe)

	World	*OECD*	*China plus East Asia*	*Others*
PRIMARY ENERGY	+3,554 (100%)	+1,072 (30.2%)	+1,262 (35.5%)	+1,220 (34.3%)
Solids	+979 (100%)	+186 (29.0%)	+602 (61.5%)	+191 (19.5%)
Oil	+1,285 (100%)	+342 (26.6%)	+427 (33.2%)	+516 (40.2%)
Gas	+963 (100%)	+411 (42.7%)	+136 (14.1%)	+416 (43.2%)
Nuclear	+151 (100%)	+66 (43.7%)	+44 (29.1%)	+41 (27.2%)
Hydroelectricity	+120 (100%)	+35 (29.2%)	+33 (27.5%)	+52 (43.3%)
Geothermal/Others	+56 (100%)	+33 (58.9%)	+9 (16.1%)	+14 (25.0%)
CO_2 Emissions (Mt)	+9,612 (100%)	+2,396 (14.9%)	+3,967 (41.3%)	+3,249 (33.8%)
GDP (billion 1987US$)	+13,265 (100%)	+8,013 (60.4%)	+2,883 (21.7%)	+2,369 (17.9%)
GDP/capita (1987US$)	+1,188	+6030	+986 or +2297	–
Energy/capita (toe)	+0.18	+0.54	+0.45 or +0.52	–
Energy Intensity (toe/1,000US$)	-0.09	-0.06	-0.69 or -0.12	–

Source: Based on the data estimated for capacity constraints case (CC case) in the IEA's 1995 World Energy Outlook.

Table 110. Energy demand scenario in East Asia, 1992 and 2010

(MTOE)	*1992*	*2010*	*2010*
Primary Energy	1,102	1,574	2,364
Solids	644	883	1,246
Oil	356	509	783
Gas	56	97	192
Nuclear	24	41	68
Hydroelectricity	18	32	51
Geothermal/Others	6	15	24

Source: Based on data from International Energy Agency: World Energy Outlook, 1995 edition. East Asia here includes China.

Table 111 presents Asian APEC's primary energy supply/demand and imports outlooks from 1992 to 2010. The Asian APEC area here includes the six ASEAN countries (Brunei Darussalam, Indonesia, Malaysia, the Philippines, Singapore and Thailand), Asian NIEs (the Republic of Korea, Hong Kong, China and Taiwan Province of China) and China.

The commercial primary energy intensity in developed and developing countries (DCs) for 1990 is indicated in table 112.

Fuel minerals production and trade in North-East Asia

Oil

China produced 2.99 million barrels of oil daily in 1995 and is the world's seventh largest oil producer after Saudi Arabia, the United States, the Russian Federation, Islamic Republic of Iran, Mexico, and Norway. In 1994, about 70 per cent of China's oil production was produced from three complexes, Daqing (nearly 38 per cent), Shengli (20.5 per cent) and Liahoe (9.8 per cent), all in northeast China.

Offshore production of oil and gas in China reached only 4.44 per cent of its total in 1994. In East China Sea, as reported by Integrated Exploration & Development Services Ltd. (IEDS) in 1995, the SPC (City of Shanghai-40 per cent, MGMR-30 per cent and CNOOC-30 per cent) would proceed alone with the development of the Pinghu field with a loan provided by the Asian Development Bank. A 14" gas pipeline will cover 400 km from the field to Shanghai and a 10" oil and condensate line will come ashore on Dai Shan Island of Zhejiang province.

In the Democratic People's Republic of Korea, the last new contracts were awarded to Australian Claremont Petroleum and Beach Petroleum on 13 August 1994. There have been no significant discoveries drilled in the country, although some wells in the West Bay Basin have recorded minor oil flows. Taurus reportedly planned to acquire seismic data over its West Bay Basin acreage during 1995 in a programme originally scheduled for 1995 (IEDS, 1995).

The Republic of Korea passed the Law for Development of Submarine Mineral Resources, which divided the offshore (continental) shelf area into seven blocks. A total of 32 wells have been drilled in the area and overal results have been most disappointing.

There has been no production of oil and gas in Mongolia since 1969, when the Zuunbayan field was shut down due to an on-site fire. Cumulative production is reported to have been about 3.85 mmbbls of 28.5 degree API, low sulphur, waxy crude. This and the Tsagaan-Els field represent the only two known potentially commercial oilfields in Mongolia.

Table 111. Outlook of APEC's primary energy supply/demand and imports, 1992-2000-2010

Units: Million tonnes of oil equivalent

China	*1992*	*2000*	*2010*	*1992-2010*
Primary energy demand	%/year			
Oil	132.7	204.0	325.8	5.1
Coal	551.7	788.6	1,156.7	4.2
Natural gas	13.6	23.0	50.8	7.6
Nuclear	0.0	3.6	37.4	
Hydro/others	11.5	19.7	34.0	6.2
Total	709.5	1,038.9	1,604.7	4.6
Primary energy output				
Oil	144.3	153.0	194.0	1.7
Coal	557.5	795.0	1,088.5	3.8
Natural gas	13.6	22.2	33.2	5.1
Nuclear	0.0	3.6	37.4	
Hydro/others	11.3	19.1	36.9	6.8
Total	726.7	992.9	1,390.0	3.7
Imports (demand minus output)				
Oil	-11.6	51.0	131.8	
Coal	-5.8	-6.4	68.2	
Natural gas	0.0	0.8	17.6	
Total	-17.4	45.4	217.6	
Dependence (Import/Demand)		4.4	13.6	
Republic of Korea	*1992*	*2000*	*2010*	*1992-2010*
Primary energy demand	%/year			
Oil	70.4	94.8	124.5	3.2
Coal	23.8	38.6	59.8	5.3
Natural gas	4.4	13.4	28.3	10.9
Nuclear	14.7	22.8	46.1	6.6
Hydro/others	0.5	0.5	0.6	1.0
Total	113.8	170.1	259.3	4.7
Primary energy output				
Oil	0.0	0.0	0.0	
Coal 5.5	4.1	2.3	-4.7	
Natural gas	0.0	0.0	0.0	
Nuclear	14.7	22.8	46.1	6.6
Hydro/others	0.5	0.5	0.6	1.0
Total	20.7	27.4	49.0	4.9

Table 111. *(continued)*

Imports (demand minus output)				
Oil	70.4	94.8	124.5	
Coal	18.3	34.5	57.5	
Natural gas	4.4	13.4	28.3	
Total	93.1	142.7	210.3	
Dependence (Import/Demand)	81.8	83.9	81.1	
Japan	*1992*	*2000*	*2010*	*1992-2010*
Primary energy demand	%/year			
Oil	260.0	249.6	249.2	-0.2
Coal	75.5	82.2	84.1	0.6
Natural gas	47.6	62.4	67.9	2.0
Nuclear	58.4	77.3	115.9	3.9
Hydro/others	11.2	22.7	45.4	8.1
Total	452.6	494.2	562.6	1.2
Primary energy output				
Oil	0.9	0.8	0.8	-0.7
Coal	4.4	3.3	3.3	-1.6
Natural gas	2.0	2.0	2.0	0.0
Nuclear	58.4	77.3	1,15.9	3.9
Hydro/others	11.1	22.7	45.5	8.2
Total	76.8	106.1	167.5	4.4
Imports (demand minus output)				
Oil	259.1	248.8	248.4	
Coal	71.1	78.9	80.8	
Natural gas	45.7	60.4	65.9	
Total	375.9	388.1	395.1	
Dependence (Import/Demand)	83.1	78.5	70.2	
Asia APEC 10 total	*1992*	*2000*	*2010*	*1992-2010*
Primary energy demand				
Oil	344.4	494.5	733.5	4.3
Coal	608.0	875.1	1,319.9	4.4
Natural gas	53.3	112.5	217.6	8.1
Nuclear	23.5	36.7	75.1	6.7
Hydro/others	19.8	45.0	71.0	7.4
Total	1,049.0	1,563.7	2,446.4	4.8
Primary energy output				
Oil	268.9	275.4	301.5	0.6
Coal	582.2	856.2	1,171.7	4.0

Table 111. *(continued)*

Natural gas	89.8	155.4	207.6	4.8
Nuclear	23.5	36.7	106.1	8.7
Hydro/others	20.0	45.0	69.5	7.2
Total	984.4	1,368.7	1,862.4	3.6
Imports (Demand minus Output)				
Oil	75.4	219.1	432.0	
Coal	25.7	18.9	148.1	
Natural gas	-36.6	-43.0	9.9	
Total	64.5	195.0	590.0	
Depandence (Import/Demand)	6.1	12.5	24.1	

Source: Kazuya Fujime, 1996. Long-term Energy Supply/Demand Outlook for Asia APEC Nations, Energy in Japan No. 137'96.

Table 112. Energy intensity in different economies

	Total primary energy		*Commercial primary energy*	
	kgoe/USS	*kgoe/PPPS*	*kgoe/USS*	*kgoe/PPPS*
OECD	0.26	0.26	0.25	0.25
USA	0.36	0.36	0.36	0.36
Japan	0.14	0.19	0.14	0.19
EE and FSU	1.61	0.69	1.58	0.67
FSU	1.57	0.68	1.54	0.67
CEE	1.77	0.69	1.77	0.69
DCs	0.87	0.32	0.62	0.23
China	1.96	0.36	1.50	0.29
India	1.02	0.33	0.61	0.20

Source: Energy Policy 1995, Volume 23, Number 8.

The former Soviet Union produced 20 per cent of the world's oil in 1990. Russia's oil industry comprised 840 oil fields with 148,000 oil wells, 48,300 km of main oil pipeline, and 28 oil refineries that can refine more than 300 Mt/y of oil. The oil industry and its auxiliary services employed about 900,000 people. Oil production in Russia was 552 Mt in 1989, 395 Mt in 1992, 357 Mt in 1993, and 315.7 Mt in 1994. About 10 per cent of the Russian oil production comes from the Russian Eastern Siberian and Pacific basins. The Russian Far Eastern region (RFE) produces less than 40,000 barrels daily of crude and condensate in Sakhalin, accounting for only 0.4 per cent of Russian oil output.

Japan consumed 268.7 million tonnes of oil equivalent in 1994, ranked the second in the world. However, domestic oil production can only provide 0.3 per cent of Japanese total consumption with the balance from imports. Japan's current production rate of crude oil and condensate is estimated to be approximately 18,200 bpd, of which 4,200 bpd comes from the Iwafune-oki field in the Sea of Japan.

China became a net oil importer in 1993, the first time since the 1960s. China imports crude from the Middle East, West Africa, and South-East Asia. About 65 per cent of China's crude imports are 30-45° gravity, low sulfur. China exported 22.1 million tonnes crude oil and its refined products of totally in 1995, of which over 52 per cent goes to Japan, nearly 35 per cent to other Asian countries and nearly 12 per cent to the United States. The major buyers of China's crude oil are Brazil, Japan, the Democratic People's Republic of Korea, the Republic of Korea, the Philippines, Singapore, and the USA. The major trading countries for China's refined products in 1993 included Singapore, the Republic of Korea and Japan.

The Democratic People's Republic of Korea's oil imports fell from 3.3 Mt in 1989 to 2.5 Mt in 1990, due to cut backs in Russia exports. The supply further dropped sharply in 1991 as Russia cut its supply by more than 90 per cent and demanded hard currency for deliveries. With this sharp cut-back in the supply, the Democratic People's Republic of Korea is desperate for oil. The oil supply is heavily dependent on imports from China now.

The Republic of Korea's energy production in 1991 decreased by 2.2 per cent to 22.7 Mt of oil equivalent (TOE). Total energy imports were 9.6 million TOE – an increase of 39.4 per cent against 1990. The Republic of Korea's energy consumption amounted to 103 million TOE in 1991, an increase of 1.1 per cent over 1990. Petroleum dominated energy consumption with 424 Mbbl in 1991, up 19.1 per cent over 1990. Imports of crude oil were approximately 399 Mbbl and 2.8 Mt of LNG from Indonesia.

Russia reportedly exported 79.7 Mt of oil outside the CIS in 1993. Russia's exports of oil to other CIS states was 40.4 Mt in 1993, only 53 per cent of its 1992 level.

The insufficient indigenous production is wholly consumed within the Far Eastern region of the Russian Federation – mainly at local power plants (53 per cent), in industry (31 per cent), and for household/municipal needs (10 per cent) in the cities of Yakutsk, Okha, Komsomolsk-on-Amur, and in neighbouring industrial towns (Petromin, Oct. 1992). In 1990, the region's dependence on imported liquid fuels amounted to 80 per cent in the case of crude oil and to 55 per cent, on an average, for all oil products, including 65 per cent for diesel fuel, 58 per cent for jet kerosene, 55 per cent for residual fuel oil, and 37 per cent for motor spirit (Petromin, Oct. 1992).

In recent years, more than 200,000 b/d of petroleum products and 130,000 b/d of crude oil have been transported there by sea and railway, at an annual cost of more than 300 million Russian roubles. Russia's plan for comprehensive development of the Far Eastern region to 2000 highlighted the need to sharply increase oil and gas prospecting. The volume of surveying activity is to increase 2.5-3 times in the next two five-year development periods (table 36, Dorian, 1993).

Oil production, consumption, import and export in North-East Asia is shown in table 113. For oil refining centres in the Asia-Pacific region, Japan, China and the Republic of Korea are ranked the first three in oil refining capacity.

Natural gas

Natural gas production in Russia was 616 billion cubic metres in 1989, 640 billion cubic metres in 1992, 617.4 billion cubic metres in 1993, and 581.02 billion cubic metres in 1994. Russia's main producing area of natural gas is West Siberia. The Gazprom controls practically all gas production in Russia. Russia exported 99 billion cubic metres of natural gas in 1992, and 96 billion cubic metres in 1993. About one third of the 1993 exports went to the countries of the former Soviet Union, and the rest to market economy countries. Several developed gas fields are located in Sakhalin island and in central Yakutia, producing around 300 mmcfd gas, or 0.5 per cent of Russian gas production.

Gas production, consumption, import and export in North-East Asia is shown in table 114.

'The Trans-Asian Natural Gas Pipeline Project' proposed by the National Pipeline Research Society of Japan, calls for the construction of six international trunk lines as shown in figure 68, including the following:

- Turkmenistan – West China – Japan, aiming to supply Shanghai and other coastal areas of China and Japan through undersea pipelines.
- Yakutsk – China – Korean peninsula – Japan, aiming to supply to China, the Korean peninsula and Japan with three possible routings, i.e. via Gulf of Pohai, via Sakhalin, and via the Korean peninsula.

Table 113. Oil production, consumption and trade in North-East Asia, Asia and the world at the end of 1995

Oil	*Production (Thousand barrels daily)*	*Consumption (Thousand barrels daily)*	*Import (Thousand barrels daily)*	*Export (Thousand barrels daily)*
China	2,990	3,310	644	447
Japan	<0.05	5,780	5,581	157
Democratic People's Republic of Korea		84 ('94)*	66 ('89), 50 ('90)*	
Republic of Korea				
Mongolia				
Russia	6,200			
Total NE Asia				
Total Asia	6,695	17,085	12,663	2,474
Total WORLD	66,695	67,930	36,239	36,239

Source: British Petroleum, 1996: Statistical Review of World Energy.
* Mining Annual Review, 1990s.

Table 114. Natural gas production, consumption and trade in North-East Asia, Asia and the world at the end of 1995

Gas	*Production (Million tonnes oil equivalent)*	*Consumption (Million tonnes oil equivalent)*	*LNG Import (Billion cubic metres)*	*LNG Export (Billion cubic metres)*
China	15.8	15.8		
Japan	2.0*	55.0	57.9	
Democratic People's Republic of Korea				
Republic of Korea			7.9	
Mongolia				
Russia				
Total NE Asia				
Total Asia	156.0	175.6	>70.8	3.5**
Total WORLD	1,908.0	1,883.6	92.5	92.5

Source: British Petroleum, 1996: Statistical Review of World Energy.
* for 1994, ** from Indonesia and Malaysia.

- (Alaska) – Sakhalin – Japan, aiming to supply Japan from northern Sakhalin, and possibly from Pridhoe Bay in Alaska.

The other three international trunk lines proposed were ASEAN-southern China, ASEAN pipelines and Australia-ASEAN pipeline.

The Trans-Eurasian pipeline network is shown in figure 69, indicating the planning pipeline to be constructed in the North-East Asia sub-region.

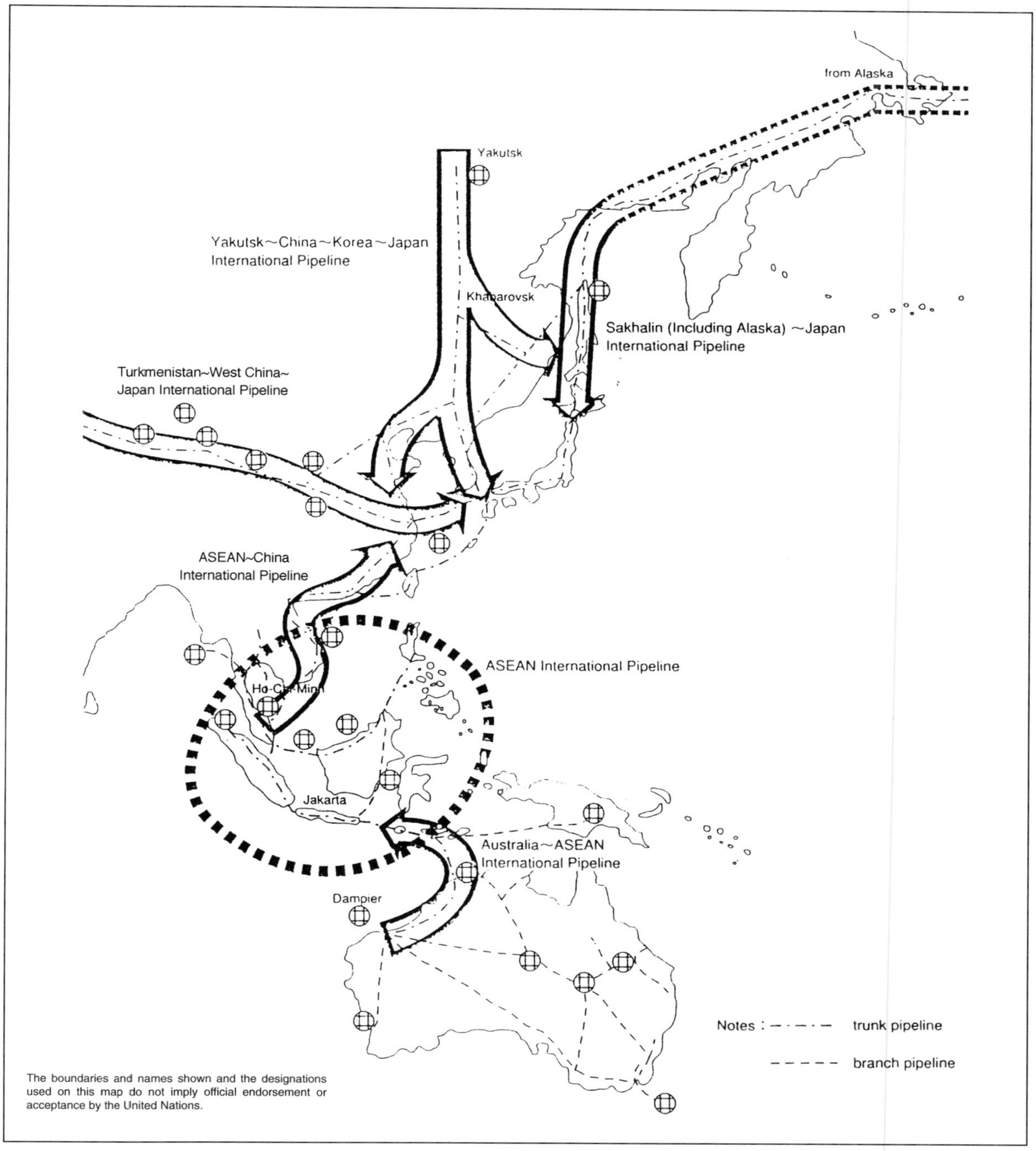

Source: Masaru Hirata, 1995.

Figure 68. Trans-Asian natural gas pipeline network

Coal

World reserves of coal at the end of 1995 were estimated to be 1,031,610 Mt (BP Statistical Review of World Energy, 1996). China had 114,500 Mt of coal reserves accounting for 11 per cent of the total and ranked the second in the world after the United States (240,558 Mt for 23.3 per cent), and followed by Australia (61,865 Mt for 8.8 per cent). Japan had 821 Mt of coal reserves accounting for 0.0796 per cent, the Democratic People's Republic of Korea with 600 Mt for 0.0582 per cent, and the Republic of Korea with 183 Mt for 0.0177 per cent. The R/P ratio for China was 88 years, Japan 130 years and the Republic of Korea 32 years.

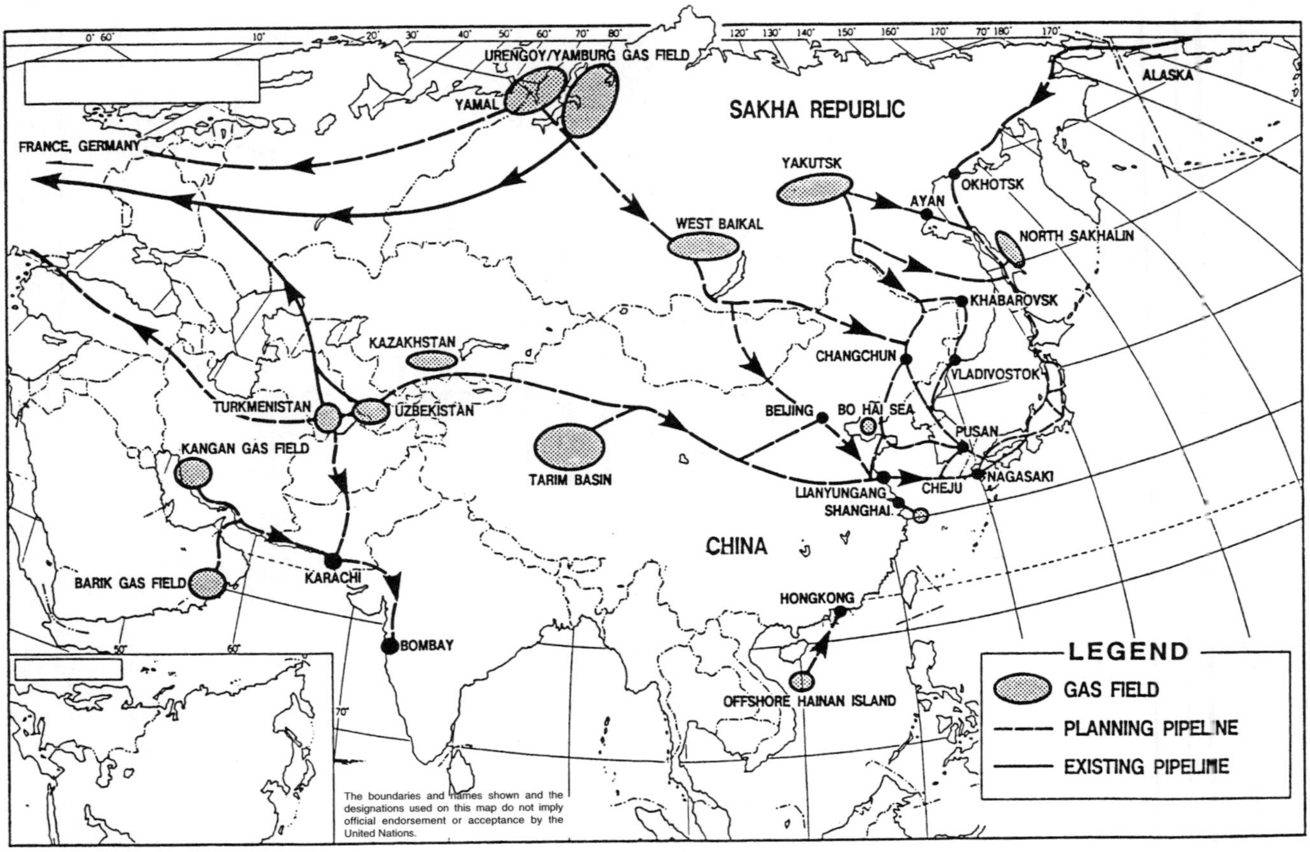

Source: Masaru Hirata, 1995.

Figure 69. Trans-Eurasian natural gas pipeline network

Acording to IEA press release in Paris, of 5 August 1996, the key features of the world coal market in 1995 were as follows:

- World coal trade was strong again in 1995, rising by 8.1 per cent or 34 million metric tonnes (Mt). Trade was estimated at 458 Mt in 1995, representing about 13 per cent of world production.
- World hard coal production rose by about 61.2 Mt or about 1.7 per cent in 1995 from 3,561 Mt in 1994 to 3,623 Mt in 1995.
- A strong recovery in coal prices, together with high freight rates was responsible for increased cost of imported coal in European and Asian markets in 1995.

The main exporters of hard coal in 1995 were Australia with 136.1 Mt, the United States with 80.3 Mt, South Africa with 59.7 Mt, Canada with 34.0 Mt, Poland with 31.9 Mt, Indonesia with 31.6 Mt and China with 27.5 Mt. Total coal imports in the Asia-Pacific area rose by 18 Mt to reach 227.3 Mt in 1995.

Total world steam coal exports rose by about 27.4 Mt to 272.7 Mt in 1995, an increase of 11.2 per cent over 1994 levels. Australia and South Africa remained the largest exporters, with about 61.2 Mt and 53.5 Mt respectively. Indonesian exports reached 29.4 Mt and China 23.2 Mt.

Steam coal imports in the Asia-Pacific market reached 127.4 Mt in 1995, about 13.8 Mt higher than in 1994, thus representing about 47 per cent of the total world steam coal market. Japan is the world's largest single market for steam coal, importing 57.1 Mt in 1995. Other major importers in the region in 1995 were the Republic of Korea with 26.7 Mt, Taiwan Province of China with 24.3 Mt and Hong Kong, China with 9.1 Mt. The major steam coal suppliers to this market in 1995 were Australia at 53.5 Mt, Indonesia at 22.2 Mt, China at 20.1 Mt and South Africa at 15.4 Mt. The U.S. steam coal exports to this region were 6.8 Mt in 1995.

The level of global coking coal trade rose in 1995 by 6.8 Mt to 184.9 Mt, mainly as a result of higher exports from Australia (up 3.1 Mt to 74.9 Mt) and the United States (up 4.4 Mt to 47.3 Mt). Australia and the United States remained the largest coking coal exporters, with about 41 per cent and 26 per cent of the total market, respectively. Canada exported 28.6 Mt, China 4.3 Mt and Russia 3.0 Mt in 1995.

Coking coal imports in the Asia-Pacific region reached 99.8 Mt in 1995, about 4.4 Mt higher than in 1994; this market represents about 54 per cent of total world coking coal imports. Within the region, Japan accounts for about two thirds of the market, importing 65.3 Mt in 1995. Other major imports in the region in 1995 were the Republic of Korea (17.2 Mt), India (9.6 Mt) and Taiwan Province of China (4.6 Mt).

World hard coal production rose by about 61.2 Mt or about 1.7 per cent in 1995, from 3,561.5 Mt in 1994 to 3,622.7 Mt in 1995. China remains the world's largest producer. Most production was for local consumption. Demand which was not met by indigenous production was satisfied by imports.

Coal production in North-East Asia in 1992 was as follows: China produced 1,110 Mt, ranked number one in the world, Russia-335 Mt (the third), the Democratic People's Republic of Korea-43.1 Mt (the 16th), the Republic of Korea-11.9 Mt (the 23rd), Mongolia-9 Mt (the 27th) and Japan-7.6 Mt (the 28th). The countries in the subregion, including the Russian Federation, produced about 1,500 Mt of hard coal, accounting for about 45 per cent of the world total.

China consumed 527.1 Mt of oil equivalent in 1992, which accounted for 24.4 per cent of the world total and ranked number one, Russia consumed 116.1 Mtoe accounting for 22.0 per cent ranked number two, Japan did 77.9 Mtoe accounting for 3.6 per cent (7th), and the Republic of Korea consumed 23.3 Mtoe accounting for 1.1 per cent (17th). The coal consumption of the Democratic People's Republic of Korea and Mongolia was less than 0.1 million tonnes of oil equivalent.

The Democratic People's Republic of Korea's annual energy consumption is about 42 Mt of oil equivalent, with coal meeting 75 per cent and oil 10 per cent of the demand (1995). According to the Democratic People's Republic of Korean press, coal production in 1988 was between 48 Mt and 70 Mt in 1988, reducing to 43.1 Mt in 1991, and increasing at 90 Mt in 1994. Output of coking coal has remained stagnant at around 1 Mt/y, coming from the Kukdong and Yangjong collieries.

The Japanese coal industry has been in a long downward trend with a production of steam coal at 6.9 Mt in 1994, comparing with about 10.4 Mt in 1988, and an average decreasing rate of 5.6 per cent. Japan produced 110,000 Mt of coking coal in 1990, falling 87 per cent from the previous year. It has not been a coking coal producer since then.

In the Republic of Korea, due to falling coal demand and rising labour costs, the Ministry of Geology and Mineral Resources implemented a Coal Mining Rationalization Plan. Tens of small coal mines were closed annually, with about 210 operating mines in 1990, 170 in 1991, 130 in 1992, and down to 26 in 1994. In connection to this, anthracite coal production has decreased sharply in recent years from total 23.8 Mt in 1988 to 20.8 Mt in 1989, 17.2 Mt in 1990, 15.1 Mt in 1991, 11.9 Mt in 1992, 9.4 Mt in 1993, and down to 7.44 Mt in 1994.

Russia's coal production was reported to be 337 Mt in 1992, 300 Mt in 1993, and 271 Mt in 1994. The Kuznetsk basin (Kuzbas) is the Russia's largest coal field.

In recent years, the Democratic People's Republic of Korea imported about 1-2 Mt/y of steam coal from China and the former Soviet Union, while Australia has supplied prime coking coal.

Due to growing demand for energy, Japan's import of steam coal has been in a long upward trend reaching 50 Mt in 1994, compared with 31 Mt in 1989, at an average increase rate of 7.6 per cent. Overall, import of coking coal has been in a downward trend, reflecting steel mills shifting to other energy sources because of increased prices. Japan imported 63 Mt of coking coal in 1994, compared with about 71 Mt in 1988.

The Republic of Korea's total imports of coal in 1992 amounted to 30.1 Mt, an increase of 3 per cent over the previous year. The Republic of Korea's imports of anthracite coal declined drastically, to 2.8 Mt in 1988, 1.5 Mt in 1989, 808,629 t in 1992 (477,014 Mt from China and 301,613 Mt from Viet Nam), 780,000 t in 1993, and 700,000 t in 1994. On the other hand, the use of imported bituminous coal, has increased significantly and it now provides around 60 per cent of primary coal consumption. Bituminous coal imports totalled 27.8 Mt (15.5 Mt for metallurgical coal and 12.3 Mt for thermal coal) in 1991, 29.3 Mt (16.4 Mt/metallurgical and 12.9 Mt/thermal) in 1992, 34.2 Mt in 1993, and 37.8 Mt in 1994.

Coking coal imports for steel use amounted to 11.6 Mt in 1989, 11.3 Mt in 1990, 17.3 Mt in 1993 and 16.9 Mt in 1994. Steaming coal imports for power generation, cement and boiler industry amounted to 11.8 Mt in 1989, 11.6 Mt in 1990 and 16.9 Mt in 1993. In 1994, the Republic of Korea imported 20.9 Mt of steaming coal for power (24 per cent).

The major coal suppliers to the Republic of Korea included Australia (14 Mt in 1992, 15.6 Mt in 1993 and 16.8 Mt in 1994), China (6.5 Mt in 1992, 6.7 Mt in 1993 and 6.3 Mt in 1994), and Canada (4.9 Mt in 1992, 5.8 Mt in 1993 and 5.4 Mt in 1994).

Russia exported 23.6 Mt of coal in 1991, 17.9 Mt in 1992, 25 Mt in 1993, and 19-20 Mt in 1994. In 1993, coal exports to countries of the former Soviet Union were about 5 Mt, and those to countries outside the former Soviet Union were 17.6 Mt. The Kuznetsk basin in Siberia accounted for most of Russia's exports.

Coal production and consumption in North-East Asia at the end of 1995 are presented in table 115.

3.2 Creation of a sound development environment

In contrast to the 1960s and 1970s, when strong national policies in many countries in the world resulted in closing doors to mineral sector foreign investment, the 1980s and 1990s have seen new and more investor-friendly mining policy in most of the countries. Now, there is a strong competition among countries to attract private sector capital for exploration and mining. Those nations that have both favourable geology and an attractive mineral sector investment environment tend to attract foreign direct investment (FDI). James Otto summarized the criteria for assessing mineral investment conditions which is presented in table 116 (Otto, 1992). Criteria for attracting foreign investment include geological, political, marketing, regulatory, fiscal, monetary, environmental, operational and profit related factors.

Table 115. Coal production and consumption of North-East Asia, Asia and the world at the end of 1995

Coal	*Production (Million tonnes oil equivalent)*	*Consumption (Million tonnes oil equivalent)*
China	655.5	640.3
Japan	4.2	85.9
Democratic People's Republic of Korea		31.5*
Republic of Korea	3.0	27.3
Mongolia		
Russia	116.9	119.4
Total NE Asia		
Total Asia	864.9	959.1
Total WORLD	2,225.1	2,210.7

Source: British Petroleum, 1996: Statistical Review of World Energy.
* Mining Annual Review, 1990s.

Table 116. List of criteria for private sector appraisal of mineral sector investment conditions

The following "investor check-list" itemizes the major criteria used for private sector appraisal of mining investment conditions abroad. It is based on a global survey and augmented from subsequent additional research.

GEOLOGICAL CRITERIA

General mineral abundance

Efficient and helpful government offices (geological survey, department of mines, ministry for lands, financial offices, central bank)

Perceived geological potential compared with competing jurisdictions

Availability of reliable geoscientific information (e.g., topographic maps, geological maps, air photo coverage, airborne geophysics, geochemical data and geological reports)

Organization and accessibility of geoscientific information, reports in common form, preferably in English, centralized database, readily available

Historical production

Ability to apply geological assessment techniques, availability of local trained personnel

Mines survey department

Availability of local laboratories

POLITICAL CRITERIA

Long-term national stability

Regime stability

Consistency and constancy of mineral policies

Internal security

Stability of neighbouring countries

Availability of foreign investment insurance

Good governance – transparency and accountability

Open versus closed market economy

Indicators in place to formalize open market economy:

Privatization

Institutionalization of secure mineral tenure embodied in mining code

Democratic linkages

Access to external dispute settlement mechanisms such as international arbitration of conflicts

Development of social policy, i.e., health services, education, land reform, income redistribution, training

Entrepreneurial skills in place

Presence of domestic investment in productive sector

REGULATORY CRITERIA

Modernized mining code

Workable mineral legislation

Stability of exploration/mining terms

Mineral ownership

Surface/land ownership

Access to water rights

Security of tenure, length of tenure exploration and mining

Quality of mineral titles system

Ability to track private deals on top of original tenure

Right to transfer ownership

Availability of a mineral agreement to supplement or in place of mining code

Dispute resolution such as international arbitration

Level of bureaucracy

Procedural efficiency and clarification of administrative competency

Simple, effective and efficient permitting system

Rating for overall simplicity

Rating for consistency

Rating for completeness

Availability of good local mineral resource lawyers

Effective and transparent legal system

FISCAL CRITERIA

Tax method and level of taxes (profit-based or not, fixed or not)

Ability to predetermine tax liability

Availability of accelerated depreciation

Availability of investment tax credits

Availability of reinvestment credits

Exploration cost treatment

Export-import credits

Stability of fiscal regime

Tax treaty with home country

Expatriate tax treatment

Competitiveness with other jurisdictions

MONETARY CRITERIA

Realistic foreign exchange regulations

Allowance for external accounts

Ability to repatriate profits

Ability to raise external financing to fund projects

Presence of strong local stock exchange

Presence of global banking institutions

Efficiency in monetary transfers

OPERATIONAL CRITERIA

- Majority equity ownership held by company
- Company has management control
- Degree of established infrastructure and utilities
- Experienced local workforce availability
- Climatic conditions
- Physical layout of land
- Altitude
- Availability of support services such as fabrication/maintenance/local geotechnical services
- Transportation infrastructure, e.g., roads, railway, air, river, deep sea ports
- Common language spoken
- Constraints on use of expatriate staff
- Restrictions on hiring, firing and negotiating wages
- Technology required
- Strength of labour unions
- Additional support services
- Effective communications systems, internal/external
- Transferability of operating permits

MARKETING CRITERIA

- Geographic location – proximity to markets including access to transportation networks
- Presence of internal markets
- Presence of local competitive smelting and refining capability
- Export/import policies and restrictions, e.g., mandatory sales of product to central bank
- Regional trade agreements
- Outlook for demand/price

ENVIRONMENTAL CRITERIA

- Legal requirements for environmental protection
- Ability to predetermine environmental-related obligations
- Anti-mining groups
- Relative sensitivity of environment

PROFIT CRITERIA

- Measure of profitability (internal rate of return, net present value, payback periods)
- Competitive cost position

OTHER CRITERIA

- Prior company experience
- Company employee prior experience
- Specialized company expertise

Experience of other firms, joint venture partner experience
Helpfulness of government offices (geological survey, department of mines, ministry for lands, finance offices, central bank)
Availability of information on government, administration, fiscal, regulatory policies, geological information
Availability of key government personnel locally and abroad
Risk containment strategies available
Availability of political risk insurance (debt and equity portion)
Bilateral/trilateral protection treaty in place or negotiable
Mine development agreement
Degree of World Bank and other multilateral agency involvement in project, in country
Legislated protection for investors against expropriation, and currency inconvertibility

Source: Annex to the Report of the Secretary-General 'Assessment of benefits accruing to host countries from the inflow of funds and technology for mineral development' (E/C. 7/1996/8) at the third session of Committee on Natural Resources on 6-17 May 1996. The initial list, prepared for the Economic and Social Commission for Asia and the Pacific by Dr. James Otto, Assistant Director, Centre for Petroleum and Mineral Law and Policy, University of Dundee, United Kingdom, 1992; subsequently revised by Diana Manson and Associates, Ltd., Canada.

The role of Government

To attract and maintain long-term private sector mining investments, the role of the Government as a regulator, promoter and facilitator, should create a sound development. Some important conditions have to be well established through efforts of the Governments at different levels, including the following:

(a) A sound macro environment should offer prudent and stable policies supportive of an open, export-oriented business environment, including the right to market products externally and the right to import equipment and supplies, assured access to foreign exchange at free market rates, and the ability to repatriate profits and keep funds offshore.

Development of open-market, investment-oriented policies need to consider sustainable development, transfer of technology, development of entrepreneurial skills, infrastructure, and long-term financial and social policy issues.

(b) A modern legal and fiscal framework should provide assured access to land for exploration, clear transparent procedures, security of tenure from exploration through production, a stable fiscal regime, equal opportunity to foreign investors and access to international arbitration.

An orderly and complete sets of laws and operating codes are required. The legislative frameworks should be complete, internally consistent and competitive with other countries. The regulatory regime should provide investors certainty of long-term rights to minerals that are to be developed, and there is a dispute mechanism in place to provide investors with fair and unbiased arbitration in case of conflicts, with emphasis on a streamlined permitting process and provisions for stability within the regulatory framework.

The tax system in place should be competitive with other jurisdictions, allow an acceptable rate of return on investment, concentrate on profit-related taxes rather than duties and royalties, and be applicable to foreign, domestic and state enterprises.

The key monetary criteria include realistic foreign exchange regulations, allowance for external accounts, the ability to repatriate profits and the ability to raise external financing to fund projects.

(c) A sound environmental framework includes the need to have predetermined environmentally related obligations, identification of anti-mining lobby groups and host Government involvement in ensuring environmentally sustainable mining practices. This can be accomplished by introducing a consultative process with involved stakeholders, introducing necessary laws and regulations in consultation with world-wide industry and multi-disciplinary groups, developing technical norms and standards, collecting environmental baseline data and protecting locally affected groups and indigenous people.

Environmental protection and guidelines are increasingly becoming an important focal point for foreign investors. Continued emphasis on working towards universal, realistic standards would be supported by the international mining community.

(d) Infrastructure conditions, especially the transportation facilities are important to development of mineral resources. Nowadays, the promising mineral reserves are usually in such remote areas that transportation of equipment, workers and ore products is very difficult. This is one of the hindrances to developing minerals and trade in Mongolia and the far eastern region of the Russian Federation. In this regard, investment efforts should be pushed in this direction.

(e) The availability of reliable, well-organized information is provided. The information regards all laws, regulations, mineral potential, support services, the ability of the skilled workforce, and the geological potential. Timely advice should be given to potential investors about changes, inter alia, in policy, mining investment codes and geological potential. This necessarily involves assistance in the design of brochures, and participation at trade shows and conferences.

Establishing and maintaining good geological information systems oriented towards both the large international mining firms and the smaller exploration companies is also essential to attract the initial interest and inflow of investment. Having geoscientific information readily available will serve to reduce the cost, time and risk of exploration.

(f) Competent public mining institutions are established which might include, under the Ministry of Mines, a mining department, geological survey office and an environmental office. Institutions should be staffed by competent individuals and be adequately funded and their mandate, authority and accountability should be clear.

Implementation takes place in a specific cultural, political and burcaucratic context. These factors will decide whether a new mineral project will proceed on time and on budget or become a victim of delays, cost overruns or similar disruptions.

In this chapter, the foreign direct investment, the transnational technology, the local expectations and the legal framework and fiscal regime will be discussed.

3.2.1 Foreign direct investment

Foreign Direct Investment (FDI) is basically a private business phenomenon reflecting flows of capital, technology and marketing/management expertise from 'source or home countries' to 'use/recipient or host countries', having a good investment environment and opportunity to save costs or making profits. FDI is affected by both the 'pushing factors' of the source countries/companies and the 'pulling factors' of the use countries/ recipients. FDI, accompanied by expansion of intraregional trade, which follows as a consequence, can play an important role to reinforce the degree of inter-dependence and inter-cooperation between the source countries and use countries (figure 70).

World FDI in 1995

According to the World Investment Report 1996: Investment, Trade and International Policy Arrangements, issued 20 September 1996 by the United Nations Conference on Trade and Development (UNCTAD), total FDI flows into developed and developing countries surged by 40 per cent in 1995, to reach $315 billion. Around $100 billion of that volume went into developing countries, a gain of 15 per cent over 1994, and China alone attracted $38 billion.

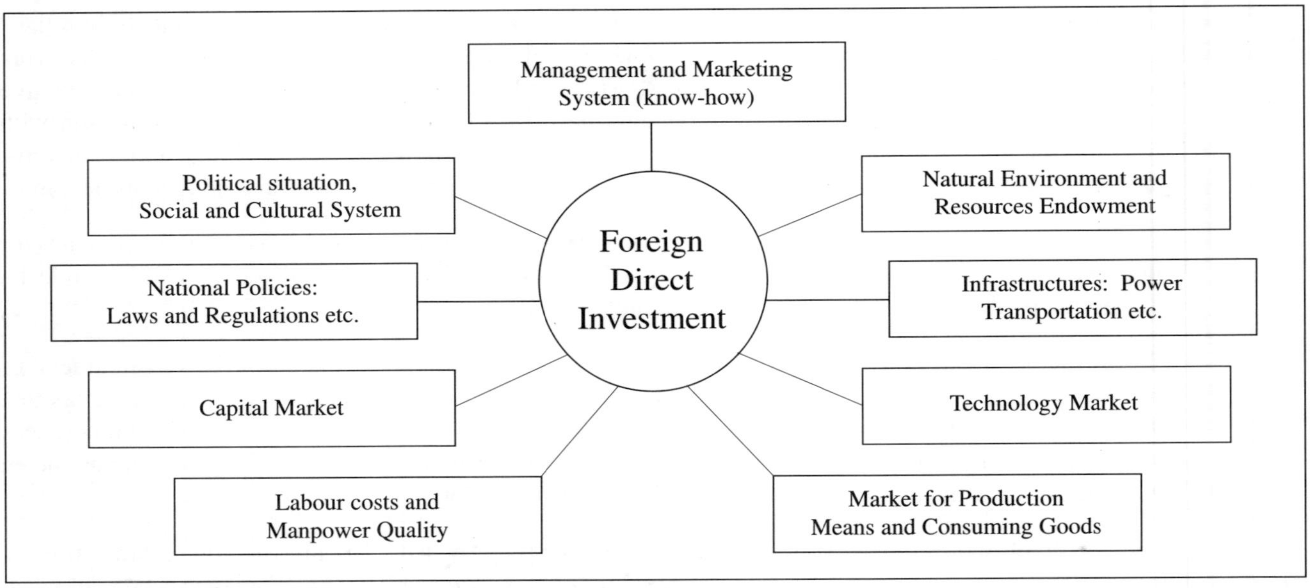

Figure 70. Factors that influence the FDI

The new FDI data show that the United States corporations as a group were the largest foreign investors ($96 billion) in 1995, and the United States was the largest single host country for FDI ($60 billion). The top 100 TNCs alone now account for about one-third of the total stock and they are all headquartered in developed countries. Firms in the United States, Germany, United Kingdom, Japan and France (in that order) dominated FDI outflows, with two-thirds of the global 1995 total of $318 billion. Japanese firms continue to expand their FDI, but the full scale of their investments is difficult to track, as a considerable volume may be originating today from Japanese affiliates in foreign countries.

The United States does more trade with Asia than with Western Europe, but there are more FDI flows from the United States to Western Europe. The FDI between Western Europe and the United States set new records and was dominated by mergers and acquisitions. Nevertheless, the rates of increase in FDI flows from the United States towards Asia are now greater than those inflows towards Western Europe, with recent annual rates of respectively 6.6 per cent and 4.3 per cent.

Inflow of FDIs are highly concentrated. Overall, FDI into the developed countries rose 53 per cent to $203 billion last year. The FDI into the United States amounted to $60 billion, 19 per cent of the total. China attacted $38 billion, 12 per cent of the total, ranked the second. The top 10 largest recipient countries accounted for 68 per cent of the $315 billion of inflows. The smallest 100 recipient countries received only 1 per cent of total investment.

Total FDI into developing countries rose 15 per cent to $100 billion. It has thus become the single largest item in net private capital flows to those countries, accounting for 54 per cent of the total. Although FDI to the 48 least developed countries rose 29 per cent in 1995, it only amounted to $1.1 billion.

An increasing amount of FDI into developing countries comes from corporations headquartered in developing countries, a trend that is likely to continue, according to UNCTAD. More uncertain is the degree to which FDI to developing countries will become more evenly spread. 65 per cent of all FDI to developing countries went to Asia. While the report suggests that FDI inflows to China may fall below $30 billion in the next few years, it finds reason to believe that this would be mainly a temporary adjustment rather than a response to a change in general economic factors. China will remain one of the top FDI destinations in the world.

More generally, 58 per cent of all remaining FDI flows to developing countries went into just 10 developing countries other than China. For example, FDI into Indonesia, Malaysia, Philippines and Thailand alone, by contrast, rose from $8.6 billion in 1994 to $14 billion in 1995. Despite increases in absolute dollar terms, levels of FDI flows into both south Asia and into west Asia remain quite modest, at $2.7 billion and $2.5 billion respectively.

UNCTAD lists the biggest TNCs in developed and developing countries in terms of their foreign assets. Again, Royal Dutch Shell of the United Kingdom/Netherlands leads the list, with foreign assets of $63.7 billion and total assets of $102.0 billion in 1994. Ford of the United States is second with $60.6 billion of foreign assets and total assets of $219.4 billion, while Exxon of the United States is third with foreign assets of $56.2 billion and total assets of $87.9 billion.

The United States firms account for one-third of the top 100 TNCs. The number of Japanese TNCs in the list is rising, with just 11 in 1990 and 19 in 1994. The world's largest 100 TNCs – just 0.3 per cent of all TNCs have dominant positions in FDI flows and international production.

The largest TNCs headquartered in developing countries, are becoming more important in global investment. Many of the largest ones are headquartered in Asia, and all of the top 50 are headquartered in either Asia or Latin America. The list is headed by Daewoo of the Republic of Korea, Hutchison Whampoa of Hong Kong, China and Cemex of Mexico.

A key element in the report is a survey of the largest 100 TNCs on their future FDI plans. Continued strong investment growth is projected. The North American TNCs continue to view Europe as the most important investment location for the future, especially in high-technology and consumer-goods industries. The European TNCs see the United States as the prime location for their FDI engagements. Meanwhile, the Japanese TNCs see Asia as their top prospective FDI priority. No slowing of FDI flows by the largest TNCs in developing countries is anticipated either.

Transnational corporations (TNCs) today are very free to choose how to service foreign markets, whether through trade or investment. As a consequence, decisions on locations by transnationals are increasingly shaping trade and investment flows at the same time. That poses a whole set of new challenges to policy makers.

FDI in North-East Asia

In the mid-1970s, foreign and private investment in the mining sector in many jurisdictions including the Asia and Pacific region, was politically identified as encroachmant on national sovereignity and an instrument of foreign domination over national economic development. Such investment was also thought to lead to an excessive drain on foreign exchange, a deterrent to industrial development and distortion of prices, actions detrimental to a developing country.

The above trends were rapidly reversed during the mid-1980s. Most countries were dissatisfied with their state owned, controlled and operated mining companies, which failed to achieve their purposes or did so at high cost. Moreover, they were faced with severe financial constraints such as inadequate local funds, difficulties in raising loans from international markets and severe debt burdens due to restrictions on commercial bank lending. Thus foreign investment oriented mineral projects became attractive, diversifying and restructuring of State-owned companies commenced and the governments of a number of countries began shifting development strategies towards relying on the private sector to lead economic growth. In line with the trends worldwide, over 75 countries have introduced or are in the process of introducing new mining laws/policies since 1985.

The majority of South-East and East Asian countries have also adopted policies designed to improve the environment for private investment. Until a decade ago, it was not possible to discuss FDI in North-East Asia because China, Mongolia, the Russian Far East and the Democratic People's Republic of Korea were more or less closed to FDI with the only exception being China which adopted an open policy in the late 1970s. Generally in the subregion, Japan and the Republic of Korea are regarded as potential investors and the rest of the four countries as potential recipients of the FDI.

China has been one of the largest recipients of FDI in the world in recent times. By the end of 1993, foreign capital accounted for 70 per cent of the total investment in China. FDI in China totalled US$61.797 from 1979 to 1993, in which US$27.515 (44.5 per cent) was contributed in 1993. About 80 per cent of foreign funded projects were in coastal cities in China. However, the FDI flows to China from within the North-East Asia subregion were relatively small. The Japanese firms have shared only 4.2 per cent of their investments in China over the period 1979 to 1993, but they accounted for almost 20 per cent of the total worldwide FDI.

For various reasons, the FDI flows in the subregion are very small. The total value of FDI in the Russian Federation as a whole by the end of 1993 was only about US$2,6 billion. The Russian Far East accounted for 10 per cent of some 6,500 operating foreign firms in Russia. Mongolia approved some 388 projects over the four year period (1990-94) after it opened its market to foreign investors but the total amount of FDI even on an approval basis reached only about US$50 million. It seems that there are almost no foreign firms operating in the Democratic People's Republic of Korea.

FDI by host region and economy in North-East Asia in the period of 1984 to 1995 is shown in table 117.

FDI In China

China has been one of the largest recipients of the FDI in the world recently. By the end of 1993, foreign capital accounted for 70 per cent of the total investment in China. About 80 per cent of foreign funded projects were in coastal cities in China.

The Chinese government enacted a regional "inclining policy", which encourages foreign investors to invest in coastal regions by providing preferential policies to these regions. By the end of 1993, 30 national Economic and Technology Development Areas had been established in China since 1979, and 13 bonded areas since 1990. The foreign funded enterprises approved in 1993 in the northeastern part of China in the Liaoning, Jilin, Helongjiang provinces are listed in table 118.

The priority sectors set by the government in the 1980s included energy exploitation, building materials, the chemicals industry, the metallurgical industry, the machine manufacture industry, and the electronic industry. However, service industries have also been emphasized since the early 1990s. It is said that less than 10 per cent of the FDI in 1995 went to the priority areas of basic infrastructure and power generation and only 1.4 per cent to agriculture.

In 1993, out of the total 83,437 projects with a total agreed upon foreign capital of US$111,435.66 million in China, those projects related to productive industry accounted for nearly 68 per cent, and those related to service industries about 30 per cent. There were 112 FDI projects with a total amount of US$743.82 million for the petroleum industry accounting for 0.664 per cent of the 1993 FDI total. There were 11 FDI projects related to geological prospecting with a total amount of US$80.53 million accounting for only 0.07 per cent of the 1993 FDI total in China (Ji, 1995).

The first five countries/regions investing in China in 1993 were Hong Kong, China, Macao (actual investment US$17.861 billion or 65 per cent), Taiwan Province of China (US$3.139 billion or 11.4 per cent), the United States (US$2.063 billion or 7.5 per cent), the Republic of Korea (US$1.5 billion or 5.5 per cent), and Japan (US$1.324 billion or 4.8 per cent). Among the investors, the Republic of Korea's operations developed very quickly with only one investment project of US$144,000 in 1985 but increasing to 2000 projects valued US$1.5 billion by the end 1993. Labour intensive and export-oriented enterprises were predominant among the investment by Hong Kong, China and Macao. Although the investment by western developed countries did not contribute a large amount in the total FDI in China, they mainly invested in capital and technology intensive projects.

Out of the total actual investment of US$61.797 billion in China from 1979 to 1993, equity joint ventures accounted for 53.23 per cent or US$32.894 billion, cooperative joint ventures for 21.93 per cent or US$13.552 billion, and wholly foreign-owned enterprises for 18.75 per cent or US$11.590 billion. The remaining 6.09 per cent or US$3.761 billion were from cooperation development (table 119). New ventures such as joint-stock companies have been actively promoted by the government recently. The BOT (Building-Operation-Transfer) method is also being researched.

The FDI has promoted the development of the Chinese socialist market economy and the reform of the economic structure. It has become one of the main channels for raising development funds. In addition to the fund, it also introduced advanced technology and management know-how to China and promoted industrial development and reformation.

Table 117. Foreign direct investment by host region and economy in North-East Asia, 1984-1995

	1984-1989	*1990*	*1991*	*1992*	*1993*	*1994*	*1995*
INFLOWS, in million US Dollars							
China	2,282	3,487	4,366	11,156	27,515	33,787	33,787
Hong Kong, China	1,422	1,728	538	2,051	1,667	2,000	2,100
Taiwan Province of China	691	1,330	1,271	879	917	1,375	1,375
Democratic People's Republic of Korea							
Japan	81	1,753	1,730	3,490	234	908	39
Mongolia							
Republic of Korea	592	788	1,180	727	588	809	1,776
Russian Federation							
OUTFLOWS, in million US Dollars							
China	581	830	913	4,000	4,400	2,000	3,467
Hong Kong, China	1,833	2,448	2,825	8,254	17,713	21,347	25,000
Taiwan Province of China	1,999	5,243	1,854	1,869	2,451	2,460	3,822
Democratic People's Republic of Korea							
Japan	20,793	48,024	42,619	21,916	15,471	18,521	21,286
Mongolia							
Republic of Korea	137	1,056	1,500	1,208	1,361	2,524	3,000
Russian Federation							

Source: Asian Economic Survey, 1996-1997, the Asian Wall Street Journal, Oct. 22, 1996.

Table 118. Foreign funded enterprises approved in northeastern provinces of China in 1993

	Total		*Equity joint-venture*		*Cooperation Joint-venture*		*Wholly foreign-owned*	
PROVINCE or city	*Number*	*Foreign capital*	*Number*	*Foreign capital*	*Number*	*Foreign capital*	*Number*	*Foreign Capital*
Total	83,423	11,113,104	54,003	5,517,427	10,445	2,549,998	18,975	3,045,679
Tianjin	3,558	225,570	1,466	122,253	65	5,098	2,007	98,219
HEBEI	2,013	192,862	1,762	148,562	90	29,641	161	14,659
INNER MONGOLIA	413	35,862	357	26,566	21	3,155	35	6,141
LIAONING	4,054	379,616	3,394	279,586	361	34,895	299	65,135
Shenyang	996	83,100	850	53,599	36	10,287	110	19,214
Dalin	1,655	178,244	1,256	113,504	261	21,596	138	43,144
JILIN	1,469	82,885	1,121	59,806	59	5,584	289	17,495
Changchun	295	22,860	236	16,834	11	1,389	48	4,637
HEILONGJIANG	1,705	98,329	986	53,936	32	5,426	688	28,967
Harbein	691	43,554	333	26,205	17	4,053	341	13,296
SHANGDONG	7,227	706,043	5,625	452,093	297	49,557	1,305	204,393
Qingdao	1,351	177,781	746	67,552	120	24,242	485	85,987
JIANGSU	10,257	1,084,308	8,994	748,244	448	134,372	815	201,692
Shanghai	3,645	698,902	2,443	441,697	691	135,490	511	121,715
ZHEJIANG	4,487	374,563	3,770	269,855	158	13,144	559	91,564
Ningbo	1,015	107,152	779	66,895	34	3,183	202	37,074

Source: Ji Hong, 1995. Foreign Direct Investment in China: Historical Trends and Investment Environment in Foreign Direct Investment and Cross Cultural Management in North-East Asia, 1995, the Sasakawa Peace Foundation.

Table 119. Investment project approved in China from 1979 to 1993

Ventures	*Number of Projects*	*Proportion (per cent)*	*Agreed Investment*	*Proportion (per cent)*	*Actual Investment*	*Proportion (per cent)*
EJV	113,041	64.88	105,529	47.60	32,849	53.23
CJV	27,245	15.64	57,061	25.72	13,552	21.93
WFO	33,847	19.43	55,442	24.99	11,590	18.75
CD	95	0.05	3,764	1.69	3,761	6.09
Total	174,228	100.00	221,896	100.00	61,797	100.00

EJV: Equity joint-venture

CJV: Cooperation joint-venture

WFO: Wholly foreign-owned

CD: Cooperation development

Source: Ji Hong, 1995. Foreign Dircet Investment in China: Historical Trends and Investment Environment in Foreign Direct Investment and Cross Cultural Management in North-East Asia, 1995, the Sasakawa Peace Foundation.

The FDI in China has also become the main power behind Chinese foreign trade. In 1993, the value of import and export trade of foreign-funded enterprises was $US67.07 billion which accounted for 34.4 per cent of the national import and export value.

The FDI in China improved employment and increased the national financial income. The FDI enterprises accounted for 9.5 per cent of national industry output value in 1993 and there were 10 million employees in the foreign-funded enterprises. The generalized view on risks of investing in China is shown in table 120.

FDI in northeast China

The northeastern provinces of China, including Heilongjiang, Jilin, and Liaoning, with a total area of 4.873 million sq km and total population of 101 million, are the thriving centres of Chinese heavy industries. The total value of its heavy industries rank first in China. But the aging equipment, obsolete technological knowledge and underdeveloped management systems of its industries need to be replaced or reformed. The output value of its iron and steel accounts for 25 per cent of the national total, oil for 50 per cent, coal for 20 per cent, machine tools for 20 per cent, and automobiles for 50 per cent.

Mineral deposits are abundant in these provinves. The Heilongjiang province has oil and gold. The Jilin province possesses 70 kinds of minerals. The Liaoning province has 115 kinds of minerals, including iron ore, talc, bentonite, and jade reserves which are the largest in China. Its Liaohe oil field is the third largest in China. Twelve kinds of non-ferrous metals have been found in the Liaoning province.

The trunk railway lines of Liaoning form a dense network with an operational mileage that ranks the highest in China. The highway between Henyang and Dailian is the largest in the country. The Dailian, Yingkou, and Dandong ports have now opened to the rest of the world. These ports have 148 berths out of which 28 could hold ten thousand tonnes of goods.

The FDI into the three provinces in 1993 is presented in the following table.

Province	*Number of signed projects*	*Actual Investment in US$*
Heilongjiang	1,329	232,320,000
Jilin	1,469	275,270,000
Liaoning	3,442	1,279,130,000

Table 120. Generalized view on risks of investing in China

Basic Risks (Country Risks)
- Political Stability
- Social stability
- Policy stability and macro-economic control
- Command economy and bureaucracy
- Legal and institutional setting

Contract Risks (Pre-investment Risks)
- Negotiation process
- Contract approval process
- Finding partners
- Evaluation of investment in kind
- Foreign exchange balance
- Local content
- Technology transfer
- Practice of market economy in China
- Export control of high technology (previous COCOM)

Operational Risks (Post-investment Risks)
- Operational approval process
- Decision making on profit sharing
- Infrastructure management
- Labour management
- Cost management (inflation)
- Procurement of local materials
- Local financing
- Domestic marketing
- Foreign exchange market
- Extra expenditures
- Protection of intellectual property
- Qualified manpower supply
- Public security

Source: Tan Ashide, 1995. "Japanese Direct Investment in China and its Implications for NEA" in "Foreign Direct Investment and Cross cultural Management in North-East Asia", 1995, the Sasakawa Peace Foundation.

Foreign capital came not only from Hong Kong, China, Macao, U.S.A., Japan, and the Republic of Korea, but also from Russia, Ukraine, Lebanon, Czech Republic, Panama and 33 other countries and regions.

FDI in the Democratic People's Republic of Korea

An initial commitment of $826 million worth of foreign investment has heralded a new era for industry in the Democratic People's Republic of Korea. The announcement of 14 tentative agreements with firms from abroad came at the end of the first investment forum for the Rajin-Sonbong Free Economic and Trade Zone in Rajin City, the Democratic People's Republic of Korea from 13 to 15 September, 1996.

Organized by the United Nations Industrial Development Organization (UNIDO), with funding from the Government of the Democratic People's Republic of Korea and United Nations Development Programme, the forum brought together some 400 foreign participants from 20 countries (mostly from Japan, China, the United States and the Russian Federation) to negotiate with 150 local project sponsors. Interest in the event was so great that some 850 potential investors from abroad wanted to participate.

More than a hundred prime investment opportunities in the above Economic Zone were on offer, which attracted collaboration worth $826 million during the forum for projects ranging from $180 million for a luxury hotel with a Hong Kong-based group to $257,000 for petrol stations with a Chinese company. Other investments announced by the end of the meeting were for a hospital, bank, tourist facilities, oil refinery, brick factory, paper mill, toy manufacturing, timber processing, vinyl plant, seaweed processing, prefabricated housing production and telecommunications equipment manufacturing.

Further agreements are expected with industries in the Rajin-Sonbong Zone as UNIDO builds on the interest generated by the forum through its network of Investment Promotion Service offices in Athens, Bahrain, Beijing, Istanbul, Milan, Moscow, Paris, Seoul, Tokyo, Vienna, Warsaw and Zurich.

FDI from Japan to China

In addition to developing a new market, securing a market already created and utilizing abundant and inexpensive manpower and raw materials in China, a new tendency has arisen since 1990 with Japanese companies starting to regard China as a promising site for relocating their manufacturing operations, serving both local and export markets.

Japanese FDI in China stayed at a marginal 4.7 per cent of the total investment outflow from Japan in 1993, well below the 8.5 per cent directed to the ASEAN five countries. According to Chinese statistics, Japanese FDI of the total investment inflow into China remained also marginal at 2.7 per cent in 1993, in contrast with it being a large ratio for Japan, approximately 20 per cent in 1993, in the total trading value of China.

The country risks in China are normally not of concern to Japanese investors. Contract risks have been greatly reduced since China has made efforts to clarify and rectify the procedures of pre-investment, but the operational risks are of more concern to Japanese investors today.

FDI in Mongolia

The present policy of the Government is to encourage the inflow of foreign investment, particularly in export-oriented and manufacturing industries. Since 1989, many new laws and regulations relating to FDI and trade have been adopted. Several bilateral agreements between Mongolia and developed countries have also been signed.

A government resolution passed at the end of 1989 allowed some large enterprises to handle foreign trade. The Law of Foreign Investment was passed and at the same time it became effective in March 1990. The Company Law was adopted in 1991, and all business, irrespective of ownership, became able to handle foreign trade. A uniform tariff rate of 15 per cent was set for most imports (Batbayar, 1995).

The main laws governing foreign investment cover organization of joint ventures, income tax, foreign exchange controls, customs duties, banking, civil procedures, trade mark protection, oil exploration, and development of minerals.

The Foreign Investment Law of Mongolia was significantly revised in May 1993 to stimulate the FDI. It permits land to be leased up to 60 years. Infrastructure projects, including highways, railways, and telecommunications, as well as power and thermal plants and their transmission network were defined as priority industries.

In addition, development of mineral resources, especially oil and coal, were treated very favorably. A number of significant incentives were permitted including the following: the mining and processing of mineral resources (except precious metals), oil and coal, metallurgy, metal processing, chemical production, machinery and electronics (which will enjoy 5 years of tax exemption and 50 per cent tax relief in the following 5 year period).

Mongolia has signed bilateral investment agreements with 12 countries including the Republic of Korea and China in North-East Asia. It has also signed double taxation agreements with China and the Republic of Korea.

The Ministry of Trade and Industry has approved 388 foreign investment projects during the period 1990-1994, totalling about US$50.299 million. Investors were from Russia (150 JVs), China (123 JVs), Hong Kong, China (19 JVs), Japan (17 JVs) and the US (17 JVs). By committed amount of foreign capital, the list included Russia with US$17.994 million, China with US$9.084 million, the United States with US$7.948 million,

Hong Kong, China with US$4.084 million and Singapore with US$1.724 million (table 121). Among the 393 JV projects, 9 were for mineral resources exploration and development, 11 for metalworking, 3 for making of dynamite (for mines) and 5 for transportation.

Table 121. Foreign investment projects in Mongolia approved by other North–East Asian countries, 1988-1994

Unit: US$1,000
1994.06.20

	1988-1990		*1991*		*1992*		*1993*		*1994*	
	A	*B*	*A*	*B*	*A*	*B*	*A*	*B*	*A*	*B*
China	3	688.5	8	1,069.2	36	3,004.1	51	3,505.8	25	816.93
Hong Kong, China	1	84.0	4	1,152.2	4	1,138.5	7	1,525.0	3	184.30
Taiwan Province of China	–	–	–	–	1	56.0	1	600.0	–	–
Democratic People's Republic of Korea	–	–	1	268.5	–	–	–	–	–	–
Republic of Korea	–	–	–	–		–	1	11.1	6	2,545.50
Japan	2	200.0	1	62.0	2	46.1	8	180.3	4	222.25
Russia	1	221.3	24	6,965.4	55	8,163.6	44	1,045.3	26	1,548.90

A – Number of approved projects
B – Capital invested by a foreign partner

Source: Ministry of Trade and Industry, Mongolia, quoted by Ts. Batbayar, 1995. 'Foreign Direct Investment in Mongolia: Problems and Perspectives' in Foreign Direct Investment and Cross Cultural Management in North-East Asia, 1995, the Sasakawa Peace Foundation.

One of the most attractive features to FDI in Mongolia is that it is rich in mineral resources. One of the most challenging problems for development of its mineral resources, however, is the extreme climate and poor infrastructure. The joint development projects could be based upon the principle that the investors would supply the required equipment on a long term deferred payment basis and in turn would be provided with an agreed quantity of minerals.

FDI in the Russian Federation

As early as 1991, a dynamic process of establishing JVs began in the RFE. By August 1991, 126 JVs were registered in the RFE, the majority of them in Primorsky and Khabarovsk territory. The most active participants were Japanese companies (43 per cent), the US (18 per cent) and China (15 per cent) of the total. Initially, these companies were more active in selling goods and services in the domestic market than in export operations. A major part of them was in services (43 per cent) and fisheries (33 per cent). The number of operating enterprises in the RFE in 1989-1991 had grown almost 15 times, while in Russia, as a whole, by only 3.7 times. By the middle of 1992, there were 445 companies registered in the region and by the beginning of 1994 over 1500. The total value of FDI in the RF by the end of 1993 was estimated at US$2.6 billion.

About 88 per cent of total ventures in the RFE were in the southern part of the RFE and its Sakhalin region. This is no surprise since the Primorsky and Khabarovsk territories, as well as the island of Sakhalin, possess a relatively well developed transportation and communications structure and good opportunities for developing cross-border trade with the neighbouring countries. In addition, the island of Sakhalin and the area around the city of Nakhodka in Primorsky territory were granted the status of Free Economic Zone.

The average value of FDI began to decrease from 1992 when the central government abolished all the preferences for foreign investment. The observation is confirmed by table 122, whereby the average foreign investment amount per joint venture in the RFE has dropped significantly during 1991-1993. Although the number of FCF increased in 1993, the average value of the FDI decreased greatly.

Table 122. Average foreign investment per joint venture in the Far Eastern region of the Russian Federation

(US$1,000)

	1991	*1992*	*1993*
Japan	1,038	163	19
China	243	71	30
USA	1,850	14	13
Republic of Korea	100	45	39

Source: Minakir, P.A., Admidin, A.G., 1995. 'Trends in Foreign Direct Investment in the Russian Far East' in Foreign Direct Investment and Cross Cultural Management in North-East Asia, 1995, the Sasakawa Peace Foundation.

In 1993, the share of foreign capital in regional exports decreased from 25.1 per cent to 13.1 per cent (the fishery exports decreased almost two folds although fish and sea products continued to dominate in the RFE exports), while imports declined from 18.1 to 10.8 per cent. The share of timber products decreased almost 20 per cent. The rise in the share of ferrous and non-ferrous metals from 4 to 7 per cent of the total exports was not able to compensate for the contraction.

The earnings from foreign trade operations was the major attraction for foreign investors. The attractiveness of the domestic market decreased in 1993 due to the introduction of limitation in foreign exchange regulations.

In 1993, Japanese companies accounted for 47.5 per cent (exports 59.5 per cent and imports 25.8 per cent) of total foreign trade Chinese companies accounted for the second position in imports, surpassing the Republic of Korean companies which accounted for almost half of total JVs in 1992. In 1993, the number of JVs with Chinese participation accounted for 40 per cent of the total JVs in RFE (27.6 per cent in 1992). The second position was Japan which accounted for 15.5 per cent (22.3 per cent in 1992).

3.2.2 Transfer of technology

Technology and development

The industrial success in East and South-East Asian countries has been contributed to by an outward-looking trade policy with export-oriented strategies. However, successful industrialization implies more than simply adding physical resources or increasing production over a brief period. It also implies that the capacity is built and utilized efficiently, and that growth is sustained over the longer term by increases in productivity and competitiveness. The East Asian NIEs have a generally higher human capital base than other developing countries. Technological efforts related to the extent and direction of local investments in technology, contribute to the extent and form of technology imports.

Recent technological breakthroughs, particularly in the fields of electronics and information, biotechnology, new materials, and renewable energy technologies, are currently reshaping the basic parameters of goods and services production. The newly emerging technologies, which are highly knowledge- and research-intensive, are certain to have a profound impact on economic and social activities in all countries, with a world-wide restructuring of production, trade, and investment. These changes will definitely affect the future scope for industrialization in developing countries and have important implications for their integration into the world economy.

In development and management of the fuel and non-fuel mineral sectors, computers, micro processors and equipment and their diverse applications have been widely used in data collection (such as remote sensing and GPS), data processing (such as 3-D seismic processing) and data presentation (such as GIS) in all of the exploration, development and exploitation stages. Although developments of bio-technology are still at an early stage, the mineral sector is among a few sectors to which its applications are confined.

New materials that are being developed include powder metallurgy materials, low-alloy steels, advanced ceramics, polymers, and composites. These developments may have far reaching negative implications for the mineral exporting Asian countries, but positive influences also in terms of substitution of some raw minerals. The demand for copper, steel, tin, and lead is likely to go down, because of the gradual replacement of copper cables by optical fibers and the increasing use of composites and synthetic materials in the automotive and other industries. The countries that are currently dependant on income from those minerals then have to find new products for their export earnings.

The development of new and renewable energy technologies, including nuclear and solar energy, mini-hydro stations, wind power, geothermal energy, and biomass will in the long run lower the demand for coal and oil.

The impacts of the new technologies on mineral resources development and management are far greater than the technological and economic aspects, affecting the whole structure of the mineral industries, from upstream to downstream. For example, the advantage of cheap labour in developing countries will be gradually eroded. Since the world is moving fast towards information economies, manufacturing will lose its importance. Thus reliance on cheaper labour for comparative advantage, a strategy that worked for a manufacturing-based economy, will no longer be sustainable in an information economy.

To successfully compete in the new information economy, one prime ingredient is higher education. Productivity convergence during the 1950-1985 period relied mainly on good primary and secondary schooling, because it was based on manufacturing and thus the need of a labour force that could function in a factory environment. During the 1990s, productivity catch-up and technology transfer will likely depend on a broad-based university trained labour force, which can manipulate an information-based software.

Space technology

The French-launched satellite SPOT images have high spatial resolution: 10 m in panchromatic mode and 20 m in multi-spectral mode. American satellite Landsat 5 thematic mapper deliver high-quality thematic mapped scenes. These, combined with SPOT P and SPOT XS scenes, provide satellite images for customers all over the world.

In the Asian and Pacific region, there have been 30 national remote sensing programmes/centres, 11 earth observation satellite ground receiving stations and three comprehensive national remote sensing satellite programmes, as well as mushroom development of the spatial information industry.

After more than two decades of development, remote sensing technology has now become operational and indispensable for providing information regarding natural resources and environment. Remote sensing data sets can offer spatial resolutions varying from 10 to 1,000 metres in various spectral bands. The costs range from a few dollars per sq km in digital form of raw data to a few hundred dollars per sq km in the form of value-added information.

Major developments of remote sensing technology in recent years include high spatial resolution, hyperspectral imaging, and imaging radar technology. Most of earth observation satellites planned for the next decade will provide data with a spatial resolution of 1-3 metres, an improvement of one order of magnitude as compared to that available in current satellite systems. Geographic information systems (GIS), developed independently but in parallel with remote sensing, have become another important tool. The integration of remote sensing and GIS will soon lead to the true operationalization of spatial information infrastructure for natural resources management and environmental monitoring in the next decade.

Thermal infrared remote sensing data provide information notably for silicate and carbonate minerals and expand the capability for delineating lithologic units. Imaging spectrometers allow for detailed lithologic identification based on mineral content, such as iron oxide and hydroxide minerals, used especially for hydrothermal alteration zonation mapping. Radar imagery shows variations in fold geometry and fractures in the rock that are controlled by the fault of basement rocks.

Remote sensing data have been used in the mapping of anomalous limonite concentrations for finding hydrothermally altered rocks, identifying clays and other OH-bearing minerals, and interpreting linear structures to help characterize structure setting. Multispectral and thermal remote sensing data as well as radar images have been used together to interpret geological structure important in the accumulation of petroleum deposits, sedimentary facies affecting petroleum accumulation and possible carbonate accumulations.

The new generation of remote sensors will be able to identify subtle changes in rock composition. Cloud-penetration and whole-weather capability make it possible to acquire information over tropical regions.

Earth resource satellites, unlike meteorological or other environmental satellites which are operated as a public service with data made available at low cost, provide earth resources data which are expensive. Developing countries have often pointed out that commercialization policies substantially limit the use of remote sensing technology in those countries, particularly for such purposes as environmental monitoring and protection. They have also expressed concern that changes in satellite technology can require expensive upgrades to ground stations and processing equipment. Although a number of proposals for a system of concessionaire prices or access fees for developing countries have been put forward, there have so far been no results.

Heap leach technology for gold extraction

While heap leach technology is well proven as a means of extracting gold from low grade deposits, its application has, until recently, been resident with companies operating in North America and Australia. The transfer of this technology to developing countries and countries in transition, as a result of change to the respective business environments, is significant.

The availability of a competent workforce and the ability to utilize technology advances at all levels of operations is critical to the ongoing ability of companies to operate efficiently in developing countries and economies in transition.

The policy of TNCs in technology transfer

Although there already exists a range of commercial channels through which mine operators can purchase capital goods, engineering services and design specifications, the market for knowledge and expertise, including training programmes, is less mature. Therefore, the new technologies are transferred mainly through the investment mechanism which make them embodied and diffused. Advanced technologies can either be imported from developed countries or equipped through the involvement of transnational corporations. Foreign direct investment (FDI) and joint ventures with foreign investors including transnational corporations (TNCs) have already significantly brought the new technologies into developing countries, and they are likely to play a greater role in the future.

The TNCs undertake a major part of the world's research and development (R & D) efforts and produce, own, and control most of the advanced production technology. R & D is crucial for TNCs, since such efforts create ownership-specific advantages that enable firms to create advantages in foreign countries. They have to protect their intangible assets and other similar advantages needed to make foreign investment possible. The more modern and complex the technology, the less willing the TNCs are to accept any arrangements other than wholly-owned subsidiaries, in order to avoid leakages. The cost of transferring specific technologies decreases with increasing capabilities in the host economies.

The main technology exporter to the Asian countries is Japan. As a newly industrialized economy (NIE), the Republic of Korea made big domestic efforts to gain competence in science and technology that to a large extent explain its impressive economic performance during the last decades. For example, computer companies from the Republic of Korea, together with the Taiwan Province of China, have obtained market shares in the United States during 1980s.

Since the newly emerging technologies are so advanced and expensive to develop, developing countries may wish to develop their capability to use rather than to produce them. Thus, less emphasis should be put on generation of new ideas and developing new, cutting-edge technologies, than on widespread dissemination of technological capacity throughout their economies and using the advanced technologies. This might be an appropriate strategy for technology development in a developing country.

The host countries will gain from the training of labour and management which takes place in the TNCs and may then become available to the economy in general. But the acquisition of industrial capabilities is not an easy, automatic or a less cost process. The full range of industrial capabilities needed by modern industry is very large and can be grouped under three broad headings – entrepreneurial, managerial, and technological. The technological capabilities at firm-level can be categorized by functions related to three general sets of activities, i.e. investment, production and linkages (table 123). The matrix also show the degree of complexity of the capabilities.

Traditionally, technology transfer has meant a transfer of capital goods, engineering services and equipment designs – the physical items of the investment, accompanied by training in the skills and know-how for operating plant and equipment. As a consequence, the innovative capacity of recipients is undeveloped and they remain purchasers and operators of imported plant and equipment. This is the case especially in developing countries, as recipients become dependent upon their suppliers to make changes or improvements in successive vintages of technology. Contractual conditions may reinforce this situation.

New forms of technology transfer need to go further so as to embrace, first, the knowledge, expertise and experience required to manage technical change – of both an incremental and a radical nature – and, second, the development of human resources to implement organizational changes to improve overall production efficiency, energy efficiency and environmental management throughout the plant and facility, from mine development through production, to waste treatment and disposal. It is the capacity to effect technical change, not just the skill to operate an item of technology, that will ultimately determine the success with which recipient firms build up and sustain competence in capabilities to implement innovation.

Table 123. Illustrative matrix of technological capabilities

			FUNCTIONAL					
			Investment		*Production*			*Linkages*
			Pre- investment	*Project execution*	*Process engineering*	*Project engineering*	*Industrial engineering*	*within economy*
DEGREE OF COMPLEXITY	BASIC	SIMPLE, ROUTINE (Experience based)	Prefeasibility and feasibility studies, site selection, scheduling of investment	Prefeasibility and feasibility studies, site selection, scheduling of investment	Debugging, balancing, quality control preventive maintenance, assimilation of procees technology	Assimilation of product-disign, minor adaption to market needs	Work flow, scheduling, time-motion studies. Inventory control	Local procurement of goods and services, information exchange with suppliers
	INTER-MEDIUM	ADAPTIVE DUPLICATIVE (Search based)	Search for technology source, Negotiation of contracts. Bargaining suitable terms. Information systems	Equipment procurement, detailed engineering, training and recruitment of skilled personnel	Equipment stretching, process/raw material adaption and cost saving, licensing new technology	Product/ quality improvement, licensing and asimilating new imported product technology	Monitoring productivity, improved coordination	Technology transfer to local suppliers; coordinated design; S and T links
	ADVANCED	INNOVATIVE RISKY (Reasearch based)		Basic process design. Equipment design and supply	In-house process innovation, basic research	In-house product innovation, basic research		Turn-key capability, cooperative R and D, licensing, own technology to others

Source: ESCAP, 1994, Transnational Technology (ST/ESCAP/1071).

Competition, particularly in international markets, provides the most basic spur or incentive to 'learning' and innovation, but it is a double-edged sword. It can destroy a new entrant, especially from a developing country. Nevertheless, protection intervention for sheltering domestic markets has to be selective and must be balanced, as far as possible, with the need to provide a competitive spur to capability acquisition.

Skill development arises partly from simple learning-by-doing, but largely it is dependent on training. Technological effort covers routine technical activity as well as the search for new information, the integration of brought-in-technology and formal R & D. Organizational investments include all the efforts needed to improve management and marketing.

At national level, a proper mix of market and government interventions and improvements both in human resources and in science and technological infrastructure may speed up the technology transfer by TNCs and the diffusion of this technology to local firms.

Technologies to small and medium enterprises (SMEs)

There are many barriers that hinder the transfer of environmentally sound technologies (ESTs) to small and medium enterprises (SMEs). A major barrier to SME technology transfer is the lack of appropriate resources to obtain information and to explore the opportunities for technology cooperation. Not many SMEs employ highly qualified scientists, engineers and professionals to address the complex issues involved. In addition, SMEs rarely have highly developed management structures and support mechanisms. Other barriers for SME technology transfer include the short term focus on production, restricted cash flow, relative immaturity and high birth and death rates. In addition, SMEs are not always fully aware of the value of technology transfer as a strategic business tool.

The Expert Group Meeting (EGM) on Transfer of ESTs among SMEs which was organized in New Delhi, India, from 22-24 January 1996, by the Asian and Pacific Centre for Transfer of Technology (APCTT) recommended for improvement of the environment for commercial EST transfer in developing countries as follows:

1) Clarifying the concept and approach to ESTs. The life cycle approach should be promoted instead of pollution control or prevention alone. ESTs are more than mere physical facilities or technologies, involving human capabilities, know-how, documented facts and organizational frameworks.

2) Creation of awareness by SMEs as well as by governments for the value of ESTs, potential threats and opportunities created by them, and options available.

3) Development of enabling policy framework for encouraging the use of ESTs by SMEs; Promotion of market-based instruments (MBIs) based on the balanced application of command & control regulations, as well as incentives.

4) Technical and managerial capacity-building in SMEs, and among technology transfer intermediaries. Consideration should be given to gaps in needs and capabilities between small enterprises and medium enterprises.

5) Information on ESTs and on their sources that is digestible by SMEs, by policy makers and by intermediaries.

6) Access to financing and other resources.

7) Support structures and the role of intermediaries need to be strengthened, as well as linkages between them and SMEs.

8) Enhance cooperation and networking among SMEs on issues related to ESTs.

9) Meeting international standards and requirements in order for SMEs to gain international competitiveness (e.g. ISO 9,000 and ISO 14,000).

10) Cooperation among national and development assistance institutions needs to be enhanced in order to avoid overlapping and to optimize resources.

3.2.3 Local expectations

Local population concerns highlight improvement of living standards, access to opportunities, increase in social equity and environmental considerations. The mining project should benefit the economy as a whole, encourage and develop local industries to meet the needs of the people and should offer an opportunity for education and training of nationals and also ensure technology transfer.

The host Government has responsibility to the local population concerns. Poverty alleviation, basic education, job training, health infrastructure, project funding for the poor, labour equity, including minimum wage and job security, and movement towards pension provisions should be government social policy, with programmes related to the mining projects.

Many companies are realizing that the "stakeholders" include workers, neighbours, investors, community and the environment itself and the best political risk insurance available is the full support of the local community in which they work. Assisting in capacity building of the community, including supporting local businesses is becoming increasingly important, with job-training, health and welfare programmes, education and assist the development of local enterprise. Another key element in attaining the goal of sustainable development is having effective participation in the decision-making process of local communities that surround a potential mine. The World Bank recently conducted a review of 20 to 25 projects and concluded that effective public participation resulted in the sustainability of long-term objectives of both communities and projects.

Small-scale mining

Small-scale mining, particularly artisanal mining, is a labour-intensive activity that has seen a world-wide resurgence over the past 25 years. It provides employment to a significant number of people, most of whom live in remote areas where there are few other well-paid job opportunities. People working in small-scale mining, particularly artisanal miners, are generally unskilled laborers with little, if any, formal education. The average incomes earned from mining, however, are usually higher than subsistence wage levels, and are often higher than incomes paid in comparatively formal employment in other sectors of the economy.

A summary of small-scale mining employment in selected countries for which estimates are available is provided in the table 124. The number of people active in small-scale mining world wide exceeds 6 million – more than 20 per cent of those active in the mining industry as a whole. Of the more than 6 million people, a significant number are women. In many high-density artisanal mining districts, the percentage of females ranges from 10 to 50 per cent of the workforce. Women often work long hours as ore and concentrate carriers, panners, ore sorters or cooks, and most of them are also engaged in child-rearing.

In China, as of 1992 there were 167,351 privately run small-scale mines, of which 82,921 cover coal, metals and building construction material, with the other 84,430 mines in clay and quarry products for the construction of highways, railway lines and irrigation projects (Jin, 1995). A large proportion of the small-scale mining is coal mining. From 1980 to 1988, the annual output of coal produced from small-scale coal mines accounted for 53.6 per cent of the total national output. During that same period, the total number of local coal mines increased from 20,000 in 1980 to over 80,000 in 1988, of which 79,000 mines were run either jointly or independently by counties, towns and individuals (see TCD/NRED/E. 13, p. 23). The growth was encouraged by a series of policies implemented during that period by the Chinese Government. More recently, however, due to environmental damage from the small-scale mines and serious concerns about mine safety, the Government has been scaling back the preferential treatment.

Based on the conclusions and recommendations made by the international round table on artisanal mining in May 1995 in Washington, D.C., the World Bank has subsequently proposed an assistance strategy, focusing on legal and technical actions, that emphasizes a partnership with international organizations, non-governmental organizations, government entities, artisanal miners and international mining companies. It was indicated that serious efforts were needed in (a) small-scale mining regulation and promotion, particularly to mitigate

Table 124. Small-scale mining employment in selected countries in the world

(Thousands of workers employed)

Country	*Estimated Employment*	*Data source*[a]
China	3,000	Jennings (1993)
Brazil	1,000	Davidson (1990)
India	500	Chakravorty (1989)
Zaire	500	Jennings (1994)
Indonesia	465	ILO (1990)
Philippines	250	Muyco (1993)
United Republic of Tanzania	100	Noetstaller (1994)
Mali	100	world Bank (1992)
Sierra Leone	100	World Bank (1992)
Bolivia	70	Priester (1996)
Burkina Faso	60	World Bank (1992)
Guinea	60	World Bank (1992)
Ghana	30	World Bank (1992)
Angola	30	World Bank (1992)
Zambia	30	World Bank (1992)
Zimbabwe	30	World Bank (1992)
Peru	20	ILO (1990)
World total	6,345	Jennings (1993)

Source: Noetstaller (1995), with additions from Muyco (1993) and Priester (1996).

environmental damage, minimize social upheavals and curb smuggling; (b) institutional and capacity-strengthening; and (c) technology transfer. The overall objective of such assistance is to transform informal mining into environmentally sustainable, legally structured, formal mining.

Recently, many large companies seeking to establish operations in developing countries have concerned themselves with the small-scale mining issue, establishing specialized divisions that deal with community relations. Conflict resolution and the practice of placing the firm's presence in a position of mutually beneficial coexistence with the local community are now being incorporated into the development plans of many large-scale ventures. In some cases, artisanal miners are given marginal parts of the company's concession and/or are allowed to rework tailings, and some companies are also willing to extend credit services and financing. Cooperation and assistance can take a variety of forms, including joint ventures with local entrepreneurs in mining and the provision of ancillary supplies and services.

3.2.4 Legal framework and fiscal regime

Legal framework

A coherent legislative framework, and a stable and reasonable tax system are essential before international mining companies will invest in exploration and/or mining projects in a country.

In most countries, the mining code defines the basic framework under which exploration and mining will be regulated. The legal requirements regulating the mining process may be all-inclusive within the mining code or they may be found in a variety of laws dealing such matters as environment, labour and safety. Substantial transactional costs may be incurred by a company in a country where many different laws are applied to mineral projects.

If the mining code is not available, bad, out of date, vague or silent on important matters, then a mineral investment agreement will mostly be needed. A simplified mineral investment agreement may attract smaller companies because they have limited management time and budget to spare on protracted negotiations. For large projects, however, companies probably prefer more flexibility, and a simplified agreement may not be adequate.

Mineral ownership is perhaps the first issue of concern to investors. Who owns the mineral? The answer may be the land owner, the state, the state in the name of the people, the ruler, the ruling council, both the local and national governments, all the private interests, and so on. Problems may occur where different owners have conflicting claims. In addition, the regulatory system should optimally identify at what point, if any, in the extraction/marketing process the ownership of the mineral becomes vested in the company.

Many companies will not invest in any project in which they do not hold majority equity ownership because they think it is doubtful that the local investment capital is available. However, others may prefer minority equity ownership because it is obvious that local partners can offer assistance in spreading risk, accessing finance, understanding the bureaucracy, deciphering the local geology etc.

One of the most important decision criteria for investment in a country by exploration/mining companies is whether the national regulatory regime provides the linkage for the transition between the exploration phase and the mining phase. Very few companies will undertake exploration unless the right to mine is also assured.

The mining regulatory regime should also clearly define the rules for the company, whether it will only provide rights to exploration, and perhaps transfer ownership to others for development and mining. Many smaller firms nowadays specialize in only a single aspect of the mineral sequence, exploration or mining. With regard to the mining phase, the regulatory system should spell out clearly under what conditions the right to mine may be cancelled and whether the right may be renewed.

Since the exploration and mining activity is often a lengthy and complex business, there is a possibility that disputes between the government and the investor may arise. In many instances, large sums are involved, and the investor may prefer or require recourse to international arbitration. Otherwise, a strong independent judiciary should be well established and the available national or local court be familiar with hearing mining cases to make sure the dispute is settled in a fair, neutral and transparent.

Fiscal regime

Taxation

Some countries legislate and publish tax rates that are uniformly applied to all companies, while others rely on negotiations to set some or all tax rates on a project-by-project basis. Some exploration companies will not undertake exploration until the rates that would apply to the development of a discovery are known.

A tax holiday, especially if a fairly long period, will be considered an added enticement to investment by almost all mining companies. Considering the high-cost/high-risk nature of exploration and the great potential of wealth creation through the development of economic mineral deposits, many governments provide a few years tax holidays.

Of importance is the net mineral tax liability imposed on the investor no matter what methods could be used. However, taxes that are not profit-based such as royalty tax will be more harshly viewed by the investor than taxes that are profit-based such as income tax.

Royalty tax

A royal tax (also called a production tax, products tax or severance tax) is levied by most nations and is not usually based on profits but is calculated on the basis of the physical quantity or value of the mineral being produced or sold.

A unit-based royalty tax, as one of the oldest types of mining taxes, either equals a prescribed fee per unit weight or per unit volume. The ease of calculation is apparent for some types of mines, particularly quarries and coal mines, which produce a more or less homogeneous product. However in many cases, it can be more difficult to monitor a unit-based tax than a tax based on some measure of sales value. Well defined reporting requirements and penalties are then necessary.

A unit-based tax is used by many countries for some types of commodities that are fairly homogeneous, such as sand, gravel, dimension stone and coal. It is less seldom applied to metallic minerals. Particular difficulties are encountered in trying to apply such a scheme to polymetallic deposits or to mines that sell the mineral at several stages of recovery as ore, concentrate and metal.

The inability of unit-based taxes to differenciate between low- and high-value materials that fall within a commodity group tax classification may result in significant loss of revenue to the government. It precludes an increase in revenue in times of rising prices although it also provides revenue stability. It also fails to keep pace with inflation.

Income tax

Income tax is profit-based. Income tax is the percentage (tax rate) of the taxable income which equals the estimated project's revenues minus the costs and allowances. Some of the tax deductions/credits are outlined below.

Deduction of exploration costs. If the exploration expenses could be carried forward as deductions against future earnings from a mine, it is an attractive feature to almost all mineral companies.

Accelerated depreciation. Although the total amount of depreciation allowable as a tax deduction does not change, the company is allowed to make parts of such deductions earlier in the project's life. This is welcome by the investors for the time value of money.

Investment credits. Many countries allow some types of investments to be credits against the tax liability. A credit of a given amount, which reduces the tax liability, has a greater benefitt than a taxable income deduction, such as accelerated depreciation, of that same amount.

Reinvestment credits. This could be a major enticement for some companies, such as those who wish to start out small then expand an operation based on internally generated cash flow.

Expatriates exempt from income tax. Exploration companies are particularly sensitive to this type of tax because no revenues are coming into the exploration stage and they have to pay the expatriate (a highly qualified employee) with a high salary.

Repatriation of profits. The ability to repatriate profit is essential for most foreign mining companies in order to make a positive decision. Countries that restrict the repatriation of profits, or tax such remittances heavily, will discourage potential mineral sector investors.

Countries have many options to choose from in designing a mineral-sector tax system. A carefully designed tax system will encourage the creation of a large tax base over the long run and will avoid heavy taxation, which discourages the exploration and long-term development of the mining industry. There is no perfect mining tax system adequate to any country. Every country has its own unique approach to taxation, one that takes into account the cultural background, level of training, administrative capability, political philosophy and expectations of the government.

Foreign exchange regulations. Countries whose currencies are not easily convertible in the world marketplace may not be attractive to investors. In many developing countries, the internal mineral markets are small, and for many mines the majority of the production is sold abroad for hard currency. The investors also need hard currency to repay loans, purchase equipment and supplies, and pay other expenses which are denominated in hard currency.

An overview of financial instruments is shown in table 125.

Table 125. An overview of financial instruments

(1) Financial Instruments in General		
Instruments	Description	Characteristics
Forward	– An agreement to purchase or sell a given asset at a future date at a preset price. – Transactions are made mostly through brokers by phone and telex. – A typical use is for locking-in a future price	– No cash transfer is needed at the beginning. Cash transfer occurs only at maturity. – Credit risk is involved. – Tailor-made contracts are avaiable for specific hedging needs. – Contracts are available primarily for short-term maturities up to one year).
Futures	– An agreement to purchase or sell a given asset at a future date at a preset price. – Transactions are made in formal exchange through clearinghouse systems. – Contract terms (amounts, grades, delivery dates, etc.) are highly stanardized. – Profits and losses are settled daily, requiring daily cash flows. – Margin (collateral) money is requied at the beginning. – A typical use is for locking-in a future price.	– Initial cash transfer is required for margin money. – Daily cash transfers are necessary. – Credit rick is minimal. – Tailor-made contracts are not available. – Contract are available primarily for short-term maturities (up to one year). – Markets are more active than forward markets for some contracts. – An original position can be closed or reversed easily and quickly.
Option	– The right to purchase or sell a certain asset at a preset price on (or before) a specified date. – Transactions are made both through brokers by phone and telex and in formal exchanges. – A typical use is for setting a ceiling or floor for prices.	– A buyer of an option contract can limit the maximum loss, but keep an opportunity to take advantage of favorable price movements. – A buyer has to pay a premium (cost of option) up-front. – A buyer faces a seller's credit risk. (A buyer has the right: a seller has the obligation). – Tailor-made contracts are available for specific hedging needs. – Contracts are available primarily for short-term maturities (up to one year).
Swap	– An agreement to exchange specified cash flows at fixed intervals. – A series of forward contracts lined up on a shedule. – Transactions are made through brokers by phone and teles. – A typical use is for locking-in future prices for a long period.	– No cash transfer is needed at the beginning. – Credit risk is involved. – Tailor-made contracts are available for specific hedging needs. – Contracts are available for medium – and long-term maturities (one to ten years).

Table 125. *(continued)*

(2) Commodity-Linked Instruments		
Instruments	Description	Characteristics
Commodity Swap	– a swap contract on a certain commodity. An agreement to pay a pre-fixed amount of cash in exchange for a variable amount of cash at fixed intervals, or vice versa. A variable amount of cash is determined by the market price for a set quantity of a commodity. A fixed amount is based on a fixed price for the same quantity of the commodity. – Contracts are procided by international banks. – A typical use is for locking-in a price of a commodity for medium–and long-term	– No deliverise of physical commodities are involved. transactions are made as purely financial, as the other swap contracts (see the swap saction above for characteristics of swap countracts in general). – The markets are not very active.
Commodity-Linked Loan	– A loan in which interest and/or repayment amount are linked to the market price of a certain commodity. – A loan can be viewed as a combination of a conventional fixed-rate loan and a commodity swap contract. – These loans are procided by international banks. – (Forward-type) A bond in Which coupons and/or principal are linked to the market price of a certain commodity. – (Option-type) A bond to which the right to buy or sell a certain commodity at a preset price is attached. – These bonds are underwritten by international banks.	– A loan can be regarded as effectively denominated in a commdity. – Credit risk of the loan is lower than that of a conventional loan, if used by a commodity producer. A producer can repay the loan even if the price of the commodity fell significantly. – (Forward-type) Characteristics are similar to commodity-linked loans. – (Option-type) this type is often useful for commodity producers to reduce the cost of financing. – The bonds have been issued primarily on gold and oil. Some are available for silver, copper, and nickel.

Source: "Modern financial Techniques: A Primier for Asset and Liability Management in the Development Countries" in United Nations Publication on Minerals and Metal Trading (ST/ESCAP/1288).

Legislation in North-East Asian countries

China

Since the Third Plenum of the 11th Congress of the Communist Party in December 1978, China has adopted the 'Open Economic Policy' which include attracting FDI. In 1979, China issued the 'Law of the People's Republic of China on Enterprises Operated Exclusively with Foreign Capital' , the 'Law of the PRC on Chinese-Foreign Cooperative Joint Ventures', and the 'Law of the PRC on Chinese-Foreign Equity Joint Ventures'. Shenzhen, Zhuhai, Shantou, and Xiamen have been opened as Special Economic Zones (SEZs). In 1984, 14 more coastal cities opened as SEZs and also Hainan Island. To develop the regions along the Changjiang River, the development of Pudong and Shanghai began. All inland provinces opened gradually.

In 1983, China promulgated the Regulations for the Implementation of the Joint Venture Law. Investors gradually increased investment. The State Council promulgated the Provisions for Encouragement of Foreign Investment (the '22 Articles') in 1986. Policies to promote the 'Economic Development Strategy for the Coastal Region of China' had been adopted since 1986.

In 1990, a grand plan of 'Pudong New Zone Development' was announced and proclaimed as an amendment of the 'Joint Venture Law'. In the early 1990s, the Planning Committee issued the 'Regulations on Guiding Foreign Investment' which divided the foreign-funded projects into ones that would be encouraged, limited, prohibited, or licensed by the state. In 1992, the concepts of 'Socialistic Market Economy' was announced. Since 1992, the regional preferential policies began to extend to the inland regions, including opening the border cities such as Heihe, Luanfenhe, Manzhouli, Earlianhaote, and others. The inland cities enjoyed the same preferential policies as those in the coastal cities. Since 1996, the central and western regions of China began to be emphasized under China's implementation of its regional "inclining policy".

The main laws and regulations related to environment protection are the Environmental Protection Law of 1989; Law of the Prevention and Control of Air Pollution, 1987; Law of the Prevention and Control of Water Pollution, 1984; and Regulations on the Prevention and Control of Noise Pollution, 1989.

In 1986, China issued the "Mineral Resources Law" which includes regulations on exploration rights management, mining rights management, mine inspection, mineral royalty collection and the rules for the implementation of the Mineral Resources Law. In 1982 and 1983, The State Council issued regulations on Chinese Foreign Cooperation for the exploration and development of oil and gas, onshore and offshore.

In May, 1996, the Amendment of Mineral Resources Law was passed by The State Council, which includes usage problems from mineral resources state ownership, privately operated mining enterprises and permits foreign company investment in exploration, and mining of mineral resources. This amendment has been submitted to the National People's Congress Standing Committee for discussion. The department responsible for geology and mineral resources management directly under The State Council is drafting a complete set of regulations. These laws and regulations enable mining in China to gradually enter the legal system.

However, as the programme of establishing a legal system for a socialist market economy system, for mining is not yet complete, the organization of mining activities in several mining areas has not yet improved. Distribution, import and export of mineral products, and profit distribution from mining development, as well as introduction of foreign investment into the mining industries, need to be further cleared through mining legislation. It is necessary to draft a "Mining Law" which will cover all aspects of the mining activities and adjust the interests of all parties concerned.

Japan

Japan promulgated a new Mining Law in March 1991, and abolished the Mining Law (Law No. 45 of 1905) and the Alluvial Mining Law (Law No. 13 of 1909) when the new Mining Law was enforced. Article 1 of Chapter 1 of the new Mining Law, states' the purpose of this Law is to establish the fundamental system of mining in order to contribute to the promotion of public welfare by rationally developing mineral resources.

Democratic People's Republic of Korea

By Decision Number 74 of 28 December 1991, the Administration Council of Democratic People's Republic of Korea decreed the establishment of the 621 square kilometre Rajin/Sonbong Free Economic and Trade Zone. The city of Chongjin, which is outside of the zone to the south was also declared a free port. Legislation governing direct investment in the Zone was enacted. Among the Laws were:

- Law on Foreigners' Investment
- Law on Foreign Enterprises
- Law on Foreign Exchange Administration
- Law on Contractual Joint Ventures
- Law on Foreign Investment-Business Enterprises and Foreign Individual Tax
- Law on Free Economic and Trade Zone

Mongolia

Following the adoption of its new Constitution in January 1992, a new Foreign Investment Law came into force on 1 July 1993. Its purpose is to encourage foreign investment, to protect the investor's rights and property in Mongolia and to regulate matters related to the operation of a foreign investment business entity. According to the Foreign Investment Law, foreign companies are granted five years tax holiday and 50 per cent tax relief in the subsequent five years when they invest in certain areas including mining. Equal investment conditions exist for both Mongolian and foreign mining companies.

The Law on Mineral Resources was approved on 20 September 1994 and became effective as of 1 January 1995. Exploration licenses will be given for three years, with two renewals for another two years, if requested, and exploration license holders have the first right to change their exploration license to mining licenses which will be given for a period of 30 years with renewals for another 20 years if requested. The law allows the royalty rates for strategic minerals (e.g. gold) to be set between 1 1/2 per cent and 12 per cent. A new regulation establishing the actual royalty rates is expected soon. The law gives the exploration title holder the "priority" right rather than the exclusive right.

A series of thirteen Environmental Laws have been approved and enacted in the period of 11 November 1994-22 May 1995 and EIA procedures and enforcement for any development projects in Mongolia is coming into force.

A total of thirteen Environmental Laws has been enacted by the National Assembly during the last two years, which includes the Mongolian Law on Land, the Mongolian Law on Special Protected Areas, the Mongolian Law on Environmental Protection, the Mongolian Law on Air, the Mongolian Law on Hunting, the Mongolian Law on Water, the Mongolian Law on Forests, the Mongolian Law on Natural Plants, the Mongolian Law on Protection from Toxic Chemicals, the Mongolian Law on Hunting Reserve Use Payments, and on Hunting and Trapping Authorization Fees, the Mongolian Law on Water and Mineral Water Use Fees, the Mongolian Law on Fees for Harvest of Forest Timber and Fuelwood and the Mongolian Law on Natural Plant Use Fees.

In 1995, the Mongolian Environmental Impact Assessment Procedures (MEIAP) were worked out by the Ministry, and will be enforced them and will be an integral part of any development project in Mongolia. The document clearly spells out the objectives and goals of MEIAP, the EIA procedure enforcing and reporting, the roles and responsibilities of individuals and organizations involved in MEIAP, the role of public consultation and participation as well as environmental standards and guidelines.

Mongolia, however, has not yet established a national capacity for EIA procedures.

Mongolia is making an effort to privatize some of its mining industry. The government has undertaken a world-wide campaign to invite foreign firms to take part in the exploration and development of the nation's metal and mineral resources.

The current economic reforms, recent improvements of the regulatory framework for foreign investment and structural adjustment have resulted in an upsurge of ODA assistance from US$12 million in 1990 to US$260 million in 1995, multilateral bank loans (two coal projects worth US$75 million) and an influx of foreign investment to the mineral sector.

A large number of mining companies are now showing a keen interest in carrying out mineral exploration programmes in Mongolia. The target areas for such activity are spread over the entire territory of Mongolia and target minerals are gold, copper, lead, zinc, molybdenum, fluorite and coal.

The gold mining industry is now booming in Mongolia to generate quick cash revenue desperately needed by the country. From 167 applications, 76 mining companies have been awarded gold mining licenses in recent years. This government's policy has resulted in a notable increase of gold production from 850 kg in 1992 to 4,650 kg in 1995.

A 5-year Mongolian-Japanese joint project on geological investigation and exploration was started in 1993 as part of the Japanese technical development aid to Mongolia. Reconnaissance exploration conducted by the JICA funds encompassed an area of 300,000 km^2 mainly in eastern and southern Mongolia.

The Mining Journal Ltd. has printed a book on Mongolia's mineral resources and mining opportunities. The traditional journal's country supplement on mining investment opportunities in Mongolia has been issued in April 1997. These activities of the Mining Journal Ltd. culminated in organization of a Round Table Conference on Investment Opportunities in Mining held in Ulaanbaatar in July 1997.

Russian Federation

The first legislation for foreign capital companies (FCC) appeared in USSR in 1987. Only minority-owned joint ventures were allowed with up to 49 per cent equity participation.

A strategic goal was declared to transform the Soviet Far East into a zone of economic prosperity and cooperation with the countries of Asia Pacific and a new state programme of economic development of the Far East adopted. The RFE received some preferences in 1988 for attracting more FDI to its territory.

In August 1991, the Russian Parliament passed the "Law on Foreign Investment in the Russian Federation". Foreign companies were allowed to establish, not only JVs, but also fully-owned subsidiaries, equity participation in the Russian national enterprises, and to conclude concession agreements for the exploitation of natural resources in the Russian Federation. From 1 January 1992, the right to register foreign-affiliated (with the value of foreign investment below 100 million roubles) was designated to local territorial authorities.

However, the peculiarities of the economic policies of the central government of the Russian Federation strongly influenced the development pattern of FDI. The Government rejected any claims for economic preferences from the regions and imposed the national regime for foreign investors.

3.3 Environmentally sound development

The most important result of the 1992 United Nations-sponsored Rio de Janeiro Earth Summit was the world-wide change in environmental awareness, leading to a de-emphasis on constant economic growth because of its accompanying unlimited pollution. There is a trend towards the globalization of environmental legislation and standards in industrialized countries, developing countries and economies in transition.

Increasingly international standards are being adopted by developing countries. This is related to the fact that where mining and metals processing ventures are owned and operated by multinational mining companies, the companies are using the cleanest available production technology – in response to good corporate management, meeting project financing investment guarantees in response to the global pressures of highly organized environmental groups.

3.3.1 Environmental impact assessment (EIA)

Environmental impacts of human production activities are shown in table 126. It is clear that the energy and manufacturing industries are responsible for most of the CO_2 and SO_2 emissions.

Mining related pollution

The mining and mineral processing sectors cover an enormous variety with a myriad of potential environmental effects and management requirements. The environmental impacts may be categorized into physical-chemical, ecological, and socio-economic ones. The chemical and physical pollution of land, water and air from mining and mineral processing activities is the primary effect of the pollution. The environmental hazards in base metal mining and processing are shown in table 127.

Chemical pollution

Chemical water pollution in the mining industry is caused primarily by toxic metals and acids. Run-off from open strip mines, drainage from tailings storage areas, effluent from milling or processing plants may be contaminated with heavy metals or acid. There are certain pathways by which the groundwater contamination may be severe. These include mine water percolation, borehole percolation, and tailing areas percolation. Chemical air pollution can be produced by two general activities in the mining sector: smelting of ores and fossil-fuel-burning in industrial processes.

Table 126. Environmental impacts of human production economic activities

Affected quality	*Human natural baseline*	*Disruption index*	*Share of human disruption caused by*			
			Industrial energy	*Traditional energy*	*Agriculture*	*Manufacturing/others*
Lead flow	25,000 tonnes/year	15.00	63 per cent fossil-fuel burning, including additives	Small	Small	37 per cent metals processing, manufacturing, refuse burning
Oil flow to oceans	500,000 tonnes/year	10.00	60 per cent oil harvesting, Production, processing, transport	Small	Small	40 per cent disposal of oil waste
Cadmium flow	1,000 tonnes/year	8.00	13 per cent fossil-fuel burning	5 per cent burning traditional fuels	12 per cent agricultural burning	70 per cent metals processing, manufacturing, refuse burning
SO_2 flow	50 million tonnes/year	1.40	85 per cent fossil-fuel burning	0.5 per cent burning traditional fuels	1 per cent agricultural burning	13 per cent smelting, refuse burning
Methane stock	800 parts per billion	1.10	18 per cent fossil-fuel harvesting and processing	5 per cent burning traditional fuels	65 per cent rice paddie domestic animals, land clearing	12 per cent landfills
Mercury flow	25,000 tonnes/year	0.70	20 per cent fossil-fuel burning	1 per cent burning traditional fuels	2 per cent agricultural burning	77 per cent metals processing, manufacturing, refuse burning
Nitrous oxide flow	10 million tonnes/year	0.40	12 per cent fossil-fuel burning	8 per cent burning traditional fuels	80 per cent fertilizer, land clearing, aquifers	Small
Particle flow	500 million tonnes/year	0.25	35 per cent fossil-fuel burning	10 per cent burning traditional fuels	40 per cent agricultural burning, wheat handling	15 per cent smelting, non agricultural land clearing, refuse burning
CO_2 stock	280 parts per million	0.25	75 per cent fossil-fuel burning	3 per cent net deforestation for fuel wood	15 per cent net deforestation for land	7 per cent net deforestation for lumber, cement manufacturing

Source: Allen Clark, South-East Asia Energy and Minerals Development: Environment and Economics, United Nations publication on Minerals and Metals Trading (ST/ESCAP/1268).

Table 127. Environmental hazards in base mental mining and processing

1. Mining – Dust – Noise – Vibrations – Mining waste disposal/spoil heaps – Mine waters (acid mine drainage)	4. Primary/secondary cooper refining and winning – Materials handling – Gaseous emissions – Hydrometallurgical wastes and residues
2. Milling – Dust – Noise – Tailings/slime disposal – Effluent disposal – Groundwater contamination	5. Primary/secondary lead smelting – Sulphur dioxide emissions – Metallic particulate emissions – Dumping – Volatile organic chemical emissions – Other off-gas components
3. Primary/secondary cooper smelting – Sulphur dioxide emissions – Metallic particulate emissions – Slag dumping – Volatile organic chemical emissions – Other off-gas components (e.g. chloride, fluorine)	6. Primary/secondary zinc smelting – Sulphur dioxide emissions – Metallic particulate emissions – Slag dumping – Volatic organic chemical emissions – Other off-gas components – Hydrometallurgical wastes and residues (e.g. jarostie)

Source: ESCAP Secretariat, 1992, 'An Environmental Audit of the Mining Industry in the Asia-Pacific Region' in 'Mineral Resources Development and the Environment' (ET/ESCAP/1192).

Mining and milling suddenly exposes large surface areas of the minerals to oxygen, causing oxidation of the minerals and a consequent change in their chemical nature. The oxidation of various metals, such as lead, zinc, arsenic, and mercury, cause them to become soluble and enter the surrounding environment through drainage water.

Many of the important metals are mined as sulfides, and sulfides occur with many of the minerals. Following the oxidation of the sulfide and contact with water, sulfuric acid can be formed. This is the main process by which acidic mine water is formed. In an acid environment, metals become even more soluble, hence acidic effluent often occurs with release of toxic heavy metals.

Hydro-metallurgy, which is the extraction of metals from their ores, by leaching with aqueous solutions, and the recovery of the metals by cementation or electrolysis, represents an important source of chemical water pollution. The waste water can contain heavy metals, acids, alkalis and cyanides if not properly treated.

The consequences of acidic effluent and toxic trace compounds can be disastrous. The contamination of surface water and ground water may result in the loss of beneficial uses, such as drinking water supply, fisheries and irrigation. If the pH value of stream or lake water falls below 5, then the aquatic ecosystem may suffer irreparable damage. Heavy metals can bio-accumulate in organisms and cause serious chronic illnesses, cancer, and mutagenicity.

Physical pollution

Physical pollution, mainly taking the form of dust in air, suspended solids in water and solid waste, is the most obvious environmental effect of mining activities. Small particulate matter is formed by excavating ores through drilling, blasting, crushing, grinding, processing and from tailings storage. If this particulate matter comes in contact with run-off or drainage water, it causes turbidity from the suspended solids.

Dust emanating from wind erosion of tailings and mining areas, blasting operations, haulage roads, and dry processing operations can reduce visibility. High concentrations of suspended particulate matter in the air can have adverse effects on human health, particularly by causing respiratory problems. The dusty wind can also erode buildings and cause equipment maintenance problems.

Excessive turbidity can result in the silting of rivers and lakes. Fisheries can be destroyed or harmed by the excessive turbidity, as it alters the aquatic ecosystem by affecting light penetration and algae, fish migration patterns and the covering of benthic communities. Suspended particles which settle out in the river or lake may also induce flooding because they affect the land form and natural drainage pattern. Furthermore, the excessive turbidity can damage the water for irrigation purpose by resulting in the silting of farming land and excessive wear on irrigation distribution systems.

Coastal/offshore dredging can have serious effects. Direct physical disruption of marine environment can destruct benthic fauna in the path of a dredge, and destroy the natural marine habit, swamps, coral reefs and gravel bars, where many species breed, spawn and spend the larval and early stages of their life. Settling of turbid water from the tailings discharge can alter the ocean floor texture and ecosystems. Fish migration patterns and primary production may also be affected.

Tremendous amounts of solid waste are produced by the mining sector. The average per capita consumption of all minerals has been estimated at close to 10 tonnes per annum, the tonnage of which the industrial minerals, especially construction sand, gravel and stone are mainly accounted for. In Germany, over 20 tonnes per annum is estimated to be consumed. In China, with a population of 12 billion, the figure would be 240 billion tonnes consumed per annum. It is hardly possible to argue that such consumption rates are sustainable, either in terms of their environmental impact or in terms of the availability of resources.

Waste rock from surface and underground mines and overburden from surface mining represent the largest sources of solid waste generation. It can also be generated from mill benefication in the form of mill tailings and metal processing wastes, mainly as smelter slag. Mining has the potential to leave many scars on the landscape and make the land become biologically inactive and unsightly. Mining projects often encroach into forests, swamplands and wildlife resources, in the form of the project site, roads, pipelines and transmission lines.

The concept of EIA

EIA is a crucial part of environmental impact management for a mineral resources development project. The EIA could be used before, during and after a mining project is executed. However, it is strongly recommended that EIA be done at the planning stage, along with or even before prospecting. Effective reclamation requires consideration from project inception, with the establishment of reclamation land-use objectives and with each phase of the project life-cycle (design, construction, start-up, operation and decommissioning) carried out in such a manner as to facilitate economical and effective rehabilitation.

Comprehensive knowledge of the local pre-project environmental conditions must be generated during the course of the EIA study to ensure that impacts can be reliably forecast. The relevant data and information collected in this pre-project stage establishes the baseline data which will be used for comparison with those data monitored during and after project implementation, so the environmental impacts of the resources development throughout the project can be estimated or evaluated.

The environmental impacts of a mining activity may relate to all kinds of natural and man-made capital including land, surface/ground water and coastal/marine water, the atmosphere, the flora and fauna in the terrestrial and aquatic environments, and the people, their socio-economic activities and infrastructures, aesthetic and cultural values. The impacts on the mining site may be serious, but may also be serious in remote areas by the circulation of pollutants through atmospheric, river and marine ecosystems.

At the initial stages of project development, simpler and more routine types of information will be gathered, with more highly technical and expensive data collection phased in as the geological results and economic prospects become more promising. It should be noted that about one year of field studies is needed to establish a reliable environmental baseline and several years of data will be required for many aspects of detailed project planning such as hydrological design and air dispersion modelling.

Estimating the value of natural resources and the costs of their degradation, associated with the mineral exploitation, is a complex task. Even more complicated is how to distribute those costs among polluter, State and community. In the past, environmental costs were measured largely in terms of the expense involved in the remedial treatment of degraded water quality, investment in environmental control technologies, or compensation for damage caused to local farm land by toxic emissions. More recently, environmental costs have been estimated in terms of extensive rehabilitation programmes that transform the previous mine and plant site with alternative resource uses such as re-vegetation and recreation (Kopp and Smith, 1989).

The EIA procedure identifies the possible positive and negative impacts to the natural and 'human' environments. The EIA provides for a plan which, upon implementation, will reduce or offset the negative impacts and may also utilize the positive impacts for enhancement measures to offset negative impacts. The EIA also provides a monitoring programme to measure the level of plan implementation and the degree of effectiveness of the environmental protection provisions.

The Battelle environmental evaluation system was developed by the Council on Environment Quality of the United States of America (figure 71). This method has been the basis for the development of several other methods and it clearly presents a more complete item-by-item picture of the factors that should be considered in an EIS. Although it was conceived for use with water resource development projects, the method can also be adapted for use with other development sectors including the mining industry. The figures indicated in the matrix shown in figure 71 indicate the environmental resources values (ERVs), which were estimated by a group of experts and may change from project to project.

An important requirement for EIA is the preparation of an environmental base map (EBM), or a set of maps, which shows the salient information and data collected or monitored so that the reviewer can readily interpret the EIA and especially the conclusions and recommendations. For a mining project EIA theme maps may include the following,

(a) geoscientific information such as meteorological, marine and geological status and trends including surface/ground water and soil conditions and natural hazards;

(b) information on living resources such as forests, local wildlife, fisheries and specific endangered ecological resources;

(c) socio-economic environment such as demography, areas of cultural, archaeological and tourist interests, and transportation and communication infrastructure;

(d) human activities such as performances of agriculture and industries and their waste streams; and

(e) institutions and communities such as NGOs, and Green Peace that are concerned or affected by the project activities.

The difficulty encountered in making these maps lies in how to present the maps so as not only the planner or manager for the project, but the decision and policy makers of the relevant authorities could understand the conclusions and recommendations of the EIA report and make an appropriate decision resulting in a sound environmental development. In this regard, ESCAP secretariat has strived for many years through its urban projects to bridge the gap between geoscientists and decision makers. Based on professional geological maps, a set of thematic maps are produced, which are easily understood by the economists and policy makers (ST/ESCAP/1374).

The methodologies and techniques used for EIA (table 128) include checklists, network, overlay, environmental index using factor analysis, cost-benefit analysis and simulation modelling workshops (ST/ESCAP/351 and 784).

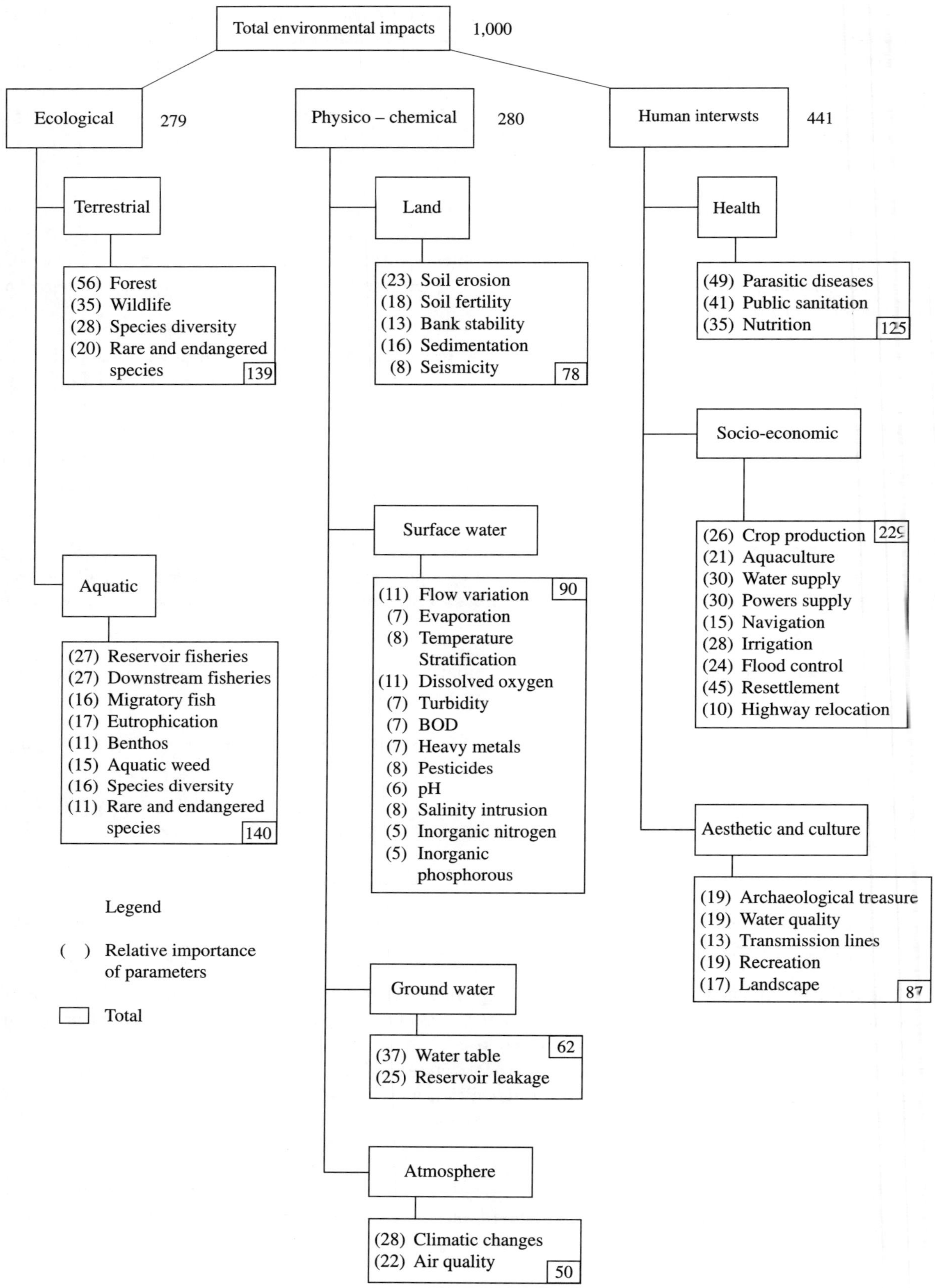

Source: Environmental Evaluation for Water Resources in Thailand, Water Resources Development vol. 1, No. 3, Lohani, B.N. and Khan, S.A. 1983. (11).

Figure 71. Battelle environmental evaluation system

Table 128. Summary of Current EIA Methodology Evaluation

Criteria		Checklist	Overlay	Network	Matrix	Environmental Index	Cost/benefit Analysis	Simulation Modeling Workshop
1.	Comprehensiveness	S	N	L	S	S	S	L
2.	Communicability	L	L	S	L	S	L	L
3.	Flexbility	L	S	L	L	S	S	L
4.	Objectivity	N	S	S	L	L	L	S
5.	Aggregation	N	S	N	N	S	S	N
6.	Replicability	S	L	S	S	S	S	S
7.	Multi-function	N	S	S	S	S	S	L
8.	Uncertainty	N	N	N	N	N	N	S
9.	Space-dimension	N	L	N	N	S	N	S
10.	Time-dimension	S	N	N	N	S	S	L
11.	Data requirement	L	N	S	S	S	N	N
12.	Summary format	L	S	S	L	S	L	L
13.	Alternative comparison	S	L	L	L	L	L	L
14.	Time requirement	L	N	S	S	S	S	N
15.	Manpower requirement	L	S	S	S	S	S	N
16.	Economy	L	L	L	L	L	L	N

Source: ESCAP, 1990, Environmental Impact Assessment, Guidelines for Industrial Development. (ET/ESCAP/784).

3.3.2 Environmental impacts management

Water pollution

Liquid wastes generally emanate from three main sources: mine de-watering, process water, and surface water run-off. Mine de-watering water is usually low in suspended solids but may contain high concentrations of dissolved minerals and metals. Process water, which is used in the transportation, washing, benefication, and separation of ores, usually contains heavy loading suspended solids and may contain high concentrations of dissolved metals. In some cases it may also be very acidic. Surface water run-off contains high levels of suspended solids and may contain high concentrations of metals.

The essential elements of a successful water management programme revolve around water use minimization and control. The amount of water entering the facilities through surface runoff should be minimized by proper civil engineering drainage techniques to prevent erosion, water ingress into mines, flooding, and destruction of waste/tailings detention areas. The lack of attention to proper drainage control represents one of the primary causes of water pollution at mining facilities. The use of process water should be minimized through efficient utilization of water and recycling where possible.

Dissolved metals are normally precipitated and settled out. Acidity is removed through addition of lime or some other neutralization method. Suspended solids are removed through sedimentation.

An ongoing waste water effluent monitoring programme should be established to asses the performance or the need for the treatment system, although the specific compounds which should be monitored depend on the mining process and ore. In general, at least the following groups should be considered: suspended solids, pH, metal particulate in suspension, toxic metallic salts in solution and BOD (bio-chemical oxygen demand). The monitoring should take place at all significant discharge points and, when a waste water treatment system exists, before and after treatment.

The design and location of the tailings detention area is extremely important. It should not be placed in a location where drainage from a large watershed area is received. Provision should be made for decant from the surface and drainage from the bottom of the tailings detention area. The decant and drainage water can then be treated, if required, and recycled or discharged. The tailings retention dam should be properly constructed, with a spillway for heavy rains and appropriate drainage around the waste retention area and dam. The percolation of water from the tailings retention dam into the ground water should be avoided by proper engineering design. The collapse of retention dams produces one of the most common and severe water pollution problems.

Surface and ground water pollution from abandoned mine sites can pose serious problems. The ore in the ground or in the waste rock can be exposed to either oxygen or water. In underground mines all surface openings to the mine should be completely sealed. Inundation of the underground mine is acceptable if the mine is able to hold water and does not leak into the groundwater, as it provides an effective seal against the entry of oxygen.

Solid waste

Waste management technologies may involve a combination of physical treatment for solids removal and reagent degradation; biological treatment for domestic sewage and the reduction of nutrients; and chemical treatment for metals removal, pH adjustment and reagent destruction. It should always be recognized, however, that wastes are often concentrated rather than destroyed, which means that we have not solved the problem, only changed its form. The end-product of many waste treatment processes, particularly pH adjustment and metals removal, will be a solid residue or sludge that must be disposed of. Many sludges are less than 5 per cent solids and must be dewatered before disposal. Some toxic solids may be classed as hazardous wastes and require special, licensed disposal.

Mine waste rock and tailings from processing plants must be placed in such a manner that they will not endanger the public or worker safety and will remain stable and environmentally benign in perpetuity, with little or no maintenance, and can be effectively and economically reclaimed to a useful state.

Accomplishing these various objectives, particularly where the possibility of acid mine drainage is involved, may require detailed technical information. This might include the estimated rate and volume of wastes to be generated over the life of the mine, the configuration and physical properties of the foundation soils, the surface and ground water hydrology, and the precipitation and run-off data, to allow an estimation of a designated storm event for design purposes.

In addition to site information, an effective waste management programme would require field characterization and specialized handling of large volumes of material, continuous supervision of, and design improvements to waste disposal procedures, regular maintenance and environmental monitoring of waste disposal procedures, and progressive stabilization and reclamation.

Air pollution

Emissions of fugitive dust and dust generated by mining activities are affected by many factors such as moisture content of the ore, the type of ore, the amount processed, the type of equipment, operating factors and a variety of geographical and seasonal factors.

Almost all fugitive dust controls involve one or a combination of three techniques: watering, chemical stabilization, and reduction of surface wind speed crossing exposed surfaces. Watering costs the least but also provides the least permanent dust control. Chemicals may be applied to piles of overburden, waste, and tailings in the mining industry, although the primary use of chemical stabilizers is for land reclamation after mining. Common methods of reducing wind speed are construction of windbreaks and enclosures or coverings of dust generation sources and the planting of vegetation on or adjacent to the exposed surfaces.

Noise pollution associating with drilling, blasting, loading, hauling and crushing operations in mining activities can represent a serious environmental quality problem and could be reduced considerably if these activities are arranged properly.

Acid mine drainage

Acid mine drainage has, over the last decade, been recognized as one of the most significant environmental problems associated with the mining of minerals and coal. One of the products of sulphide oxidation is sulphuric acid, which in turn can dissolve residual metals and other minerals such as arsenic in the mine waste, to produce a toxic leachate of very high acidity (pH 1.8 to 3.0). Uncontrolled acid mine drainage from many abandoned mines has caused serious environmental problems and the need to collect and treat acid leachates, has turned several operating mines from profitable enterprises into costly liabilities.

There are four requirements for acid mine drainage: sulphide minerals, oxygen, water, and iron/sulphur-oxidizing bacteria (such as Thiobaccilus ferro-oxidants). If any of these are missing, acid mine drainage will not occur, and it is this fact that provides the basis for prevention and control technology. The key to acid mine development is the early identification of ore zones and planning to ensure the prevention of acid generation from waste rock, tailings, open pits and underground workings. Decision makers will properly require an unprecedented degree of preplanning and risk assessment for any future proposals to mine high-sulphide ore bodies (ST/ESCAP/ 1192).

Progressive companies are not closing down, reinvesting elsewhere or exporting pollution to less restrictive regulatory regimes in developing countries. Rather they are adapting to environmental regulatory pressures by innovating, improving and commercializing their environmental technology and management practices at home and abroad. This evidence challenges the "pollution haven" hypothesis.

For instance, more than 12 per cent of Inco's capital spending during the last 10 years has been related to environmental concerns (Coppel, 1992). Modernization plans include replacement of its reverberatory furnaces with a new innovative oxygen flash smelter, a new sulphuric acid recovery plant and an additional oxygen plant. By incorporating two of the flash smelters, the firm reduced sulphur dioxide emissions by over 100,000 tonnes in 1992, and by 1994 planned to achieve a government-set target level of 175,000 tonnes per year. Other environmental benefits have included a cleaner, safer work environment (Mining Journal, 23 February 1990). Inco is now one of the world's lowest-cost nickel producers. Furthermore, like other dynamic firms that are responding to environmental regulation through innovation, Inco is seeking to recoup research and development costs through an aggressive effort to license its technology with firms in other copper and nickel-processing countries.

Homestake's McLaughlin gold-mine in California is a good example of a mine and processing facility that has been designed, constructed and operated from the outset within the bounds of the world's strictest environmental regime. Environmental efficiency is built into every aspect of the gold-mining process. Innovative process design criteria, fail-safe tailings and waste disposal systems and extensive ongoing mine rehabilitation and environmental monitoring systems characterize the mine site. The mining operation, therefore, combines a myriad of innovative technologies to define "best practice in environmental management". Most of these environmental management initiatives have not resulted in any substantial extra cost; indeed many have improved the efficiency of the mine, and this has positively affected the economics of the overall operation.

Environmental impact management

Environmental management, including EIA has become a crucial element of a mine development project. In fact, it is the element of highest priority, during the licensing process and throughout the development and implementation of the development project. The EIA should include early and comprehensive environmental impact assessment, pollution control and other preventive and mitigative measures, monitoring and auditing activities and emergency response procedures.

A recent development in this respect is to establish capability in integrated economic environmental accounts (IEEA) in industry and government at the highest management and policy-making levels. The IEEA seeks to evaluate the resources endowment and the environment impacts of the resource development through an accounting system. At the national level, the IEEA is part of a system of national accounts (SNA). Efforts have be made to develop computer programmes to evaluate natural resources and environmental resources (STAT/SERA/1).

Environmental accounting is also crucial to the mining sector. A example for base metal mining, is the estimation that environmental restrictions could lead to an increase in current base metal costs by 20 per cent in the 1990s. It can also be predicted that the growing disparity between the stringency of the environmental laws in rich industrialized countries and the laws in the developing and least developed countries in the Asia-Pacific region will encourage the location of new smelters and processing plants for base metals and other minerals in the future in these latter countries. This undoubtedly will have major implications for future investment patterns affecting both international trade in metals and minerals and giving rise to price fluctuations.

For environmental considerations to be an integrated part of mining development practice, the following activities are required:

(a) to ensure participation and dialogue with the affected community and other directly interested parties on the environmental aspects of all phases of mining activities;

(b) to adopt environmentally sound technologies in all phases of mining activities and increase the emphasis on the transfer of appropriate technologies that mitigate environmental impacts, including those from small-scale mining operations;

(c) to seek to provide additional funds and innovative financial arrangements to improve the environmental performance of existing mining operations;

(d) to adopt risk analysis and risk management in the development of regulations and in the design, operation and decommissioning of mining activities, including the handling and disposal of hazardous mining and other wastes;

(e) to reinforce the infrastructure, information systems service, training and skills in environmental management in relation to mining activities; and

(f) to explore the feasibility of reciprocal agreements to reduce trans-boundary pollution.

3.3.3 Environmental legislation

International agreements

Marine environment

International conventions related to protection of the marine environment adopted in the 1970s and 1980s include the Prevention of Pollution from Ships (MARPOL), the Convention on the Prevention of Marine Pollution by Dumping from Ships and Aircraft (London Convention) (1972) and the Basel Convention on the Control of Transboundary Movements of Hazardous Wastes and their Disposal (1989).

The Chapter 17 of Agenda 21, adopted at the United Nations Conference on Environment and Development at Rio de Janeiro, 3-14 June 1992, deals with the issue of protection of the oceans, all kinds of seas, including enclosed and semi-enclosed seas, coastal areas and the protection, rational use and development of their living resources.

Considerable progress has been made since the Earth Summit in 1992 in intergovernmental negotiations related to oceans and seas. The entry into force of the United Nations Convention on the Law of the Sea in 1994 was a fundamental achievement and provides the framework for the protection of the marine environment.

Other recent successes include, inter alia, the Agreement to Promote Compliance with International Conservation and Management Measures by Vessels Fishing in the High Seas (Rome, Food and Agriculture Organization of the United Nations, 1993); the Agreement relating to the Implementation of Part XI of the United Nations Convention on the Law of the Sea of 10 December 1982 (General Assembly resolution 48/263, annex); the Agreement for the Implementation of the Provisions of the United Nations Convention on the Law of the Sea relating to the Conservation and Management of Straddling Fish Stocks and Highly Migratory Fish Stocks (Document A/50/550, annex I, see also A/CONF. 164/37); the Code of Conduct for Responsible Fisheries (Rome, FAO, 1995) and the Global Programme of Action for the Protection of the Marine Environment from Land-based Activities (UNEP, November 1995) (A/51/116, annex II).

The Global Programme of Action for the Protection of the Marine Environment from Land-based Activities was adopted at the Intergovernmental Conference held in Washington, D.C., in 1995. The Conference decided to submit to the Economic and Social Council at its substantive session of 1996 a draft resolution to be considered by the General Assembly at its fifty-first session on the institutional arrangements for the implementation of the Global Programme of Action.

The Intergovernmental Conference also adopted the Washington Declaration on the Protection of the Marine Environment from Land-based Activities, which contained the request for the Executive Director of UNEP, in close partnership with the World Health Organization (WHO), the United Nations Centre for Human Settlements (Habitat), the United Nations Development Programme (UNDP) and other relevant organizations, to prepare proposals for a plan to address the global nature of the problem of inadequate management and treatment of waste water and its consequences for human health and the environment, and to promote the transfer of appropriate and affordable technology drawn from the best available techniques, and referred to in the Global Programme of Action. Such proposals are to be considered by the Governing Council of UNEP at its nineteenth session.

The Governments participating in the Washington Intergovernmental Conference expressed an intention to take action to develop, in accordance with the provisions of the Global Programme of Action, a global, legally binding instrument for the reduction and/or elimination of emissions, discharges and, where appropriate, the elimination of the manufacture and use of the persistent organic pollutants identified in decision 18/32 adopted by the Governing Council of UNEP at its eighteenth session (see A/50/25, annex).

In completing its review of all the chapters of Agenda 21 in the context of its first multi-year thematic programme of work, the Commission of Sustainable Development at its fourth session held from 18 April to 3 May 1996, also offered the first opportunity to review events since UNCED in the important area of oceans and seas (chapter 17 of Agenda 21). The Commission welcomed the considerable progress in recent intergovernmental negotiations related to oceans and seas, endorsed the Global Programme of Action for the Protection of the Marine Environment from Land-based Activities, adopted at the Intergovernmental Conference in Washington, D.C., in November 1995. As requested by the Conference, the Commission formulated a draft resolution, for consideration by the General Assembly at its fifty-first session, on institutional arrangements for the implementation of the Global Programme of Action (table 129).

The Commission has placed special importance on ensuring more effective methods and activities, including in the United Nations system, for enhancing international cooperation and coordination of marine issues. It has therefore recommended, for approval by the Economic and Social Council and subject to the outcome of the special session in 1997, that there should be a periodic overall review by the Commission of all aspects of the marine environment, the results of which should be considered by the General Assembly under a new consolidated agenda item entitled "Oceans and the law of the sea". It further recommended a review of existing inter-agency coordination mechanisms dealing with marine issues.

Protection of the atmosphere

The close interrelationship between protection of oceans and all kinds of seas and protection of the atmosphere should be noted, in view of the exchange of matter and energy that takes place between the atmosphere and oceans and their influence on marine and terrestrial ecosystems.

It has been recognized that an essential component of measures to protect the atmosphere, environment and human health is the reduction of local emissions – especially urban air pollution – which must be dealt with at the local, regional and international levels on the basis of common but differentiated responsibilities. The importance of combating all kinds of land degradation, deforestation, forest degradation and desertification (which have adverse impacts on human health and the environment), and the importance of improved land use management have also been recognized.

The Second Assessment Report (SAR) adopted by the Intergovernmental Panel on Climate Change (IPCC) in December 1995, as the most comprehensive assessment of climate change issues to date, states among other conclusions, and in the full context of the report, that the balance of evidence suggests a discernible human influence on global climate.

Table 129. The resolution and decisions of the 4th session of the Commission on Sustainable Development

A. Draft resolution

Institutional arrangements for the implementation of the Global Programme of Action for the Protection of the Marine Environment from Land-based Activities

(* Related to Chapters 17, 33, 34, 38 and other related chapters of the Agenda 21)

B. Draft decisions

I. Matters relating to the third and fourth sessions of the Ad Hoc Intergovernmental Panel on Forests

(* Related to Chapters 11, and 10, 12, 13 and 14 of the Agenda 21)

II. Report of the Commission on Sustainable Development on its fourth session and provisional agenda for the fifth session of the Commission

(* Related to all the chapters of the Agenda 21)

C. Matters brought to the attention of the Council

Decision 4/1. Trade, environment and sustainable development

(* Related to Chapter 2 of the Agenda 21)

Decision 4/2. Combating poverty

(* Related to Chapter 3 of the Agenda 21)

Decision 4/3. Demographic dynamics and sustainability

(* Related to Chapter 5 of the Agenda 21)

Decision 4/4. Integrating environment and development in decision-making

(* Related to Chapter 8 of the Agenda 21)

Decision 4/5. Information for decision-making

(* Related to Chapter 40 of the Agenda 21)

Decision 4/6. International legal instruments and mechanisms

(* Related to Chapter 39 of the Agenda 21)

Decision 4/7. International institutional arrangements

(* Related to Chapter 38 of the Agenda 21)

Decision 4/8. Information provided by Governments and organizations

(* Related to Chapter 17 and other related chapters of the Agenda 21)

Decision 4/9. Major groups

(* Related to Chapters 23-32 of the Agenda 21)

Decision 4/10. Transfer of environmentally sound technologies, cooperation and capacity-building

(* Related to Chapter 34 of the Agenda 21)

Decision 4/11. Promoting education, public awareness and training

(* Related to Chapter 36 of the Agenda 21)

Decision 4/12. National mechanisms and international cooperation for capacity-building in developing countries

(* Related to Chapter 37 of the Agenda 21)

Decision 4/13. Changing production and consumption patterns

(* Related to Chapter 4 of the Agenda 21)

Decision 4/14. Financial resources and mechanisms

(* Related to Chpter 33 of the Agenda 21)

Decision 4/15. Protection of the atmosphere and protection of the oceans and all kinds of seas

(* Related to Chapters 9 and 17 of the Agenda 21)

Decision 4/16. Review of the implementation of the Programme of Action for the Sustainable Development of Small Island Developing States

(* Related to Chapter 17 and most of the other chapters of the Agenda 21)

Decision 4/17. Matters relating to the inter-sessional work of the Commission

(* The Inter-sessional Ad Hoc Open-ended Working Group meeting is to be held from 24 February to 7 March 1997, in New York, to assisting the Commission in undertaking the review for the special session of the General Assembly that convined for the purpose of carrying out an overall review and appraisal of progress achieved in the implementation of Agenda pursuant to its resolution 50/113 of 20 December 1995)

Decision 4/18. Proposals for the medium-term plan for the period 1998-2001

(* Related to all the chapters of the Agenda 21)

Source: E/1996/28, E/CN. 17/1996/38.
* Author's note.

The United Nations "Framework Convention on Climate Change", the "Vienna Convention for the Protection of the Ozone Layer", its Montreal Protocol and subsequent amendments and adjustments, and the United Nations "Convention to Combat Desertification in those Countries Experiencing Serious Drought and/or Desertification, Particularly in Africa", are the most important conventions related to the issues of climate change.

Acidifying substances

Acid rain is the key trans-boundary problem associated with mining and non-ferrous metals production, although groundwater and surface-water pollution can cause regional problems, for example, mercury contamination from gold-mining in the Amazon.

The European Community now relies on the use of Directives that are applicable to all member States to achieve political aims with respect to acid rain. The Canada/United States Agreement on Air Quality was signed in March 1991. The United States Congress introduced a plan, under the authority of the Clean Air Act (1990 Amendments), to reduce sulphur dioxide emissions from 1980 levels by 10 million tonnes. A national cap of 8.9 million tonnes of sulphur dioxide per year is required by the year 2000. The Agreement also committed the Government of Canada to overseeing the reduction of sulphur dioxide emissions by 40 per cent of 1980 levels by 1994. All mining operations having smelters in the seven eastern-most provinces of Canada are required, under the Canada/United States Agreement on Air Quality, to reduce their sulphur dioxide emissions to the agreed level. The provincial Governments, by entering into agreements with the federal Government, have been largely responsible for enforcing the required reductions at smelters within provincial boundaries.

To address the growing problem of transboundary air pollution, in particular the pollution is affecting the Arctic and risks caused by persistent organic pollutants, there is a need for effective transboundary air pollution agreements such as the ECE Convention on Long-range Transboundary Air Pollution and its protocols in all affected regions.

Hazardous wastes

It is highly probable that international environmental issues and agreements will have a more significant impact on the world mining industry. The current world-wide interest in making environmental issues relevant to trade in world commodities will likely require that all nations produce metals and minerals in an environmentally sound manner in order to avoid restrictive trade barriers in some form. Such agreements may have positive and negative environmental consequences for the world mining industry.

The Basel Convention on the Control of Trans-boundary Movements of Hazardous Wastes and their Disposal attempts to regulate the trade of hazardous substances. Many metals that could be recycled more effectively by countries that wish to do so are currently not transported across national boundaries because of restrictions applied by the Basel Convention. However, the Basel Convention also prevents the transport of mining slag from developed countries to the underdeveloped ones that cannot treat the waste but may wish to earn foreign exchange through payment for its disposal.

On a positive note, environmental codes or agreements applicable to key mining industry processes, enforced by an international trade regulatory body, could push world best practice in environmental management to new levels, although they might have negative competitive implications for certain developing country companies that are constrained by obsolete technology and scarce resources.

National environmental laws in Asia

Many developing countries have not yet adopted environmental regulations and standards for the mining industry. The mining code or other general environmental legislation may mention something on the subject but it is usually too general or vague. This situation should be changed urgently if the country wants to develop its mining industry responsibly.

Many international mining companies nowadays would prefer to invest in countries that have established practical and reasonable environmental protection legislation, or equivalent provisions in a standard mineral agreement. With regard to environmental protection regarding mining, they would prefer to know in advance what their obligations are than to invest heavily and then be notified that new requirements will come into force.

China

The First National Environmental Protection Conference was held in Beijing in August 1973. The Environmental Protection Leading Group was established under the State Council in 1974, and various national environmental standards were developed specifying norms for environmental quality and pollutant emissions. The Constitution of the People's Republic of China, adopted on 5 March 1978, states specifically that the State protects environment and natural resources against pollution and other public nuisances (Gao, Zhiyou and Wenmin, 1992). In September 1979, the Environmental Protection Law of the People's Republic of China was promulgated. Implementation of this Act marked the beginning of legislative environmental management in China. Table 130 outlines the environmental regulations applicable to mining in China.

Japan

A relatively small indigenous mining capacity has forced Japan to invest in mining projects abroad in order to secure mineral supplies. It has developed a very strong smelting and refining industry at home and has invested increasingly in environmentally and economically efficient mining and smelting technologies for use at home and abroad (Warhurst, 1993).

Initial environmental regulations aimed at the Japanese smelting industry tended to require the modernization of smelters and their being equipped with emission control technologies. These add-on controls were required to disperse waste gases through high stacks, recover modest amounts of sulphuric acid and remove dust through electrostatic precipitation. All new smelters required similar controls. Later, all existing and new Japanese smelters were required to install double-contact acid plants and/or sulphur dioxide scrubbing plants, fugitive gas collecting systems and waste-water treatment systems. In addition, individual firms were encouraged to develop innovative smelting technologies that improved environmental and productive performance. It was during this period that the state-of-the-art Mitsubishi smelting process was developed.

Table 130. Environmental policy and legislation applicable to mining in China

Type	*Regulations and policy*
Command-and-control standards	Environmental Protection Law (1979) Marine Protection Act (1982) Water Pollution Prevention Act (1984) Atmosphere Pollution Prevention Act (1988)
	Supplemental administrative acts and rules applicable to the mining industry: implementing Rules for Water Pollution Prevention Act, the Provisional Measures for Collecting Pollutant Discard Charges, the Atmospheric Quality Standard, the Sea-Water Quality Standard and the Implementation Rules for the Atmosphere Pollution Prevention Act.
	A large number of regional and local regulations have also been enacted by the provinces, municipalities and autonomous regions to control specific environmental problems associated with mining.
	Land Reclamation Regulations (1988) set standards for reclaimed land.
Liability legislation	Land Reclamation Regulations (1988) establish liability for reclaiming mined land, or impose penalties.
Economic incentives legislation	Environmental Protection Law and other relevant regulations have been enacted, using economic incentives rather than the old style of propagandist slogans and education.
	Mining enterprises generally benefit from reclamation work undertaken by receiving favourable taxation status and special credit.
Permitting and environmental impact assessment legislation (or process)	Environmental Protection Law requires that environmental impact assessments be carried out prior to construction of new mining operations. Also, pollution disposal facilities must be designed, built and made operational at the same time as the rest of a mining or smelting operation.

Japanese air control requirements have become increasingly stringent – in line with those developed in the United States. However, since Japan does not mine its own minerals, it faces the unique challenge of confronting a wide variety of conditions and government regulations in the countries in which its mining and smelting industries operate. The response of Japanese mining and smelting firms has therefore been to set levels of technological excellence that, in not only meeting but often exceeding government requirements elsewhere, often establishes best environmental practice in mining and smelting.

3.4 Sub-regional cooperation programmes

Unlike South-East Asia, for which many international development programmes have been formulated and implemented for decades, the North-East Asia has few subregional programmes. The situation has however begun to change since 1990.

3.4.1 United Nations system

The CSD'S 1996 resolution and recommendation – marine sector

The United Nations Commission on Sustainable Development (CSD) at its fourth session from 18 April to 3 May 1996 recommended to the Economic and Social Council (ECOSOC) the adoption of the draft resolution: Institutional arrangements for the implementation of the Global Programme of Action for the Protection of the Marine Environment from Land-based Activities. If the draft resolution would be concurred by the ECOSOC and adopted by the General Assembly, the General Assembly would call upon the United Nations Environment Programme, within its available resources, and with the aid of voluntary contributions from States for this purpose, to take expeditious action to provide for the establishment and implementation of the clearing-house mechanism referred to in the Global Programme of Action[1], and requests the Executive Director of the United Nations Environment Programme to prepare and submit to the Governing Council at its nineteenth session specific proposals on, inter alia:

(a) The establishment of an inter-organizational group to develop the basic design and structure of the clearing-house data directory and its linkages to information delivery mechanisms;

(b) The means of linking the inter-organizational group to ongoing work within the United Nations system on the identification of and access to relevant databases and the comparability of data;

(c) The outline of a pilot project on the development of the clearing-house's source category component on sewage, to be implemented in partnership with the World Health Organization;

The General Assembly would also call upon States, in relation to the clearing-house mechanism, to take action in the governing bodies of relevant intergovernmental organizations and programmes so as to ensure that these organizations and programmes take the lead in coordinating the development of the clearing-house mechanism with respect to the following source categories, which are not listed in order of priority:

(a) Sewage – the World Health Organization;

(b) Persistent organic pollutants – Inter-organizational Programme for the Sound Management of Chemicals, the International Programme on Chemical Safety and Intergovernmental Forum on Chemical Safety;

(c) Heavy metals – the United Nations Environment Programme in cooperation with the Inter-organizational Programme for the Sound Management of Chemicals;

(d) Radioactive substances – the International Atomic Energy Agency;

(e) Nutrients and sediment mobilization – the Food and Agriculture Organization of the United Nations;

(f) Oils (hydrocarbons) and litter – the International Maritime Organization;

(g) Physical alterations, including habitat modification an destruction of areas of concern – the United Nations Environment Programme.

The Commission on Sustainable Development, in order to enhance implementation of the commitment set forth in section F of chapter 17 of Agenda 21 to promote regular intergovernmental review and consideration, within the United Nations system, of general marine and coastal issues, including environment and development matters, at its fourth session in 1996, agreed on the need:

---

1 The Global Programme of Action was adopted at the Intergovernmental Conference to Adopt a Global Programme of Action for the Protection of the Marine Environment from Land-based Activities, which was held in Washington, D.C. from 23 October to 3 November 1995.

(a) To better identify priorities for action at the global level to promote conservation and sustainable use of the marine environment;

(b) For better coordination among the relevant United Nations organizations and intergovernmental financial institutions;

(c) To ensure sound scientific, environmental, economic and social advice on these issues.

The Commission therefore recommends that the Economic and Social Council approve the following conclusions as regards addressing these issues, subject to the outcome of the special session of the General Assembly in 1997 at which the Assembly will, inter alia, decide on the future work programme of the Commission:

(a) There should be a periodic overall review by the Commission of all aspects of the marine environment and its related issues, as described in chapter 17 of Agenda 21, and for which the overall legal framework is provided by the United Nations Convention on the Law of the Sea. This review should cover other chapters and provisions of Agenda 21 directly related to the marine environment. This review should draw upon reports of the United Nations Environment Programme (UNEP), and those of other relevant United Nations bodies and international organizations in their respective fields, coordinated by the Administrative Committee on Coordination (ACC) Subcommittee on Oceans and Coastal Areas. Other modalities for the review should be decided by the Commission on Sustainable Development. The results of such reviews should be considered by the General Assembly under an agenda item entitled "Oceans and the law of the sea";

(b) In order to address the need for improved coordination, the Secretary-General should be invited to review the working of the ACC Subcommittee on Oceans and Coastal Areas, with a view to improving its status and effectiveness, including the need for closer inter-agency links between, inter alia, the secretariat of the Subcommittee and UNEP;

(c) The Secretary-General and the executive heads of the agencies and organizations of the United Nations system sponsoring the Joint Group of Experts on the Scientific Aspects of Marine Environmental Protection (GESAMP) should be invited to review the Group's terms of reference, composition and methods of work, with a view to improving its effectiveness and comprehensiveness while maintaining its status as a source of agreed, independent scientific advice.

CNR'S recommendations – mineral setor

In the field of mineral resources, the Third Session of the Committee on Natural Resources held in May 1996 adopted resolutions 3/4, 3/5 and 3/6 and brought them to the attention of the Economic and Social Council.

The resolution 3/4 endorsed the need for authoritative assessment and dissemination of information on technological progress towards the sustainable extraction and use of mineral resources through improved efficiency, new technology, substitution and recycling. The CNR recommended a commission on mining and materials be established, placed in the DDSMS, or in UNCTAD.

The resolution 3/5 endorsed the need to establish a global geochemical database as an essential contribution to objective and effective environmental and resource management. The CNR recommended a plan to be formulated for cooperation with national agencies to develop the database. It has been estimated that full data acquisition will require a minimum of a decade.

The resolution 3/6 endorsed the need for a global knowledge base on mineral resource potential, in terms of potential source areas, to be developed so that it can be integrated with other land-use information as an essential part of an integrated approach to the planning and management of natural resources. The CNR recommended that the detailed requirements of such a knowledge base and the ways in which such a knowledge base might be achieved, including initiation through a regional pilot project base, to be considered.

UN's institutions

Several organizations and agencies within the United Nations system, in accordance with their mandates and areas of specialization, participate in technical cooperation activities in the field of mineral resources. These have included the Department for Development Support and Management Services (DDSMS) of the United Nations secretariat, the regional commissions and UNCTAD.

DDSMS was charged with, inter alia, the implementation of technical cooperation activities, in cooperation with other relevant agencies, in the field of mineral resources. UNCTD is mainly concerned with mineral economics and trade in primary commodities. The regional Commissions with the capacity for the execution and finance, through grants, of technical cooperation assistance in regional projects include, among the five regional Commissions, the Economic and Social Commission for Asia and the Pacific (ESCAP) and the Economic Commission for Africa (ECA), which are particularly active in the field of mineral resource development.

Since the United Nations Conference on Environment and Development in 1992 and its adoption of Agenda 21, the United Nations system has increased its emphasis on the formulation and implementation of environmentally sound mineral development policies and adopted a cross-sectional, multi-disciplinary approach in coordinating and integrating technical cooperation activities in mineral resources development, particularly upstream technical support services. Activities have been undertaken to provide for the means of facilitating complementarity between the roles and obligations of host Governments and of investors in the development of mineral resources within the framework of sustainable development.

The United Nations system, World Bank and the IFC system and other multilateral and bilateral organizations have provided invaluable assistance to industry and host Governments in respect of mineral resource technical support and advice; training of national staff in all skills required in mineral resource development; hosting of regional promotional round-table forums to facilitate dialogue between industry, Government and facilitating organizations; institutional structuring and reforms; capacity building; provision of research on trends in mineral, fiscal and legal policies; and for commodity supply and demand studies.

DDSMS

The Department for Development Support and Management Services (DDSMS) of the United Nations was the principal United Nations office responsible for the implementation and coordination, in cooperation with other relevant bodies, of technical cooperation activities in the field of mineral resources development.

The Department for Development Support and Management Services (DDSMS) of the United Nations Secretariat organized a workshop sponsored jointly with Carl Duisberg Gesellschaft of Germany at Windhoek in April 1993, which was a contribution to the practical implementation of the guidelines adopted in Berlin at the International Round Table on Mining and the Environment, held in June 1991.

The Interregional Seminar on Foreign Investment and Joint Ventures in the Mining Sector was organized at Haikou City, China, during the period 7 to 12 December 1992. It was organized by the Department of Economic and Social Development of the United Nations Secretariat, with the collaboration of the Department of Foreign Affairs of the Hainan Provincial Government. The format of the Haikou seminar is an example of how to provide another opportunity for the decision makers of developing countries and economies in transition to become better acquainted with current issues related to foreign investment.

The breakdown of the mineral projects implemented by Department of Economic and Social Development in 1992-1993 is shown in figure 72.

UNCTAD

The activities of the United Nations Conference on Trade and Development (UNCTAD) relating to minerals and metals include, inter alia, the following recurrent publications:

(a) Annual and quarterly statistical bulletins on tungsten and tin; annual and, with the assistance of a trust fund, quarterly statistical publications on iron ore; and annual statistics on bauxite/alumina/aluminium;

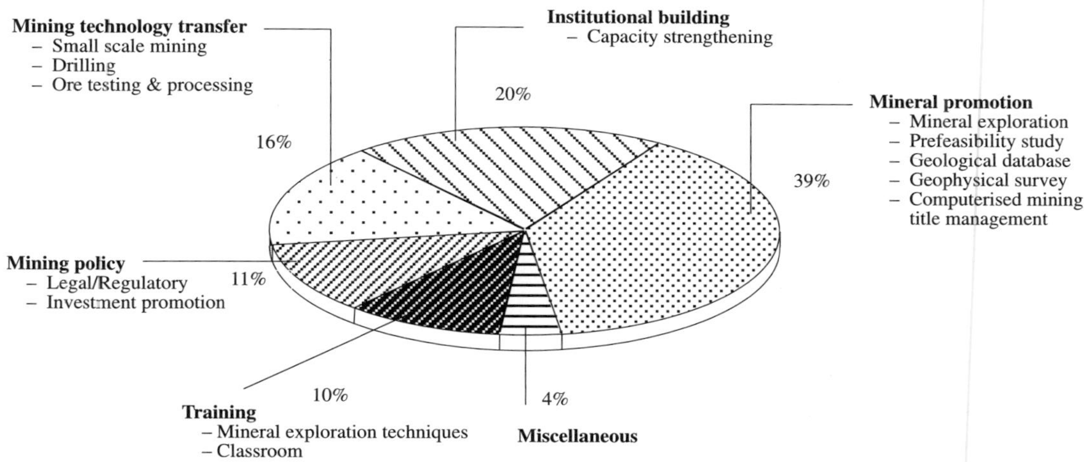

Figure 72. Project areas of the Mineral Resources Brannch of UN/DESD, 1992-1993

(b) The Monthly Commodity Price Bulletin, which includes prices and indexes in nominal and real terms for major minerals and metals;

(c) The Commodity Yearbook, which includes time-series on production, consumption and international trade of major minerals and metals.

In addition, the UNCTAD secretariat was preparing a worldwide comparative statistical analysis on minerals and metals, as well as a Minerals and Metals Yearbook, which it envisaged to start publishing in 1994.

UNIDO

The United Nations Industrial Development Organization (UNIDO) assists among other functions in developing mineral-based industrial processing operations.

UNDP/Tumen river project

The UNDP Tumen River Area Development Project formulated in 1991 is a plan for a multi-billion dollar trade and transport mega-complex in the Tumen River delta in the Sea of Japan, which is an area crossing the borders between China, Russia and the Democratic People's Republic of Korea (figure 73). It is one of the most far reaching strategic economic venues ever proposed for North-East Asia and has been supported by the countries of the Democratic People's Republic of Korea, China, Russia, Mongolia, the Republic of Korea and Japan in the subregion. The plan is to collect three hundred billion dollars in 20 years to develop the Tumen river delta area as an economic, trade and transportation centre in the world. Three agreements have been signed in 1995 among China, the Democratic People's Republic of Korea, Russia, and Mongolia to establish a consultative commission and a coordinative commission to deal with the projected development issues of the area.

UNEP/NOWPAP

The Governing Council of UNEP at its Fifteenth Session approved the 'preparation of new action plans for seas not yet covered by the regional seas programme (North-West Pacific, Black Sea)' as one of the activities listed within the 'Supplementary of Environment Fund Activities for the Bennium 1990-1991', attached as an Annex to the Decision 15/1 'Strengthening the role and effectiveness of the United Nations Environmental Programme'. The

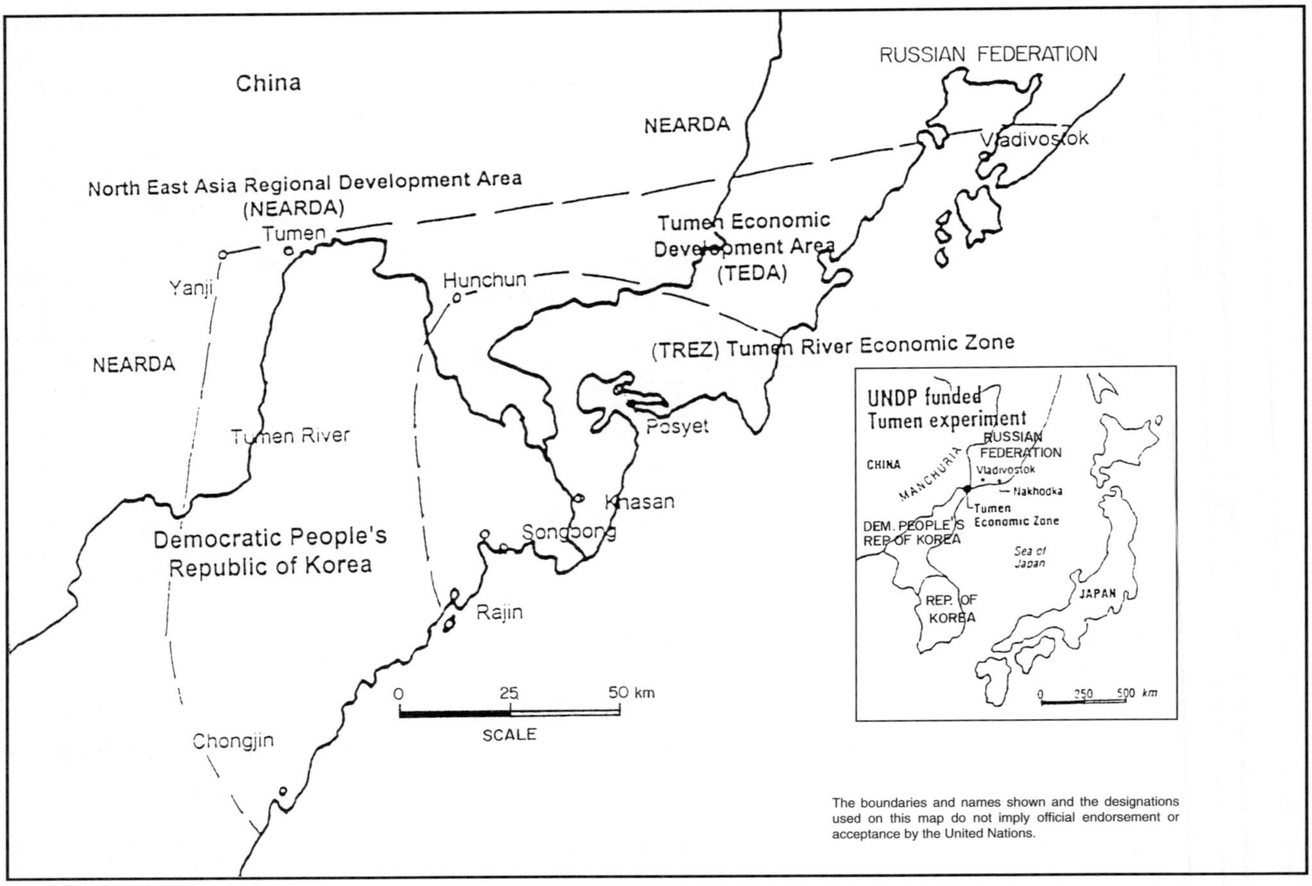

Source: Marton, McGee and Paterson, 1995.

Figure 73. Tumen river delta area with large/small development zones

countries that make up the North-West Pacific region are: the Democratic People's Republic of Korea, Japan, China, the Republic of Korea, and the Russian Federation.

The First Meeting of Experts and National Focal Points on the Development of the North-West Pacific Action Plan was held in Vladivostok from 28 to 31 October 1991, the Second Meeting in Beijing from 26 to 30 October 1992 and the third Meeting in Bangkok from 25 to 29 October 1993. In the Third Meeting, it was recommended to convene an intergovernmental meeting in 1994 to adopt a format framework for cooperation in the region in the form of the North-West Pacific Action Plan (NOWPAP) within UNEP's Regional Seas Programme and to get agreements reached and temporary solutions in the form of resolutions.

To protect the coastal/marine environment, the Washington Conference held in November 1995 adopted the Global Programme of Action (GPA) and requested UNEP to prepare a proposal setting forth a specific plan for implementing the institutional arrangements contained in the GPA. The Organizational Committee of ACC held a meeting on 30 September - 4 October, 1996 and drafted a report to UNEP Governing Council for its endorsement on the implementation of the Global Programme of Action for the protection of the marine environment from land-based activities.

ESCAP

ESCAP resolution 50/8 for North-East Asia

The Economic Social Commission for Asia and the Pacific (ESCAP) at its 50th Session in 1994 adopted resolution 50/8 which called for sustainable development in North-East Asia and called upon the secretariat to provide the disadvantaged economies in transition with, inter alia, technical assistance during their economic reform and structural adjustment phase. The Commission also urged the secretariat to undertake activities for the evaluation of the economic potential and possibility of exploration of mineral resources on the continental shelf of North-East Asia.

In response to the resolution and subsequent request by the Commission, the ESCAP secretariat organized an international conference held in 1994 with a theme of Economic Cooperation in North-East Asia: Capacity Building Towards Policy Research and Development Management. The secretariat further organized a Roundtable on Economic Cooperation Possibilities through Exploitation of Trade and Investment Complementarities in the North-East Asian Subregion in Seoul on 10-12 July 1995. In cooperation with International Trade Centre of UNCTAD/WTO, the secretariat organized a Workshop on International Market Research for the Democratic People's Republic of Korea Trade Officials held in Bangkok on 13-24 May 1996.

To promote the international cooperation in mining sector, the ESCAP secretariat is pursuing a study on development of the mineral resources in North-East Asia. A workshop on the subject has been organized in Changchun, China in February 1997.

Mineral resources

The systematic collection of regional knowledge on mineral resources has been one of the primary goals of the ESCAP secretariat. The secretariat has assisted member countries since 1985 in the compilation of national atlases of mineral resources, and the countries covered so far are Afghanistan, Cambodia Bhutan, Lao People's Democratic Republic, Malaysia, Myanmar, New Zealand, Nepal, the Republic of Korea, Solomon Islands, Sri Lanka, and Viet Nam. Each atlas consists of an explanatory brochure in English with two atlas sheets – a geological map and a mineral resources map. The atlases provide an essential overview to international mining companies for assessing the geological potential for detailed exploration and investment in mining.

During the past decade, requests from developing countries for United Nations technical expertise to assist in the formulation and design of foreign investment policy and regulatory and tax legislation have increased. A Regional Advisor on Mineral Policy and Mineral Economics was recruited in August 1995 and continued advise in this sector is now being provided. Two mineral investment oriented projects, namely, review of mineral policies for development of the mining industry and investment promotion in selected countries of the Asia-Pacific region and training on negotiating mineral investment agreements in Asian LDC's and South Pacific island states were formulated in 1996 for extra-budgetary funding from bilateral donors.

The ESCAP secretariat also focussed its activities on the assessment of geological and mineral resource potential and policy for development in Asian least developed countries and on subregional/regional assessment of important mineral commodities such as gold, base metals, industrial and non-metallic minerals and construction materials.

The issues related to environmental management of mineral resources development were also pursued, and to this end, two regional studies on exploitation and use of mineral waste and confinement of waste in underground space have been carried out. The findings of these studies are contained in a publication on mineral recovery, recycling, waste prevention and confinement for sustainable development in the Asia-Pacific region which has been released in 1996 (ST/ESCAP/1554).

Land-use and urban geology

The application of geological principles for land-use planning and environmental geology is becoming increasingly important in the Asia-Pacific region mainly due to rapidly expanding urban centres as well as increased mining activity centred around human habitations. The ESCAP secretariat is actively involved in compilation and issuance of a series of urban and environmental geology maps. A recently released publication is the environmental and urban geology of Ningbo city, Zhejiang province, China (ST/ESCAP/1374). Two regional expert group meetings and training activities through the implementation of a project on Environmental and Urban Geology for Sustainable Development of Fast-growing Cities were also carried out in 1995.

Marine non-living resources and coastal management

ESCAP secretariat has pursued studies on development and management of coastal/marine non-living resources such as the potential for use of ocean energy resources (ST/ESCAP/827) and the potential for development of offshore construction sand and gravel in East Asia (ST/ESCAP/1341). A workshop on review of mineral and petroleum resources potential of North-East Asia in support of sustainable development of coastal and offshore areas was held in February 1997 and the proceedings with relevant studies and country reports is contained in this volume. ESCAP secretariat has also been actively involved in the study of integrated coastal zone management and practices in the Asia-Pacific region (ST/ESCAP/1137 and 1577).

ESCAP secretariat also actively pursued training activities related to legal, technical and financial aspects for removal of obsolete offshore installations and platforms on the Exclusive Economic Zone and the Continental Shelf of the Asia-Pacific region. With the conducting of the third seminar in September 1995, four volumes of publication have been issued (ST/ESCAP/1389, 1424, and 1664). Such activity will be pursued during 1996 where it is expected to hold a final fourth seminar with extrabudgetary financial resources.

Assistance to member states in implementing the new regime of the sea established under the 1982 United Nations Convention on the Law of the Sea has been provided by the secretariat. In this respect, a series of publications have been released (ST/ESCAP/826, 1097 and 1332).

Information system

ESCAP initiated action to establish a mineral information system with a workshop held in Colombo, Sri Lanka, at the Institute of Computer Technology in July 1992. The database was specifically designed for use by earth scientists, mining engineers and governmental policy makers. Collaboration with the UNCTAD Micas System, a computer-based commodity analysis and information system for world-wide distribution has been developed. The ESCAP database is now functional in six countries – namely, Bhutan, Nepal, Philippines, Sri Lanka, Thailand and Viet Nam. ESCAP has also initiated a project for digital compilation of geological and related thematic maps covering East and South-East Asia in collaboration with CCOP and the Geological Survey of Japan.

Space technology

In cooperation with the Chinese Government, ESCAP organized a Ministerial Conference on Space Applications for Development in Beijing in September 1994. To implement the recommendations of the Ministerial level conference, an Expert Group Meeting on Space Technology Applications for Sustainable Development in Asia and the Pacific was organized by ESCAP in Bangkok from 30 January to 3 February 1995. ESCAP's space technology programme is one of the most successful programmes executed by ESCAP and has received continued broad support from ESCAP's member countries.

Transportation facilities

As a priority activity for Phase II (1992-1996) of the Transport and Communications Decade for Asia and Pacific, ESCAP has been implementing the 'Asian Land Transport Infrascture Development' (ALTID) project which was endorsed at its 48th session in 1991 as a regional project comprising the Asian Highway, the Trans-Asian Railway and the facilitiation of land transport projects.

The participating countries of the project entitled 'A Study for the Development of Asian Highway Network' implemented in 1992 included 18 Asian countries including China and Mongolia from East Asia. To include seven Asian republics namely Armenia, Azerbaijan, Kazakhstan, Kyrgyzstan, Tajikistan, Turkmenistan and Uzbekistan in the development of the highway network, ESCAP initiated in July 1994, a project study on the development of highway network in Asian Republics, as part of the ALTID project with financial support from the Government of Japan. The Asian Highway route criteria are: a) capital-to-capital links; b) connections to main industrial and agricultural centres; c) connections to major sea and river ports; and d) connections to major container terminals.

The Trans-Asian Railway (TAR) projects originally consisted of a southern corridor going through South-East Asia, Bangladesh, India, the Islamic Republic of Iran, Pakistan and Turkey, but was later expanded to include a northern corridor. The northern corridor of the project involved China, the Democratic People's Republic of Korea, Mongolia, the Russian Federation, and Kazakhstan. The Trans-Asian Railway system and its relevant part in North-East Asian countries is shown in figure 74.

To connect the sourthern and northern Asian railway network, the Asian Development Bank (ADB) funded a project in 1993 to develop a railroad between China and Thailand (figure 74). The railroad in Yunan province of China has been extended to Chuxiong, 100 km west to Kunming city. It will extend to Daili in China, then down to the south through Luang Prabang and Vientiane of Lao PDR, to Nong Khai of Thailand. The total length of the

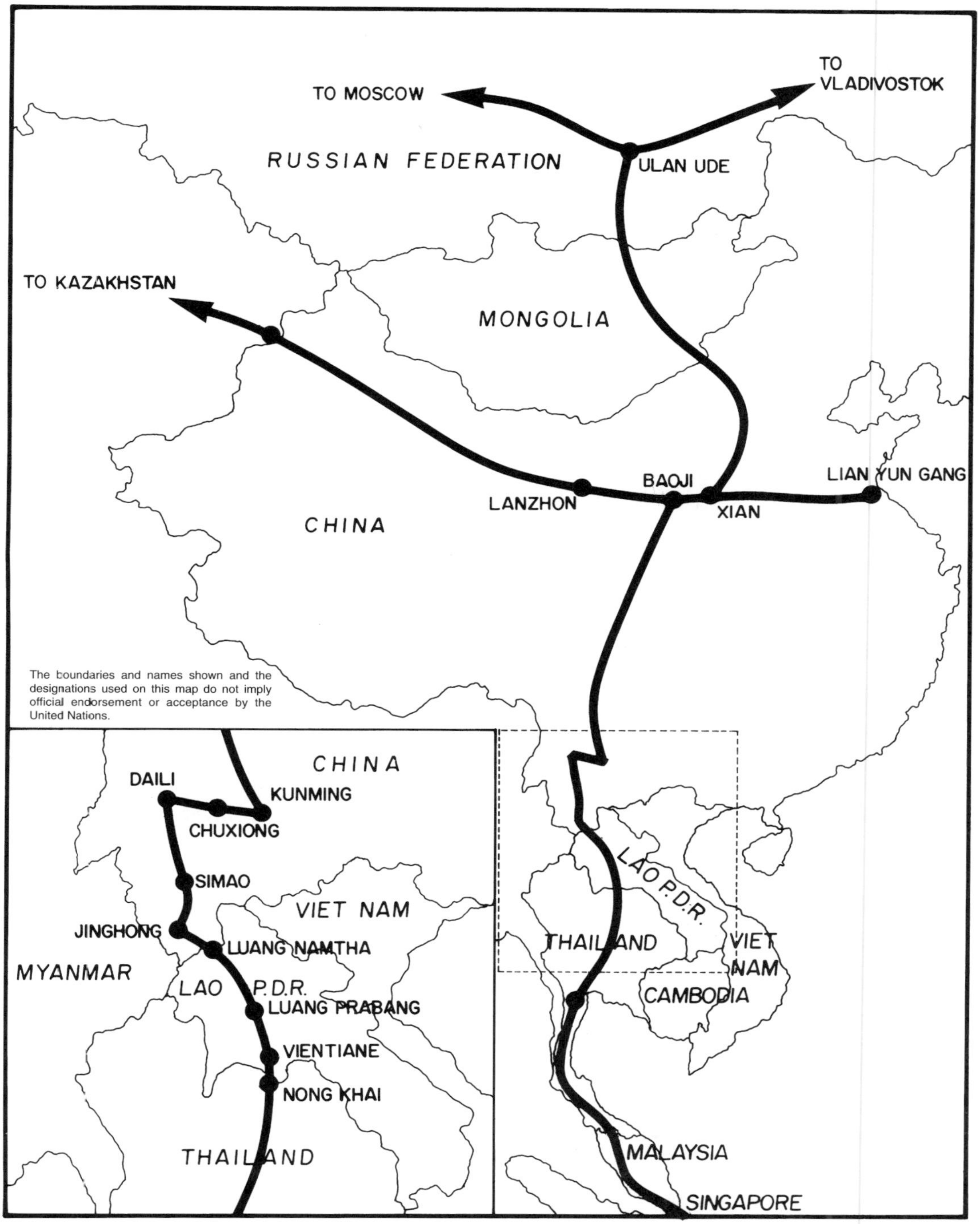

Source: World Daily, 13 July 1996.

Figure 74. Connecting railroad systems of North-East Asia and South-East Asia

railroad between China through Lao PDR to Thailand is 2,119 km. The rail road could also connect the railroad of so-called new Asia-Europe trans-continent bridge from Lianyun Gang in the east of China to Amsterdam in the west of Europe. This Asia-Europe trans-continent bridge was completed in September of 1990 when the railroad in Xinjiang province of China connected to that in Kazakhstan.

3.4.2 Geoscientific communities

CCOP'S Geoscientific projects

The technical activities of Coordinating Committee for Coastal and Offshore Geoscience Programmes in East and South-East Asia (CCOP) include four programme areas covering energy, minerals, coastal zone management and natural disaster mitigation.

Hydrocarbon resources assessment has been one of the most important technical activities in the energy sector of CCOP over the years. As an important output of the project, 'The Total Sedimentary Isopach Maps Offshore East Asia' and its explanatory text were published in 1991. Of the six maps, the seas surrounding northeast China, Korean peninsula and Japan are presented in sheet 1 and 2 on the scale of 1:4 million. CCOP also cooperated with the U.S. Geological Survey (USGS) and the Circum-Pacific Council for Energy and Mineral Resources (CPCEMR) for publishing the East Asia Geographic Map Series. Another important project named Project on Oil and Gas Resources Management was completed by the end of 1994, with a main output of a CD-ROM produced. A new project on Dynamic Basin Analysis and Description in the CCOP area was approved by the Danish government in 1995.

In the mineral sector, CCOP has worked on a project in recent years on Digital Compilation of Geoscientific Maps in East and South-East Asia (DCGM). The compilation, digitization and editing of the geological maps cover the CCOP member countries but also cover the northern part of the Korean Peninsula and the Russian Far East. The final outputs of the phase I of the DCGM has been made available in the form of CD-ROM at the end of 1996. The second phase of the project includes digital compilation of thematic maps. A new proposal of CCOP, in cooperation with the Geological Survey of Japan, is to compile a marine geological and natural resources map during the period of 1996-1997.

The International Council on Metals and Environment (ICME)

The International Council on Metals and the Environment (ICME) comprises the majority of the major companies involved in the international mining industry. The group was initially formed as a central voice for the international mining community on environmental issues, and it provides information to governments and multilateral institutions on issues related to the development and coordination of world-wide environmental standards for the mining resource industry. It responds to the global pressures of highly organized environmental groups.

International Strategic Minerals Inventory (ISMI)

The International Strategic Minerals Inventory (ISMI) started in 1981. Earth science agencies which have participated in ISMI's resource inventory compilation include the Bureau of Mines and the Geological Survey of the United States Department of the Interior; the Geological Survey of Canada and the Mineral Policy Sector of the Canadian Department of Energy, Mines and Resources, and the British Geological Survey, a component of the Natural Environment Research Council. Other ISMI participants include the Bureau of Resource Sciences of the Australian Department of Primary Industries and Energy, the Institute for Geosciences and Natural Resources of Germany, and the Geological Survey and Minerals Bureau of the Department of Minerals and Energy Affairs of South Africa.

ISMI decided to begin with commodity studies on chromium, manganese, nickel, and phosphate. These studies, plus others on platinum-group metals, cobalt, titanium, graphite, lithium, tin, vanadium, and zirconium, have now been published. Additional studies on niobium (columbium) and tantalum, tungsten, and rare-earth oxides and yttrium have subsequently been undertaken. A regional survey of the strategic minerals of sub-Equatorial Africa has been published, and a survey on eastern Europe is under way.

ISMI reports are publicly available, in convenient form, non-proprietary data and characteristics of major deposits of strategic mineral commodities for policy considerations in regard to short-term, medium-term and long-term world supply. They provide summary statements of the data compiled and overviews of supply aspects in a format designed to benefit policy analysts and geologists.

The Circum-Pacific Map Project

The Circum-Pacific Map Project is a collaborative global effort to compile and publish geologic, geophysical and resources maps of the Pacific Basin and surrounding continental areas. A series of more than 60 maps have been designed, and about two thirds have been published by the United States Geological Survey. Maps are designed to illustrate the relationship of known energy and mineral resources to the geology, tectonics, and crustal dynamics of the Pacific region. Geologic, geophysical, mineral, and energy data are being compiled at a scale of 1:10,000,000 on equal-area map projections. Where possible, available data are complemented by new, project-developed data sets such as magnetic lineations, earthquake first-motion solutions, lithospheric stress, seafloor mineral deposits, and seafloor sediments. Geologists and earth scientists from throughout the Asia/Pacific region are participating in this work.

Covering more than half of the earth's surface, the project area extends from the Indian Ocean (lat 90°E) eastward across the Pacific to include most of North and South America (lat 50°W). It also includes the Arctic Ocean and the entire continent of Antarctica. Six overlapping regional maps at a scale of 1:10,000,000 form the cartographic base for the project. In most of the eight map series there is also a Pacific basin map at 1:17 million scale for depicting basin-wide data. These maps cover 220° of latitude, more than half of the earth's surface. Maps are compiled on a Lambert azimuthal equal area projection to minimize distortion. Base map series include geographic and base maps, the latter with 2° grids for plotting data. The thematic map series include plate-tectonic, geologic, geodynamic, energy resources, mineral resources, and tectonic maps.

The Circum-Pacific Map Project is currently organized under six panels of geoscientists representing national earth-science organizations, universities and natural-resource companies. Forty-one map sheets were published as of 1992. Four map series have now been issued: geographic, base, plate-tectonic, and geodynamic. Publication of the geologic map series began in 1983 and was scheduled for completion in 1994. The first map in the mineral resource series was published in 1984. The first map of the energy resources series was published in 1986. The first map of the tectonic series was published in 1991. The Northwest Energy Resources map was suppose to be issued in 1992.

Preparation of eight 1:2,000,000 scale base maps for a basin evaluation/resources assessment study of South-East Asia was initiated in late 1983 in cooperation with the International Union of Geological Sciences (IUGS) and the Committee for Coordination of Joint Prospecting for Mineral Resources in Asian Offshore Areas (CCOP). Other regional mapping projects in Central and South America are in the planning stage.

United States Geological Survey (USGS)

The resource assessment activities undertaken by USGS assess the distribution, quantity, and quality of the mineral resources of the United States, particularly those on public lands, by studying the geology of known mineral occurrences and potentially mineralized areas; by developing and improving exploration techniques and mineral-occurrence models necessary in the continuing search for new deposits; and by enhancing knowledge and understanding of domestic and world resources of non-fuel minerals. Particular emphasis is placed on strategic and critical minerals, minerals that are largely or entirely imported and those that are necessary to the economy of the United States.

In 1993, the Office of Mineral Resources (OMR) of USGS proposed to undertake a seven-year plan to provide probabilistic quantitative assessment of the undiscovered non-fuel mineral resources of the United States and its public lands. For the first time, a consistent, usable, minimum level of current mineral-resource information, together with estimates of total undiscovered mineral endowment, will be provided for the entire United States.

The plan calls for a two-year preliminary national assessment, followed by an iterative national assessment recurring about every five years. The two-year preliminary assessment, serving as a springboard, will consider a limited number of types of deposit based on existing data. The report of this preliminary effort will include maps showing the outlines of tracts that are prospective for the types of deposits considered and a broad description of the basis for the delineation of the tracts. Tables will list the deposit types assessed, the identified resources, estimates of numbers of undiscovered deposits, and a summary of quantities of estimated undiscovered mineral resources. Maps will be provided to land-management agencies and resource-planning organizations in paper and digital formats at whatever scales they request for use in land- and resource-planning and decision-making. USGS scientists will visit these users to identify their future planning needs.

Integrated Exploration & Development Services LTD (IEDS)

It should be noted that the most updated and valuable data and information on mineral resources including oil and gas are with private companies, especially TNCs. Those data and information are very well organized, clearly presented and very costly. The most valuable data/information handled by the TNCs is at high security level and not for sale.

A North-East Asia Monthly Activities Report issued by the Integrated Exploration & Development Services LTD (IEDS) contains detailed updated data and information on contracts, current exploration and development activities, future exploration and development programmes onshore and offshore in China, Democratic People's Republic of Korea, Republic of Korea, Mongolia and Japan in the North-East Asia. This monthly publication is for sale.

4.0 CONCLUSIONS AND RECOMMENDATIONS

General Principles

1. In the context of **sustainable development**, the issue of energy and minerals development is considered as a link of sustainable supply with energy and minerals to the society. To address this issue, consideration should be given to the whole energy/mineral circle from upstream exploration/mining to downstream extracting/manufacturing industries, the activities from production, consumption, recycling and substitution, and all the development factors including resources potential, technology, commodity market, capital, trained manpower and infrastructure facilities. The interaction of the developmental activities of the energy and minerals sectors with those of the other sectors should also be considered. The sustainable extraction and use of energy and mineral resources is expected to be achieved through improved efficiency, new technology, substitution and recycling, and integration into land development and use.

2. **Environment protection** is a key component of sustainable development and it would be inappropriate to relax environmental laws, regulations and standards or their enforcement in order to encourage foreign direct investment and promote development. A more 'clean' approach for development of energy and minerals should be adopted and EIA for energy and mineral development projects should be performed at the early planning stage. The CO_2, SO_2, NO_x and other wasteful/hazardous gases emission, the discharged wasteful/ hazardous water and the disposed wasteful/hazardous solid matters from energy/minerals exploration, production, transportation and consumption must be controlled.

3. **Metallic** minerals consumption in developed countries has declined in recent decades because of increased efficiency in the use of the metals, the increase in value-adding in finished products and substitution by non-metallics. Exploration for metallic mineral resources has been increasingly dominated by international mining companies. In contrast, **non-metallic** minerals require less capital investment than the metallic minerals and often contribute quickly and inexpensively to a country's development. Many non-metallic minerals also lend themselves to processing and marketing inside the country, which encourages entrepreneurship, creates local value added and leads to the improvement of regional transportation systems.

4. The **development and export** of energy and minerals provides developing countries with valuable government revenues, much needed foreign currency, employment (particularly for rural workers), infrastructure

investment, technology transfers and establishment of ancillary industries. On the other hand, the petroleum exploration and modern mining industry are high risk, capital and technologically intensive, and often remotely located. Oil/gas projects and mines have finite lives and also have potential to cause environmental impacts. These basic characteristics should be considered by both investor and recipient.

5. To achieve energy and material efficiency, and to meet sound environment requirements in the development of energy and minerals, cost-effective, energy/material-saving and environment-friendly **technologies**, including improved GHGs (Greenhouse gases) technologies should be adopted, so as to formulate more efficient systems of production that emphasize pollution prevention and waste minimizing and recycling, with particular attention to the adoption of such systems by small and medium-sized enterprises.

6. To attract and maintain long-term private sector **investments** in energy and minerals development, the role of Government should be as regulator, promoter and facilitator. Some important conditions need to be established, including the following:

(a) A sound macro investment environment offering prudent and stable policies, which are supportive of an open, export-oriented business environment;

(b) A modern legal and fiscal framework, providing assured access to land for exploration, clear transparent procedures, security of tenure from exploration to production, and a stable fiscal regime;

(c) A sound environmental framework, including the need to have predetermined environmentally related obligations, identification of stakeholder groups and host government involvement in ensuring environmentally sustainable mining practices;

(d) Infrasctructure conditions – transportation facilities are especially important to the development of mineral resources. Promising mineral reserves are usually located in remote areas, which makes the transport of equipment, workers and ore products difficult. This is a hindrance to mineral projects and trade especially in Mongolia and the Far Eastern region of the Russian Federation. Infrastructure investment efforts should be a priority;

(e) Competent public energy and mining institution(s) should be established to survey and/or assess resource potential, to provide reliable, well-organized information, to manage environmental issues and to deal with foreign investment issues;

(f) A broad-based higher education system including a university trained technical and management team is a prime ingredient to successfully compete in an information-based industry, and with implementing the productivity and technology gains of the 1990s elsewhere;

(g) A knowledge base of energy and minerals potential which could be integrated with other land/marine resources data bases, and with relevant environment and socio-economic information.

7. The role of indigenous communities and provincial or local government entities will largely strengthen the decision making process. **Local population concerns** usually highlight: improvement of living standards, access to opportunities, increase in social equity and environmental considerations. Energy and mining projects should benefit the whole economy, encourage the development of local industries to meet the needs of people and should provide opportunities for education and training of nationals which will facilitate technology transfer.

The North-East Asia

1. According to the World Bank, the East Asian nations have achieved spectacular success in raising incomes and reducing the incidence of poverty from 30 per cent of the population in 1960, to 10 per cent in 1990. North-East Asia one of the fastest growing areas, with a strong potential to sustain its **economic growth** during the 1990s – and to become a major growth centre in the world economy during the 21st century.

The Far East region of the Russian Federation, Mongolia, the Democratic People's Republic of Korea and China are richly endowed with mineral resources and with low-cost labour, but are lacking capital, modern technology and managerial skills. On the other hand, Japan and the Republic of Korea are endowed with much capital, high technology and modern managerial skills, but lack mineral resources and low-cost labour.

2. GDP percentage figures for **mining** in 1993 were as follows: 9.39 per cent for China's three northeastern provinces, 3.8 per cent for the Far East part of the Russia Federation, 26.44 per cent for Mongolia, and 8.2 per cent for the Democratic People's Republic of Korea. Their mining sectors have played an important role in the development of their respective economies and will continue to do so in the future.

GDP figures in US dollars for mining in 1993 were nearly 11 billion for Japan and over 1.1 billion for the Republic of Korea, although their GDP percentage figures were only 0.26 per cent for Japan and 0.34 per cent for the Republic of Korea. Their manufacturing industries are well developed and they are large consumers and importers of energy and minerals. In the Republic of Korea, the value of non-metallic mineral products was $1,021 million in 1994, accounting for 64 per cent of its total mineral value. However, the recent strong increase of domestic demand for non-metallic minerals has caused a sharp increase in imports.

3. The Far East region of the Russian Federation and Mongolia are endowed with abundant **energy** and **mineral resources**, and is one of the most prospective areas, yet still relatively unexplored. The region continues to attract many of the world's largest mining companies including US-based Cyprus Minerals Company, Canada's Placer Dome, Inc. and Teck Corporation, Australia's Broken Hill Proprietary Co. Ltd. and CRA Ltd. Many experts believe that exploration and development expenditures in the region will increase in the early 21st century.

China is also well endowed with energy and mineral resources, and has a large domestic market for its commodities. China imports both ferrous and some of non-ferrous metal ores for years, and became a net importer of oil in 1993. China will be a net-importer of natural gas by 2000 and is expected to become a coal importer by 2010. China will continue import some of the metal and non-metallic minerals.

The Democratic People's Republic of Korea is endowed with some minerals and there is potential for exports. The country is rich in coal but no oil resources have been discovered. The shortage of energy is a serious problem for it's economy.

Japan and the Republic of Korea are poorly endowed with energy and mineral resources. Their demand has long been met through imports. They import oil, coking coal, iron ore and some of non-ferrous metal ores such as copper and zinc.

North-East Asia will continue to depend on imports of crude oil from other regions, especially from the Middle East. The Democratic People's Republic of Korea, Japan, Mongolia and the Republic of Korea have long been dependent on imports of oil from abroad. The Far East region of the Russian Federation imports 80 per cent of its crude oil from abroad, while offshore areas may have potential for discoveries. The Far East region of the Russian Federation is endowed with rich resources of natural gas, while coal is abundant in the North-East Asian countries with the exception of the Republic of Korea and Japan.

4. **Energy structure** is destined to change. Substitution of coal and especially natural gas for oil in the energy mix is likely to be a long-term goal for countries in North-East Asia. The R/P ratio of oil is much shorter than for coal and gas. The carbon released by natural gas is much less than that from coal and oil.

Japan and the Republic of Korea have been among the leading producers of non-ferrous and metal products. North-East Asian countries' continue to seek direct investments in industrial minerals including cement, aggregates and dimension stones.

5. While combating the shortage of energy/minerals supply through developing and importing energy/minerals, the conservation of energy/minerals through raising production efficiencies and consumption of energy/minerals should be an important policy for sound development of energy/minerals in North-East Asia.

The energy/minerals intensity (ratio of energy/minerals consumption to GDP) of the developing countries is usually rather high, indicating a capacity for developing countries to conserve energy/minerals. A conservation policy will contribute to the sustainable development of resources.

6. An **open doors policy** and a 'market-friendly' approach to development has been viewed by many developing countries and economies in transition as the basis for economic growth, technological progress, and job creation. The economies of China, Hong Kong, China, the Republic of Korea and Taiwan Province of China in East Asia are today among the fastest-growing economies in the world. On the other hand, the economies of Mongolia the Democratic People's Republic of Korea, and the Russian Federation are still experiencing difficulties. However, the Mongolian economy appears to be improving with a previous decline in its GDP in 1990 having now been reversed in 1994.

7. To facilitate the development of energy and minerals, the mining policy and legislation is undergoing changes in the North-East Asian countries since the 1980s. New mining laws have been implemented by Mongolia, the Russian Federation and China. Legislation will continue to evolve in the future to accommodate the need for attracting foreign direct investment, dealing with environment protection, social-cultural issues, and improving fiscal regimes. To attract foreign direct investments **(FDI)** special incentives need to be adopted by host governments.

8. To attract foreign investment for developing its non-ferrous minerals sector, China opened its Metals Exchange market at Shenzhen and Shanghai in 1992. The Shenzhen metal exchange lists eight metals for trade – aluminium, antimony, copper, lead, magnesium, nickel, tin and zinc. China's CNNC focuses on diversifying its business and tapping overseas resources and capital to secure the supply of raw materials needed. China's oil industry also invests in countries abroad, with an approach of "many channels, many blocks and small shares".

A recent increase in the involvement of Japanese mining companies in world-class mining projects is apparent – Nippon Mining in Australia, Samitomo Metal Mining in Chile, Dowa Mining in Mexico are examples. MMAJ is also active in worldwide geological surveys and exploration through its governmental grant system.

9. Lack of **transportation facilities**, especially in the Far East region of the Russian Federation and Mongolia, is one of the difficulties in developing energy and mineral projects in the North-East Asia. The policy and planning framework of countries in the North-East Asia at macro-economic and sectoral levels should address the issue of infrastructure services. Partnership between the public and private sectors should be examined to identify mechanisms for encouraging private-sector involvement in project financing, management, operations and risk-sharing.

10. **Further assessment** of resources potential, environmental conditions and ecoñómic and social requirements is needed. Detailed and comprehensive nation-wide geoscientific survey of mineral and energy resources and mines has been undertaken by Japan, the Republic of Korea and China. However, the widespread investigation of mineral occurrences in geologically promising areas in the Far East region of the Russian Federation and Mongolia has been limited, thus the current knowledge of the potential for new discoveries is also limited.

11. A **development challenge** remains for the Asian developing countries. Over one billion people in East Asia live in poverty. In addition, the deterioration of the ecological environment caused by unsustainable development practices continues to cause alarm.

Rasing living standards by promoting growth and development and protecting the environment from pollution in developing countries alleviate poverty, expands trade, job opportunities, and incomes, and reduces transnational boundary pollution. **Win-win strategies** for sustainable development should be adopted in North-East Asia.

12. The intra-regional (subregional) and inter-regional economic and technological **cooperation** should be encouraged to facilitate social-economic development and sound environment management in countries of North-East Asia. Cooperation should include capacity-building programmes in the areas of technological, economic and environmental assessment of mineral resources and the protection of the environment. The exchange of information, especially to enable information to be more accessible to decision makers is important.

REFERENCES

1. Asian Economic Survey, 1996-1997, *Asian Wall Street Journal, 22 October 1996.*

2. *Asian Journal of* Mining, May/June 1995.

3. Batbayar, Ts., 1995. "Foreign Direct Investment in Mongolia: Problems and Perspectives" in the SPF publication *of Foreign Direct Investiment and Cross Cultral Management in North-East Asia, 1995.*

4. *British Petroleum Statistical Review of World Energy, 1995, and, 1996.*

5. China, 1994. "The Petroleum resources and cooperation prospectivity in China", paper presented at the 1st Exploration Promotion Forum, 2-4 June, Stavanger, Norway.

6. China Statistic Publishing House: Energy Statistical Yearbook, 1989.

7. Chinese Institute of Geology and Mineral Resources Information, 1992. Mineral Resources of China, published by the China Building Materials Industry Press.

8. Clark, A.L., 1982. The importance of data bank for resource assessment, in New Paths to Mineral Exploration. Proceedings of the Third International Symposium, Hannover, Federal Republic of Germany, 27-29 October 1982.

9. Clark, A.L., and G.J. Jeon, 1990. Metal consumption trends in the Asia-Pacific region: 1960-2015. Proceedings of the Minerals and Energy Forum, Pacific Economic Cooperation Conference, Manila, Philippines.

10. Committee on Natural Resources, 1996. Assessment of benefits accruing to host countries from the inflow of funds and technology for mineral development (E/C. 7/96-8), reported on the Third Session, 5-7 May 1996.

11. Dorian, James P., 1993. Mining in the CIS, Comercial opportunities abound. Financial Times Management Report published and distributed by Financial Times Business Information, 152 pp.

12. Dorian, James P., 1994. Minerals, Energy and Economic Development in China. Oxford University Press, 288 pp.

13. Emery, K.O. and others, 1969. "Geological structure and some water characterisitics of the East China Sea and Yellow Sea", *United Nations ECAFE, CCOP Tech. Bull., vol. 2, p. 343.*

14. Economic and Social Council. *Towards the sustainable supply of minerals in the context of Agenda 21, Inter-sessional strategy paper of the Committee on Natural resources* (E/C. 7/1996/11).

15. ESCAP, 1990. Environmental Impact Assessment, Guidelines for Industrial Development (ST/ESCAP/784).

16. ESCAP, 1992. Mineral Resources Development and the Environment (ST/ESCAP/1192).

17. ESCAP, 1992. Minerals and Metal Trading (ST/ESCAP/1268).

18. ESCAP, 1992. Negotiating Mineral Industry Agreements and Contracts (ST/ESCAP/1267).

19. ESCAP, 1994. Transnational Technology (ST/ESCAP/1071).

20. ESCAP, 1995. Foreign Trade Statistics of Asia and the Pacific, 1988-1993 (ST/ESCAP/1589).

21. ESCAP, 1996. Studies in Trade and Investment (ST/ESCAP/1640).

22. *International Energy Agency, 1995. "Oil Energy Outlook".*

23. Ji Hong, 1995. "Foreign Direct Investiment in China: Historical Trends and Investment Environment" in the SPF publication of *'Foreign Direct Investiment and Cross Cultral Management in North-East Asia, 1995'.*

24. KIGAM, 1995. Geological Investigation and Mineral Resources Exploration & Development in the Republic of Korea, presented in the Expert Group Meeting organized by ESCAP in May 1995.

25. Minakir, P.A. and Admidin, A.G., 1995. "Trends in Direct Investment in the Russian Far East" in the SPF publication of *'Foreign Direct Investiment and Cross Cultral Management in North-East Asia, 1995'.*

26. *Mining Annual Review, 1989, 1990, 1991, 1992, 1993, 1994, 1995 and 1996, published by Mining Journal, London.*

27. *Oil & Gas Journal Special, 9 May 1994. "China's downstream Boom".*

28. Otto, J., 1992. "Criteria for Assessing Mineral Investment Conditions". In the ESCAP publication 'Mineral investment conditions in selected countries of the Asia-Pacific Region' (ST/ESCAP/1197).

29. Pechey, R., 1992. World salt resources, Industrial Minerals, February 1992.

30. *Petroleum Economist, February 1995, vol. 62, No. 2.*

31. *Petroleum Economist, May 1995.* "Russia: Far East finding its feet?" vol. 62, No 2.

32. *Petroleum Economist, June 1995, "China: More help needed to boosy output" Vol. 62, No. 6.*

33. *Petroleum Economist, July 1995, vol. 62, No. 7.*

34. *Petromin, August 1995.*

35. Qiu Da Xiong, 1991. "The role of petroleum in sustainable development in China". In Proceedings on the Role of Petroleum in Sustainable Development, January 1991, Penang, Malaysia.

36. Government of the Republic of Korea, WTO Secretariat Report on Trade Policy Review of the Republic of Korea.

37. Tan Ashide, 1995. Japanese direct investiment in China and its implication for NEA. Paper in Foreign Direct Investment and Cross Cultural Management in North-East Asia, published by the Sasakawa Peace Foundation.

38. UNCTAD, 1995. Market situation and outlook for tin, 1994 (UNCTAD/COM/54).

39. UNCTAD. Commodity Yearbook, 1994, and 1995.

40. UNCTAD, 1996. Recent and planned changes in production capacity for bauxite, alumina and aluminium.

41. UNDP, 1994. Annual Development Cooperation Report for the Democratic People's Republic of Korea.

42. UNESCAP/CCOP, "The offshore hydrocarbon potential of East Asia: A decade of investigations in 1966-1975". *UNESCAP/CCOP Technical Bulletin, vol. 11, October 1977.*

43. *World Daily, 13 July 1995.*